Chemistry of Ozone in Water and Wastewater Treatment

Chemistry of Ozone in Water and Wastewater Treatment
From Basic Principles to Applications

Clemens von Sonntag and Urs von Gunten

IWA Publishing
London • New York

Published by   IWA Publishing
                Alliance House
                12 Caxton Street
                London SW1H 0QS, UK
                Telephone: +44 (0)20 7654 5500
                Fax: +44 (0)20 7654 5555
                Email: publications@iwap.co.uk
                Web: www.iwapublishing.com

First published 2012
Reprinted 2013
© 2012 IWA Publishing

Cover image: Ozone Generator Ozonia, Degrement Technologies, with permission.
Photograph: Urs von Gunten.
Cover Design: Timo von Gunten and Sixteen Design (www.sixteen-design.co.uk)

Apart from any fair dealing for the purposes of research or private study, or criticism or review, as permitted under the UK Copyright, Designs and Patents Act (1998), no part of this publication may be reproduced, stored or transmitted in any form or by any means, without the prior permission in writing of the publisher, or, in the case of photographic reproduction, in accordance with the terms of licenses issued by the Copyright Licensing Agency in the UK, or in accordance with the terms of licenses issued by the appropriate reproduction rights organization outside the UK. Enquiries concerning reproduction outside the terms stated here should be sent to IWA Publishing at the address printed above.

The publisher makes no representation, express or implied, with regard to the accuracy of the information contained in this book and cannot accept any legal responsibility or liability for errors or omissions that may be made.

Disclaimer
The information provided and the opinions given in this publication are not necessarily those of IWA and should not be acted upon without independent consideration and professional advice. IWA and the Author will not accept responsibility for any loss or damage suffered by any person acting or refraining from acting upon any material contained in this publication.

British Library Cataloguing in Publication Data
A CIP catalogue record for this book is available from the British Library

ISBN 9781843393139 (Paperback)
ISBN 9781780400839 (eBook)

Printed in Great Britain by Bell & Bain Ltd., Glasgow

Contents

About the Authors xi

Chapter 1
Historical background and scope of the book 1

Chapter 2
Physical and chemical properties of ozone 7
2.1 Introductory Remarks 7
2.2 Generation of Ozone 8
2.3 Ozone Solubility in Water 9
2.4 UV–VIS Spectrum of Ozone 10
2.5 Determination of the Ozone Concentration 12
    2.5.1 The *N,N*-diethyl-*p*-phenylenediamine (DPD) method 12
    2.5.2 The indigo method 13
2.6 Methods for Measuring Ozone Kinetics 14
    2.6.1 Ozone decay measurements 15
    2.6.2 Quenching of ozone with buten-3-ol 16
    2.6.3 Reactive absorption 16
    2.6.4 Competition kinetics 17
2.7 Reduction Potentials of Ozone and Other Oxygen Species 18
2.8 Stability of Ozone Solutions 19
2.9 Reactivity of Ozone 19
    2.9.1 pH dependence of ozone reactions and the "reactivity pK" ... 20
    2.9.2 Multiple reaction sites within one molecule 21

Chapter 3
Ozone kinetics in drinking water and wastewater 23
3.1 Stability of Ozone in Various Water Sources 23

3.2	Molecular Weight Distribution of Dissolved Organic Matter	31
3.3	Mineralisation and Chemical Oxygen Demand	33
3.4	Formation of Assimilable Organic Carbon	33
3.5	Formation and Mitigation of Disinfection By-products	35
3.6	UV Absorbance of Dissolved Organic Matter	36
3.7	Relevance of Ozone Kinetics for the Elimination of Micropollutants	37
3.8	Hydroxyl Radical Yield and •OH-Scavenging Rate of Dissolved Organic Matter	39
3.9	Elimination of Ozone-Refractory Micropollutants by the •OH Route	40
3.10	Ozone-based Advanced Oxidation Processes	42
	3.10.1 Peroxone process	42
	3.10.2 UV photolysis of ozone	45
	3.10.3 Reaction of ozone with activated carbon	46

Chapter 4
Inactivation of micro-organisms and toxicological assessment of ozone-induced products of micropollutants ... **49**

4.1	Disinfection Kinetics	49
4.2	Inactivation Mechanisms: Role of Membranes and DNA	52
4.3	Reactions with Nucleic Acid Components	53
4.4	Reaction with DNA	54
4.5	Application of Ozone for Disinfection in Drinking Water and Wastewater	55
4.6	Toxicological Assessment of Ozone Induced Transformation Products	55
4.7	Endocrine Disrupting Compounds	56
	4.7.1 Laboratory studies	58
	4.7.2 Full-scale studies	59
4.8	Antimicrobial Compounds	60
4.9	Toxicity	62

Chapter 5
Integration of ozonation in drinking water and wastewater process trains ... **65**

5.1	Historical Aspects	65
	5.1.1 Drinking water	65
	5.1.2 Municipal wastewater	65
5.2	Drinking Water Treatment Schemes Including Ozonation	66
5.3	Micropollutants in Water Resources, Drinking Water and Wastewater	70
5.4	Enhanced Wastewater Treatment with Ozone	72
5.5	Energy Requirements for Micropollutant Transformation in Drinking Water and Wastewater	73
5.6	Source Control	74
5.7	Reclamation of Wastewater	75
5.8	Comparison of the Application of Ozone in the Urban Water Cycle	77

Chapter 6
Olefins **81**
6.1 Reactivity of Olefins 81
6.2 The Criegee Mechanism 84
6.3 Partial Oxidation 87
6.4 Decay of the Ozonide via Free Radicals 88
6.5 Detection of α-Hydroxyalkylhydroperoxides 88
6.6 Ozone Reactions of Olefins – Products and Reactions of Reactive Intermediates 89
    6.6.1 Methyl- and halogen-substituted olefins 89
    6.6.2 Acrylonitrile, vinyl acetate, diethyl vinylphosphonate, vinyl phenyl sulfonate, vinylsulfonic acid and vinylene carbonate 91
    6.6.3 Acrylic, methacrylic, fumaric, maleic and muconic acids 92
    6.6.4 Muconic acids 96
    6.6.5 Cinnamic acids 98
    6.6.6 Dichloromaleic acid 99
    6.6.7 Pyrimidine nucleobases 99
6.7 Micropollutants with Olefinic Functions 102

Chapter 7
Aromatic compounds **109**
7.1 Reactivity of Aromatic Compounds 109
7.2 Decay of Ozone Adducts 116
7.3 Ozone Reactions of Aromatic Compounds – Products and Reactions of Reactive Intermediates 118
    7.3.1 Methoxylated benzenes 118
    7.3.2 Phenols 121
7.4 Micropollutants with Aromatic Functions 124

Chapter 8
Nitrogen-containing compounds **131**
8.1 Reactivity of Nitrogen-containing Compounds 131
8.2 General Mechanistic Considerations 138
    8.2.1 Aliphatic amines 138
    8.2.2 Aromatic amines (anilines) 143
    8.2.3 Nitrogen-containing heterocyclic compounds 145
8.3 Micropollutants with Nitrogen-containing Functions 146
    8.3.1 The *N*-nitrosodimethylamine (NDMA) puzzle 156

Chapter 9
Reactions of sulfur-containing compounds **161**
9.1 Reactivity of Sulfur-containing Compounds 161
9.2 Thiols 162

9.3   Sulfides, Disulfides and Sulfinic Acids 163
9.4   Sulfoxides 165
9.5   Micropollutants Containing an Ozone-reactive Sulfur 166

Chapter 10
Compounds with C–H functions as ozone-reactive sites **169**
10.1 Reactivity of Compounds with C–H Functions as Ozone-reactive Sites 169
10.2 General Mechanistic Considerations 171
10.3 Formate Ion 173
10.4 2-Methyl-2-Propanol (Tertiary Butanol) 175
10.5 2-Propanol 176
10.6 Carbohydrates 180
10.7 Dihydrogen Trioxide – Properties of a Short-lived Intermediate 182
10.8 Saturated Micropollutants Lacking Ozone-reactive Heteroatoms 184

Chapter 11
Inorganic anions and the peroxone process **185**
11.1 Introductory Remarks 185
11.2 Hydroxide Ion 187
11.3 Hydroperoxide Ion – Peroxone Process 188
11.4 Fluoride 189
11.5 Chloride 190
11.6 Hypochlorite 191
11.7 Chlorite 192
11.8 Bromide 192
11.9 Hypobromite 193
11.10 Bromite 194
11.11 Iodide 194
11.12 Nitrite 195
11.13 Azide 196
11.14 Hydrogen Sulfide 197
11.15 Hydrogen Sulfite 198
11.16 Bromate Formation and Mitigation in Water Treatment 198
11.17 Bromide-catalysed Reactions 201
11.18 Mitigation of Iodide-related Problems 202

Chapter 12
Reactions with metal ions **205**
12.1 Reactivity of Metal Ions 205
12.2 Arsenic 206
12.3 Cobalt 207
12.4 Copper 207

12.5	Iron	207
12.6	Lead	208
12.7	Manganese	208
12.8	Selenium	209
12.9	Silver	209
12.10	Tin	211
12.11	Metal Ions as Micropollutants	211

Chapter 13
Reactions with free radicals ... 213
13.1	Reactivity of Radicals	213
13.2	Ozone Reactions with Reducing Radicals	214
13.3	Ozone Reactions with Carbon-centered Radicals	215
13.4	Ozone Reactions with Oxygen-centered Radicals	217
13.5	Ozone Reactions with Nitrogen- and Sulfur-centred Radicals	219
13.6	Ozone Reactions with Halogen-centred Radicals	220

Chapter 14
Reactions of hydroxyl and peroxyl radicals ... 225
14.1	Introductory Remarks	225
14.2	Hydroxyl Radical Reactions	225
	14.2.1 Addition reactions	225
	14.2.2 H-abstraction reactions	227
	14.2.3 Electron transfer reactions	228
14.3	Determination of •OH Rate Constants	229
14.4	Detection of •OH in Ozone Reactions	230
14.5	Determination of •OH Yields in Ozone Reactions	232
14.6	Formation of Peroxyl Radicals	233
14.7	Redox Properties of Peroxyl Radicals and Reaction with Ozone	233
14.8	Unimolecular Decay of Peroxyl Radicals	234
14.9	Bimolecular Decay of Peroxyl Radicals	235
14.10	Reactions of Oxyl Radicals	236
14.11	Involvement of •OH Radicals in Chlorate and Bromate Formation	237
	14.11.1 Chlorate formation	237
	14.11.2 Bromate formation	238
14.12	Degradation of Ozone-refractory Micropollutants by •OH/Peroxyl Radicals	241
	14.12.1 Saturated aliphatic compounds	241
	14.12.2 Aromatic compounds	243
	14.12.3 Chlorinated olefins	245
	14.12.4 Perfluorinated compounds	248

References ... 249

Index ... 287

About the Authors

Prof. Dr Clemens von Sonntag
Max-Plack-Institut für Bioanorganische Chemie, D-45470 Mülheim an der Ruhr, Germany, (retired: postal address: Bleichstr. 16, D-45468 Mülheim an der Ruhr)

and

Institut für Instrumentelle Analytik, Universität Duisburg-Essen, D-45117 Essen.
(email: clemens@vonsonntag.de)

Prof. Dr Urs von Gunten
Department of Water Resources and Drinking Water, Eawag, Swiss Federal Institute of Aquatic Science and Technology, CH-8600 Dübendorf, Switzerland

and

Institute of Environmental Engineering, School of Architecture, Civil and Environmental Engineering, ENAC, Ecole Polytechnique Fédérale Lausanne, EPFL, CH-1012 Lausanne.
(email: vongunten@eawag.ch)

Chapter 1
Historical background and scope of the book

The discoverer of ozone, *Christian Friedrich Schönbein* (1799–1869, cf. Figure 1.1) was born in Metzingen (Germany) as the son of a dyer. With only one exam in his life, which he passed on his own, without a regular education, he became one of the leading chemists in Europe. Before being nominated Professor at the University of Basel, he studied in Germany, England and France. In 1830, he received an honorary doctoral degree from the University of Basel. He also became an honorary citizen of the City of Basel in 1840 and later on was politically active in the legislative and executive government of this city (Nolte, 1999). He is best known for his discovery of ozone (1839), but he also discovered the principle of the fuel cell (1839), and gun cotton (1846). The test for ozone that he had developed on the basis of the guajac resin led to the discovery of peroxidases (1855) and is still in use as a simple screening test for colon cancer (the haemoglobin in the faeces act like peroxidases). He also was the first (von Sonntag, 2006) to use the Fe^{2+} plus H_2O_2 reaction (Schönbein, 1859), which was later termed the Fenton reaction after *Henry John Horstman Fenton* (1854–1929) who nearly forty years later looked into the reaction in more detail (Fenton, 1894; Fenton & Jackson, 1899). *Schönbein* gave the new oxygen species the name "ozone" because of its strong smell [taken from Greek "ὄζειν" (ózein): to smell (see Chapter 2) (Schönbein, 1840)] and was very close to deducing the right structure (Schönbein, 1854). He also described the reaction with iodide and the most sensitive indigo assay (Schönbein, 1854). This assay is still in use today (Chapter 2). His famous 1854 review was requested by *Justus von Liebig* (1803–1873) to be published in his "Annalen", in which he writes: "*Herr Professor Schönbein hat auf meinen Wunsch seine Untersuchungen über diesen Gegenstand für die Leser der Annalen zusammengestellt. Ich betrachte die Erscheinungen und Beobachtungen, welche dieser ausgezeichnete Forscher beschreibt, für eben so wichtig wie bedeutungsvoll für die Wissenschaft, denn es ist von jeher die Entdeckung einer neuen Eigenschaft der Materie die Quelle neuer Naturgesetze und die Quelle der Einsicht in bis dahin unerklärliche Erscheinungen gewesen.* – On my request, professor Schönbein has compiled his studies on this subject. I consider the phenomena and observations described by this distinguished scientist as important as well as significant for science, since the discovery of a novel property of matter has always been the source of new laws of nature and the source of comprehension of hitherto unexplainable phenomena."

Schönbein was not only an excellent scientist but must also have been good company (Oesper, 1929a; Oesper, 1929b). *Justus von Liebig* wrote to *Friedrich Wöhler* (1800–1882): "*Schönbeins Humor ist unschätzbar; wenn ich nur seinen Magen hätte.* – Schönbein's sense of humour is invaluable; I wish I had his stomach."

Figure 1.1 *Christian Friedrich Schönbein* (1799–1868). University Library Basel, Portrait Collection, with permission.

The history of the first hundred years of ozone chemistry has been reported in eight excellent articles by M.C. Rubin (Rubin, 2001, 2002, 2003, 2004, 2007, 2008, 2009; Braslavsky & Rubin, 2011), and here we can give only a very short account.

Schönbein had discovered ozone when he electrolysed dilute sulfuric acid and observed that it was also formed in the autoxidation of white phosphorus ("the phosphorus smell"). The latter was the standard method for obtaining ozone in the first years of ozone chemistry. He reported his discovery to the Basel Natural Science Society on 13 March 1839: "*Über den Geruch an der positiven Elektrode bei der Elektrolyse des Wassers* – On the odour at the positive electrode during electrolysis of water." *Schönbein* had already realised that low concentrations of carbon, iron, tin, zinc and lead hindered ozone production (Schönbein, 1844). For *Schönbein*, this was proof of the oxidising properties of ozone. Yet, it was more difficult at the time to derive the structure of ozone. Originally, *Schönbein* thought that ozone is related to halogens, because of its smell, which is similar to chlorine and bromine. Later, he hypothesised that it contained oxygen and hydrogen (Schönbein, 1844). Only years later, did he accept that ozone was a modification of oxygen as was described by *Jacob Berzelius* (1779–1848) in 1846 (Nolte, 1999). He writes to *Michael Faraday* (1791–1861): "*Wir können nicht länger an der Tatsache zweifeln, dass Sauerstoff in zwei verschiedenen Zuständen, in einem aktiven und einem inaktiven, in dem ozonischen und dem normalen Zustand existiert.* – We can no longer doubt the fact that oxygen exists in two different states, an active and an inactive one, in the ozonic and normal state."

The ozone generator that we use today for its production was invented by *Werner von Siemens* (1816–1882) in 1857, and only this invention made industrial applications of ozone possible.

Applications often very rapidly follow technical progress. It was less than a fortnight after the discovery of x-rays by *Wilhelm Konrad Röntgen* (1845–1923), when a physicist in Chicago realised that this biologically active radiation might be used in cancer therapy, and the first patient was treated (Grubbé, 1933; von Sonntag, 1987). Also, when reliable UV-lamps became available (Perkin, 1910), the first plant providing UV-disinfected drinking water to a community of 20,000 people was installed in the same year (von Sonntag, 1988). Similarly, very shortly after the discovery of the pathogenic agents of anthrax in 1876 and of cholera in 1884 by *Robert Koch* (1843–1910), the disinfecting power of ozone was reported (Sonntag, 1890) in the same issue as that of chlorine (Nissen, 1890). The implementation of ozone in water treatment followed about one decade later (see below and Chapter 5).

The ozone chemistry of organic compounds was first studied systematically by *Carl Friedrich Harries* (1866–1923), professor at the University of Kiel and son-in-law of *Werner von Siemens*, and it was he who coined the name "ozonide" for compounds formed in the reaction of ozone with organic compounds, notably olefins (Rubin, 2003).

A breakthrough in the understanding of ozone reactions mechanistically was achieved by *Rudolf Criegee* (1902–1975, Figure 1.2), with experiments starting in the late 1940s, and the reaction of olefins with ozone (Criegee, 1975) rightly carries his name. One of us (CvS) knew Criegee quite well, as Criegee had been his PhD examiner in Organic Chemistry at the Technical University of Karlsruhe, but, at the time, trained as a photochemist and as a radiation chemist; the candidate would never have dreamt that, one day, ozone chemistry may find his own interest as well.

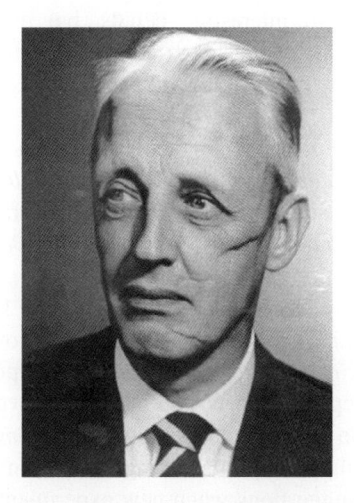

Figure 1.2 *Rudolf Criegee* (1902–1975). Chemistry Department of the Karlsruhe Institute of Technology (formerly the Technical University of Karlsruhe), with permission.

In those times, ozone chemistry was carried out largely in organic solvents (Bailey, 1978; Bailey, 1982) [for aqueous solutions see (Bailey, 1972)]. *Werner Stumm* (1924–1999) (Giger & Sigg, 1997), director of Eawag (1970–1992) realised the high potential of ozone in water treatment, but also the very limited knowledge of its reactions in aqueous solution (Stumm, 1956). He thus enforced his group by asking *Jürg Hoigné* (Giger & Sigg, 1997) to join in, and due to *Hoigné's* pioneering work on ozone chemistry in aqueous solution, the topic of this book, found more than a little interest. It was he who showed that ozone reactions in aqueous solution may induce free-radical reactions (Hoigné & Bader, 1975), reactions that seem not to occur in organic solvents. *Hoigné* also started off as a radiation chemist, and a friendship with one of us (CvS) dates back to the mid-1960s, when *Hoigné* was still an active member of the radiation chemistry community. With this background knowledge, he introduced radiation-chemical tools for elucidating aspects of ozone chemistry in aqueous solution (Bühler *et al.*, 1984; Staehelin *et al.*, 1984). The other author of this book (UvG) joined the *Hoigné* group at Eawag in 1992 and later became his successor. We (CvS and UvG) profited greatly from discussions with *Jürg Hoigné*, and it is our great pleasure to dedicate this book to him.

The disinfecting power of ozone (Sonntag, 1890) and chlorine (Nissen, 1890) were realised practically at the same time in the late 19th century. The first ozone disinfection unit was installed in 1906 in Nice (France)

(Kirschner, 1991). Not much later (1911), a UV-disinfection plant was built in nearby Marseille (von Sonntag, 1988). Despite this very early start of ozone- and UV-disinfection technologies, chlorination dominated over many decades, and it was only in the 1970s and even later, when the shortcomings of chlorination became apparent (chlorination by-products, lack of inactivation of the cysts of *Giardia* and oocysts of *Cryptosporidium*) that disinfection with ozone and UV gained in importance. Later on, the oxidation of micropollutants also became an important field of ozone application (Chapter 5).

Based on the increasing importance of ozone in drinking water and wastewater, a number of books appeared on this topic (Evans, 1972; Langlais *et al.*, 1991; Beltrán, 2004; Rakness, 2005; Gottschalk *et al.*, 2010). They often dealt with technical aspects or, when ozone chemistry was in the foreground, they no longer covered the recent developments in this area of research. Most scientific papers at conferences and in publications report interesting details, but they are not embedded in a general mechanistically based concept of ozone chemistry in aqueous solution. The present book intends to fill this apparent gap and should enable researchers to sharpen their research by applying basic mechanistic principles. Mechanistic considerations "hypotheses" are the basis of scientific progress: "*Hypothesen sind Netze, nur der wird fangen, der auswirft.* – Hypotheses are nets, only those who cast will catch." (*Friedrich Philipp Freiherr von Hardenberg* (1772–1801), "*Novalis*", German poet and scholar). Yet, there is a *caveat* that we should not stick to these concepts slavishly: "*Hypothesen sind Wiegenlieder, womit der Lehrer seine Schüler einlullt, der denkende treue Beobachter lernt immer mehr seine Beschränkung kennen, er sieht: je weiter sich das Wissen ausbreitet, desto mehr Probleme kommen zum Vorschein.* – Hypotheses are lullabies, by which the teacher lulls his pupils; the thinking and careful observer increasingly realises his limitations; he sees: the further knowledge expands, the more problems appear." (*Johann Wolfgang von Goethe* (1749–1832), German poet and scholar). Mechanisms are always open to revision, since concepts in science can never be proven and must contain the potential of falsification – otherwise they are too general and useless [*Karl Raimund Popper* (1902–1984), Austrian philosopher]. The reader will see this principle operating in relation to our work also; we had to revise our already published mechanistic suggestions when new experimental data became available. Here, we are in accord with *Schönbein* who is reported (Oesper, 1929a) to have said: "As for me, the determination of the truth is far more important than the maintenance of my views, for why should one hold fast to notions that will not withstand the criticisms of facts. The sooner they fall, the better, even though prima facie they appear ever so ingenious."

Yet, mechanistic considerations are not hatched in the ivory tower for the amusement of physical chemists, but are of great predictive value. As in analytical chemistry, one typically only finds what one is looking for. Mechanistic considerations lead to more detailed and in-depth studies.

With this concept in mind, *Maggie Smith*, responsible for IWA Publishing, and the authors agreed to launch this book at as low a price as possible to make it not only affordable for senior scientists but also for students of environmental sciences and engineering. For expanding the knowledge in ozone chemistry and application or finding an entry into the field, as many references as possible were included and updated in early 2012. Ozone rate constants in aqueous solution are compiled, updating an earlier compilation (Neta *et al.*, 1988). Managers of water supplies and wastewater treatment plants will find here the state-of-the-art in disinfection and pollution abatement using ozone and ozone-based advanced oxidation processes and a discussion of certain limitations that may be caused by problematic by-products such as bromate. Furthermore, examples of the incorporation of ozone into water and wastewater treatment schemes are given. Finally, as our drinking water resources become scarcer, notably in arid countries, a paragraph is devoted to the contribution of ozone treatment in reclamation technologies.

While writing this book, we spent much of our limited spare time in front of the computer screen or correcting drafts. This was a considerable burden on our families, and in particular on our wives, Ilsabe and Birgit, who had to miss activities that would have been fun to share. We were in the most fortunate situation that despite these sacrifices Ilsabe and Birgit gave us their loving support, which is reflected in the successful termination of this project. We are more than just most thankful for this.

Chapter 2
Physical and chemical properties of ozone

2.1 INTRODUCTORY REMARKS

Schönbein gave ozone its name because of its strong smell (Chapter 1). In his 1854 review, he calculated that ozone should be detectable by its smell at a concentration of 1 ppm (Schönbein, 1854). He also raised the question of why the nose is that sensitive in detecting ozone. One of us (CvS) hypothesises that the receptors in the nose do not record ozone as such but a strongly smelling as yet unknown product formed upon the reaction of ozone with some material contained in the skin. Evidence for this comes from a typical lab experience. When one spills some ozone water on one's hands, they smell like ozone, even after a time, when all the ozone must have evaporated/reacted completely. The smell of iron is due to the formation of unsaturated aldehydes and ketones on the skin which are in contact with iron (Glindemann et al., 2006). Similarly, it might be possible that strongly smelling volatile compounds form when ozone is in contact with skin.

The nose is a most sensitive instrument for warning that some ozone must be in the air with an indicative level of about 15 µg/m^3 and a clear detection at around 30–40 µg/m^3 (Cain et al., 2007). However, the sensitivity soon fades away, and one has the impression that ozone is no longer present to the same extent. Thus measures have to be taken *immediately* to ventilate the room and free it from toxic ozone.

The toxicity of ozone has already been described by *Schönbein*, and he mentions that about 2 mg kills a large rabbit (Schönbein, 1854). Prolonged exposures should hence be avoided. The maximum daily allowance in air at work is 200 µg/m^3 (8-h-value) in most industrialised countries (Rakness, 2005). Information on the human toxicity limits for ozone exposure is available (Kirschner, 1991; Rakness, 2005).

Some physical properties of ozone are compiled in Table 2.1.

Table 2.1 Compilation of some physical properties of ozone

Property	Value
Molecular weight	48 Da
Dipole moment	0.537 Debye
Bond length	1.28 Å

(Continued)

Table 2.1 Compilation of some physical properties of ozone (*Continued*)

Property	Value
Bond angle	117°
Melting point	−192.7°C
Boiling point	−110.5°C
Solubility in water at 0°C	2.2×10^{-2} M
Solubility in water at 20°C	1.19×10^{-2} M
Henry constant at 0°C	35 atm M^{-1}
Henry constant at 20°C	100 atm M^{-1}
Explosion threshold	10% Ozone

2.2 GENERATION OF OZONE

Schönbein discovered ozone, when he electrolysed dilute sulfuric acid. Electrolysis of sulfuric acid (20%) with gold or platinum anodes at high current density and cooling may yield oxygen with an ozone content of 4–5%. Yields may be even further increased using a platinum wire anode and by cooling to −14°C.

Electrolysis, albeit with other electrodes and ozone-resistant membranes, continues to be a convenient means for producing ozone in aqueous solution (McKenzie *et al.*, 1997). Such equipment is commercially available, but most commercial ozone generators work on the basis of a silent discharge first developed by *Werner von Siemens* in 1857. Dry air or oxygen (dew point minimum −65°C) may be used. With air, an ozone concentration of 1–5% (by weight) and with oxygen 8–16% may be reached (Rakness, 2005). Depending on pressure and energy of ignition, ozone concentrations >10% can be explosive (Koike *et al.*, 2005). Thus, high-ozone-yield systems have to be operated according to the guidelines of the suppliers. The energy requirement for ozone production is about 12–15 kWh/kg for ozone including oxygen production, transport and destruction (Hollender *et al.*, 2009; Katsoyiannis *et al.*, 2011). Based on this approach to generating ozone, the cost/energy requirements of ozone in technical applications has been evaluated (Ried *et al.*, 2009) and will be discussed in more detail in Chapter 3.

The chemistry in the plasma of the microdischarge columns is quite complex with about 300 reactions that may have to be considered (Eliasson *et al.*, 1987; Okazaki *et al.*, 1988). Under optimised conditions, the major fraction of the energy of the electrons gained in the electric field leads to excited atomic and molecular states of oxygen (feed gases: O_2 and air) and nitrogen (feed gas: air). The excited states of O_2 (O_2^*, $A^3\Sigma_u^+$, $B^3\Sigma_u^-$) dissociate according to reactions (1) and (2) (Kogelschatz *et al.*, 1988; Kogelschatz, 2003).

$$O_2 + e^- \longrightarrow O_2^* + e^- \tag{1}$$

$$O_2^* \longrightarrow 2\,O \tag{2}$$

In this oversimplified scheme, there is no differentiation between triplet O atoms, $O(^3P)$, and singlet O atoms, $O(^1D)$. Since O_2 has a triplet ground state, reaction (3) proceeds only readily with $O(^3P)$ (spin conservation rule). Ozone formation is facilitated through a three-body reaction (3) with M (O_2, O_3, O, or in case of air, N_2) being a collision partner.

$$O + O_2 + M \longrightarrow O_3^* + M \longrightarrow O_3 + M \tag{3}$$

O_3^* is the initial transient excited state of ozone. Ozone formation in reaction (3) is in competition with reactions (4)–(6) which also consume O atoms (Kogelschatz, 2003).

$$O + O + M \rightarrow O_2 + M \qquad (4)$$
$$O + O_3 + M \rightarrow 2 O_2 + M \qquad (5)$$
$$O + O_3^* + M \rightarrow 2 O_2 + M \qquad (6)$$

When air is used as the feed gas, several nitrogen species such as N^+, N_2^+, N and excited atomic and molecular species increase the complexity of the reaction system. This leads to the additional reactions (7)–(10), involving nitrogen atoms and excited molecular states of N_2 (A: $^3\Sigma_u^+$, B: $^3\Pi_g$) (Kogelschatz, 2003).

$$N + O_2 \rightarrow {}^\bullet NO + O \qquad (7)$$
$$N + {}^\bullet NO \rightarrow N_2 + O \qquad (8)$$
$$N_2(A) + O_2 \rightarrow N_2O + O \qquad (9)$$
$$N_2(A, B) + O_2 \rightarrow N_2 + 2 O \qquad (10)$$

Approximately 50% of the ozone formed in air-fed systems is produced from these nitrogen-based processes. Ozone formation through these processes is slower (ca. 100 μs) than in O_2 (10 μs), and a significant part of the electron energy which is lost through collisions with nitrogen molecules can be recovered for ozone formation by reactions (7)–(10) followed by reaction (3) (Kogelschatz, 2003). In addition to reactions (7)–(10), several other nitrogen oxide species, ${}^\bullet NO_2$, ${}^\bullet NO_3$, N_2O_5, are formed that consume ozone [reactions (11)–(14)] (Kogelschatz & Baessler, 1987).

$${}^\bullet NO + O_3 \rightarrow {}^\bullet NO_2^* + O_2 \qquad (11)$$
$${}^\bullet NO_2^* \rightarrow {}^\bullet NO_2 + h\nu \qquad (12)$$
$$2\, {}^\bullet NO + 3\, O_3 \rightarrow N_2O_5 + 3\, O_2 \qquad (13)$$
$${}^\bullet NO_2 + O_3 \rightarrow {}^\bullet NO_3 + O_2 \qquad (14)$$

${}^\bullet NO_2^*$ is an excited form of ${}^\bullet NO_2$. In typical air-fed ozone generators, ${}^\bullet NO_x$ formation is nearly 2% of ozone formation.

Feed-gases for ozone generators should be dry to avoid undesired effects on ozone generation. The singlet O atom, $O(^1D)$, reacts very quickly with water vapour by insertion [reaction (15)], cf. (Taube, 1957). The thus-formed H_2O_2 retains much vibrational energy and readily decomposes into two ${}^\bullet OH$ radicals [reaction (16)], which induce the decomposition of ozone (Chapter 13).

$$O(^1D) + H_2O \rightarrow H_2O_2^* \rightarrow 2\, {}^\bullet OH \qquad (15)/(16)$$

Furthermore, ${}^\bullet OH$ radicals react readily with ${}^\bullet NO$ and ${}^\bullet NO_2$ to HNO_2 and HNO_3. N_2O_5 also hydrolyses to HNO_3 in water (Kogelschatz & Baessler, 1987). Therefore, when the air-feed is not dry, formation of nitrous and nitric acids can lead to corrosion of metal parts in ozone generators and tubing (Kaiga et al., 1997).

Even in large-scale ozone generators where typically pure oxygen is used, the presence of nitrogen (1%) has a beneficial effect on ozone generation by increasing the ozone yield compared to pure oxygen owing to the O-forming reactions (7)–(10) (Kogelschatz, 2003).

2.3 OZONE SOLUBILITY IN WATER

Ozone is about ten times more soluble in water than oxygen (Figure 2.1), and this allows one to obtain rather high ozone concentrations by saturating water with an ozone/oxygen mixture from an ozone generator that is still rich in oxygen. For the solubility of ozone at high ozone concentrations in the gas phase, see Mizuno & Tsuno (2010).

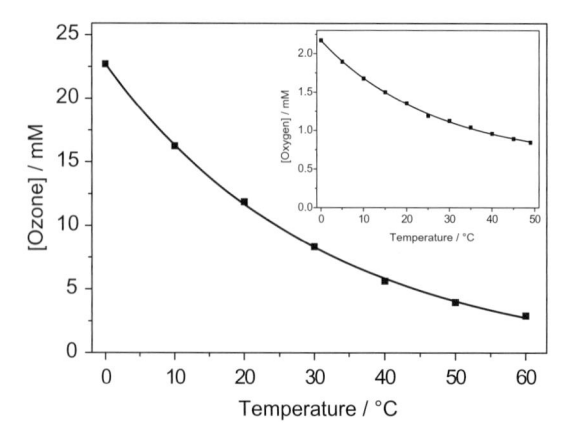

Figure 2.1 Solubility of ozone and oxygen (inset) in water as a function of the temperature for pure gases. The maximum aqueous ozone concentration for a given ozone/oxygen gas mixture can be calculated by Henry's law according to the ozone partial pressure which is achieved by a given ozone generation system.

Ozone solubility strongly depends on temperature. Ozone solubility is about twice as high at 0°C than at room temperature (Figure 2.1). Hence, cooling with ice can be used with advantage when ozone-rich stock solutions are desired. Ozone concentrations then range near 1–1.5 mM as is often required for kinetic studies (Ramseier et al., 2011).

2.4 UV–VIS SPECTRUM OF OZONE

The first absorption band of ozone in aqueous solution is at 590 nm. Its absorption is only weak [$\varepsilon = 5.1 \pm 0.1$ M^{-1}cm^{-1} (Hart et al., 1983)], and this weak absorption in the visible region causes the blue colour of concentrated solutions. The second absorption band is in the UV region, and its maximum centres at 260 nm (Figure 2.2).

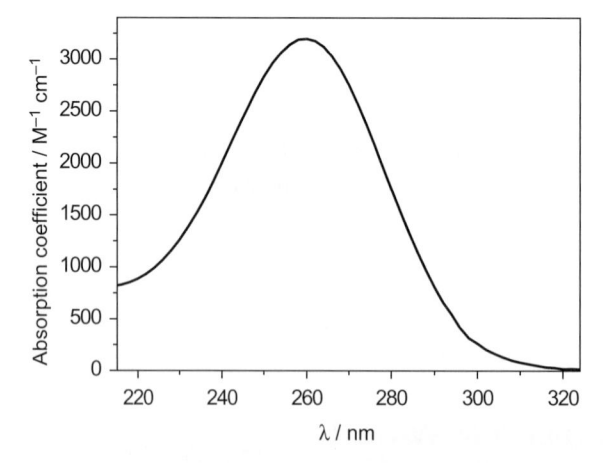

Figure 2.2 Ozone spectrum in the UV region taking a molar absorption coefficient of 3200 M^{-1}cm^{-1} at the maximum (260 nm) (courtesy A. Tekle-Röttering).

Its absorption coefficient at the maximum is a matter of continuing debate. This is not an academic question, as the ozone concentration in water is often determined by measuring the ozone absorption. The difference between the highest and the lowest reported values is about 20% (Table 2.2). This may not seem much, but when a complete material balance is attempted, 20% is non-negligible.

Table 2.2 Molar absorption coefficient of ozone at 255–260 nm in aqueous solution

Molar absorption coefficient/M^{-1} cm^{-1}	Reference
3600	Taube, 1957
2930 ± 70	Kilpatrick et al., 1956
2000	Boyd et al., 1970
2900	Bader & Hoigné, 1982
3314 ± 70	Forni et al., 1982
2950	Gilbert & Hoigné, 1983
3292 ± 70	Hart et al., 1983
3150	Hoigné, 1998

For two of those (Hoigné, 1998; Forni et al., 1982), no information is given as to how these values have been obtained. Except for the value that was based on the oxidation of Fe^{2+} (Hart et al., 1983), all values were determined by reacting ozone with I^- and measuring the iodine formed. It has been mostly assumed that one mole of ozone produces one mole of iodine according to reactions (17) and (18).

$$O_3 + I^- \rightarrow IO^- + O_2 \tag{17}$$

$$IO^- + I^- + H_2O \rightarrow I_2 + 2\,OH^- \tag{18}$$

Yet, a ratio of 1.5 (without explaining the chemistry behind this higher value) has also been given (Boyd et al., 1970); reported ratios range between 0.65 and 1.5 mol/mol (Rakness et al., 1996). Experiments confirming a somewhat higher value have as yet not been carried out, but a speculation may still be at place here. It is recalled that in reaction (17) oxygen is released to some extent as singlet oxygen ($^1O_2$) (Muñoz et al., 2001). I^- reacts moderately fast with $^1O_2$ ($k = 7.2 \times 10^6\,M^{-1}\,s^{-1}$) in competition with a decay to the ground state ($k \approx 3 \times 10^5\,s^{-1}$, in H_2O) (Wilkinson et al., 1995). At an I^- concentration near 4×10^{-2} M about 50% of $^1O_2$ has reacted with I^-. This reaction is reported to lead to I_3^- (Wilkinson et al., 1995). Thus, reaction (19), which gives rise to peroxoiodide ($\Delta G^0 = -11$ kJ mol^{-1}, Naumov and von Sonntag, 2010, unpublished results) may take place.

$$I^- + {}^1O_2 \rightleftarrows IOO^- \tag{19}$$

Peroxoiodide is in equilibrium with its conjugate acid [reaction (20)].

$$IOO^- + H^+ \rightleftarrows IOOH \tag{20}$$

The free acids of this group of compounds have very low O–O BDEs [cf. ONOOH: 92 ± 8.5 kJ mol^{-1} (Brusa et al., 2000)], and undergo rapid homolysis such as reaction (21).

$$IOOH \rightarrow IO^\bullet + {}^\bullet OH \tag{21}$$

IO$^\bullet$ provides three oxidation equivalents and $^\bullet$OH provides one leading to the overall reaction (22).

$$IOO^- + 3\,I^- + 4\,H^+ \longrightarrow 2\,I_2 + 2\,H_2O \qquad (22)$$

Reaction (19) is in competition with the reversion of $^1O_2$ to the ground state ($t_{1/2} = 5\,\mu s$ in H_2O) and hence its importance should depend on the I^- concentration used in such experiments. Reactions (19) and (20) have to occur only to a small extent to increase the value of the absorption coefficient, as two molecules of iodine are formed per $^1O_2$ reacting according to reaction (22). It is noteworthy, that Hoigné used increasingly higher absorption coefficients as his experience with ozone reactions increased (Table 2.2). There is also some information that may be drawn from the ozonolysis of olefins (Chapter 8). Without steric hindrance by bulky substituents, they give rise to a carbonyl compound and an α-hydroxyalkylhydroperoxide. Taking an absorption coefficient of 3314 $M^{-1}\,cm^{-1}$, a material balance (mol product per mol ozone) is obtained with a tendency of an excess of up to 5% (typically less, near 2%, cf. Table 6.3). This seems to indicate that the chosen absorption coefficient may be somewhat on the high side. Correcting for this, the value would come very close to the most recent value chosen by Hoigné, and it is suggested here to use a molar absorption coefficient of 3200 $M^{-1}\,cm^{-1}$ for the determination of the ozone concentration in water.

2.5 DETERMINATION OF THE OZONE CONCENTRATION

Analytical methods for determining ozone concentrations in water and the gas phase have been reviewed (Gottschalk *et al.*, 2010). The most straightforward method is measuring its absorption at 260 nm (the spectrum near the maximum is relatively broad, and an absorption maximum at 258 nm is also found in the literature. Due to this broadness, measurements at either 260 nm or 258 nm give practically identical results). We suggested above basing such measurements on an absorption coefficient of $\varepsilon(260\,nm) = 3200\,M^{-1}\,cm^{-1}$. It is well suited for the determination of ozone in stock solutions (Ramseier *et al.*, 2011b) and in waters with low UV absorbance [$A(258\,nm) < 1\,m^{-1}$] (Hoigné & Bader, 1994). Such measurements require, however, that there is no other material such as dissolved organic matter (DOM), turbidity and iron that absorb at this wavelength (Hoigné, 1994). For coping with such conditions, assays have been developed that are discussed below (Hoigné & Bader, 1994).

For on-line measurements of ozone, amperometric electrodes without and with membranes can be used (Stanley & Johnson, 1979; Langlais *et al.*, 1991; Rakness, 2005; Gottschalk *et al.*, 2010). Many systems are commercially available and are not discussed any further.

2.5.1 The *N,N*-diethyl-*p*-phenylenediamine (DPD) method

The first studies with *p*-phenylenediamines for the determination of ozone used tolidine as a substrate (Zehender, 1952; Zehender & Stumm, 1953).

The yellow colour that is formed was measured at 440 nm, but this colour faded away too rapidly, and quenching ozone first with Mn^{2+} in sulfuric acid solution has been suggested. This reaction gives rise to MnO_2 solution (Chapter 12). The subsequent addition of tolidine gave a more stable colour. *p*-Phenylenediamine and its *N*-alkylated derivatives have low reduction potentials. For the parent, a value of +309 mV and for N,N,N',N'-tetramethyl-p-phenylenediamine (TMPD) a value of +266 mV is given (Wardman, 1989). The value for DPD must lie in-between. In its reactions with one-electron oxidants, stable radical cations are formed [reaction (23)]. For TMPD this is commonly known as Wurster's blue. DPD is widely used for the determination of free and combined chlorine in drinking water (Eaton et al., 2005). The DPD radical cation is red and absorbs strongly at 551 nm.

DPD has also been proposed as an agent for the determination of ozone (Gilbert, 1981; Gilbert & Hoffmann-Glewe, 1983; Gilbert & Hoigné, 1983). Based on $\varepsilon_{app}(260\,nm) = 2.950\,M^{-1}\,cm^{-1}$ for ozone, an absorption coefficient of $\varepsilon = 19.900 \pm 400\,M^{-1}\,cm^{-1}$ has been derived (Gilbert & Hoigné, 1983). This value is different from that given for the H_2O_2 assay: $\varepsilon_{app}(551\,nm) = 21.000 \pm 500\,M^{-1}\,cm^{-1}$ (Bader et al., 1988). This difference may be due to an error in the absorption coefficient of ozone (see above) and/or in a more complex reaction.

In the reaction of TMPD with ozone, $^{\bullet}OH$ radicals are generated (via $O_3^{\bullet-}$) in a yield near 70% (Chapter 8, Table 8.4) [reaction (24)].

$$TMPD + O_3 \rightarrow TMPD^{\bullet+} + O_3^{\bullet-} \qquad (24)$$

Based on the DMSO test for $^{\bullet}OH$ formation (Chapter 14), the $^{\bullet}OH$ yield in the reaction of ozone with DPD is only 23% [Jarocki & von Sonntag (2011), unpublished results], that is, its precursor $O_3^{\bullet-}$ and hence primary $DPD^{\bullet+}$ is also only 23%. A detailed study that would have elucidated other potential reactions giving rise to $DPD^{\bullet+}$ has still to be carried out. From radiation-chemical studies it is known that $^{\bullet}OH$ in its reaction with the stronger reductant TMPD gives rise to $TMPD^{\bullet+}$ (partially via an adduct), and the reduction potential of TMPD is so low that even peroxyl radicals can also undergo this reaction. Rate constants range from 1.1×10^6 to $1.9 \times 10^9\,M^{-1}\,s^{-1}$ depending on the nature of the substituents (Neta et al., 1989; Schuchmann & von Sonntag, 1988). To what extent all this also holds for the less reducing DPD is not yet known. If not, $^{\bullet}OH$ scavenging by the water matrix may lower $DPD^{\bullet+}$ yields.

In Mn(II)-containing waters, the MnO_2 colloids that are formed upon ozonation (Chapter 12) also readily oxidise DPD. This may result in an overestimate of residual ozone concentrations when such waters are assayed by the DPD method. At an extremely high Mn(II) concentration of 8.5 mg/L (154 µM) converted to MnO_2, an ozone equivalent of 1.01 mg/L (21 µM) has been reported (Gilbert, 1981). For more typical Mn(II) concentrations of the order of 1 mg/L, the interference would be much smaller but still mimic an ozone residual. Additionally, the presence of Br^- and its oxidation to HOBr during ozonation (Chapter 10) can result in a false positive ozone residual because HOBr also reacts with DPD (Pinkernell et al., 2000). Similar interferences are observed with the indigo method (see below).

2.5.2 The indigo method

For the quantification of ozone, *Schönbein* developed the indigo method, and in his 1854 review (Schönbein, 1854) he writes at the end of it: "*Um die in einem gegebenen Luftvolumen vorhandene Menge ozonisirten*

Sauerstoffes dem Gewichte nach zu bestimmen, bediene ich mich schon seit Jahren der Indigolösung, und vielfache Versuche haben mich überzeugt, daß dieses Mittel rasch zum Ziele führt; denn mit Hülfe desselben läßt sich der Gehalt einiger Liter Luft an ozonisirtem Sauerstoff in wenigen Minuten bis zu einem kleinen Bruchtheil eines Milligrammes bestimmen, wie sich aus nachstehenden Angaben ergeben wird. – For the determination of the amount per weight of ozonised oxygen in a given volume of air, I have been using a solution of indigo for years, and many experiments have convinced me that this agent leads quickly to the goal; with its help the content of ozonised oxygen can be determined up to a fraction of a milligram within a few minutes, as can be seen from the ensuing description."

The indigo solution that *Schönbein* used has also been sulfonated. At present, the bleaching of indigotrisulfonate by ozone is measured to determine ozone concentrations. As a decrease (base: 100% with a given uncertainty) rather than an increase (base: 0%, no uncertainty) is measured, there is an intrinsic analytical uncertainty. The indigotrisulfonate that is commercially available is a technical product with an unknown purity (possibly near 85%). This material reacts very quickly with ozone, $k = 9.4 \times 10^7 \, M^{-1} \, s^{-1}$ (Muñoz & von Sonntag, 2000a). Details of the reaction have not yet been investigated, but if the site of ozone attack is the central C–C double bond, sulfonated isatine and the corresponding α-hydroxyhydroperoxide should be the primary products (Chapter 6). In contrast to the reaction of ozone with DPD, no •OH is generated in the reaction of ozone with indigotrisulfonate [Jarocki & von Sonntag (2011), unpublished results].

Indigotrisulfonic acid   **Isatine**

The indigo method, now a kind of standard method, is not a primary method, and the extent of bleaching has been based on the molar absorption coefficient of ozone (for its value see above). The purity of the indigo sample, the ozone absorption coefficient and the reaction efficiency thus determine the value of $\varepsilon(600 \, nm)$ to be used for the indigo assay. A value close to $20,000 \, M^{-1} \, cm^{-1}$ has been found (Bader & Hoigné, 1982; Muñoz & von Sonntag, 2000a).

Indigotrisulfonate is also readily oxidised by some products that may be generated by ozone with water containing impurities such as Mn(II). The rate constant with the MnO_2 colloids is $k > 10^7 \, M^{-1} \, s^{-1}$ and with permanganate, which is also formed to some extent, it is $k = 1.3 \times 10^3 \, M^{-1} \, s^{-1}$. The Mn(III) species dominating in acid solution reacts at $2 \times 10^4 \, M^{-1} \, s^{-1}$ (Reisz et al., 2008). HOBr may also react with indigo. Currently, the extent of interference is, however, not entirely clear.

The use of indigotrisulfonate for the determination of ozone residual concentrations in drinking water plants has been assessed. Indigo stock solutions are not stable and using solutions that have been standing for several weeks can cause a major underestimate of ozone residual concentrations (Rakness et al., 2010).

2.6 METHODS FOR MEASURING OZONE KINETICS

To measure rate constants for the reaction of ozone with a substrate under first-order conditions, experiments can be performed in excess of ozone or the selected substrate. Typically, a substrate concentration is chosen in excess of ozone (e.g. tenfold), and the ozone decrease is measured as a function of time. Because, the stoichiometry of the ozone–substrate reaction may deviate from 1.0, more than 1 mol of ozone may be consumed per mol of degraded substrate. Therefore, under first-order conditions, the determined rate constant for the decrease of ozone or the decrease of the substrate may deviate by more than a factor of 3 (e.g. phenol, triclosan and diclofenac, paragraph 2.6.4; monochloramine, Chapter 8). For water treatment, this has to be considered, because in real systems ozone is typically in large excess over the substrate. Thus, the second-order rate constant determined by the decrease of the substrate should be used for the assessment of substrate abatement.

There are several methods for determining the rate constant of ozone with a given compound. The most reliable ones are the direct methods. A larger error may be involved in the method that uses competition kinetics, as there is already an uncertainty, albeit typically small, in the rate constant of the competitor. Direct methods, however, may also have their problems, but these are not as straightforward. In all cases, the determination of rate constants with ozone requires extreme care to avoid reactions with •OH, which may be formed during ozonation. Therefore, kinetic measurements should be carried out at low pH, where ozone is more stable and/or in presence of •OH scavengers (Hoigné & Bader, 1983a). Methods based on reactive ozone absorption are not easy to perform and some have led to results not compatible with more straightforward methods (see below) and should be avoided if possible.

2.6.1 Ozone decay measurements

Following ozone decay as a function of time is a direct method and thus possibly the most reliable one. Here, the compound whose rate constant is to be determined is typically present in large excess (e.g. tenfold) over ozone. The other way round, ozone in a large excess over the substrate is also feasible but often not as convenient. Under such conditions, the reaction is kinetically of (pseudo-) first order. For the substrate (**M**) in excess, one may write equations (25) and (26).

$$O_3 + M \longrightarrow P \tag{25}$$

$$\frac{-d[O_3]}{dt} = k_1 [O_3] \times [M] \tag{26}$$

As the concentration of **M** does not significantly change during the reaction, [**M**] becomes a constant and equation (26) can be integrated to equation (27).

$$\ln\left(\frac{[O_3]}{[O_3]_0}\right) = -k_1[M] \times t = -k_{obs} \times t \tag{27}$$

A plot of $\ln([O_3]/[O_3]_0)$ vs. the time (t) yields a straight line from the slope of which k_{obs} is calculated and division by [**M**] yields the bimolecular rate constant k_1 (unit: $M^{-1} s^{-1}$). The ozone decay can be followed spectrophotometrically at 260 nm. The absorption coefficient of ozone at this wavelength is high (3200 $M^{-1} cm^{-1}$; for a discussion see above), but its exact value is not required here as only the absorption ratios are of relevance. Absorption of **M** in the same wavelength region as ozone does usually not affect the determination of the rate constant by this method as the same kinetics are followed even if **M** is bleached or an absorption due to the formation of **P** builds up. Strong absorptions by **M** may impede such measurements. This is typically avoided in the batch quench method (see below).

For low rate constants, kinetics can be followed in a UV-spectrophotometer set at the time-drive mode. A variation of the direct determination of ozone rate constants is the batch quench method. Here, a solution of indigotrisulfonate is added at different times, and the remaining ozone concentration is determined by the bleaching of the indigotrisulfonate (Bader & Hoigné, 1981). Alternatively, the reaction solution is dispensed into sampling tubes containing indigotrisulfonate, which quenches the residual ozone (Hoigné, Bader, 1994). The reaction of ozone with indigo is so fast [$k = 9.4 \times 10^7 \, M^{-1} \, s^{-1}$ (Muñoz, von Sonntag, 2000)] that it occurs practically instantaneously.

For high rate constants, the stopped-flow technique is of advantage. Here, the available time range allows the determination of rate constants near $10^6 \, M^{-1} \, s^{-1}$. Alternatively, quench flow techniques can be used, in which the ozone consumption is measured for various predetermined reaction times by quenching the solution with indigotrisulfonate. The bleaching of indigo, a measure for the ozone residual concentration, can then be measured off-line by spectrophotometry. The determination of rate constants with this method is in a similar range as stopped-flow – in the order of 10^5–$10^6 \, M^{-1} \, s^{-1}$ (Buffle et al., 2006b). For higher rate constants, methods based on competition kinetics are required.

For dissociating compounds where the base reacts too fast to be monitored, kinetics may be carried out in a more acidic environment. Sufficiently far from the pK_a, the observed rate constant, k_{obs}, drops by one order of magnitude per pH unit as does the concentration of the more reactive base in equilibrium. This allows one to measure the rate of reaction on a convenient timescale. Taking the pK_a of the substrate into account, extrapolation to high pH allows the calculation of the rate constant of the highly reactive base (Hoigné, Bader, 1983b). Typical examples are amines and phenols, where this difference in the rate constants is several orders of magnitude. At lower pH, the poorly reactive conjugate acid (BH^+) is present in excess, but the base (B) dominates the rate of reaction. Under such conditions, the pH-specific rate constant (k_{obs}) can be conveniently determined by equation (28).

$$k_{obs} = k(BH^+) + k(B) \times 10^{(pH - pK_a)} \tag{28}$$

2.6.2 Quenching of ozone with buten-3-ol

There may be conditions where spectral interference does not allow following the 260 nm absorption as a function of time and quenching with indigotrisulfonate cannot be used because oxidising species build up during ozonation, the progress of the reaction may then be followed by quenching ozone with buten-3-ol (Chapter 6) and measuring formaldehyde [e.g. spectrophotometrically (Nash, 1953)] generated in a 100% yield according to reaction (29) (Dowideit & von Sonntag, 1998).

$$CH_2 = CH_2C(OH)HCH_3 + O_3 \rightarrow CH_2O + H_2O_2 + HC(O)C(OH)HCH_3 \tag{29}$$

2.6.3 Reactive absorption

Ozone rate constants are sometimes also determined by making use of reactive absorption measurements. In a typical setup, 0.5 ml of a solution containing the compound whose rate constant is to be determined is placed in a polystyrene tube (12 mm i.d.) (Kanofsky & Sima, 1995). An ozone/oxygen flow passes 1.2 cm above the solution at $1.25 \, ml \, s^{-1}$. The difference between the ozone concentration in the gas inlet and outlet is measured, and the fraction of ozone absorbed after 2 min is plotted against the logarithm of the substrate concentration. Such data are evaluated on the basis of the *Reactive Absorption Theory* discussed in detail in the given reference. Another approach has also been described (Utter et al., 1992). In some cases, reliable (supported by more direct methods) rate constants were obtained. This approach has been extended to ozone and substrate uptake measurements in a stirred bubble column (Andreozzi et al., 1996). As long as there is a 1:1 ratio of ozone uptake and substrate disappearance, this approach may also yield acceptable rate constants. But when this prerequisite is not met, the method may fail (typically, values may come out too low). For example, the rate constant of diclofenac by this method gave a value of $1.8 \times 10^4 \, M^{-1} \, s^{-1}$ (Vogna et al., 2004), while the more reliable determination by competition kinetics yielded 6.8×10^5 M^{-1} s^{-1} (Sein et al., 2008) (Chapter 8). Therefore, we recommend that, whenever possible, one should stick to more direct methods including determination by competition kinetics. These methods are addressed in the next paragraph.

2.6.4 Competition kinetics

The determination of ozone rate constants of a given compound **M** requires that the ozone rate constant of the competitor **C** is known to a high accuracy, that is, it should have been determined by a reliable direct method.

In competition kinetics, two substrates **M** and the competitor **C** react with ozone [reactions (30), rate constant k_m and (31), rate constant k_c] (Dodd, 2008).

$$M + O_3 \longrightarrow \text{oxidation products of M} \qquad (30)$$

$$C + O_3 \longrightarrow \text{oxidation products of C} \qquad (31)$$

The relative degradations as a function of the ozone concentration are then given by equation (32).

$$\ln\left(\frac{[M]}{[M]_0}\right) = \ln\left(\frac{[C]}{[C]_0}\right) \times \frac{k_m}{k_c} \qquad (32)$$

For this approach, it is required that **M** and **C** are degraded by ozone with the same efficiency, for example, unity efficiency. An efficiency of unity is often found, for example, with olefins (Chapter 6). But, with some aromatic compounds, marked deviations from an efficiency of unity have been reported, for example, phenol [~0.42 (Mvula & von Sonntag, 2003)], triclosan [0.41 (Suarez et al., 2007)] and diclofenac [~0.4 (Sein et al., 2008)] (Chapters 7 and 8). The reasons for such deviations are not yet fully understood. Apparently, there are fast side reactions that compete with the destruction of the substrate. These will continue to occur under the conditions of the competition kinetics as well. Thus, such deviations will result in an under/overestimation of the rate constant when determined according to equation (32). The error will be typically not more than a factor of two or three, and this is often quite acceptable.

The second approach is based on the measurement of just the competitor **C**. While the product of the reaction with **C** can be monitored, the reaction with **M** remains silent. Detection can be by bleaching of **C** or build-up of absorption or by the formation of a specific product due to the formation of **C***.

$$C + O_3 \longrightarrow C^* \text{ (detected)} \qquad (33)$$

$$M + O_3 \longrightarrow P \text{ (not detected)} \qquad (34)$$

At a given ozone concentration ($[O_3]_0 \ll$ [**M**] and [**C**]) relationship (35) holds ([**C***]$_0$ is the concentration of **C*** in the absence and [**C***] in the presence of **M**).

$$\frac{[C^*]}{[C^*]_0} = \frac{k_c[C]}{k_c[C] + k_m[M]} \qquad (35)$$

This can be rearranged into equation (36).

$$\frac{[C^*]_0}{[C^*]} = \frac{k_c[C] + k_m[M]}{k_c[C]} = 1 - \frac{k_m[M]}{k_c[C]} \qquad (36)$$

Plotting $([C^*]_0/[C^*] - 1)$ vs. $[M]/[C]$ yields a straight line with a slope of k_m/k_c. Since k_c is known, k_m can be calculated.

Various potential competitors have been discussed (Muñoz & von Sonntag, 2000a). A most convenient one is buten-3-ol. Its solubility in aqueous solution is high, as is its ozone rate constant [$k = 7.9 \times 10^4$ M^{-1} s^{-1} (Dowideit & von Sonntag, 1998)]. One of its ozonation products, formaldehyde, can be readily determined (cf. Paragraph 2.6.2). The use of competitors with pH-dependent rate constants, for example, phenol or olefinic acids should be avoided, as a small uncertainty in the pH changes the observed rate constant significantly.

2.7 REDUCTION POTENTIALS OF OZONE AND OTHER OXYGEN SPECIES

The reduction potential of ozone is of relevance, whenever one-electron transfer reactions have to be considered. Although other ozone reactions dominate (e.g. Chapters 6, 7 and 8), one-electron transfer reactions often take place in competition. There are two potential routes, an inner sphere type electron transfer, that is, when an adduct is formed first that subsequently decays into $O_3^{\bullet-}$ and the corresponding radical cation. Alternatively, an outer-sphere electron transfer may take place. The energetics of the latter is given by the reduction potential shown in Table 2.3. Reduction potentials [indicated by (g)] corrected for the solubility of ozone and oxygen can be calculated by equation (37).

Table 2.3 Reduction potentials, $E°$ (vs. NHE) at pH 7 in V at 25°C, of O_2 and water, partially reduced intermediates, ozone, and singlet dioxygen according to (Koppenol et al., 2010). Reduction potentials of gases are based on their saturated solutions, that of other solutes on their 1 M solution. For making reduction potentials comparable with the reduction potentials of solids, such standard reduction potentials [indicated by (g)] were converted in the table to the basis of 1 M according to the solubility of ozone and O_2 (see Figure 2.1)

Reaction	Reduction potential/V
$O_2 + e_{aq}^- \rightleftharpoons O_2^{\bullet-}$	−0.18
$O_2(g) + e_{aq}^- \rightleftharpoons O_2^{\bullet-}$	−0.35
$^1O_2 + e_{aq}^- \rightleftharpoons O_2^{\bullet-}$	+0.81
$^1O_2(g) + e_{aq}^- \rightleftharpoons O_2^{\bullet-}$	+0.64
$O_3 + e_{aq}^- \rightleftharpoons O_3^{\bullet-}$	+1.03
$O_3(g) + e_{aq}^- \rightleftharpoons O_3^{\bullet-}$	+0.91
$O_2^{\bullet-} + e_{aq}^- + 2H^+ \rightleftharpoons H_2O_2$	+0.91
$HO_2^{\bullet} + e_{aq}^- + H^+ \rightleftharpoons H_2O_2$	+1.05
$O_2 + 2e_{aq}^- + 2H^+ \rightleftharpoons H_2O_2$	+0.36
$O_2(g) + 2e_{aq}^- + 2H^+ = H_2O_2$	+0.28
$H_2O_2 + e_{aq}^- + H^+ \rightleftharpoons {}^{\bullet}OH + H_2O$	+0.39
$H_2O_2 + 2e_{aq}^- + 2H^+ = 2\,H_2O$	+1.35
$O_2 + 4e_{aq}^- + 4H^+ \rightleftharpoons H_2O$	+0.85
$O_2(g) + 4e_{aq}^- + 4H^+ \rightleftharpoons H_2O$	+0.81
${}^{\bullet}OH + e_{aq}^- + H^+ \rightleftharpoons H_2O$	+2.31

$$E = E^0 - 0.059 \log[\text{red}]/[\text{ox}] \tag{37}$$

For O_2 reduction to $O_2^{\bullet-}$ ($[O_2] \approx 1.2$ mM) this leads to equations (38)–(40).

$$E = E^0 - 0.059 \log[O_2^{\bullet-}]/[O_2] = E^0 + 0.059 \log[O_2] - 0.059 \log[O_2^{\bullet-}] \tag{38}$$

$$E = -0.18 - 0.059 \times 2.92 - 0.059 \log[O_2^{\bullet-}] \tag{39}$$

$$E = -0.35 - 0.059 \log[O_2^{\bullet-}] \text{ (i.e., } E^0[O_2]/[O_2^{\bullet-}] = -0.35\text{V for a saturated oxygen solution)} \tag{40}$$

2.8 STABILITY OF OZONE SOLUTIONS

Aqueous ozone solutions are unstable. Many effects contribute to this instability, but not all of them are fully elucidated. In basic solutions, ozone is especially unstable. This is due to the formation of $^{\bullet}$OH by OH^- (Chapter 11) and the reaction of $^{\bullet}$OH with ozone (Chapter 13). This reaction proceeds even in neutral solutions, where the OH^- concentration is very low (1×10^{-7} M). Acidification and the addition of $^{\bullet}$OH scavengers such as bicarbonate further increase the ozone stability in aqueous solutions. In acid solutions and at 31°C, the rate constant of ozone decomposition has been reported at 3×10^{-6} s^{-1} ($E_a = 82.5 \pm 8.0$ kJ mol^{-1}) (Sehested et al., 1992). Mechanistic details of the 'spontaneous decomposition' are not yet fully understood.

In natural waters, the dissolved organic matter (DOM) contributes significantly to ozone decay, and waters that have a low DOM and high bicarbonate content show relatively high ozone stability (Chapter 3), which is of relevance for the disinfection efficiency of ozone (Chapter 4).

2.9 REACTIVITY OF OZONE

In micropollutant abatement, for example, the reactivity of a micropollutant determines the efficiency of its elimination by an ozone treatment. Ozone rate constants may vary 8–10 orders of magnitude even within one group of compounds. Cases in point are olefins (Chapter 6) and aromatic compounds (Chapter 7) and also compounds which carry C–H functions as only ozone-reactive sites (Chapter 10). In general, ozone rate constants depend on temperature, but there are only very few cases, where details have been measured. The temperature dependence of the second order rate constants can be expressed by the Arrhenius equation [A: pre-exponential factor, E_a: Activation energy, R: Universal gas constant, T: absolute temperature (K)] (41).

$$k = A \times e^{-\frac{E_a}{RT}} \tag{41}$$

To determine the parameters A and E_a, equation (41) can be logarithmised, yielding equation (42).

$$\log k = \log A - \frac{1}{2.3} \frac{E_a}{RT} \tag{42}$$

A plot of $\log k$ versus $1/T$ allows the determination of $\log A$ and E_a. Available data are compiled in Table 2.4 and some plots are shown in Chapters 9 and 10.

All these compounds react only slowly with ozone. For the more reactive ones, much lower activation energies are expected and hence the temperature dependence of the rate constants will be less pronounced.

There may be a dramatic effect of pH when the site of ozone attack can be deprotonated/protonated, such as in the case of phenols/amines or inorganic ions (Chapters 7, 8, 11 and 12).

Table 2.4 Compilation of available reaction parameters (rate constants at 20°C) for ozone reactions in aqueous solution

Compound	$k/M^{-1}s^{-1}$	log A	$E_a/kJ\,mol^{-1}$	Reference
Cl$^-$	1.4×10^{-3}	10.3	74	Yeatts & Taube, 1949
Br$^-$	160	8.8	37	Haag & Hoigné, 1983a
ClO$^-$	35	12.2	59.8	Haag & Hoigné, 1983b
BrO$^-$	505	13.4	60	Haag & Hoigné, 1983a
(CH$_3$)$_2$SO	1.8	11.5	63.1	Reisz & von Sonntag, 2011*
H$_2$O$_2$	0.036	11.5	73.5	Sehested et al., 1992
HC(O)O$^-$	82	10.9	50.3	Reisz & von Sonntag, 2011*
HC(O)O$^-$	46**	11.4	54.6	Reisz & von Sonntag, 2011*
HC(CH$_3$)$_2$OH	0.83	12.5	70.7	Reisz & von Sonntag, 2011*
(CH$_3$)$_3$COH	0.0011	9.3	68.7	Reisz & von Sonntag, 2011*

*Reisz & von Sonntag, 2011 (unpublished results)
**In the presence of tBuOH

2.9.1 pH dependence of ozone reactions and the "reactivity p*K*"

Whenever ozone reacts with a compound that can be present in different protonation states, there will be a pH dependence of the rate constant. When deprotonated, the electron density within a given molecule is higher, and due to the electrophilicity of ozone the rate constant for the reaction with ozone is also higher. The magnitude of this pH effect strongly depends on the type of ozone reaction. With amines, for example, where ozone adds to the lone pair at nitrogen, the high reactivity of the free amine (in the order of $10^6\,M^{-1}\,s^{-1}$) drops to nearly zero when this reaction site is blocked upon protonation or is significantly lowered by complexation with a transition metal ion (Chapter 8). With phenols, on the other hand, there is already a marked activation of the aromatic ring by the OH group ($k = 1.3 \times 10^3\,M^{-1}\,s^{-1}$). This is, however, strongly increased upon deprotonation of the phenol, resulting in a rate constant of $1.4 \times 10^9\,M^{-1}\,s^{-1}$ for the phenolate ion (Chapter 7). In olefinic acids the anion supplies some additional electron density to the C–C double bond, and in acrylic acid, for example, the rate constant of the free acid is $k = 2.8 \times 10^4\,M^{-1}\,s^{-1}$, while that of the anion is $1.6 \times 10^5\,M^{-1}\,s^{-1}$ (Chapter 6). This is only a factor of 5.7, while deprotonation of phenol increases the rate constant by a factor of about one million. Quantum-chemical calculations discussed in Chapter 7 can account for this dramatic difference.

pH effects are not restricted to organic compounds but are also observed with inorganic compounds. A case in point is As(III). With As(OH)$_3$, ozone reacts with $5.5 \times 10^5\,M^{-1}\,s^{-1}$, while As(OH)$_2O^-$ reacts with $1.8 \times 10^8\,M^{-1}\,s^{-1}$ (Chapter 12). Similarly, HS$^-$ reacts five orders of magnitude faster than H$_2$S (Chapter 11).

Such differences in the rate constants may result in marked effects on product distribution as a function of pH, which does not follow the pK_a of the starting material. This effect has been termed "reactivity p*K*".

If compound **MH** is an acid that is in equilibrium with its anion **M**$^-$ [equilibrium (43)], it reacts with ozone according to equation (44), where α is the degree of dissociation and $k(O_3+$**MH**$)$ and $k(O_3+$**M**$^-)$ are the ozone rate constants of **M** and **M**$^-$, respectively.

$$\text{MH} \rightleftarrows \text{M}^- + \text{H}^+ \tag{43}$$

$$d[\text{MH}]_{total}/dt = (\alpha \times k(O_3 + \text{MH}) + (1-\alpha)\,k(O_3 + \text{M}^-)) \times [\text{MH}]_{total} \times [O_3] \tag{44}$$

The degree of dissociation (α) can be calculated on the basis of the dissociation constant (K) and the pH [equation (45)].

$$\alpha = 1/(1 + K/[\text{H}^+]) \tag{45}$$

The situation for phenol is shown in Figure 2.3. On the basis of the above rate constants, a reactivity pK_a of 4 is calculated. This means that despite the fact that phenol has a pK_a of 10, a predominant reaction of ozone with the neutral form will only occur below pH 4.

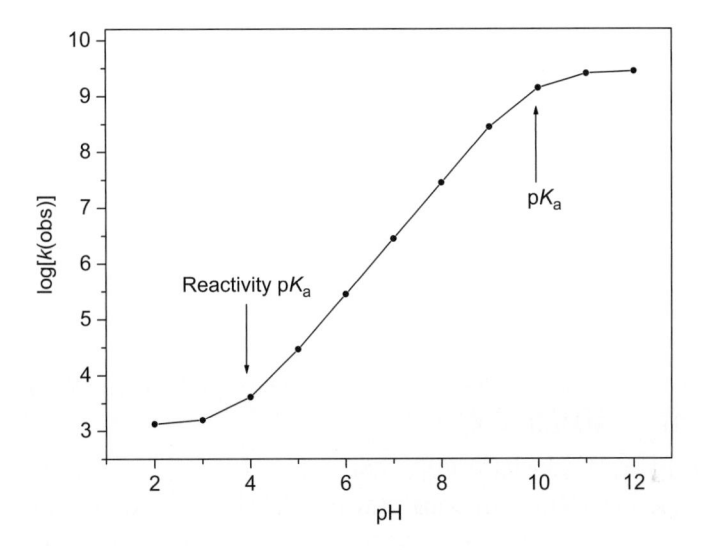

Figure 2.3 Plot of the logarithm of the observed rate constant of the reaction of phenol with ozone as a function of pH. pK_a(phenol) and its reactivity pK_a are indicated by arrows.

2.9.2 Multiple reaction sites within one molecule

There may be different ozone-reactive sites within a molecule such as an aromatic ring and an aliphatic amino group. This situation is found in some pharmaceuticals such as in the beta blocker metoprolol. For other compounds with multiple sites of attack, the situation may become more complicated.

In metoprolol, a compound containing an amino group and an aromatic ring (for its structure see Chapter 8), the amino group loses its ozone reactivity upon protonation. Thus at low pH, the overall reactivity will only be due to the rate constant of the aromatic ring [in the case of metoprolol: $k(\text{metoprolol–H}^+) = 330 \text{ M}^{-1} \text{ s}^{-1}$, (Benner et al., 2008)], while at high pH the ozone reactivity is determined by the reaction of the free amine [pK_a(metoprolol) = 9.7; $k(\text{metoprolol}) = 8.6 \times 10^5 \text{ M}^{-1} \text{ s}^{-1}$ (Benner et al., 2008)]. The percentage of ozone reacting with the free amine in equilibrium as a function of pH is shown in Figure 2.4.

The pH at which the two reaction sites, aromatic ring and amino group, react equally fast is again a kind of "reactivity pK". Due to the much higher rate constant of the free amine, the reactivity pK of metoprolol is much lower than its pK_a value. The reactivity pK for metoprolol (Meto) is calculated at 6.3 on the basis of equations (46–52).

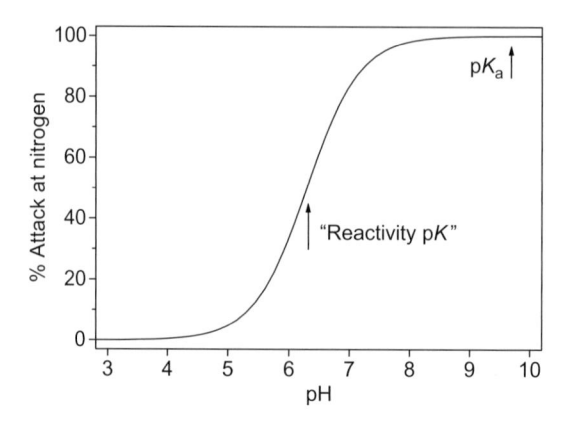

Figure 2.4 Reaction of ozone with metoprolol (pK_a = 9.7). Percentage of ozone attack at nitrogen as a function of the pH. The "reactivity pK" is at 6.3.

$$k(\text{Meto} - \text{H}^+) \times [\text{Meto} - \text{H}^+] = k(\text{Meto}) \times [\text{Meto}] \tag{46}$$

$$[\text{Meto}] = [\text{Meto}]_{\text{total}} \times K_a/(K_a + [\text{H}^+]) \tag{47}$$

$$[\text{Meto} - \text{H}^+] = [\text{Meto}]_{\text{total}} \times [\text{H}^+]/(K_a + [\text{H}^+]) \tag{48}$$

$$k(\text{Meto} - \text{H}^+) \times [\text{Meto}]_{\text{total}} \times [\text{H}^+]/(K_a + [\text{H}^+]) = k(\text{Meto}) \times [\text{Meto}]_{\text{total}} \times K_a/(K_a + [\text{H}^+]) \tag{49}$$

$$k(\text{Meto} - \text{H}^+) \times [\text{H}^+] = k(\text{Meto}) \times K_a \tag{50}$$

$$\log[(k(\text{Meto} - \text{H}^+))] - \text{pH} = \log[k(\text{Meto})] - \text{p}K_a \tag{51}$$

$$\text{"reactivity p}K\text{"} = \text{p}K_a - \log[k(\text{Meto})] + \log[k(\text{Meto} - \text{H}^+)] \tag{52}$$

This is of a practical consequence for the formation of transformation products. In wastewater that has typically a pH of 7–8, metoprolol is almost entirely degraded via a reaction of ozone at the amino group despite the fact that the pK_a of protonated metoprolol is 9.7. Therefore, under these conditions, mainly transformation products resulting from an ozone attack on the amino group will be formed.

Chapter 3
Ozone kinetics in drinking water and wastewater

3.1 STABILITY OF OZONE IN VARIOUS WATER SOURCES

The stability of ozone in drinking water and in wastewater is largely determined by its reaction with the dissolved organic matter (DOM). The nature of DOM varies among waters of different origin as does its concentration. For example, in drinking waters, DOM, measured as DOC, is typically below 4 mg/L, while in wastewaters it ranges between 5 and 20 mg/L. The nature of DOM has an influence on the rate of its reaction with ozone and thus on the ozone lifetime in these natural waters, drinking waters and wastewaters. Carbonate alkalinity influences ozone stability by scavenging •OH (see below). This is of major importance, as the two desired effects of ozone, disinfection and micropollutant abatement, depend on the lifetime of ozone in these waters (see below). Therefore, it will be necessary to discuss the properties of aquatic DOM as much as they are known today and then discuss the lifetime of ozone in different waters.

An aspect of considerable consequence in this context is •OH production, resulting from a side reaction of ozone with DOM. The •OH radical is an important intermediate in the decomposition of ozone in water (Hoigné & Bader, 1975). In a study on wastewater that may be generalised on this point to other DOMs as well, it has been suggested that ozone reacts with the electron-rich aromatic components of DOM by electron transfer [reaction (1)] (Nöthe et al., 2009; Pocostales et al., 2010).

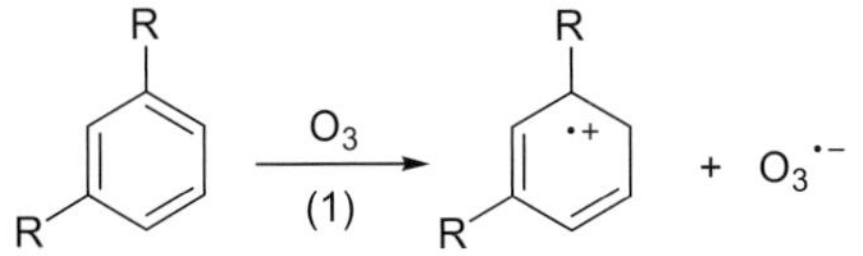

DOM model

The $O_3^{•-}$ radical gives rise to •OH [reactions (2) and (3)] (Merényi et al., 2010a) (Chapter 11).

$$O_3^{•-} \rightleftarrows O_2 + O^{•-} \tag{2}$$

$$O^{•-} + H_2O \rightleftarrows {}^{•}OH + OH^- \tag{3}$$

Production of •OH does not cease when the electron-rich aromatics present in DOM have reacted with ozone (Nöthe et al., 2009). Hence, new electron-rich sites must be created upon the action of ozone. A typical reaction of ozone with aromatic compounds is hydroxylation, and it has been suggested that phenols thus formed are responsible for the continuing •OH production (Nöthe et al., 2009). In this context, it may be recalled that ozone adducts to aromatic compounds [reaction (4)] can eliminate singlet oxygen [$^1O_2$, reaction (5)], and the resulting zwitterion rearranges into phenol [reaction (6)] (Mvula et al., 2009) (cf. Chapter 7).

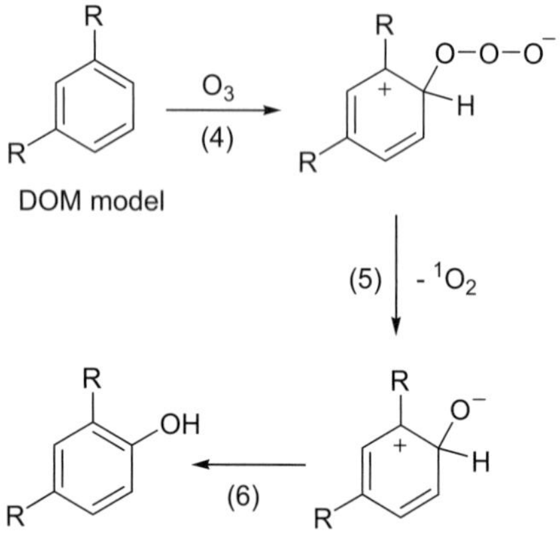

Phenols belong to the group of electron-rich aromatics that undergo electron transfer to ozone, cf. reaction (1) (Mvula & von Sonntag, 2003) and thus are capable of continuing •OH production according to reactions (2) and (3). Moreover, •OH adds to aromatic compounds of DOM [reaction (7)].

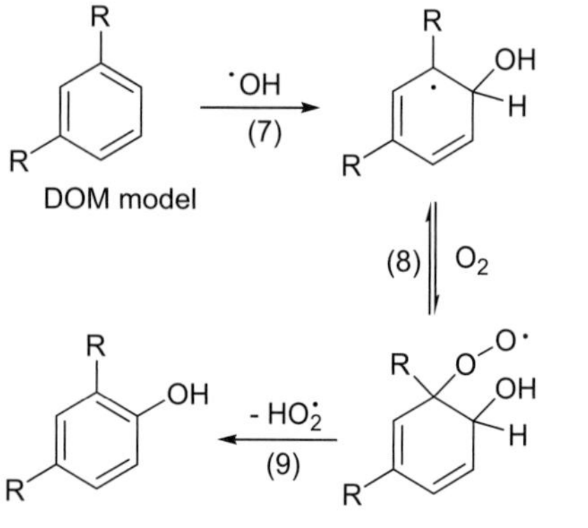

In the presence of O_2 (available at high concentrations during ozonation), the thus-formed hydroxycylohexadienyl radicals are in equilibrium with the corresponding peroxyl radicals [equilibrium (8)], which eliminate $HO_2^•$ (in competition with other reactions) (Pan et al., 1993; Fang et al., 1995; Naumov & von Sonntag, 2005) [reaction (9)]. In wastewater but also in drinking water, $HO_2^•$ is deprotonated to $O_2^{•-}$ [equilibrium (10), $pK_a(HO_2^•) = 4.8$ (Bielski et al., 1985)], and the latter reacts rapidly with ozone [reaction (11)] (Chapter 13). There are indications of the formation of $O_2^{•-}$ in the reaction of ozone with DOM [cf. reaction (9) and (10)] (Staehelin & Hoigné, 1982).

$$HO_2^• \rightleftharpoons O_2^{•-} + H^+ \tag{10}$$

$$O_2^{•-} + O_3 \rightarrow O_2 + O_3^{•-} \tag{11}$$

In these reactions, the aromatic DOM subunits are not yet mineralised but only hydroxylated. This is the likely reason why •OH generation does not cease during ozonation even at elevated ozone doses. In addition, other moieties of DOM such as aliphatic C—H bonds also lead to superoxide and eventually •OH through the above reactions (see below and Chapter 14). These reactions also apply to DOM in drinking waters.

Figure 3.1 shows the ozone stability in five Swiss waters with various compositions (DOC and alkalinity).

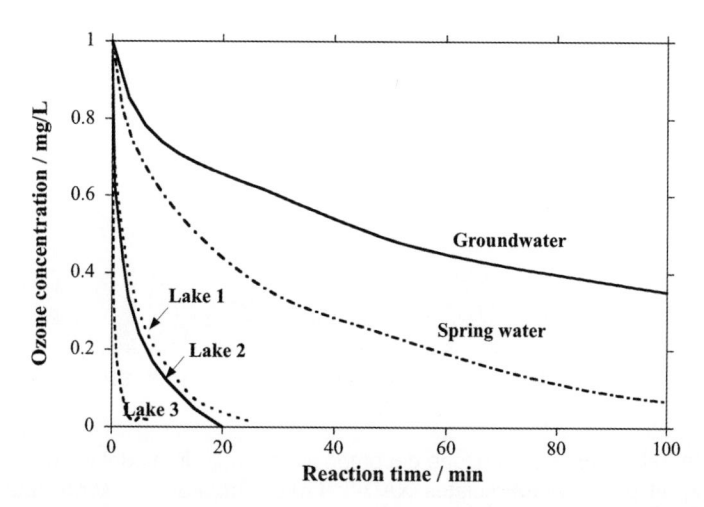

Figure 3.1 Stability of ozone in various Swiss natural waters at pH 8 and 15°C (ozone dose 1 mg/L). Water quality data: Groundwater (DOC 0.7 mg/L, carbonate alkalinity 6.7 mM); Spring water (DOC 0.9 mg/L, carbonate alkalinity 5.4 mM); Lake 1 (DOC 1.3 mg/L, carbonate alkalinity 2.5 mM); Lake 2 (DOC 1.6 mg/L, carbonate alkalinity 3.6 mM); Lake 3 (DOC 3.2 mg/L, carbonate alkalinity 3.4 mM). From Urfer et al., 2001 with permission.

At pH 8, ozone stability decreases in the sequence groundwater > spring water > lake water 1, 2 > lake water 3. This corresponds to an increasing trend in DOC concentration and a decreasing trend in alkalinity. Ozone has a very similar stability in lake waters 1 and 2, even though the DOC is higher in lake water 2. Carbonate alkalinity which has a stabilising effect on ozone is, however, higher in lake water 2. The effect of carbonate alkalinity has been systematically tested in Lake Zurich water, by varying the carbonate/bicarbonate concentration from 0 to 2.5 mM at pH 8. While keeping the DOC constant, an increase in carbonate alkalinity leads to a significantly lower rate of ozone decomposition (Elovitz et al., 2000a).

In a survey of eleven DOM isolates, a hundredfold variation of the approximate pseudo-first-order rate constant, k_{DOC}, for the ozone decrease was observed for synthetic waters containing 2 mg/L of the DOM isolate and 2.5 mM HCO_3^- at pH 7 (Figure 3.2) (Elovitz et al., 2000b). This figure shows that Suwannee River fulvic and especially humic acids are outliers and do not represent ozone consumption kinetics of typical DOM in surface waters. Therefore, results from the wide application of these sources of organic matter to simulate drinking water ozonation have to be interpreted with caution and do not allow generalisations on other water sources.

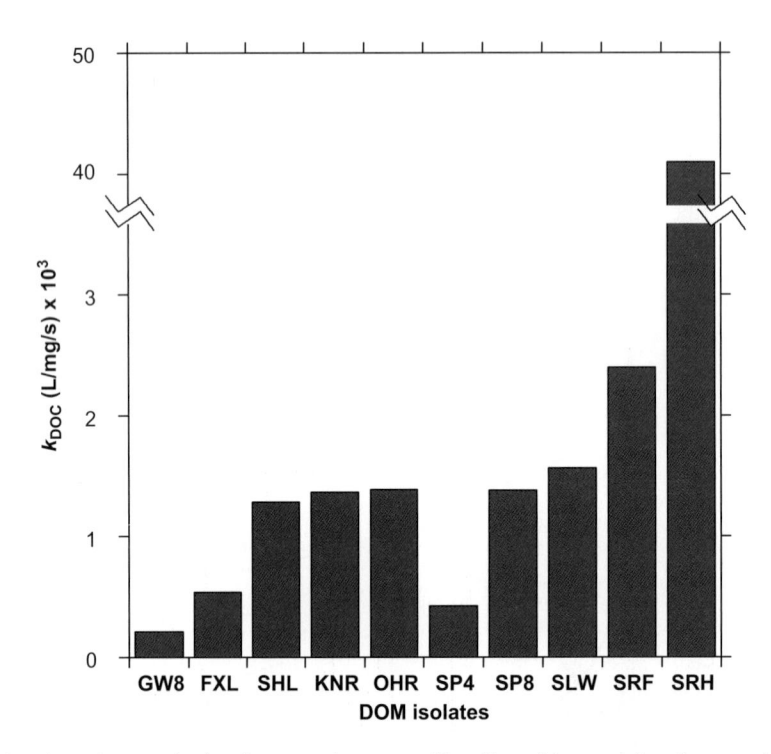

Figure 3.2 First-order rate constants of ozone decomposition (k_{DOC}) in model waters containing $\cong$ 2 mg/L DOC, 2.5 mM HCO$_3^-$ at pH 7. DOM isolates XAD-8: GW8, Groundwater Minnesota, USA; FXL, Lake Fryxell, Antarctica; SHL, Lake Shingobee, Minnesota; KNR, Yakima River, Washington; OHR, Ohio River, Ohio; SP8, California State Project Water, California; SLW, Silver Lake, Colorado. DOM isolate XAD-4: SP4, California State Project Water, California. SRF and SRH: Suwannee River, Georgia fulvic and humic acid, respectively. Reprinted with permission from Elovitz et al., 2000b. Copyright (2000) American Chemical Society.

The role of DOM and alkalinity for ozone stability can be explained by (i) the direct reaction of DOM with ozone and (ii) •OH scavenging by DOM and carbonate/bicarbonate. Whereas the type of DOM is highly relevant for ozone reactions, it is of minor importance for •OH scavenging. Second order rate constants for the reaction of various DOM sources with •OH are reported to vary within about a factor of two with an average value of $(2.3 \pm 0.77) \times 10^4$ (mgC/L)$^{-1}$ s^{-1} (Brezonik & Fulkerson-Brekken, 1988; Reisz et al., 2003). This value is very close to a more recent study with DOM isolates with an average value of 1.9×10^4 (mgC/L)$^{-1}$ s^{-1} (Westerhoff et al., 2007). The DOM isolate with the highest rate constant is a wastewater [3.9×10^4 (mgC/L)$^{-1}$ s^{-1}] which might have significantly different properties, that is, less coiling and a higher fraction of the carbon available for reaction with •OH compared to natural DOM sources. In other wastewater studies, values of 3×10^4 (mgC/L)$^{-1}$ s^{-1} and 3.5×10^4 (mgC/L)$^{-1}$ s^{-1} were found (Nöthe et al., 2009; Katsoyiannis et al., 2011).

The scavenging of •OH by DOM and bicarbonate has different effects on ozone stability. Note that at the typical pH values of drinking water and wastewater (pH $\leq$ 8.5) the contribution of •OH scavenging by carbonate is smaller despite the fact that the rate constant of carbonate with •OH ($k = 3.9 \times 10^8$ M^{-1} s^{-1}) is higher than that of bicarbonate ($k = 8.5 \times 10^6$ M^{-1} s^{-1}). The reason for this is the high pK_a of bicarbonate [$pK_a(HCO_3^-) = 10.3$; the corresponding reactivity pK (cf. Chapter 2) is 8.6]. The reaction of $^\bullet OH$ with bicarbonate leads to carbonate radicals, $CO_3^{\bullet-}$ [reaction (15); $pK_a(HCO_3^\bullet) < 0$ (Czapski et al., 1999)], Part of the $^\bullet OH$ reaction with DOM leads to $O_2^{\bullet-}$ (see above), and $O_2^{\bullet-}$ reacts quickly with ozone to $^\bullet OH$, whereas $CO_3^{\bullet-}$ further reacts by self-reaction [reaction (16)] and with DOM [Suwannee River Fulvic acid, $k = 280 \pm 90$ (mg of C/L)$^{-1}$ s^{-1} (Canonica et al., 2005)].

Because of the fast reaction of $O_2^{\bullet-}$ with ozone, leading to $^\bullet OH$ (Chapter 13), this pathway leads to a destabilisation of ozone. Hoigné introduced the concept of "promoter" and "inhibitor" for compounds that accelerate or reduce the rate of ozone decay (Staehelin & Hoigné, 1985). This concept, often cited in the literature and repeated like a prayer wheel, is sometimes misunderstood, and it seems to be helpful to discuss it here in some detail. A typical promoter is methanol at pH 8. The reaction may be induced by the slow reaction of ozone with methanol, which may generate $^\bullet CH_2OH$ radicals (cf. Chapter 10). These react readily with O_2 present in excess [reaction (12)]. The resulting peroxyl radical can eliminate $HO_2^{\bullet-}$, but this reaction is slow ($k < 3$ s^{-1}) (Rabani et al., 1974). At pH 8, the OH$^-$ concentration is 1×10^{-6} M. OH$^-$ induces an $O_2^{\bullet-}$ elimination with a bimolecular rate constant close to 1.8×10^{10} M^{-1} s^{-1} (Rabani et al., 1974). Thus at pH 8, the rate of $O_2^{\bullet-}$ formation is near 1×10^4 s^{-1} [reaction (12)–(14), for details see Chapter 14].

$$^\bullet CH_2OH + O_2 \rightarrow {}^\bullet OOCH_2OH \tag{12}$$

$$^\bullet OOCH_2OH \rightarrow CH_2O + HO_2^\bullet \tag{13}$$

$$^\bullet OOCH_2OH + OH^- \rightarrow CH_2O + H_2O + O_2^{\bullet-} \tag{14}$$

The product of reaction (16) gives rise to percarbonate upon hydrolysis. This sequence induces a very efficient chain reaction, that is, ozone decay is promoted. Scavenging of $^\bullet OH$ by bicarbonate [reactions (15) and (16), $2k_{16} = 1.25 \times 10^7$ M^{-1}s^{-1} (Weeks & Rabani, 1966)] or carbonate, interrupts the chain and enhances the lifetime of ozone. It acts as an inhibitor.

$$HCO_3^- + {}^\bullet OH \rightarrow CO_3^{\bullet-} + H_2O \tag{15}$$

$$2\, CO_3^{\bullet-} \rightarrow (CO_3^-)_2 \text{ (precursor of percarbonate)} \tag{16}$$

The formation of percarbonate according to reaction (16) deserves a note. Percarbonate is in equilibrium with H_2O_2 and bicarbonate (Richardson et al., 2000), and is a stronger oxidant than H_2O_2 (Bennett et al., 2001; Yao & Richardson, 2000). It has been suggested as a suitable agent for pollution control, at least in a limited number of cases (Xu et al., 2011).

Other $^\bullet OH$ scavengers such as tertiary butanol (tBuOH) and acetate also interrupt the chain, although some $O_2^{\bullet-}$ is also formed in the bimolecular decay of their peroxyl radicals [for tBuOH see Schuchmann & von Sonntag (1979), details in Chapter 14, for acetate see Schuchmann et al., (1985)], the dominant decay routes do not give rise to $O_2^{\bullet-}$. Thus, these compounds also act effectively as inhibitors, although with a small promoting contribution. Other organic compounds act in a similar way, promoting and inhibiting as well. A case in point is DOM as discussed above. Thus, a careful look at the concept of "promoter" and "inhibitor" is required, and its discussion should take into consideration the advantages and the limitations of this approach.

To mimic the interaction between DOM moieties leading to $O_2^{\bullet-}$ formation and components (promoters) that scavenge $^\bullet OH$ without further reaction with ozone (inhibitors), ozone stability was investigated in a synthetic system containing methanol which releases $O_2^{\bullet-}$ after H-abstraction by $^\bullet OH$ and oxygen addition and either acetate, tBuOH or bicarbonate, which do not react further with ozone (Elovitz & von Gunten, 1999).

Table 3.1 shows the results for the synthetic systems at pH 8. For a methanol concentration of 70 μM (scavenging rate $k(^\bullet OH + MeOH) \times [MeOH] = 7 \times 10^4$ s^{-1}), the pseudo-first-order rate constant for ozone decrease k_d is very similar for all scavengers, which act as inhibitors in this case.

Table 3.1 Pseudo first-order rate constants for ozone decomposition (k_d) and calculated R_{ct} values for model systems containing 70 µM methanol and the inhibitors acetate, tertiary butanol (tBuOH) and bicarbonate at pH 8 [according to Elovitz & von Gunten, 1999]

Inhibitor S	Conc./µM	$k(^{\bullet}OH + [S]) \times [S]/s^{-1}$	k_d (s^{-1})	R_{ct}
None	–	–	~1.4 × 10^{-2}	–
Acetate	350	2.8 × 10^4	1.9 × 10^{-3}	1.4 × 10^{-8}
tBuOH	46	2.7 × 10^4	1.9 × 10^{-3}	1.3 × 10^{-8}
HCO$_3^-$	2600	2.7 × 10^4	2.4 × 10^{-3}	1.9 × 10^{-8}

This shows that the interplay between promotion and inhibition of ozone decomposition is independent of the inhibitor and leads to similar ozone stability. Furthermore, the R_{ct}, the ratio between the concentrations of $^{\bullet}$OH and ozone also is very similar in the range of (1.3–1.9) × 10^{-8} (see below). Due to their high reactivity towards the water matrix and ozone, $^{\bullet}$OH radicals have very low steady-state concentrations during ozonation, typically below 10^{-12} M. Because direct measurement of $^{\bullet}$OH concentrations during ozonation is impossible, there are basically two ways to tackle this problem: (i) Modelling ozone decay and thereby calculating the $^{\bullet}$OH concentrations (Chelkowska et al., 1992; Westerhoff et al., 1997). The application of these models to natural waters is difficult due to the varying reactivity of DOM (see above). Usually, some rate constants have to be fitted to mimic kinetics of ozone decay in real water systems (Chelkowska et al., 1992; Westerhoff et al., 1997). Therefore, model predictions for transient $^{\bullet}$OH concentrations generally disagree with experimental observations. (ii) Experimental calibration of $^{\bullet}$OH formation by ozone in natural waters with the help of an ozone-resistant probe (Hoigné & Bader, 1979; Elovitz & von Gunten, 1999; Haag & Yao, 1992; Haag & Yao, 1993). Here, the decrease of an added probe (in low concentrations, ≤1 µM) which does not react with ozone but reacts quickly with $^{\bullet}$OH is measured either as a function of the ozone dose (Hoigné & Bader, 1979) or continuously during ozonation (Elovitz & von Gunten, 1999; Haag & Yao, 1992; Haag & Yao, 1993). The first approach leads to the integral $^{\bullet}$OH exposure and the corresponding ozone dose required for eliminating a particular compound to a certain percentage. The latter approach yields information on the dynamics of the oxidation by ozone and $^{\bullet}$OH. Haag & Yao (1993) used continuous ozonation experiments for arriving at steady-state concentrations of $^{\bullet}$OH. Part of the scavenging capacity of DOM was lost during these experiments and the mean steady-state concentration of $^{\bullet}$OH became very high, resulting in a ratio of the concentrations of $^{\bullet}$OH and O$_3$ of about 10^{-7}. In single ozone dosage experiments, ratios in the order of 10^{-8}–10^{-9} were obtained (Elovitz & von Gunten, 1999). Such ratios are more typical for ozonation and ozone-based AOPs under more realistic conditions.

The kinetics of the decrease of a probe (e.g. p-chlorobenzoic acid, pCBA) can be expressed as follows [equations (17) and (18)] (Elovitz & von Gunten, 1999).

$$-\frac{d[\text{pCBA}]}{dt} = k(^{\bullet}\text{OH} + \text{pCBA}) \times [\text{pCBA}][^{\bullet}\text{OH}] \tag{17}$$

Rearranging and integrating equation (17) leads to equation (18).

$$-\ln\left(\frac{[\text{pCBA}]}{[\text{pCBA}]_0}\right) = k(^{\bullet}\text{OH} + \text{pCBA}) \times \int [^{\bullet}\text{OH}]\,dt \tag{18}$$

The term ∫ [•OH]dt represents the time-integrated concentration of •OH, which is equal to •OH exposure or •OH-ct. The •OH exposure can therefore be determined by measuring the relative decrease of, for example, pCBA as a function of the reaction time. The term R_{ct} defined in equation (19) describes the ratio of •OH exposure to ozone exposure (or •OH-ct and O_3-ct) (Elovitz & von Gunten, 1999).

$$R_{ct} = \frac{\int [\text{•OH}]\,dt}{\int [O_3]\,dt} \tag{19}$$

Substitution of equation (19) into equation (18) results in equation (20).

$$-\ln\left(\frac{[\text{pCBA}]}{[\text{pCBA}]_0}\right) = k(\text{•OH} + \text{pCBA}) \times R_{ct} \int [O_3]\,dt \tag{20}$$

Ozone exposure can be calculated from the integral of "ozone concentration versus time data" (von Gunten & Hoigné, 1994).

For many natural waters, a plot of the logarithm of the relative decrease of pCBA vs. ozone exposure showed basically two phases for which fairly good linear correlations of the two parameters were obtained (Elovitz et al., 2000a, b). This is quite similar to the behaviour of ozone (see below). This means that the ratio R_{ct} of the exposures (ct values) of •OH and ozone remains constant, and for these conditions the ratio of the concentrations of •OH and ozone ([•OH]/[O_3]) can be considered constant. This empirical concept is based on the observation that during ozonation the pseudo-first-order rate constant for transforming ozone into •OH and scavenging of •OH or their ratios remain constant. R_{ct} is typically in the range of 10^{-8}–10^{-9} (M/M) for various waters and varying water quality parameters such as pH, alkalinity, DOC and temperature (Elovitz et al., 2000a). During the initial fast decomposition of ozone, the ratio is usually higher (by about a factor of 10) and varies as a function of the ozone dose (Pinkernell & von Gunten, 2001). Even during the initial phase of ozonation, R_{ct} can be approximately expressed by just one value without significant deviations in calculated micropollutant degradation (Acero et al., 2000, 2001). Therefore, measurement of the ozone concentration and the oxidation of a probe allow prediction of the oxidation of a particular micropollutant during ozonation when rate constants for its reaction with ozone and •OH are known (Acero et al., 2000; Peter & von Gunten, 2007).

Ozone kinetics in most waters is multi-phasic (e.g. Figure 3.3). In drinking waters and wastewaters, kinetics may be adequately described by breaking it down in to three components following first-order kinetics (Nöthe et al., 2009) (Figure 3.3) or to two components when the very fast and comparatively little ozone consuming component is disregarded (Schumacher et al., 2004b; Buffle et al., 2006b).

At a DOC of 7.2 mg/L and an ozone dose of 10 mg/L, the first component consumes ~1 mg/L ozone at $k = 0.071$ (mg DOC)$^{-1}$ s^{-1}. The second component ($k = 0.011$ (mg DOC)$^{-1}$ s^{-1}) consumes 5 mg/L ozone, while the third component consumes 4 mg/L ozone at 0.0019 (mg DOC)$^{-1}$ s^{-1}. These data are normalised to the DOC in Table 3.2.

It seems that the variability of the ozone decomposition rate in wastewaters with similar DOC concentrations is much smaller than in natural waters which contain DOM of quite variable composition (see above). Figure 3.4 shows results for ozone decrease in Opfikon wastewater (DOC 4.5 mg/L) and Berlin wastewater (DOC 8.5 mg/L). For the same ozone dose (2.5 mg/L) and pH 8, ozone consumption in undiluted Berlin wastewaters is much faster than in Opfikon wastewater. However, for a 1:1 dilution of Berlin wastewater, adjusting the DOC to a similar level, ozone consumption rate is very similar for both waters. Thus, the concentrations of the ozone-reactive moieties must be very similar.

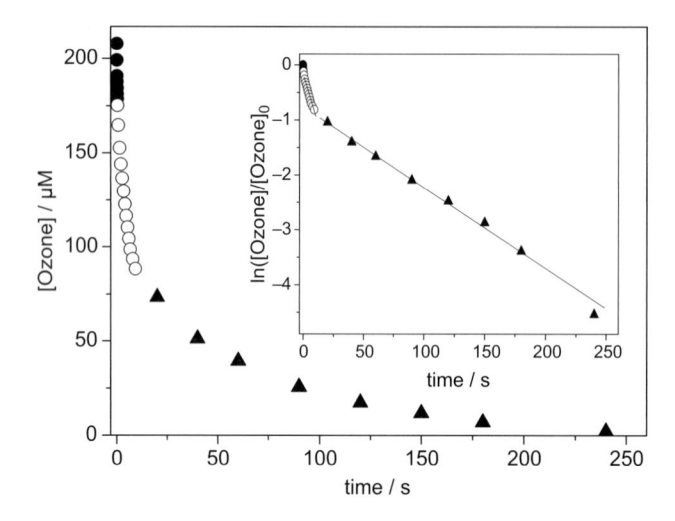

Figure 3.3 Ozone decay as followed by stopped flow (circles) and batch-quench (triangles) at 17.3 °C and pH 8. Wastewater (WWTP Bottrop) and ozone solution were mixed in a 4:1 ratio leading to concentrations of [DOC] = 7.2 mg/L and [O$_3$] = 10 mg/L (208 μM); [HCO$_3^-$] = 5.08 mM. Inset: data plotted as ln([ozone]/[ozone]$_0$) vs. time. Reprinted with permission from Nöthe et al., 2009. Copyright (2009) American Chemical Society.

Table 3.2 Rate of ozone decay and ozone consumption in wastewater according to (Nöthe et al., 2009)

Process	Rate/(mg DOC)$^{-1}$ s^{-1}	Ozone consumption/mg O$_3$ (mg DOC)$^{-1}$
Initial	0.071	0.15
Fast	0.011	0.7
Slow	0.0019	>0.85

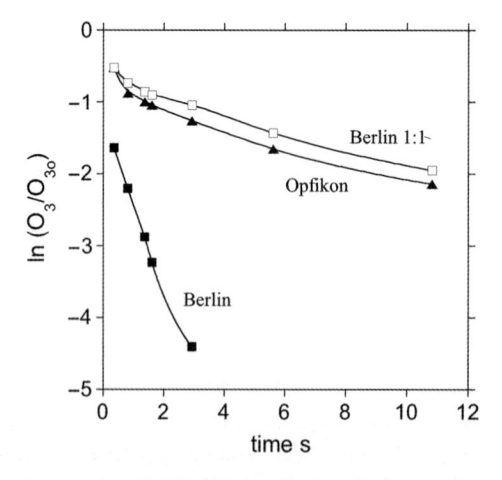

Figure 3.4 Ozone consumption in Berlin (DOC 8.5 mg/L) and Opfikon (DOC 4.5 mg/L) wastewater for an ozone dose of 2.5 mg/L at pH 8. Data for a 1:1 dilution of Berlin wastewater (DOC 4.25 mg/L) with ultrapurified water are also shown. According to Buffle et al., 2006a, with permission.

To reflect the continuum of the reactive moieties in DOM, the reactivity of ozone with DOM can conceptually be described by a model in which DOM is divided into, for example, five classes of compounds with second-order rate constants ranging from 10 to 10^7 $M^{-1}s^{-1}$ (Buffle et al., 2006a). For a wastewater, these individual moieties were assigned with fictitious concentrations in the range between 10 and 70 µM. With this approach it was possible to describe the ozone decrease in a particular wastewater (Buffle et al., 2006a). A similar approach allowed the modelling of reactions occurring in a bubble column (Nöthe et al., 2010).

3.2 MOLECULAR WEIGHT DISTRIBUTION OF DISSOLVED ORGANIC MATTER

Gel permeation chromatography (GPC), also called size exclusion chromatography (SEC), has been widely used for the determination of the distributions of molecular weights of the DOMs in various waters. In brief, a solution of the analyte (here DOM) is injected onto a chromatographic column containing the separation gel. Low-molecular-weight material can penetrate into the pores of the gel, wherefrom it is eluted slowly while high-molecular-weight material that cannot penetrate the pores elutes faster. Lacking exact reference material, it is not possible to correlate a given retention time with the exact molecular weight of that fraction, but the order of elution, high-molecular-weight material first, low-molecular-weight material later will continue to be approximately correct. In wastewater, for example, the various GPC fractions have a different specific UV absorbance (SUVA). This is a clear indication, that one deals with polymers of different chemical properties and that the above caveat as to the uncertainties of the molecular weights of the various fractions is justified.

The development of detection systems that record not only UV absorbance but also DOC have been essential for the present topic (Huber et al., 1990; Huber & Frimmel, 1991; Huber & Frimmel, 1992; Huber & Frimmel, 1996; Her et al., 2002). Figure 3.5 shows an example of a SEC-OCD chromatogram of a wastewater before and after ozonation.

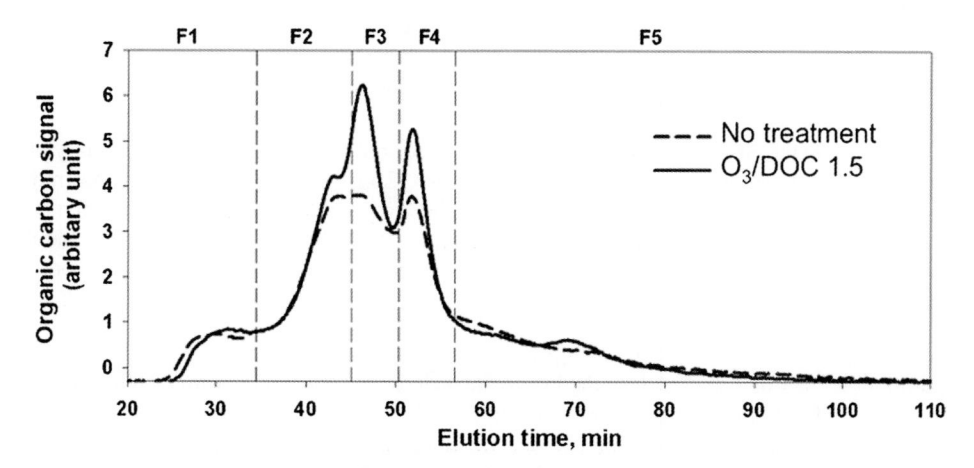

Figure 3.5 Changes in the SEC-OCD chromatogram due to ozonation of 8-µm-filtered secondary effluent of the Kloten-Opfikon wastewater treatment plant (Switzerland). SEC-OCD chromatogram before and after ozonation (specific ozone dose 1.5 gO$_3$/g DOC). Regensdorf wastewater: TOC 5 mg/L, DOC 4.7 mg/L, HCO$_3^-$ 2.86 mM, pH 7.0. F1, Biopolymers; F2, Humics; F3, Building blocks; F4, Low-molecular-weight humics and acids; F5, Low-molecular-weight neutrals (Lee & von Gunten, unpublished).

Irrespective of the origin of a water, there are always various partially-separated fractions detected. In natural waters and wastewaters, there is barely a fraction that may be associated with really low-molecular-weight material (<150 Da). SEC-OCD chromatograms of natural waters, drinking waters, soluble microbial products and wastewater have been published (Fuchs, 1985a, b, c; Allpike et al., 2005; Meylan et al., 2007; Jiang et al., 2010). As expected, waters from different origins such as groundwaters, surface waters and wastewaters show substantial differences. Waters contained in barrages that are fed by nutrients from nearby agricultural activities may be rich in extracellular organic matter (EOM) of algae (Hoyer et al., 1985) or groundwaters in peaty areas may be enriched in humics. Such differences are reflected in differences in ozone reactivity (Figure 3.2) but (obviously) also in SEC-OCD chromatograms. Common study objects are fulvic and humic acids, and data as to their structures are becoming increasingly available (Reemtsma & These, 2005; Reemtsma et al., 2006a, b, 2008; These & Reemtsma, 2003, 2005; These et al., 2004)

Figure 3.6 shows the changes of the various DOM fractions caused by ozone reactions. The so-called hydrophobic DOM fraction is retained by the chromatographic column, that is, not the entire DOC is accounted for by SEC-OCD.

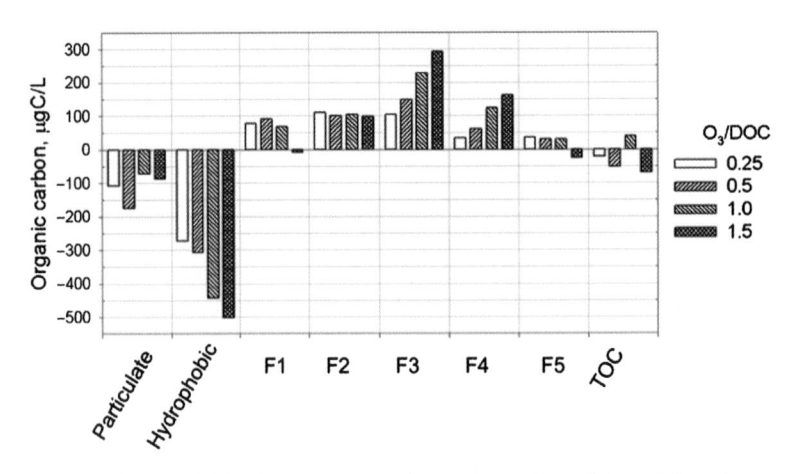

Figure 3.6 Changes in the SEC-OCD chromatogram due to ozonation of 8-μm-filtered secondary effluent of the Kloten-Opfikon wastewater treatment plant (Switzerland). Changes of individual fractions (mgC/L) for various specific ozone doses between 0.25–1.5 gO$_3$/gDOC. Regensdorf wastewater: TOC 5 mg/L, DOC 4.7 mg/L, HCO$_3^-$ 2.86 mM, pH 7.0. F1, Biopolymers; F2, Humics; F3, Building blocks; F4, Low-molecular-weight humics and acids; F5, Low-molecular-weight neutrals (Lee & von Gunten, unpublished).

The hydrophobic fraction can be determined from the difference between measured DOC and the sum of all fractions. This hydrophobic fraction decreases significantly with increasing ozone dose. This means that this part of DOC becomes detectable upon ozone treatment as fractions denoted 'building blocks and low-molecular-weight humics and acids'. The higher molecular weight fractions 'biopolymers' and 'humics' are only marginally changed. Note that at the very high ozone dose of 7.5 mg/L only about 10% of the DOM-DOC is converted to low-molecular-weight compounds.

The specific ozone doses in Figures 3.5 and 3.6 are given on a gO$_3$/gDOC basis. This unit has been chosen, because it is very practical for water treatment purposes, notably in wastewaters (Hollender et al., 2009).

For any calculations of the reaction of ozone with DOM on the molecular level, the ozone concentration has to be given in units of M instead of mg/L and for the DOM (unit: mgC/L) an equivalent unit also has to be found. Considering that DOM is polymeric, one may express its concentration, as conveniently done with other polymers, in subunits. In DOM, which is not made up of repeating subunits of known molecular weight, this is not possible, but one may make the assumption that such subunits are made up of ten carbon atoms on average (Nöthe *et al.*, 2009). For aromatic subunits (see below), this would imply six for the core benzene ring and four for substituents and linking these subunits to one another. Potential polymeric carbohydrate-based polymers would have six carbon atoms for the core unit and two further carbon atoms when *N*-acetylated. This is not far from the ten carbon atoms assumed for the aromatic units. Heteroatoms (O, N, S) and hydrogens do not have to be included in this simplified approach, as it is only based on carbon atoms (DOC). With this assumption, water that has a DOC of 12 mgC/L is 100 μM in DOM subunits. This approach now allows us to discuss many aspects of ozone chemistry in drinking water and wastewater on the molecular level. For example in a wastewater that contains a DOC of 10 mg/L (83 μM in subunits), at an ozone dose of only 4 mg/L (83 μM), each DOC subunit has reacted with ozone – on average. This approach will be used for a semi-quantitative assessment of certain DOM properties [Chemical Oxygen Demand (COD), absorption spectra, etc.].

3.3 MINERALISATION AND CHEMICAL OXYGEN DEMAND

The degree of mineralisation of DOM is typically very small, of the order of 10% for an ozone/DOC mass ratio of 1.0 (Nöthe *et al.*, 2009). For the determination of the chemical oxygen demand (COD), another bulk method to characterise the organic matter in water treatment systems, two oxidants, namely dichromate and permanganate, are in use. Dichromate is the stronger oxidant and is capable of oxidising >95% of the DOC. Thus the COD value as determined with dichromate is typically 1.5 times higher than the permanganate value.

It has been observed in a wastewater that the COD drops by ~23% at an ozone to DOC mass ratio of 1.0. Typically, one out of the three O-atoms in ozone is transferred in ozone reactions. This indicates that the COD is reduced by about 0.65 mol O_2 (1.3 mol O) at an ozone/DOC mass ratio of 1.0. From the above, one would expect a value of 0.5 mol O_2 (1.0 mol O). Thus, it seems that the oxidation of wastewater is somewhat more efficient than suggested by the above stoichiometry. This is most likely due to the formation of •OH radicals which induce peroxyl radical reactions, where O_2 serves as the oxidant (Chapter 14). In comparison, the DOC concentration is much less affected (Nöthe *et al.*, 2009). This is understood, as the DOC concentration only decreases when decarboxylation reactions set in, that is, when already substantially oxidised material is further oxidised (note that naturally occurring DOM is already partially oxidised). Mechanistic aspects of ozone-induced decarboxylation reactions are discussed in Chapters 6 and 14.

3.4 FORMATION OF ASSIMILABLE ORGANIC CARBON

As a consequence of the reaction of ozone with DOM, assimilable organic carbon (AOC) or biodegradable organic carbon (BDOC) is formed. This leads to the formation of smaller oxygen-rich molecules, such as carboxylic acids, aldehydes, ketones, etc. (Richardson *et al.*, 1999b), which are typically more biodegradable and can be summarised under AOC or BDOC (Hammes *et al.*, 2006; van der Kooij *et al.*, 1989; Siddiqui *et al.*, 1997). The mechanisms and kinetics of the formation of these compounds is governed by the reaction of ozone with DOM moieties such as phenols or other activated aromatic systems (Chapter 7). Upon ozonation, this leads to various olefins and eventually small organic acids and aldehydes (Chapter 6). Even though this is only a model approximation of the interaction between DOM and ozone, similar processes occur under realistic conditions (Figure 3.7).

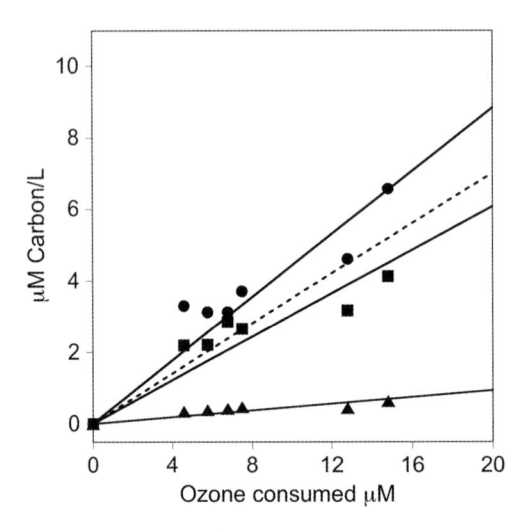

Figure 3.7 Formation of AOC (circles), organic acids (squares) and aldehydes (triangles) during ozonation of Lake Zurich water. Dashed line: sum of acids, ketones and aldehydes. DOC 1.4 mg/L, T = 22 °C, pH 8, bicarbonate = 2.6 mM. Adapted from Hammes *et al.*, 2006, with permission.

In presence of tBuOH (as an •OH scavenger), formation of organic acids is almost identical to that in absence of tBuOH in natural waters (Hammes *et al.*, 2006). Therefore, the formation of AOC which is mostly composed of carboxylic acids and aldehydes in the investigated waters is most likely due to direct ozone reactions with fast reacting moieties present in the humic fraction of DOM. In the case of Lake Zurich water, AOC can be mostly explained by carboxylic acids, whereas aldehydes and ketones contribute little. The results look different when algae are present during (pre-) ozonation. In this case, a significant portion of the AOC results from the lysis of algal cells and consists of intracellular cytoplasmic DOM (Hammes *et al.*, 2007). This material is released very quickly, even though the algae are not completely destroyed. Significant AOC formation of up to 740 µg C/L has also been observed during ozonation of wastewater (Zimmermann *et al.*, 2011).

The removal of these compounds, measured as AOC or BDOC, is one of the major tasks in water treatment plants containing an ozonation step. Their removal can be achieved in a biological filtration step (sand, biological activated carbon filtration) after ozonation (cf. Chapter 5). This avoids regrowth of micro-organisms in the distribution systems, which is particularly important in countries where no or only limited disinfectant residual is used in the distribution systems (e.g. Germany, Netherlands, Austria, Switzerland).

Figure 3.8 shows the formation and removal of AOC, carboxylic acids and aldehydes/ketones during ozonation and biological filtration, respectively, in a drinking water treatment plant. Even though the AOC concentration at the effluent of the plant is similar to that of Lake Zurich water, the treatment train including pre- and intermediate ozonation and several biological filtration steps leads to a significant overall reduction of DOC which is very beneficial for the general water quality. Similarly, during post-sand filtration of ozonated wastewater, AOC concentration was reduced significantly (up to 50%) (Zimmermann *et al.*, 2011).

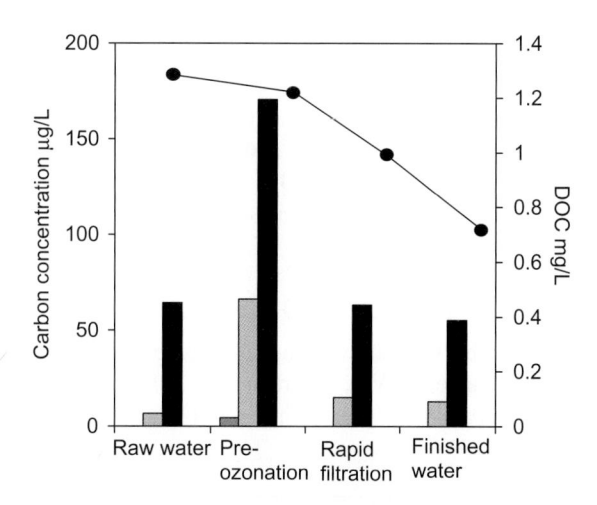

Figure 3.8 Formation and removal of AOC, carboxylic acids, aldehydes and ketones in the Lake Zurich drinking water treatment plant (Lengg, Zurich, Switzerland). Shaded bar: aldehydes and ketones; grey bars: total organic acids; black bars: assimilable organic carbon (AOC); circles: DOC. Adapted from Hammes *et al.*, 2006, with permission.

3.5 FORMATION AND MITIGATION OF DISINFECTION BY-PRODUCTS

As shown above, the application of ozone leads to a transformation of electron-rich moieties of the DOM. This has consequences for the formation of disinfection by-products (DBPs) during post-disinfection in, for example, the distribution system, and mostly leads to a reduced formation of DBPs. Nevertheless, for certain DBPs an increased formation has also been found after ozonation. One of the early applications of ozone was its significant potential to mitigate the formation of chlorophenols and bromophenols which are potent taste and odour compounds that can be formed during post-chlorination of phenol-containing waters (Bruchet & Duguet, 2004; Acero *et al.*, 2005; Piriou *et al.*, 2007). Under typical ozonation conditions, phenol is efficiently destroyed and therefore, halo-phenols cannot be formed any more (Chapter 7). Related to this, ozonation of natural waters generally reduces the formation of trihalomethanes (THMs) and haloacetic acids (HAAs) during post-chlorination (Reckhow *et al.*, 1986; Chaiket *et al.*, 2002; Chang *et al.*, 2002; Gallard & von Gunten, 2002; Chin & Bérubé, 2005; Meunier *et al.*, 2006; Hua & Reckhow, 2007; Li *et al.*, 2008). HAA formation was found to increase when chlorination was replaced by chloramination (Hua & Reckhow, 2007). The formation of trichloronitromethane (chloropicrin) and other halonitromethanes increased if natural waters were treated with ozone/chlorine compared to chlorine alone (Hoigné & Bader, 1988; Krasner *et al.*, 2006). In Br^--containing waters, a shift to bromonitromethanes was observed (Krasner *et al.*, 2006; Krasner, 2009).

These studies did not use a biological filtration step between ozonation and post-chlorination/chloramination. It can be expected that some of the halonitromethane-precursors are biodegradable and that biofiltration would reduce formation of halonitromethanes during post-chlorination/chloramination (Krasner, 2009).

In Br$^-$-containing waters, bromate is the main ozone disinfection by-product of concern (Chapters 11 and 14). However, numerous organic by-products such as bromoform, bromopicrin, (di)bromoacetic acid, dibromoacetonitriles, bromoacetone, cyanogen bromide, bromoketones, bromonitriles, bromoalkanes and bromohydrins can also be formed during ozonation or in combination with chlorine or chloramine (Richardson *et al.*, 1999a). In I$^-$-containing waters, ozonation rapidly oxidises I$^-$ to iodate, which hinders the formation of iodo-organic compounds (Chapter 11).

NDMA precursors are generally oxidised during ozonation which leads to a reduction of NDMA formation during post-chlorination (Lee *et al.*, 2007b; Krasner, 2009; Shah *et al.*, 2012). Ozonation, however, may also enhance NDMA formation, as discussed in Chapters 8 and 11.

3.6 UV ABSORBANCE OF DISSOLVED ORGANIC MATTER

All natural DOMs have in common that they show strong UV absorptions. There is, however, a considerable variation in the specific UV absorbance (SUVA, absorbance at 254 nm per mg DOC) (Huber & Frimmel, 1992) The UV absorption of a typical wastewater is shown in Figure 3.9.

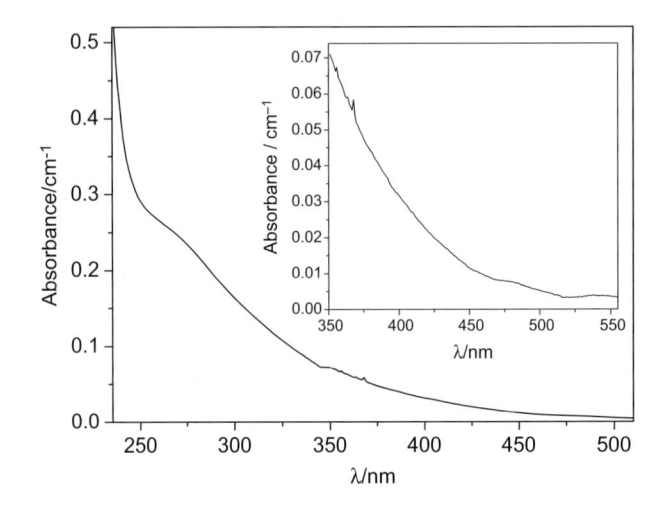

Figure 3.9 UV–Vis spectrum of a typical wastewater (effluent of the municipal WWTP Neuss, Germany). Inset: blow-up of the long-wavelength side. At the short-wavelength side the absorbance rises steeply (not shown). Courtesy T. Nöthe.

The absorption of natural water, drinking water and wastewater DOM is characterised by a continuous increase in absorption on going to shorter wavelength without any noticeable peak. This is most uncommon for any isolated UV-absorbing (e.g. aromatic) compound (Nöthe *et al.*, 2009), and it has been convincingly shown that absorptions generally observed with aquatic DOMs must be due to charge transfer transitions of electron-donating (e.g. phenolic or alkoxylated) subunits to electron-accepting (e.g. quinoid) subunits present in close neighbourhood within this polymeric material (Del Vecchio & Blough, 2004). This is schematically shown in Figure 3.10.

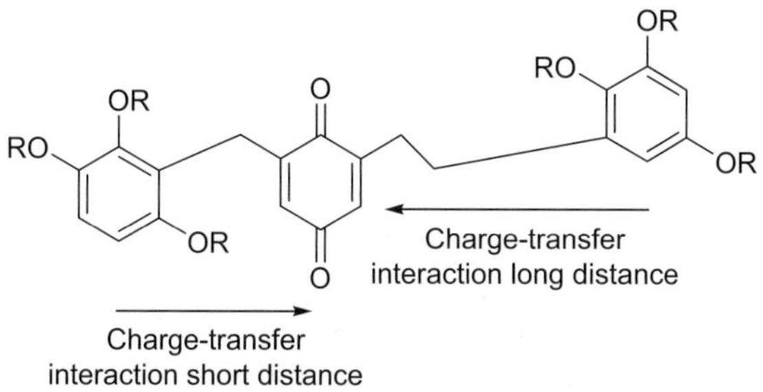

Figure 3.10 Schematic drawing of a quinone (electron acceptor) and alkoxylated benzenes (electron donors) at different distances. Charge transfer interactions between these sites in DOM give rise to electronic transitions (absorptions). The longer the distance the weaker the transition and thus the longer the wavelength of absorption.

The first electronic transition of isolated aromatic compounds that is near 250–260 nm is a forbidden transition with molar absorption coefficients of only a few hundred. Charge transfer transitions such as suggested by the Blough model (Del Vecchio & Blough, 2004) (Figure 3.10) are, however, allowed transitions and are characterised by high absorption coefficients. In a wastewater containing a DOM of 11 mg/L, the absorption at 254 nm was 30 m^{-1} (Schumacher, 2006; Nöthe et al., 2009). Based on the above estimate of the concentration of subunits of this polymeric material, this relates to a molar absorption coefficient of the DOM subunits near 3300 M^{-1} cm^{-1}, which is a much higher value than that of a typical aromatic compound. The featureless absorption, rising from the visible light into the UV region (Figure 3.9) can also be explained on the basis of this model. The absorption maxima of such charge transfer transitions vary with the distance between donor and acceptor (Figure 3.10). Structural conditions as depicted in Figure 3.10 and fluctuations of these distances induced by thermal motions are the reason for this featureless and very broad absorption.

Since the UV-active moieties in DOM (activated aromatic systems) have a certain reactivity with ozone, the UV absorption and SUVA have been used as empirical surrogate parameters to predict the ozone consumption rate in natural waters (Elovitz et al., 2000b). A reasonable correlation between the pseudo-first-order rate constant for ozone decrease and SUVA$_{254}$ with a correlation coefficient R^2 = 0.79 was found for nine DOM isolates (Elovitz et al., 2000b). In these correlations, especially Suwannee river humic acid, but also Suwannee river fulvic acid, DOM isolates, often used to simulate drinking water DOM, proved to be outliers with significantly higher ozone reactivity (see Figure 3.2).

The change in the UV absorption of DOM has been used as an empirical parameter for predicting micropollutant removal during ozonation of wastewaters (Nanaboina & Korshin, 2010; Buffle et al., 2006a; Wert et al., 2009b; Bahr et al., 2007), but such changes cannot be used for mechanistic interpretations as attempted by Nanaboina & Korshin (2010).

3.7 RELEVANCE OF OZONE KINETICS FOR THE ELIMINATION OF MICROPOLLUTANTS

For the elimination of micropollutants, competition between DOM of the water matrix and pollutant (P) for ozone has to be taken into account [reactions (21) and (22)].

$$\text{DOM} + \text{O}_3 \rightarrow \text{DOM}_{ox} \tag{21}$$
$$\text{P} + \text{O}_3 \rightarrow \text{P}_{ox} \tag{22}$$

The kinetics of the reaction of ozone with DOM thus strongly affects the efficiency of micropollutant transformation by ozone (Katsoyiannis *et al.*, 2011). Only those micropollutants that react rapidly stand a chance of being oxidised directly by ozone. Here, we only consider the very first step in the sequence of events that eventually will lead to mineralisation. This is justified, because in most cases, the biological activity of the micropollutant is practically fully suppressed by this very first oxidation step (Chapter 4). For a wastewater (DOC = 10 mg/L, pH 8), one can neglect the slow process for an ozone dose of, for example, 5 mg/L. Based on the example shown in Figure 3.3, the ozone decay can be simulated by breaking up the process into several kinetic phases. A simulation has shown (Nöthe *et al.*, 2009) that during the consumption of the first mg ozone per L all micropollutants that have ozone rate constants $>10^5$ $M^{-1}s^{-1}$ are completely (>90%) eliminated, micropollutants that react with 10^4 $M^{-1}s^{-1}$ are eliminated to about 25% remaining, and micropollutants that react an order of magnitude more slowly are practically not affected. This situation is substantially improved by the reaction of the next 4 mg/L ozone. Here, the lower rate constant of ozone consumption by the wastewater matrix and the higher ozone dose cooperate. Complete elimination is now achieved for all micropollutants that react with 3×10^3 $M^{-1}s^{-1}$ and micropollutants that react with 1×10^3 $M^{-1}s^{-1}$ are eliminated to about 50%. For the elimination of less reactive micropollutants, higher ozone doses would be required. For an ozone dose of 10 mg/L (1 gO$_3$/gDOC), micropollutants that react with 1×10^3 $M^{-1}s^{-1}$ are fully eliminated, and the concentrations of even less reactive ones are noticeably lowered ($k = 300$ $M^{-1}s^{-1}$ to 25%; $k = 100$ $M^{-1}s^{-1}$ to 65% remaining). The elimination of micropollutants with varying second-order rate constants for their reaction with ozone is illustrated in Figure 3.11 as a function of the specific ozone dose.

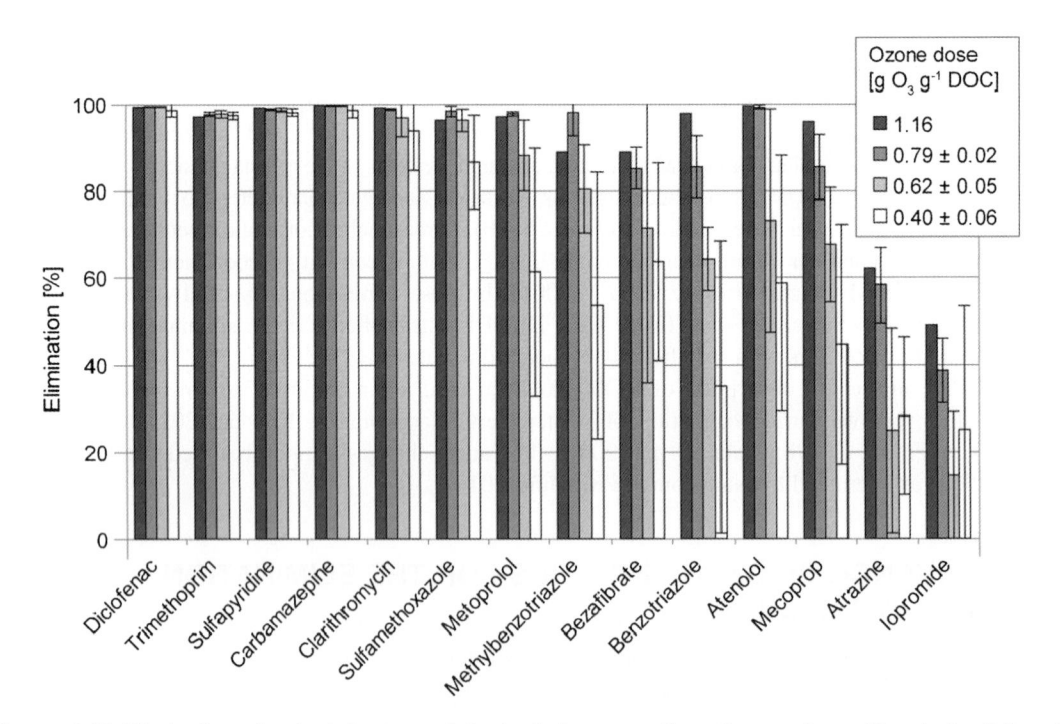

Figure 3.11 Elimination of selected micropollutants during ozonation of secondary effluent of a full-scale ozonation plant for various specific ozone doses (Regensdorf, Switzerland). DOC 5.2±0.6 mg/L, pH 7. Reprinted with permission from Hollender *et al.*, 2009. Copyright (2009) American Chemical Society.

Fast reacting compounds from diclofenac to sulfamethoxazole ($k > 10^4$ M^{-1}s^{-1}) are fully eliminated at specific ozone doses of 0.4 g ozone per g DOC. For compounds with intermediate ozone reactivity, (metoprolol to atenolol, 2×10^2 M^{-1}s$^{-1} < k < 10^4$ M^{-1}s^{-1}) significantly higher specific ozone doses of about 1 g ozone per g DOC are necessary for complete elimination. For compounds with second order rate constants near 100 M^{-1}s^{-1} and below (mecoprop to iopromide), specific ozone doses of 1 gO$_3$/gDOC lead to eliminations of 50–90%. For these compounds, elimination is dominated by •OH. The dependence of the degree of elimination of micropollutants in a wastewater on the reactivity with ozone and •OH is further illustrated in Figure 3.12 for compounds that cover the full range of ozone and •OH rate constants.

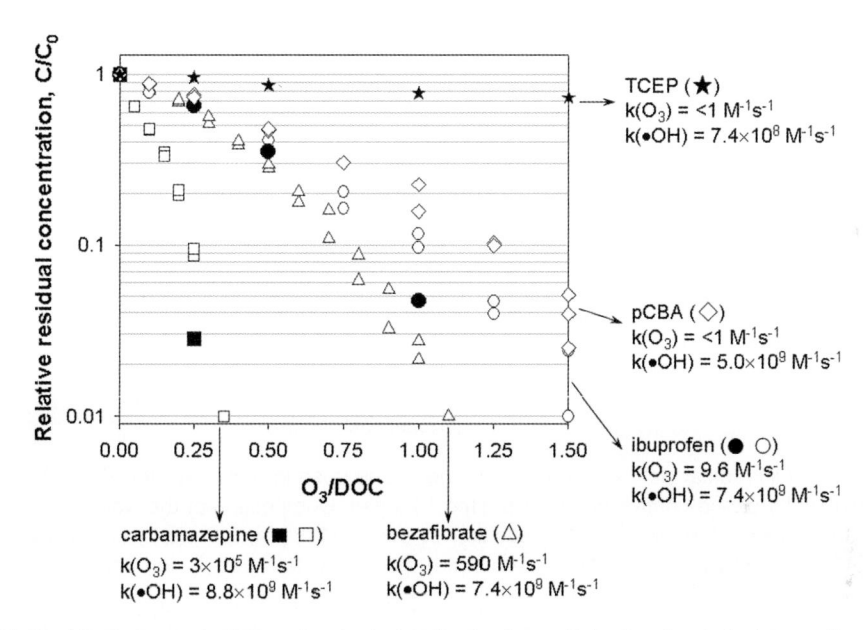

Figure 3.12 Residual concentrations of selected spiked micropollutants at sub to low µg/L ranges after ozonation of Kloten-Opfikon wastewater effluent with variable ozone doses up to ratios of ozone : DOC of 1.5 (g/g). Filled and open symbols represent experiments with low (~1 µg/L) and high dosage (~200 µg/L) of the micropollutants. pCBA: p-chlorobenzoic acid, TCEP: Tris-(2-chloroethyl) phosphate. Second order rate constants for the reaction with ozone and •OH are given. According to Lee, Gerrity, Snyder & von Gunten unpublished.

Micropollutants such as the chlorinated trialkylphosphates have considerably lower •OH rate constants than the aromatic x-ray contrast media such as iopromide (Table 3.3). These are hence degraded to an even lesser extent (cf. Figure 3.14).

3.8 HYDROXYL RADICAL YIELD AND •OH-SCAVENGING RATE OF DISSOLVED ORGANIC MATTER

Ozone reacts with a number of organic compounds, notably amines (Chapter 8) and electron-rich aromatic compounds (e.g. phenols and alkoxylated benzenes; Chapter 7) by giving rise to •OH in side reactions, and hence •OH is always formed when drinking water and wastewater are treated with ozone. The •OH yield in wastewater has been determined making use of the tertiary butanol (tBuOH) assay (Chapter 14). In brief, tBuOH is added in large excess to overrun the •OH scavenging rate of the wastewater (see above), and one of the products of the reaction of •OH with tBuOH, formaldehyde, is determined. Formaldehyde is not formed in the reaction of ozone with wastewater in the absence of tBuOH. As a rule of thumb, the •OH yield is twice the formaldehyde yield (Flyunt et al., 2003a). Using this approach, the data shown in Figure 3.13 were obtained. In the inset of this figure, a competition plot of tBuOH and the water matrix for •OH is shown (for competition kinetics see Chapter 2).

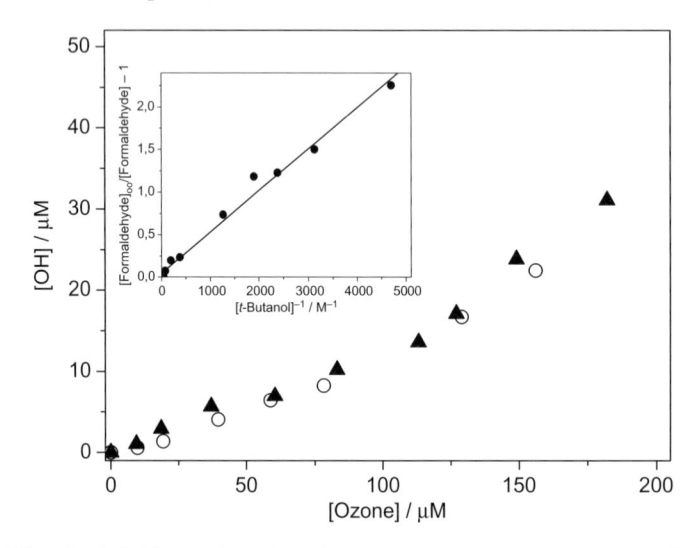

Figure 3.13 The •OH radical yield as a function of the ozone dose in a wastewater (WWTP Bottrop, pH 8, DOC after dilution by ozone-containing water 8.0 mg/L). Inset: Competition of the water matrix and tertiary butanol for •OH radicals formed upon the addition of ozone (112 µM, 5.6 mg/L) to wastewater (WWTP Neuss, pH 8, DOC = 9.05 mg/L, 3.8 mM bicarbonate). Reprinted with permission from Nöthe et al., 2009. Copyright (2009) American Chemical Society.

As seen from Figure 3.13, the •OH yield continues to remain constant or even slightly increases up to an ozone dose of 190 µM. At a DOC of 8 mg/L (67 µM in subunits) the ozone concentration is 2.8 times that of the DOM subunits. Thus, the capacity of the DOM for •OH production does not become exhausted but rather new •OH-generating sites are formed in the reaction of ozone with DOM, as discussed above.

The •OH yield can vary significantly from one water source to another. By increasing the ozone:DOC ratio (g/g) to 2.0 the •OH yield may rise to ~30% (Lee & von Gunten, in preparation). Apparently, there is no general •OH yield, and for each water to be treated with ozone •OH formation has to be assessed.

3.9 ELIMINATION OF OZONE-REFRACTORY MICROPOLLUTANTS BY THE •OH ROUTE

Ozone-refractory micropollutants are eliminated to a certain extent by the •OH route. There is not such a high variation in •OH rate constants [for a compilation see (Buxton et al., 1988)] as for ozone. Yet for ozone-refractory micropollutants that we are concerned with in drinking water and wastewater, a variation of the •OH rate constant by a factor of ten (4.5×10^8 to 7×10^9 M^{-1}s^{-1}) is typical (Table 3.3), and this is of relevance for the elimination efficiency and therefore the ozone dose required for achieving a desired elimination.

Table 3.3 Compilation of ozone and •OH rate constants (unit: $M^{-1}s^{-1}$) of selected ozone-refractory micropollutants in drinking water and wastewater at pH 7

Compound	Source/application	k (ozone)	k(•OH)	Reference
Atrazine	Herbicide	6	3×10^9	Acero et al., 2000
Iopromide	X-ray contrast agent	<0.8	3.3×10^9	Huber et al., 2003
Diazepam	Tranquiliser	0.75	7.2×10^9	Huber et al., 2003
NDMA	Oxidation by-product	5×10^{-2}	4.5×10^8	Lee et al., 2007a
Tris-(2-chloro-isopropyl) phosphate (TCPP)	Flame retardant	≪1	7×10^8	Pocostales et al., 2010
Tris-(2-chloro-ethyl) phosphate (TCEP)	Flame retardant	≪1	7.4×10^8	Watts & Linden, 2009
Tri-n-butyl phosphate (TnBP)	Plasticiser	≪1	$\approx 2.8 \times 10^9$	Pocostales et al., 2010

For the elimination of ozone-refractory micropollutants, one may write the competing reactions (23) and (24), where M denotes the water matrix (DOC plus bicarbonate, see above) and P the micropollutant.

$$M + {}^{\bullet}OH \rightarrow M\text{-ox} \quad (23)$$
$$P + {}^{\bullet}OH \rightarrow P\text{-ox} \quad (24)$$

The •OH scavenging rate that determines the rate of reaction (24) and •OH yields (in wastewater) has been given above. Based on this, the elimination efficiency of a given ozone-refractory micropollutant can be calculated. This is shown in Figure 3.14 for two organic phosphates.

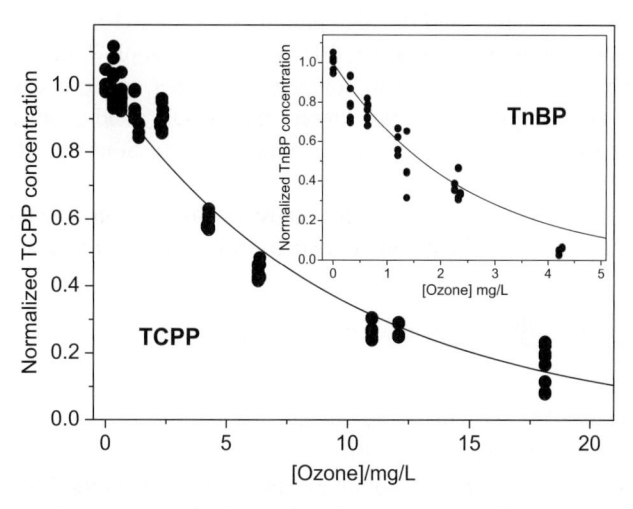

Figure 3.14 Experimental data (symbols) and simulation (solid lines) of the degradation of TCPP and TnBP (inset) in diluted Neuss wastewater (DOC = 5.5 mg/L). Reprinted with permission from Pocostales et al., 2010. Copyright (2010) American Chemical Society.

The •OH scavenging rate of a water can be determined by measuring the rate of the apparent rate constant for the decrease of a probe compound [e.g. 4-chlorobenzoic acid (pCBA)] as a function of varying tBuOH concentrations (Katsoyiannis *et al.*, 2011). Without knowing the •OH scavenging rate of a given water, one can also follow the elimination of an ozone-refractory test compound such as pCBA for which the •OH rate constant is known ($k = 5 \times 10^9$ M^{-1}s^{-1}, Figure 3.12), and with the knowledge of the •OH rate constants of other micropollutants, their degradation can be calculated (Acero *et al.*, 2000; Acero & von Gunten, 2001; Huber *et al.*, 2003; Peter & von Gunten, 2007). The •OH rate constants of selected ozone-refractory micropollutants are compiled in Table 3.3.

3.10 OZONE-BASED ADVANCED OXIDATION PROCESSES

There are three ozone-based Advanced Oxidation Processes (AOPs) that lead to the formation of •OH, (i) the reaction of ozone with H_2O_2 or, more correctly, with its anion, HO_2^- (peroxone process), (ii) photolysis of ozone with UV light (UV/ozone) and (iii) reaction of ozone with activated carbon (carboxone process). In principle, this can also provide the means for coping with ozone-refractory pollutants, as •OH radicals are much more reactive than ozone. To what extent, this expectation can be realised in practice, will be discussed below.

3.10.1 Peroxone process

The best-known of the three ozone-based AOPs is the peroxone process. Details of its chemistry are discussed in Chapter 11. Here, it is recalled that H_2O_2 does not react efficiently with ozone, but that the reaction with HO_2^- is fast, and that the HO_2^- reaction dominates over a wide pH range that extends even into the acid pH region. This pH dependence is given by equation (25).

$$k_{obs} = k(HO_2^- + O_3) \times 10^{(pH - pK_a)} \quad (25)$$

Due to this marked pH dependence and the high pK_a of H_2O_2 [$pK_a(H_2O_2) = 11.8$], this reaction is fast only at high pH. For example at pH 8, k_{obs} is 1.5×10^3 M^{-1} s^{-1} and at pH 7 one order of magnitude lower, $k_{obs} = 150$ M^{-1} s^{-1}. This slow reaction has to compete with other ozone decay processes such as the reaction with DOM (for an example see below). Moreover, the •OH yield is only half of the value that a simplified stoichiometry (Staehelin & Hoigné, 1982) suggests (von Sonntag, 2008) (for the reason for this, see Chapter 11).

Increasing the rate by increasing the H_2O_2 concentration may under certain conditions not be an advantage as H_2O_2 also acts as an •OH scavenger [reaction (26), $k = 2.7 \times 10^7$ M^{-1} s^{-1} (Buxton *et al.*, 1988)].

$$•OH + H_2O_2 \rightarrow H_2O + HO_2^• \quad (26)$$

In a sequence of reactions, $HO_2^•$ reacts with ozone [reactions (2)/(3) and (10)/(11)] giving again rise to •OH (for details. see Chapter 13). As a result, ozone and H_2O_2 are destroyed without the desired effect of pollutant degradation. Yet compared to other •OH reactions, the rate constant of reaction (26) is very low, and this prevents this chain reaction from becoming of importance when the •OH scavenging rate of the matrix is high, for example, in surface waters or wastewaters.

For a wastewater, the effect of H_2O_2 on the formation of •OH has been simulated on the basis of known rate constants (Figure 3.15).

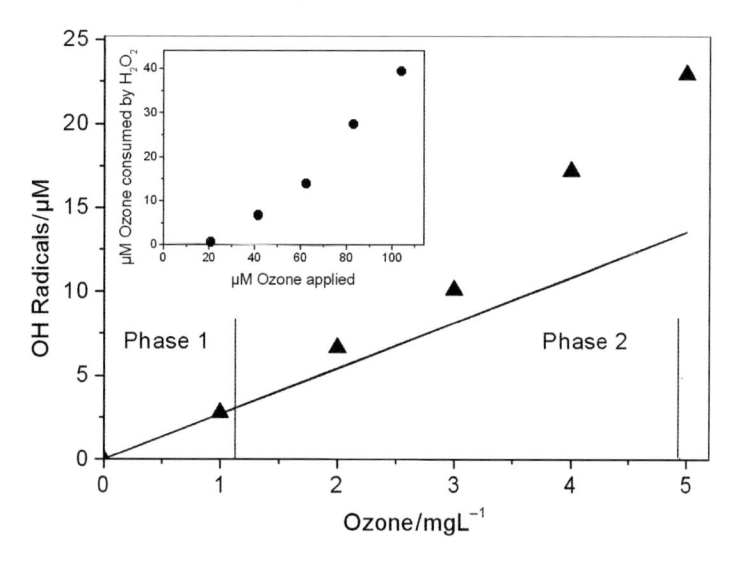

Figure 3.15 Simulation of the •OH yield for a wastewater containing a DOC of 5.5 mgC/L (diluted Neuss wastewater). The solid line is based on 13% •OH yield in the reaction of ozone with the DOC in the absence of H$_2$O$_2$ (cf. Figure 3.14). The symbols indicate the total •OH yields formed in the reaction of ozone with the DOC and in the peroxone process (molar ratio ozone/H$_2$O$_2$ = 2.0). Inset: ozone consumption from the peroxone process as a function of the applied ozone concentration (units: µM, note that 1 mg/L ozone = 21 µM). Lines near 1 and 5 mg/L ozone indicate the ranges of first and second phase in ozone kinetics of this wastewater (cf. Table 3.2). Reprinted with permission from Pocostales *et al.*, 2010. Copyright (2010) American Chemical Society.

As •OH formation in the Neuss wastewater is 13% of ozone consumption and in the peroxone process only 50% (see above), the excess gain in •OH formation is only 37% when ozone reacts with H$_2$O$_2$ instead of reacting with the wastewater DOM. Because ozone reacts rapidly with the water matrix at the early stage, H$_2$O$_2$ can barely compete. Only when the matrix reaction becomes slower and the H$_2$O$_2$ concentration higher, is there a surplus of •OH. This surplus of •OH has to be paid for by a considerable depletion of ozone (inset in Figure 3.15) and a shortening of ozone lifetime, with the effect that disinfection and micropollutant destruction by the direct ozone reaction are now much less effective. This mainly affects micropollutants that have a low to medium ozone rate constant. Whether this loss in micropollutant destruction by ozone is offset by an •OH-induced destruction has not yet been tested experimentally.

In water treatment practice, AOPs are often considered as processes with a "magic" touch. In the case of the AOP ozone/H$_2$O$_2$, a better elimination of micropollutants is suggested compared to the conventional ozonation process. In Figure 3.16, the ozone stability is shown for a groundwater (DOC 1 mg/L, alkalinity 5.2 mM) and a surface water (DOC 3.2 mg/L, alkalinity 3.8 mM). Ozone stability is affected by the addition of H$_2$O$_2$ in both waters even at pH 7 (Figure 3.16).

The effect is more pronounced in the groundwater than in the surface water. The larger effect in the groundwater is due to the lower DOC which results in a higher stability of ozone in the absence of H$_2$O$_2$. When H$_2$O$_2$ is added, the relative contribution of H$_2$O$_2$ to ozone decomposition is very high in the groundwater, whereas in the surface water it is small. The relative residual concentration of pCBA, an ozone-refractory compound, is shown in Figure 3.17 for the same experimental conditions.

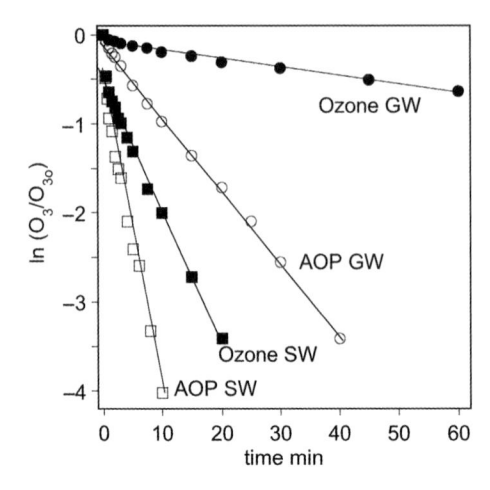

Figure 3.16 Ozonation of a groundwater (GW) (DOC 1 mg/L, alkalinity 5.2 mM) and a surface water (SW) (DOC 3.2 mg/L, alkalinity 3.8 mM). First order kinetic representation of the ozone decrease. Experimental conditions: ozone dose 2.1×10^{-5} M, H_2O_2 dose 1.0×10^{-5} M, pCBA dose 2.5×10^{-7} M, pH 7, 11 °C. Adapted from Acero & von Gunten, 2001. Reprinted by permission *Journal AWWA*. Copyright American Water Works Association.

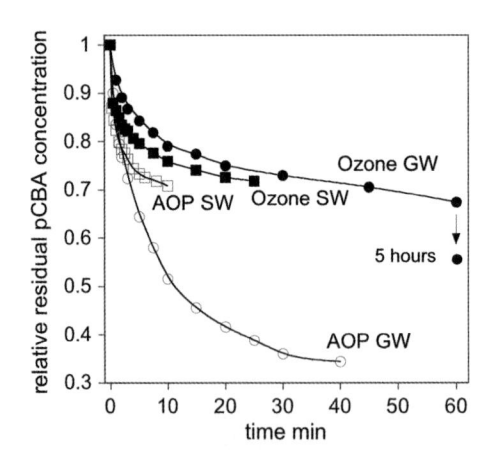

Figure 3.17 Ozonation of a groundwater (GW) (DOC 1 mg/L, alkalinity 5.2 mM) and a surface water (SW) (DOC 3.2 mg/L, alkalinity 3.8 mM). Kinetics of the transformation of pCBA. Experimental conditions: ozone dose 2.1×10^{-5} M, H_2O_2 dose 1.0×10^{-5} M, pCBA dose 2.5×10^{-7} M, pH 7, 11°C. From Acero & von Gunten, 2001, reprinted by permission *Journal AWWA*. Copyright American Water Works Association.

Two features should be highlighted: (i) the kinetics of the transformation of pCBA and (ii) the extent of pCBA transformation. In both waters, the rate of pCBA transformation is higher in the presence of H_2O_2, with a more pronounced effect in the groundwater. For a hypothetical contact time in a reactor of 15 min, the relative transformation in the absence and in the presence of H_2O_2 is 20% and 55%, respectively. This shows that the addition of H_2O_2 allows a higher degree of transformation for a given hydraulic residence time. The extent of transformation has to be compared for the complete depletion of ozone in the absence and presence of H_2O_2. In the surface water, the overall extent of transformation is fairly similar for both scenarios, whereas in the groundwater, H_2O_2 addition improves the elimination of pCBA from about 40 to 65%. Thus, depending on water quality, a similar or only small increase of the extent of transformation is observed in presence of H_2O_2.

3.10.2 UV photolysis of ozone

In his pioneering work, Taube concluded that only H_2O_2 is formed upon the photolysis of ozone in water (Taube, 1957). This conclusion was based on the apparent 1:1 stoichiometry of ozone consumption and H_2O_2 formation in the presence of acetate as an •OH scavenger. At this time, it was not yet known that in its •OH-induced reactions in the presence of O_2, acetate gives rise to relatively large amounts of H_2O_2 (Schuchmann et al., 1985). Later on, when it was realised that •OH radicals are generated in this reaction, the ozone/UV system was widely discussed among potential AOPs (Glaze et al., 1982; Peyton et al., 1982; Peyton & Glaze, 1987, 1988; Takahashi, 1990; Gurol & Vatistas, 1987; Ikemizu et al., 1987; Morooka et al., 1988). The reactions in this rather complex system are now reasonably well understood (Reisz et al., 2003).

Upon photolysis, ozone is decomposed into O_2 and oxygen atoms $O(^1D)$ (excited state) and $O(^3P)$ (ground state) (Wayne, 1987; Schriver-Mazzuoli, 2001; Bauer et al., 2000; Smith et al., 2000; Taniguchi et al., 2000). In the gas phase and below 300 nm, the main processes (quantum yield, $\Phi \approx 0.9$) are reactions (27) and (28), that is, the formation of $O(^1D)$ and singlet oxygen, $O_2(^1\Delta_g)$, as well as oxygen in its ground state, $O_2(^3\Sigma_g^-)$. Reactions that yield $O(^3P)$ [reactions (29) and (30)] are of lower importance ($\Phi \approx 0.1$) (Wayne, 1987; Hancock & Tyley, 2001; Wine & Ravishankara, 1982).

$$O_3 + h\nu \longrightarrow O(^1D) + O_2(^1\Delta_g) \tag{27}$$

$$O_3 + h\nu \longrightarrow O(^1D) + O_2(^3\Sigma_g^-) \tag{28}$$

$$O_3 + h\nu \longrightarrow O(^3P) + O_2(^1\Delta_g) \tag{29}$$

$$O_3 + h\nu \longrightarrow O(^3P) + O_2(^3\Sigma_g^-) \tag{30}$$

The quantum yield of $O(^1D)$ formation falls to a value near 0.1 above a wavelength of about 320 nm but not to zero. This shows that the spin-forbidden formation of $O(^1D)$ and $O_2(^3\Sigma_g^-)$ [reaction (28)] is possible (Bauer et al., 2000; Smith et al., 2000; Taniguchi et al., 2000; Jones & Wayne, 1970). $O(^1D)$ is very energetic [heat of formation, 437 kJ mol^{-1} (Taniguchi et al., 1999)] and therefore reacts rapidly even with water [reaction (31), $k = 1.8 \times 10^{10}$ M^{-1}s^{-1} (Biedenkapp et al., 1970), by insertion (Taube, 1957)].

$$O(^1D) + H_2O \longrightarrow H_2O_2 \tag{31}$$

In the gas phase, the excess energy of the H_2O_2 molecule so formed results in the fragmentation of the O–O bond [reaction (32); BDE = 210 kJ mol^{-1} (McKay & Wright, 1998)].

$$(H_2O_2)_{hot} \longrightarrow 2\,^{\bullet}OH \tag{32}$$

In aqueous solution, the rapid thermalisation of the "hot" H_2O_2 [reaction (33)] competes with reaction (32).

$$(H_2O_2)_{hot} \rightarrow H_2O_2 \qquad (33)$$

In addition, the solvent-cage effect inhibits the escape of $^\bullet OH$, that is, recombination competes with diffusion into the bulk. Hence, most of the $O(^1D)$ will be converted into H_2O_2.

In contrast, the reaction of $O(^3P)$ with water is endergonic by about 75 kJ mol^{-1} (Amichai & Treinin, 1969) and can be neglected here. Most of the $O(^3P)$ disappears by reacting with ground state oxygen $[O_2(^3\Sigma_g^-)]$ regenerating ozone [reaction (34); $k = 4 \times 10^9$ M^{-1} s^{-1} (Kläning et al., 1984)].

$$O(^3P) + O_2(^3\Sigma_g^-) \rightarrow O_3 \qquad (34)$$

Reactions of $O(^3P)$ with organic material are slow in comparison (Bucher & Scaiano, 1994; Herron & Huie, 1969) and may be neglected for typical AOP conditions.

The quantum yield of *direct* ozone photolysis is $\Phi = 0.5$, and, using tBuOH to scavenge $^\bullet OH$ radicals, it has been shown that only about 10% of the photolysed ozone furnish $^\bullet OH$ ($O_3 + H_2O + h\nu \rightarrow 2\,^\bullet OH$) (Reisz et al., 2003). Thus, the quantum yield of $^\bullet OH$ production is relatively low, $\Phi \approx 0.1$, when compared with the photolysis of H_2O_2 [$\Phi(H_2O_2) = 1.0$ (Legrini et al., 1993)].

The UV/ozone process has never gained much practical application in water treatment. In comparison with the UV/H_2O_2 process, it is not capable of minimising the bromate problem. Yet, there are many lab-scale studies. A considerable number of these do not take into account that ozone reacts rapidly with the substrate molecules resulting in an ozone steady-state concentration which is so low that ozone photolysis cannot contribute to product formation. The reported differences in product yields can thus not be related to ozone photolysis but rather to a photolysis of starting material and products.

3.10.3 Reaction of ozone with activated carbon

The reaction of ozone with activated carbon (AC) leads to the formation of $^\bullet OH$, and it has been believed that this is a catalytic reaction (Jans & Hoigné, 1998). Yet, a more detailed study has shown that electron-donating residues within AC, notably its nitrogen content, cause this $^\bullet OH$ production, and when this source is exhausted, $^\bullet OH$ production comes to a halt (Sánchez-Polo et al., 2005). XPS measurements indicated that the pyrrole content of the AC decreases during ozonation. Based on this characterisation and the measurement of $O_2^{\bullet-}$ formation (Sánchez-Polo et al., 2005) by the tetranitromethane method [$O_2^{\bullet-}$ reacts quickly with ozone giving rise to $^\bullet OH$ (Chapter 13)], a hypothetical mechanism for the ozone reaction with AC has been formulated [reaction (35)].

Figure 3.18 shows the kinetics of the transformation of pCBA in the absence and presence of AC. The presence of AC leads to a significant increase of the rate of pCBA transformation (adsorption is much slower and can be excluded). The presence of DOC does not affect the rate of pCBA transformation in the AC experiment.

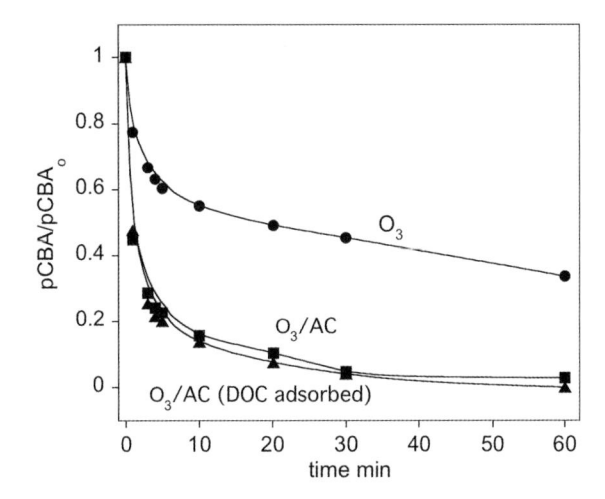

Figure 3.18 Kinetics of the transformation of pCBA during ozonation of Lake Zurich water in absence and presence of activated carbon. All experiments: pH 7, 5 mM phosphate buffer, ozone dose 1 mg/L, activated carbon 0.5 g/L. Circles (O_3), ozone only; squares (O_3/AC), ozone in presence of activated carbon F400; triangles (O_3/AC, DOC adsorbed), ozone in presence of activated carbon F400 pre-equilibrated with DOM. From Sánchez-Polo et al., 2005, with permission.

In this process, AC does not serve as a catalyst. When the pyrrole groups are exhausted upon repeated ozonation, the effectiveness of the AC decreases significantly (Sánchez-Polo et al., 2005). In the case of compounds that adsorb well on AC, adsorption may be the main removal mechanism (Sánchez-Polo et al., 2006). An experiment in the presence of AC and in the absence of ozone shows that the main removal mechanism of atrazine is its adsorption. Therefore, for compounds with high affinity to AC, a combined process has only a limited advantage over AC alone. However, the combination of activated carbon with ozone is an optimal process for removing adsorbable and non-adsorbable ozone-refractory compounds simultaneously in a one treatment step if AC is continuously renewed.

Chapter 4

Inactivation of micro-organisms and toxicological assessment of ozone-induced products of micropollutants

4.1 DISINFECTION KINETICS

The disinfecting power of ozone has already been recognised in the 19th century (Chapter 1). Ozone has been applied for primary disinfection in drinking water treatment since the beginning of the 20th century (Chapter 5). Ozone is the best chemical disinfectant currently applied in drinking water treatment (Katzenelson *et al.*, 1974; Ellis, 1991; von Gunten, 2003b; Dahi, 1976; Hoff & Geldreich, 1981; Trukhacheva *et al.*, 1992; Bünning & Hempel, 1999). It readily copes with viruses (Katzenelson *et al.*, 1979; Kim *et al.*, 1980; Thurston-Enriquez *et al.*, 2005; Katzenelson *et al.*, 1974; Kim *et al.*, 1980; Roy *et al.*, 1980; Nupen *et al.*, 1981; Roy *et al.*, 1981a, b; 1982a, b; Finch & Fairbairn, 1991; Hall & Sobsey, 1993; Botzenhart *et al.*, 1993; Lin & Wu, 2006), with bacteria and their spores (Scott & Lesher, 1963; Broadwater *et al.*, 1973; Katzenelson *et al.*, 1974; Finch *et al.*, 1988; Botzenhart *et al.*, 1993; Finch *et al.*, 1993; Hunt & Marinas, 1997; Driedger *et al.*, 2001; Larson & Marinas, 2003; Facile *et al.*, 2000; Cho *et al.*, 2002; Jung *et al.*, 2008; Komanapalli & Lau, 1996) as well as with protozoa (Wickramanayake *et al.*, 1984a, b; Rennecker *et al.*, 1999).

For bacteria, spores and protozoa, the inactivation kinetics is typically characterised by a shoulder as schematically depicted in Figure 4.1 (right curve). But in many cases, the shoulder is quite small and a fit according to a mono-exponential decay is adequate (Figure 4.1, left curve).

This is analogous to the inactivation of cells by short-wavelength UV-radiation (UVC) and by ionising radiation. In these cases, the target is definitely DNA, whereas with long-wavelength UV-radiation (UVA), a spectral range where DNA absorbs only very little, protein damage, a much less efficient process, is the molecular basis of (solar) disinfection (Bosshard *et al.*, 2010). From UVC and ionising radiation studies, it became apparent that inactivation is not due to a single lesion, but that many DNA lesions are required for preventing reproduction [reproductive cell death, for a study on ozone-induced DNA damage and repair see Hamelin & Chung (1989)]. The shoulder arises from a competition of repairing these lesions with the help of repair enzymes and the attempt to generate a second set of complete double-stranded DNA and for dividing into two daughter cells. Strains that lack such repair enzymes no longer show the shoulder. Moreover, starving the cells ("liquid holding") prevents cells from undergoing rapid reproduction, and repair becomes more efficient. When this competition is not very pronounced, the inactivation curve may show up as a straight line (Figure 4.1, left).

To a certain extent, viruses may also make use of the repair enzymes provided by their host cells.

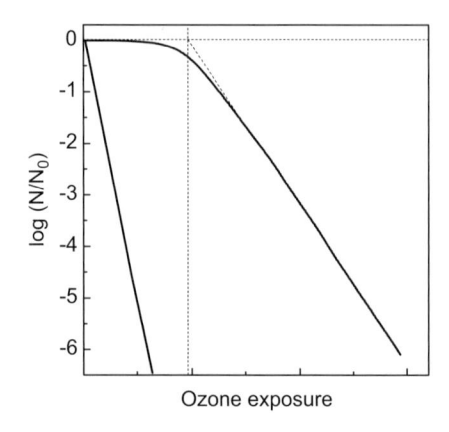

Figure 4.1 Schematic representation of the two types of inactivation curves. Mono-exponential (left), mono-exponential with shoulder (right).

The importance of repair processes may be illustrated by some numbers that are available for the lesions set by ionising radiation. The lesions set by ozone must be different, but some lesions will be similar such as single-base lesions. The more severe lesions such as DNA double-strand breaks and DNA crosslinks are possibly of minor importance in the case of ozone as a damaging agent. With ionising radiation, a dose of 1 Gy induces 0.2–0.8 lethal events in (mammalian) cells and about 1000 DNA single-strand breaks. For the same conditions, many more single-base lesions (for a typical yet not the most abundant damage, 8-oxo-adenine, 700 such lesions were estimated), 40 DNA double-strand breaks and 150 DNA-protein crosslinks occur (von Sonntag, 2006). These extraordinary large numbers show the high efficiency of the cellular repair enzymes, and it is concluded that also in the case of damages set by ozone, the majority can be repaired. It thus does not come as a surprise that some 10^8 ozone molecules are required for the inactivation of a bacterium (Scott & Lesher, 1963; Finch et al., 1988).

Ionising radiation induces chromosome aberration to a similar extent as lethal effects (von Sonntag, 2006), and mutations are also observed with ozone (Rodrigues et al., 1996; Dubeau & Chung, 1982; Dillon et al., 1992).

Starting with Chick and Watson in 1908, there were many attempts to fit experimental data (Zhou & Smith, 1995) and to model inactivation of micro-organisms by disinfectants. A compilation and discussion of current models is given by Gyürek & Finch (1998). All of these models have in common that they neglect repair processes. For the inactivation of micro-organisms showing a shoulder, kinetics can be formulated by an empirical approach (Rennecker et al., 1999; Gujer & von Gunten, 2003) [Equations (1)–(4)].

$$CT_{lag} = \frac{1}{k} \ln\left(\frac{N_1}{N_0}\right) \tag{1}$$

$$\text{if } CT \leq CT_{lag} \text{ then } \frac{N}{N_0} = 1 \tag{2}$$

$$\text{else } \frac{N}{N_0} = \frac{N_1}{N_0} \exp(-k \times CT) \tag{3}$$

$$\text{or } \frac{N}{N_0} = \exp(-k \times [CT - CT_{lag}]) \tag{4}$$

N: number of micro-organisms per unit volume; N_0: initial number of micro-organisms; N_1: intercept with ordinate resulting from extrapolating pseudo-first order line; CT: ozone exposure (see below); CT_{lag}: ozone exposure without measureable inactivation of micro-organisms; k: disinfection rate constant.

The surviving fraction (N/N_0) of a given micro-organism population when plotted as $\log(N/N_0)$ vs. the ozone CT or ozone exposure (for definitions see below) shows typically a shouldered curve (right curve in Figure 4.1). There are cases, where the shoulder is so little pronounced that this plot turns into a straight line (left curve in Figure 4.1).

Repair takes place during ozonation and any post-ozonation period including the time required for the assay measuring the surviving fraction. Table 4.1 gives an overview of inactivation parameters for various micro-organisms with ozone.

Table 4.1 Selected kinetic parameters for the inactivation of micro-organisms by ozone at 20–25°C

Micro-organism	k L mg^{-1}min^{-1}	CT_{lag} mg min L^{-1}	Ozone exposure mg min L^{-1} for inactivation of 2-log	4-log	Reference
E. coli	7800	*	0.0006	0.0012	Hunt & Marinas, 1997
B. subtilis spores	2.9	2.9 ~3 (varying)	4.5	6.1	Driedger et al., 2001 Larson & Marinas, 2003
Rotavirus[a]	76	*	0.06	0.12	Langlais et al., 1991
G. lamblia cysts	29	*	0.16	0.32	Wickramanayake et al., 1984b
C. parvum oocysts	0.84	0.83	6.3	11.8	Rennecker et al., 1999

[a]5°C; *very small, not detectable by applied methods

The rate of inactivation of micro-organisms by ozone depends on the type of organism and varies over about four orders of magnitude (Table 4.1). Although, compared to the reactivity of organic compounds, this is a narrower distribution, it is very important for the design of disinfection systems. Disinfection parameters (k, CT_{lag}) depend strongly on temperature (data in Table 4.1 are given for 20–25°C only), with a higher disinfection efficiency at higher temperature (Gallard et al., 2003; Rakness et al., 2005). Reactor hydraulics are critical for disinfection because inactivation of micro-organisms over several orders of magnitude is required (Gujer & von Gunten, 2003; Do-Quang et al., 2000). This is only possible if disinfection systems approach plug-flow behaviour (Roustan et al., 1992). In practice, this can be achieved by a series of completely stirred tank reactors (CSTRs), for example by dividing a reactor into chambers with baffles (Roustan et al., 1991). It is the most ozone-resistant target micro-organism that determines the required ozone exposure. Lag phases are most important for the required ozone exposure as may be seen from columns 4 and 5 in Figure 4.1. Stochastic modelling indicates that the largest uncertainty in predicting inactivation of C. parvum oocysts lies more in the experimental determination of the lag-phase than in the inactivation rate constant (Neumann et al., 2007).

There is only scarce information in the literature concerning the inactivation mechanisms of micro-organisms by ozone. During chlorination, the inactivation of E. coli proceeds in the following order of viability indicators: (i) loss of culturability, (ii) loss of substrate responsiveness, (iii) loss of membrane potential, (iv) loss of respiratory activity, and finally (v) loss of membrane integrity (Lisle *et al.*, 1999). Today, culturability is the main parameter for the assessment of disinfection. With a better understanding of other endpoints, (ii)–(v), and the development of new analytical tools [e.g. flow cytometry (Hammes *et al.*, 2011)], other parameters might gain in importance for the assessment of disinfection efficiency in practice.

In waters containing significant concentrations of bromide, the required ozone exposures for a certain degree of inactivation may lead to high bromate concentrations (Driedger *et al.*, 2001; Kim *et al.*, 2004, 2007a; Buffle *et al.*, 2004). Thus bromate formation may be a limiting factor, and measures have to be taken to comply with the drinking water standard (cf. Chapters 11, 14).

4.2 INACTIVATION MECHANISMS: ROLE OF MEMBRANES AND DNA

The relatively narrow distribution of inactivation rate constants for various micro-organisms may be caused by the similar types of constituents that are attacked during inactivation of all micro-organisms. Main targets for ozone may be the nucleic acids; to what extent membrane damage contributes is as yet not known.

The kinetic parameters for the inactivation of viruses are given for rotavirus in Table 4.1. In viruses, the genetic information, represented by the nucleic acids DNA or RNA, is only protected against ozone attack by a thin protein coat. Generally, only a few protein constituents react with ozone with high rate constants. The aliphatic amino acids and also phenylalanine (note that benzene reacts at $2\ M^{-1}\ s^{-1}$) are barely reactive. The basic amino acids, incorporated in the protein, will only show ozone reactivity to the extent present as free bases. Intrinsically high reactivity can only be expected from histidine (Chapter 8) and the sulfur-containing amino acids methionine, cysteine and cystine (Chapter 9). It is thus conceivable that the viral coat is not an efficient barrier for ozone diffusing to the nucleic acids. Therefore, the high rate constant (Table 4.1) is probably due to the attack of ozone on DNA/RNA.

With bacteria, the situation is more complex. Here, DNA is attached to the bacterial membrane. Some 10^8 ozone molecules are required for the inactivation of a bacterium (Scott & Lesher, 1963; Finch *et al.*, 1988). Based on this assumption for a lake water with a total cell count of 10^9 cells/L (Hammes *et al.*, 2008), 10^{17} ozone molecules/L will be required for their inactivation. This corresponds roughly to 0.2 µM (0.01 mg/L) ozone which is far below typically applied ozone doses (>10 µM $= 0.5$ mg/L). Therefore, disinfection processes do not contribute significantly to the ozone consumption under typical treatment conditions. It has been suggested that inactivation of planktonic bacterial cells is due to a destruction of the bacterial cell wall and subsequent leakage of cellular contents (Scott & Lesher, 1963). Even though there is membrane damage in the early stage of ozonation, cell viability of *E. coli* is only affected by the oxidation of DNA (Komanapalli & Lau, 1996). This is supported by the fact that the inactivation rate constant for *E. coli* (Table 4.1) is only a factor of two different from that of rotavirus, suggesting that DNA might still be the main target for ozone. Upon ozonation of a natural consortium of bacteria, membrane damage occurs simultaneously to the loss of cell numbers. For other disinfectants such as chlorine dioxide, chlorine and monochloramine, membrane integrity decreases before cell numbers are reduced (Ramseier *et al.*, 2011a). The kinetics of membrane damage of natural bacteria is slower than the kinetics of inactivation of pure cultures as determined by cultivation methods. Even though this points towards DNA damage, the two experimental systems cannot be compared directly (Ramseier *et al.*, 2011a). Therefore, further studies are needed to fully elucidate the importance of membrane damage in disinfection processes.

In addition to inactivation, ozone also causes mutations (Rodrigues *et al.*, 1996; Dubeau & Chung, 1982; Dillon *et al.*, 1992). This may be taken as evidence for DNA damage (with the cell remaining adequately intact), but the possibility that ozone by-products (e.g. formed in the reaction with the cell wall) have caused the mutagenic effect cannot be excluded. For example, hydroperoxides and H_2O_2 are typical ozonation by-products, and the latter is known to be weakly mutagenic (Thacker, 1975; Thacker & Parker, 1976). With this in mind, ozone by-products derived from membrane damage may also contribute to cell mortality.

4.3 REACTIONS WITH NUCLEIC ACID COMPONENTS

The ozone-reactive components of the nucleic acids are the nucleobases: thymine, cytosine, adenine and guanine in DNA. Some viruses contain RNA instead of DNA, and here thymine is replaced by uracil.

Uracil (Ura) Thymine (Thy) Cytosine (Cyt)

Adenine (Ade) Guanine (Gua)

There is a remarkable spread of ozone rate constants among the nucleic acid constituents (Chapter 6, Table 6.1). Notably, the rate constants of the Ade nucleosides, adenosine (Ado) and deoxyadenosine (dAdo), are very low. This is in great contrast to •OH reactions that are close to diffusion-controlled for all nucleobases (von Sonntag, 2006).

The ozone chemistry of thymine and thymidine has been elucidated in quite some detail (Flyunt et al., 2002) (Chapter 6). The reactive site is the quasi-olefinic C(5)–C(6) double bond. This may also hold for uracil and cytosine. The substantial increase in the rate of reaction of the pyrimidines with ozone upon deprotonation may be largely due to an increase in the electron density in the reacting double bond. With Cyt and cytidine (Cyd)/deoxycytidine (dCyd), protonation at the exocyclic nitrogen acts in the opposite direction, that is, it lowers the rate of reaction.

With the pyrimidine free bases, the formation of some singlet oxygen, notably at high pH is observed, but not with the corresponding nucleosides (Table 6.7) (Muñoz et al., 2001). Mechanistic aspects are also discussed in Chapter 6.

The rate of reaction of Ado and dAdo are very low, and this may be due to the fact that the aromatic ring contains two nitrogens. Nitrogen-containing heteroaromatics react only very slowly with ozone (Chapter 8). The electron density at the exocyclic amino group must be also low, but the formation of singlet oxygen (20%, Table 6.7) and of •OH radicals [43% (Flyunt et al., 2003a)] point to this nitrogen as the preferred site of attack. The N-oxide at N1 is formed upon the reaction of H_2O_2 with dAdo (Mouret et al., 1990). This N-oxide is not formed in the reaction of ozone with dAdo (Muñoz et al., 2001). A more detailed product study is still missing, but reactions (5)–(8) may be tentatively suggested to account for the formation of $^1O_2$ and of •OH. Potential precursors of •OH are $O_3^{•-}$ [reaction (7)] or $O_2^{•-}$ [reaction (8); for the ensuing chemistries see Chapter 8].

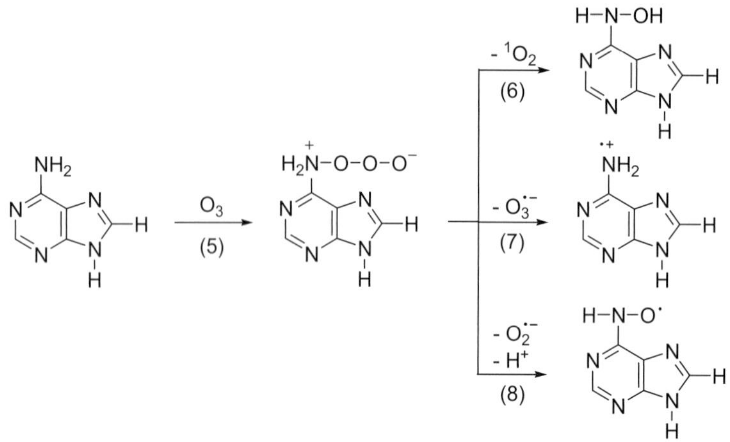

Singlet oxygen (40%, Table 6.7) but no •OH (Flyunt et al., 2003) are formed with Gua derivatives. The latter is surprising as the reduction potential of the guanine riboside Guo is lower ($E_7 = 1.29$ V) than that of adenine riboside Ado ($E_7 = 1.56$ V) (von Sonntag, 2006). This may exclude reaction (7) and favour reaction (8) as the precursor of •OH.

4.4 REACTION WITH DNA

In the reaction of ozone with DNA, •OH plays an important role (Van der Zee et al., 1987; Theruvathu et al., 2001). This •OH formation must be due to the reaction of ozone with the adenine moiety (Ishizaki et al., 1984; Theruvathu et al., 2001). For the determination of the intrinsic ozone rate constant with DNA, tertiary butanol has to be added. Under such conditions, the rate of reaction of DNA is only 410 $M^{-1}s^{-1}$ (in the absence of tertiary butanol $k_{obs} = 1.1 \times 10^3$ $M^{-1}s^{-1}$), i.e. much lower than that of the weighted average of the nucleobases. In the case of •OH, which reacts with the nucleobases and their derivatives at close to diffusion-controlled rates [$k \approx 3 \times 10^9$ $M^{-1}s^{-1}$ (Buxton et al., 1988)], the rate constant of •OH with DNA is considerably lower [$k = 2.5 \times 10^8$ $M^{-1}s^{-1}$ (Udovicic et al., 1994)], since in this non-homogeneous reaction with the macromolecule DNA two terms, a diffusion term (k_{diff}) and a reaction term (k) have to be considered (Udovicic et al., 1991). The observed overall rate constant (k_{obs}) is the harmonic mean of these two rate constants [cf. Equation (9)].

$$\frac{1}{k_{obs}} = \frac{1}{k} + \frac{1}{k_{diff}} \tag{9}$$

In contrast to •OH, ozone reacts with the nucleobases at rates much below the diffusion-controlled limit, and the second term must fall away. Hence, the rate of reaction of ozone with the nucleic acids is only given by the first term, that is, it should be close to that of the weighted average of the concentrations of the various nucleobases in the nucleic acid times their rate constants with ozone. This is not observed. The reason for this is as yet not understood. As on this basis, the dAdo moiety can barely contribute (cf. the low rate constant given in Table 6.1) and the explanation that •OH production must be due to an ozone reaction with this moiety must fall away. It is tentatively suggested that in double-stranded DNA, hydrogen bonding between the nucleobases and base stacking may be the reason for these unexpected effects.

Corresponding experiments with RNA are as yet not available. From the rate constants given in Table 6.1, one would assume that in RNA the guanine moiety is the most likely one to become degraded upon ozone treatment. This has indeed been observed (Shinriki *et al.*, 1981). In DNA, the situation might be somewhat different. Besides guanine, thymine may be the other preferred target.

4.5 APPLICATION OF OZONE FOR DISINFECTION IN DRINKING WATER AND WASTEWATER

To assess the disinfection efficiency in water treatment, the CT-concept is applied. C stands for the concentration of disinfectant (ozone) and T for the contact time; CT is the product of the aqueous disinfectant concentration and contact time. In laboratory systems where ozone can be directly measured, the CT can be expressed as ozone exposure. This corresponds to the integral under the ozone decay curve of an ozone vs. time plot (von Gunten & Hoigné, 1994). In real reactor systems, the CT value is not easily accessible. Often, time-resolved ozone concentration profiles and contact times are not readily available. To overcome this problem, several concepts have been developed (Rakness, 2005; Rakness *et al.*, 2005). A conservative approach is the calculation of CT_{10}, where C is the reactor effluent concentration (or concentration at last sampling point) and T_{10} is the travel time of the first 10% of the water going through the reactor. T_{10} is typically much shorter than the hydraulic retention time τ; the T_{10}/τ ratio is often around 0.5 (Roustan *et al.*, 1993). This approach does not take into consideration that the ozone concentration is significantly higher near the influent of the reactor, and ozone exposure is underestimated. Especially for ozone-resistant micro-organisms such as *C. parvum* oocysts (Table 4.1), this approach may lead to higher required ozone doses and hence to an increased formation of disinfection by-products such as bromate (Chapters 11 and 14). To make a better approximation of the real ozone concentration in the reactor, calculating the geometric mean of the ozone concentrations at the inlet and outlet [$C = (C_{in} \times C_{out})^{0.5}$] has been suggested (Rakness *et al.*, 2005). This method is an improvement compared to the standard CT_{10} approach. Yet, it is still far from the real ozone exposure. To further improve the prediction of disinfection efficiency, combined models including reactor hydraulics (determined by tracer experiments), ozone decay kinetics and disinfection kinetics have to be used (Roustan *et al.*, 1993; von Gunten *et al.*, 1999; Do-Quang *et al.*, 2000; Gallard *et al.*, 2003; Kim *et al.*, 2004; Smeets *et al.*, 2006). To include the variability of parameters such as inactivation rate constant, CT_{lag}, ozone decay rate, temperature, changes in water quality and hydraulics, "uncertainty modelling" is a powerful tool for assessing the variability in the disinfection efficiency (Neumann *et al.*, 2007). With increasing computing power, methods based on computational fluid dynamics can make even more accurate predictions and are a valuable tool in reactor design (Wols *et al.*, 2010).

In wastewater, ozone decay is typically too fast for using measured ozone concentrations to calculate CT values (Chapter 5). Parameters for process design in wastewater disinfection by ozone have been discussed (Xu *et al.*, 2002).

4.6 TOXICOLOGICAL ASSESSMENT OF OZONE INDUCED TRANSFORMATION PRODUCTS

The abatement of organic micropollutants during ozonation does typically not lead to their mineralisation but to the formation of transformation products (Huber *et al.*, 2004; McDowell *et al.*, 2005; Radjenovic *et al.*, 2009; Benner & Ternes, 2009a, b; Dodd *et al.*, 2010; Lange *et al.*, 2006; Schumacher *et al.*, 2004a). Thus, ozonation products may contain more or fewer structural similarities to the original compounds, and the question arises, whether the original biological effects (e.g. antimicrobial activity, oestrogenicity, herbicidal properties, etc.) are lost even when the molecules are only slightly modified. There is also some concern about new biological effects resulting from the transformation of micropollutants. Since these questions can rarely be answered by elucidating the structures of the transformation products only, ozonated solutions of a biologically active compound may have to be tested for remaining or new biological activity (Mestankova *et al.*, 2011; Escher & Fenner, 2011). In the following, some of the most important classes of biologically active compounds and the change in biological activity upon ozonation will be discussed.

4.7 ENDOCRINE DISRUPTING COMPOUNDS

Compounds behaving like hormones and disturbing the hormonal status of an organism are called endocrine disrupting compounds (EDCs). Among many other bioactive compounds, they are considered the most important class in terms of adverse effects to aquatic life (Runnalls *et al.*, 2010). The general implication for the water industry of the presence of EDCs and other micropollutants and their removal has been addressed (Snyder *et al.*, 2003, 2006; Broséus *et al.*, 2009). The simplest way to disturb the hormonal system of an organism is to interact with a receptor in a way similar to the hormone itself. The receptor of the female hormone oestrone is a typical example and is called an oestrogen receptor, because it also binds other oestrogenic compounds such as the natural hormone oestradiol and the synthetic hormone 17α-ethinyloestradiol. These oestrogenic compounds are phenols, and the phenolic group is essential for binding to the oestrogen receptor (Lee *et al.*, 2008). They are found in WWTP effluents in concentrations of up to several ng/L (Andersen *et al.*, 2003; Ning *et al.*, 2007a). One of the main concerns of the release of oestrogenic compounds is the feminisation of male fish (Sumpter & Johnson, 2008). In an experimental lake in north-western Ontario, Canada, the fish population was almost extinct after a seven-year exposure to 5–6 ng/L 17α-ethinyloestradiol (Kidd *et al.*, 2007).

Oestradiol (E2)   Oestron (E1)   17α-Ethinyloestradiol (EE2)

Besides the phenol function, there are hydrophobic binding sites that influence the equilibrium constant of equilibrium (10).

EDC + oestrogen receptor $\rightleftarrows$ bound EDC   (10)

Because of structural similarities to oestrogenic compounds, many industrial and natural compounds can also bind to oestrogen receptors with different equilibrium constants of equilibrium (10) and hence exert different endocrine disrupting potentials (Bonefeld-Jörgensen *et al.*, 2007). A critical review on *in vitro* and *in vivo* effects of synthetic organic chemicals including phenol-containing compounds is available (Tyler *et al.*, 1998). Some examples of such compounds will be discussed in the following.

Bisphenol A, *t*-butylphenol, octylphenol and nonylphenol are technical products and abundant in wastewaters and surface waters (Ahel *et al.*, 1994; Voutsa *et al.*, 2006; Ning *et al.*, 2007a).

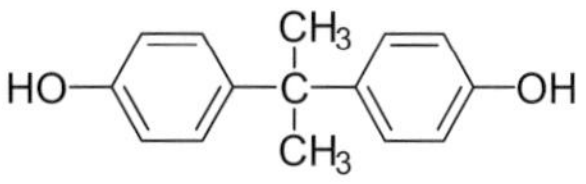

Bisphenol A

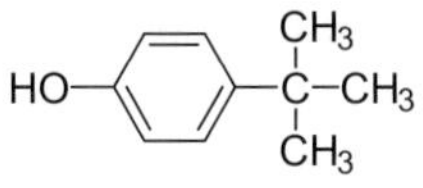

t-Butylphenol

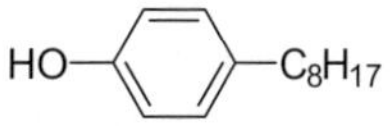

Octylphenol

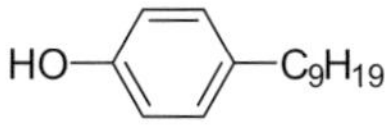

Nonylphenol

The isoflavone family (formononetine, daidzein, equol, biochanin, genistein) is typically found in surface waters (Hoerger et al., 2009).

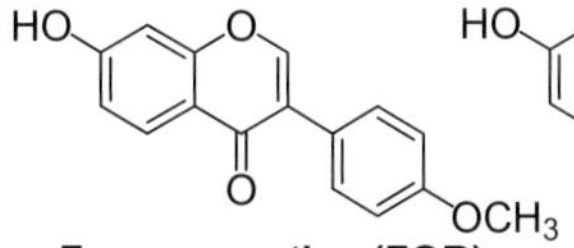

Formononetine (FOR)

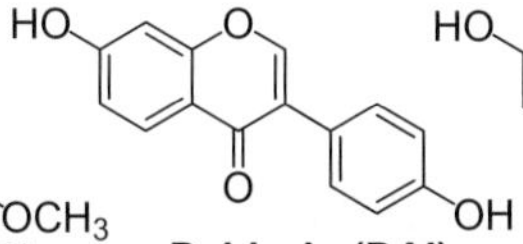

Daidzein (DAI)

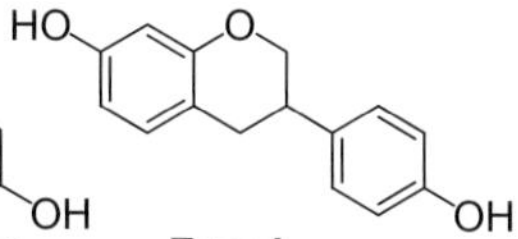

Equol

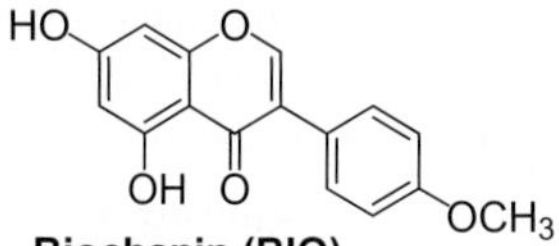

Biochanin (BIO)

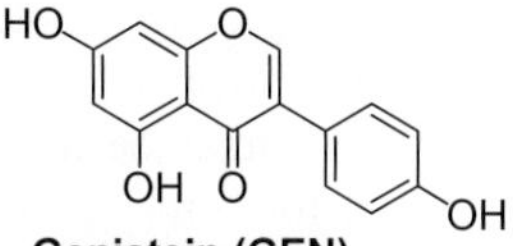

Genistein (GEN)

Parabenes, esters of the *p*-hydroxybenzoic acid, also belong to the group of EDCs. They seem to be mainly taken up by cosmetics (Darbre & Harvey, 2008).

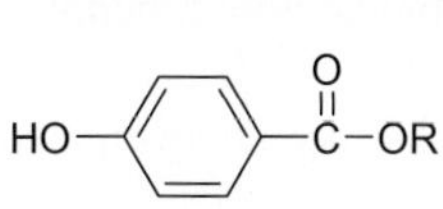

Parabene

A typical example of an endocrine disruptor is nonylphenol. There are more than 500 isomers, including stereoisomers, conceivable (Günther, 2002). To visualise this, four of them are shown below.

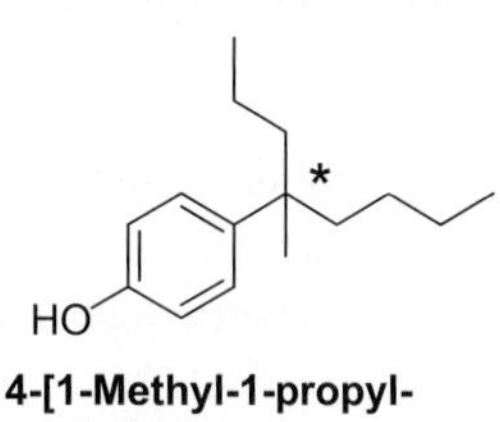

4-[1-Methyl-1-propyl-pentyl]-phenol

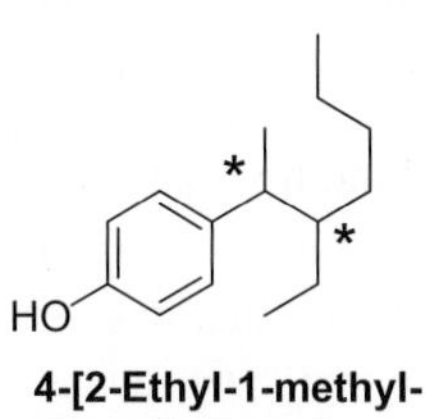

4-[2-Ethyl-1-methyl-phexyl]-phenol

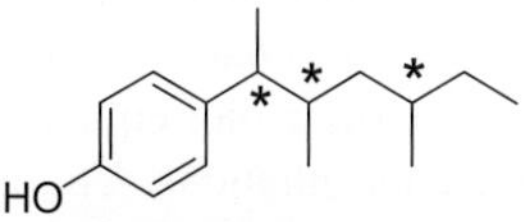

4-[1,2,4-Trimethyl-hexyl]-phenol

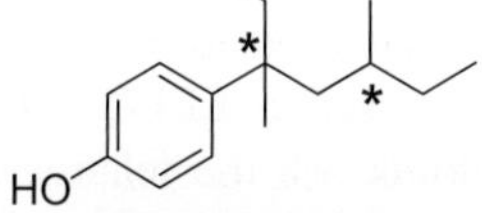

4-[1-Ethyl-1,3-dimethyl-pentyl]-phenol

They differ by orders of magnitude in their binding constants and hence in their endocrine disrupting potential. The daily intake of nonylphenols by food has been estimated at 7.5 µg for an adult in Germany (Günther, 2002). This high value indicates that drinking water may not be the major source of EDCs to man, but the main concern of these EDCs is related to their adverse effects on aquatic life (Oehlmann *et al.*, 2000, 2006; Kidd *et al.*, 2007; Sumpter & Johnson, 2008). Nonylphenols lead to feminisation of aquatic organisms and a decrease in male fertility and the survival of juveniles at concentrations below 10 µg/L (Soares *et al.*, 2008). Prosobranch snails have been suggested as test organisms (Duft *et al.*, 2007; Oehlmann *et al.*, 2007). For a comparison of prosobranch snails and fish see Jobling *et al.* (2004). Similar concerns are related to bisphenol A. Its mode of action and potential human health effects have been reviewed (Vandenberg *et al.*, 2009).

Tin compounds show a strong endocrine disrupting activity for aquatic life (Duft *et al.*, 2003a; Duft *et al.*, 2003b; Wirzinger *et al.*, 2007). Tributyl- and triphenyl tin have a very different mode of action as xeno-androgens (Schulte-Oehlmann *et al.*, 2000).

Even drugs such as carbamazepine (Oetken *et al.*, 2005) or the herbicide atrazine (Hayes *et al.*, 2002) seem to have endocrine disrupting properties.

As there are so many different compounds that give rise to endocrine disrupting activity, *in vivo* and *in vitro* test systems have been developed to assess water samples experimentally. Two test systems that are widely used for *in vivo* and *in vitro* oestrogenicity assessment are based on *in vivo* measurement of the blood plasma vitellogenin (VTG) concentrations in male rainbow trout (*Oncorhynchus mykiss*) and *in vitro* measurement of the oestrogen binding to a human oestrogen receptor (yeast oestrogen screen, YES) [for a review see (Sumpter & Johnson, 2008), for some recent developments requiring shorter reaction times (LYES) see (Schultis & Metzger, 2004)]. For the YES assay, two plasmids have been introduced into a yeast cell. The first plasmid generates the human α-oestrogen receptor. Upon addition of the EDC to be tested, it binds to the receptor according to equilibrium (10) and changes its structure. This receptor complex now binds to the second plasmid and triggers the formation of a marker enzyme. The resulting enzyme activity is measured. In these assays, bisphenol A and the mixture of technical nonylphenols are about four orders of magnitude less potent than oestradiol.

The effect of ozonation on the oestrogenic activity of natural and synthetic EDCs has been investigated in laboratory and full-scale studies.

4.7.1 Laboratory studies

Ozonation of nonylphenols leads to an intermediate increase in the oestrogenicity (Sun *et al.*, 2008). Hydroxylation is a major process in the ozone chemistry of phenols (Chapter 7), and 4-nonylcatechol has a higher oestrogenic activity than 4-nonylphenol itself. When all phenolic compounds are degraded, oestrogenic activity disappears. In contrast to this study, it has been reported that there is still some residual oestrogenicity (E-screen assay with MCF-7 cells) even after full transformation of bisphenol A, E1 and EE2 (Alum *et al.*, 2004). This is in contrast to studies that have shown a stoichiometric loss of the oestrogenicity with the transformation of EE2 by ozone and •OH radicals (Huber *et al.*, 2004; Lee *et al.*, 2008). Figure 4.2 shows the decrease of the relative EE2 concentration (open circles) as a function of the ozone dose (in presence of tBuOH as a scavenger for •OH radicals) and as a function of the fluence in the UV/H_2O_2 process (oxidation by •OH radicals; direct photolysis can be neglected) (Lee *et al.*, 2008).

Figure 4.2 also shows the oestrogenic activity expressed as EEEQ (17α-ethinyloestradiol equivalents, open circles). In the insets, the relative EEEQ is plotted vs. the relative EE2 concentrations. The good correlation between the two parameters with a slope of unity indicates that both oxidants lead to a loss of oestrogenicity by the first attack on the EE2 molecule. Loss of oestrogenicity upon •OH attack was also confirmed for E2 and EE2 (Linden *et al.*, 2007). Other oxidants such as chlorine, bromine, chlorine dioxide and ferrate (VI) also efficiently destroy the oestrogenicity of EE2 (Lee *et al.*, 2008).

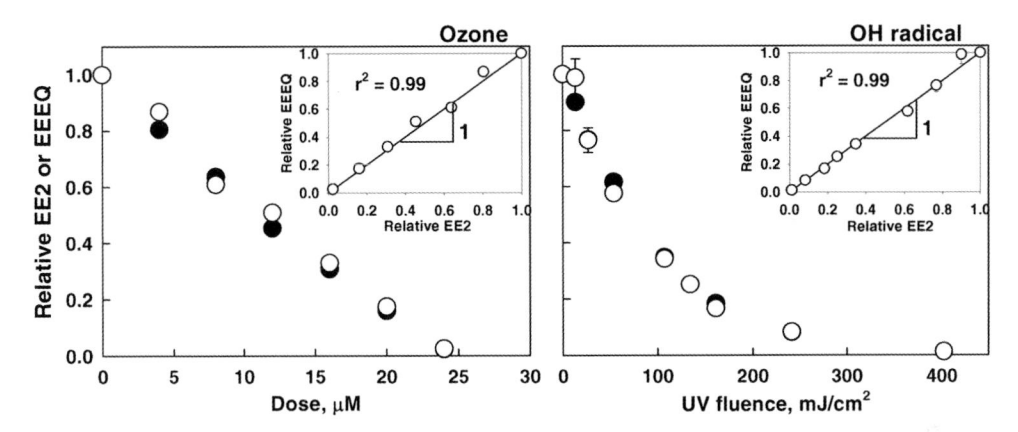

Figure 4.2 Decrease of the relative EE2 concentration (filled circles) and oestrogenic activity (open circles) EEEQ, 17α-ethinyloestradiol equivalents due to the oxidation by ozone and ˙OH radicals. Insets show plots of the relative EEEQ versus the relative EE2 concentration. Experimental conditions: [EE2]$_0$ = 10 µM, pH = 8, T = 23°C, ozonation in presence of tBuOH (5 mM). When filled circles are invisible, data overlap with open circles. Reprinted with permission from (Lee *et al.*, 2008). Copyright (2008) American Chemical Society.

4.7.2 Full-scale studies

A considerable number of EDCs have been detected in WWTPs (Spengler *et al.*, 2001). In many WWTPs, oestrogenicity is controlled by oestrogenic compounds (E1, E2 and EE2) with concentrations in the ng/L range rather than industrial compounds such as alkylphenols, alkylphenolmonoethoxylates and alkylphenoldiethoxylates, even though present in µg/L levels (Aerni *et al.*, 2004). However, this might be different in WWTPs with a high contribution of industrial wastewater. Oestrogenicity in wastewater is eliminated well by activated sludge processes (> 90% removal) (Escher *et al.*, 2009). Since many EDCs are phenols, they are readily eliminated by an ozonation step and lose their hormonal activity upon attack by chemical oxidants (see above). This was demonstrated in a full-scale WWTP in Switzerland where a > 95% elimination of oestrogenicity (YES assay) was found upon ozonation (Escher *et al.*, 2009). In another study, the oestrogenicity was reduced by 90% for an ozone dose of about 0.4 mgO$_3$/mg DOC (Stalter *et al.*, 2011). The effective removal of oestrogenic activity by ozonation has been confirmed by an additional test with yolk-sac larvae (Stalter *et al.*, 2010b). A significant reduction of vitellogenin levels was observed in fish exposed to ozonated wastewater compared to fish reared in conventionally treated wastewater.

In other WWTPs, oestrogenicity (YES assay) decreases in parallel to the degradation of bisphenol A by ozone (Figure 4.3).

The same effect is also apparent in the effluent of two other WWTPs where bisphenol A and EEQ were 10% (Köln-Stammheim) and 1% (Bottrop) of the given example. In these wastewaters, there is a very close correlation between the presence of the technical product bisphenol A and oestrogenicity. This points to the predominance of industrial sources (contraceptives were below detection) for the observed oestrogenicity in these wastewaters. However, bisphenol A and alkylphenols (data not shown) can only account for about 10% of the observed oestrogenicity. Therefore, there must be other, as yet unknown, oestrogenic micropollutants that give rise to the YES assay response. They may belong to the phenol family, as bisphenol A and YES assay show the same ozone response (Figure 4.3). Such a discrepancy between YES assay and detected micropollutants with ED activity is not uncommon. Also in estuarine sediments the oestrogenic activity is not adequately reflected (<1%) by the concentrations of known oestrogens (Thomas et al., 2004). Nevertheless, in other municipal wastewaters with less industrial influence the oestrogenicity could be reasonably well predicted by summing up the effects of individually measured compounds such as E1, E2 and EE2 (Aerni et al., 2004).

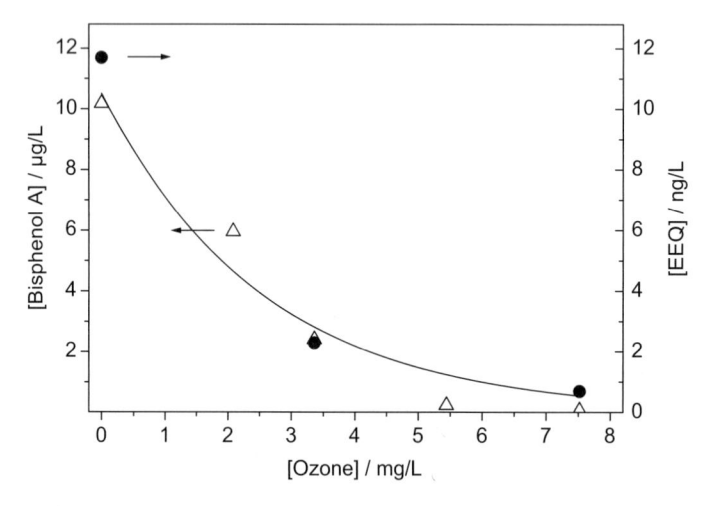

Figure 4.3 Decrease of the bisphenol A concentration (triangles, left axis) and of the oestrogenicity (YES assay, E2 equivalent (EEQ), circles, right axis in the effluents of the WWTPs of Düsseldorf-Süd (DOC = 16 mg/L). Adapted from (Nöthe, 2009), with permission.

4.8 ANTIMICROBIAL COMPOUNDS

A quasi-definition of a micropollutant is that the compound in question must show some biological activity. Two commonly asked questions are whether ozonation decreases or increases the biological activity, and when it decreases the biological activity, how many moles of ozone are required for reducing the biological activity to an insignificant level. A group of compounds that lose their biological activity are the phenolic EDCs discussed above. The fact that after ozonation ≤ 3% of the starting material is slowly regenerated after ozonation (Chapter 7) is considered insignificant in the present context. Clinically important antibiotics are found in individual concentrations from 0.5 to 3 µg/L in raw and primary wastewaters, and a reduction of 60–90% of their concentration typically occurs during activated sludge treatment [(Dodd et al., 2006a) and references therein]. In general, resulting effluent concentrations are below levels affecting bacteria and aquatic life, but there might still be effects of certain antibiotics on aquatic organisms and on bacteria in the activated sludge process (Dodd et al., 2006a). The main concern related to antibiotics is the development of antibiotic resistance in microbial consortia (Kümmerer, 2009a, b). To date it is, however, not yet clear, whether this can happen in activated sludge processes or in the aquatic environment (Kümmerer, 2009b). Antibiotics comprise many classes of compounds containing ozone-reactive moieties: macrolides (tertiary amines, Chapter 8), sulfonamides (anilines, Chapter 8), fluoroquinolones (piperazines, Chapter 8), lincosamide (thioether, Chapter 9), β-lactams (olefins, Chapter 6; thioethers, Chapter 9), tetracycline (olefins, Chapter 6;

phenols, Chapter 7), glycopeptides (phenols, methoxytoluene, Chapter 7), aminoglycosides (primary amines, Chapter 8) (Dodd et al., 2006a). With the exception of β-lactams, all investigated antibiotics (9 classes, 14 compounds) lost their antimicrobial activity during ozonation in parallel to the loss of the parent compound (Dodd et al., 2009; Lange et al., 2006). This means that the primary attack of ozone (exclusion of •OH reactions) on the parent compound efficiently removes its antimicrobial activity. This is shown in Figure 4.4 for macrolides and Figure 4.5 for fluoroquinolones.

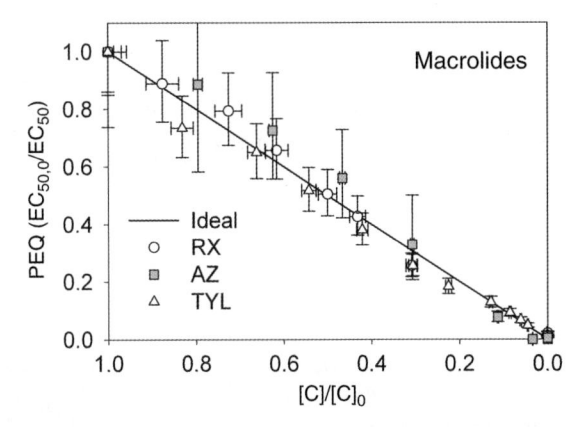

Figure 4.4 Deactivation stoichiometries of macrolides by ozone at pH 7 (tBuOH = 5 mM). PEQ: Potency equivalent derived from growth inhibition tests; RX: roxithromycin; AZ: azithromycin; TYL: tylosin. Reprinted with permission from (Dodd et al., 2009). Copyright (2009) American Chemical Society.

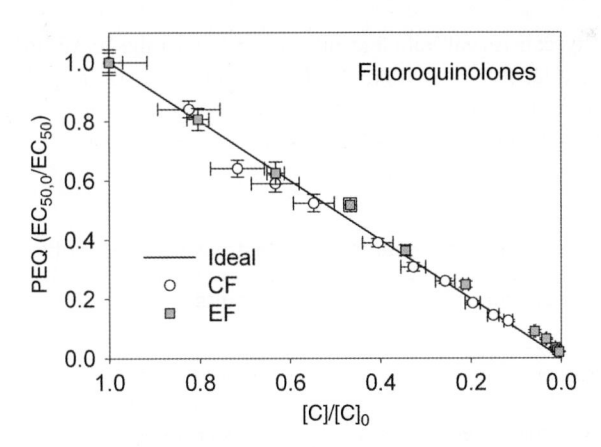

Figure 4.5 Deactivation stoichiometries of fluoroquinolones by ozone at pH 7 (tBuOH = 5 mM). PEQ: potency equivalent derived from growth inhibition tests; CF: ciprofloxacin; EF: enrofloxacin. Reprinted with permission from (Dodd et al., 2009). Copyright (2009) American Chemical Society.

Similar data as shown in Figure 4.4 have been reported for the macrolide antibiotic clarithromycin (Lange et al., 2006). For the attack of •OH on antibiotics, a slight deviation of the ideal loss of antibacterial activity was observed for β-lactams (Dodd et al., 2009). This was attributed to the formation of hydroxylated analogues of the parent compounds which are known to exert an antibacterial activity similar to or higher than the parent compounds themselves (Kavanagh, 1947). For higher degrees of transformation, however, the residual antibacterial activity was destroyed.

A discussion of ozone-reactivity of the two β-lactams penicillin G and cephalexin and the products formed upon ozone attack (Dodd *et al.*, 2010) is given in Chapter 9. The (*R*)-sulfoxides have some residual antimicrobial activity. With cephalexin, this is of no importance, because an attack of ozone on the remaining olefinic function leads to an efficient loss of the antibacterial activity. In the case of penicillin G, however, the (*R*)-sulfoxide is ozone-resistant and retains about 15% of the antibacterial activity of penicillin G. Yet during wastewater ozonation, penicillin-G-(*R*)-sulfoxide is efficiently oxidised by •OH, and the antibacterial activity is removed at typical ozone doses (Dodd *et al.*, 2010).

The antiviral drug oseltamivir acid (active metabolite of Tamiflu®, see Chapter 6) may be applied in high quantities during pandemic influenza outbreaks and significant concentrations can be expected in wastewaters, even more so because this compound is not efficiently removed in biological wastewater treatment (Prasse *et al.*, 2010). The loss of antiviral activity due to treatment with ozone and •OH was investigated by measuring neuraminidase inhibition of two viral strains (Mestankova *et al.*, 2012). For both oxidants and low doses, an increased activity was produced that disappeared at high degrees of transformation of the parent and also when solutions were analysed after 24h. Primary unstable products exerting a higher antiviral activity than the parent compound must thus be formed (Mestankova *et al.*, 2012).

Biocides are also ubiquitous compounds found in wastewater effluents. Triclosan, a common additive to soaps and personal care products reacts quickly with ozone (Chapter 7). Its antibacterial properties are lost upon the first attack of ozone (Suarez *et al.*, 2007) while some dioxin-like activity is formed (Mestankova *et al.*, in preparation.). The dioxin that is formed in the photolysis of triclosan has been characterised as 2,8-dichlorodibenzo-*p*-dioxin (Latch *et al.*, 2005).

4.9 TOXICITY

Tests on biological activity can reveal valuable information as to the feasibility of ozone application in micropollutant abatement (Gerrity & Snyder, 2011). Additional tests in real water systems might be necessary for ascertaining further endpoints and the role of mixtures of micropollutants and of the water matrix (e.g. DOM). Table 4.2 summarises effect-oriented studies on ozone-treated secondary wastewater effluents.

Table 4.2 Biological test results from *in vivo* and *in vitro* tests on secondary effluents treated with ozone

Treatment system	Test systems	Results	References
Full-scale ozonation wastewater	Pre-concentration of samples –Bioluminescence inhibition –Oestrogenicity –Arylhydrocarbon receptor response –Genotoxicity –Neurotoxicity –Phytotoxicity	Significant decrease of all effects upon ozonation step	Macova *et al.*, 2010; Reungoat *et al.*, 2010

(*Continued*)

Table 4.2 Biological test results from *in vivo* and *in vitro* tests on secondary effluents treated with ozone (*Continued*)

Treatment system	Test systems	Results	References
Full-scale ozonation of secondary effluent	Pre-concentration of samples –Bioluminescence inhibition –Growth inhibition –Inhibition of photosynthesis –Oestrogenicity –Inhibition of acetylcholinesterase –Genotoxicity	Significant removal of all effects during ozonation –No genotoxicity formation during ozonation	Escher et al., 2009
Ozonation of secondary effluent	In vitro tests after pre-concentration –*In vivo* tests with whole effluent –Genotoxicity –Retionic acid receptor (RAR) agonist activity –Acute ecotoxicity (*Daphnia magna*) –Japanese medaka embryo exposure tests	Significant removal of genotoxicity, RAR agonist activity and acute ecotoxicity – Higher ozone doses lead to a reduction in hatching success rate of Japanese medaka embryos	Cao et al., 2009
Tertiary treated sewage effluent (ozonation)	*In vivo* tests with juvenile rainbow trout *O. mykiss* in liver and kidney tissues –Glutathione S-transferase (GST) –Total glutathione (GSH) –Glutathione peroxidase (GPX) –Lipid peroxidase (LPO) –Haem peroxidase	Liver: Increased haem peroxidase, LPO and GST –Total GSH depleted –Kidney: Increased LPO, GPX observation shows oxidative stress of organism –Coagulation after ozonation reduces these effects	Petala et al., 2009
Ozonation of secondary effluent	*In vivo* tests –*Lemna minor* growth inhibition –Chironomid toxicity test with the non-biting midge *Chironomus riparius* –*Lumbriculus variegatus* toxicity –Genotoxicity –Oestrogenicity	Growth inhibition –Removal of oestrogenicity –Increased genotoxictiy –Enhanced toxicity for *Lumbriculus variegates* –Effects disappear after rapid sand filtration after ozonation	Stalter et al., 2010a

(*Continued*)

Table 4.2 Biological test results from *in vivo* and *in vitro* tests on secondary effluents treated with ozone *(Continued)*

Treatment system	Test systems	Results	References
Ozonation of secondary effluent	Fish early life stage toxicity test (rainbow trout, O. mykiss)	Development retardation –Effect disappears after post-sand filtration –Removal of oestrogenicity	Stalter et al., 2010b

Toxicity is not as well-defined as, for example, endocrine disruption. Endocrine disruption can be well-described by a relatively simple assay, for example, the YES assay that provides a reasonable answer. Other tests may be used for confirmation, but are not strictly required. In contrast for describing toxicity, many different test systems may have to be utilised depending on the relevant endpoints for ecosystems (Stalter *et al.*, 2010a). Different toxicity tests have been carried out with ozonated wastewater (Table 4.2). For example, the *Lumbriculus variegatus* test, based on the development of this worm within 28 days, revealed a significantly enhanced toxicity after ozonation compared to conventional treatment (Stalter *et al.*, 2010a). Moreover, a significantly increased genotoxicity was observed, detected with the comet assay using haemolymph of the zebra mussel (Stalter *et al.*, 2010a). The comet assay, originally developed for radiation-induced DNA strand breakage caused by ionising radiation in cells (Ostling & Johanson, 1984), has been later applied to the assessment of DNA-reactive agents (Collins *et al.*, 1997). Also the fish early life stage toxicity test (FELST) using rainbow trout (*Oncorhynchus mykiss*) revealed a considerable developmental retardation of test organisms exposed to ozonated wastewater (Stalter *et al.*, 2010b). All these effects were removed by subsequent sand filtration to the level of conventional treatment. Activated carbon treatment even resulted in a significant reduction of genotoxicity. The build-up of toxicity upon ozonation and its subsequent removal during post-sand filtration points to the formation of biodegradable organic compounds such as aldehydes and ketones from the reaction of ozone with DOM (Chapter 3). Apparently, these compounds show toxicity in certain test systems but not in others. It is very unlikely that the increased toxicity is caused by transformation products from micropollutants.

The above statement, that toxicity is not a simple parameter, is illustrated by the fact that other parameters that can measure toxicity such as the *Lemna minor* growth inhibition test and the *Chironomid* toxicity test did not give a response on ozonated wastewater (Stalter *et al.*, 2010a).

It seems fair to conclude that ozone-induced toxicity is mostly transient and can be eliminated by biological sand filtration or biological activated carbon filtration. Hence, toxicity may not be a major obstacle for introducing ozonation as a polishing step in wastewater treatment.

Chapter 5
Integration of ozonation in drinking water and wastewater process trains

5.1 HISTORICAL ASPECTS
5.1.1 Drinking water
In France, the earliest test with ozonation for disinfection dates back to 1886. In 1906, ozonation for full-scale drinking water disinfection, after slow sand filtration, was installed in Nice (France) (Le Palouë & Langlais, 1999). In the early applications of ozone in water treatment, ozonation was basically a replacement for chlorine disinfection. Especially in water supplies treating groundwater, there was a substantial carryover of ozone into reservoirs and the distribution systems, because in these waters ozone is quite stable due to the low DOC concentration and the high carbonate alkalinity (Chapter 3). In Germany, ozonation of groundwaters and surface waters also started around 1900. Several plants (Wiesbaden, Paderborn, Hermannstadt) were closed down, however, after only a few years of operation, mainly due to the lower costs of chlorination (Böhme, 1999). In the USA, the first ozone installations for taste and odour or colour removal were established in the early 1900s. Significant capacity was only installed in the mid-1980s (Rice, 1999). In other countries such as Japan, Canada, UK, The Netherlands, Belgium and Switzerland, ozone application for drinking water treatment started between the 1940s and the 1960s (Matsumoto & Watanabe, 1999; Lowndes, 1999; Kruithof & Masschelein, 1999; Geering, 1999; Larocque, 1999). A compilation of the estimated number of drinking water treatment plants in Europe and North America is shown in Table 5.1. From this comparison, it is evident that the number of ozonation plants per capita is very high in France and Switzerland, whereas it is rather low in the USA and Japan. This reflects the high affinity of many water suppliers to chlorine and related products, despite the many disadvantages of these oxidants compared to ozone (Sedlak & von Gunten, 2011).

5.1.2 Municipal wastewater
So far, there is only a limited number of wastewater treatment plants that use ozonation. Most of these plants are located in Canada, Germany, Japan, South Korea and the USA (Paraskeva & Graham, 2002) with the main objective of disinfection. Disinfection of wastewater effluent is mandatory in some states in the USA. In Europe, it is only applied occasionally for achieving bathing water quality goals. Disinfection of wastewaters is typically also applied for irrigation or other reuse purposes. Disinfection of wastewaters, however, is often achieved with chlorine or UV rather than ozone. Nevertheless, the growing importance of water reuse and the discussion on enhanced treatment of wastewaters for micropollutant removal may

render ozone more attractive because of its dual role as disinfectant and as oxidant (Ternes *et al.*, 2003; Reungoat *et al.*, 2010; Hollender *et al.*, 2009; Oneby *et al.*, 2010; Zimmermann *et al.*, 2011).

Table 5.1 Estimated number of drinking water plants using ozone in Europe, North America and Japan (numbers from period 1997–2011)

Country	Number of plants	Number of plants per million capita	References
Switzerland	108	13.8	von Gunten & Salhi, 2003
France	700	10.6	Langlais *et al.*, 1991
Canada	68	2	Larocque, 1999
Germany	>100	1.2	Böhme, 1999; Loeb *et al.*, 2011
United Kingdom	50	0.8	Lowndes, 1999
BENELUX	ca. 20	0.72	Kruithof & Masschelein, 1999
USA	200	0.64	Rice, 1999
Japan	>50	>0.39	Loeb *et al.*, 2011

5.2 DRINKING WATER TREATMENT SCHEMES INCLUDING OZONATION

Even though ozone was originally used in one- or two-step processes, today it is widely accepted that ozone should at least be combined with a biological treatment step for removal of AOC/BDOC (Chapter 3). Nowadays, ozonation is mostly used for treatment of surface waters and is integrated into multi-barrier treatment systems (Kruithof & Masschelein, 1999). The development of the integration of ozonation into water treatment trains is shown schematically in Figure 5.1(a) for the evolution of lake water treatment in Switzerland from the 1950s to 2005.

This is representative for the development of treatment trains in other industrialised countries as well. Figure 5.1(b) shows the evolution of the phosphate concentration in Lake Zurich which is an indicator of the degree of eutrophication with a peak in the early 1970s. Drinking water treatment had to follow this development (additional treatment steps) to cope with the various problems related to eutrophication (turbidity, high DOC concentrations, taste and odour issues, etc.). The phosphate concentration in Swiss lakes was reduced by rigorous measures in water pollution control (phosphate elimination in wastewater treatment plants, phosphate ban from textile washing detergents). In the 1950s, lake water treatment started with conventional treatment (sand filtration followed by chlorination) which was supplemented with a pre-chlorination followed by a flocculation process. The introduction of activated carbon was a consequence of increasing taste and odour problems but in other contexts also a barrier against micropollutants which started to emerge in the 1970s. In the mid-1970s, the discovery of trihalomethanes and the formation of chloro- and bromophenols [potent taste and odour compounds (Acero *et al.*, 2005)] which are formed during chlorination was a motivation for moving away from chlorination to ozonation. First, an intermediate ozonation was introduced followed by biological activated carbon filtration (see also Chapter 3). Then, pre-chlorination was replaced by pre-ozonation. The combination of ozone with biological activated carbon filtration is also known as the Mülheim process which was developed in 1974 (Sontheimer *et al.*, 1978; Heilker, 1979). Mülheim is one of the cities in the most densely populated industrial area in Germany, the Ruhr area (ca. 4 million people), and draws its water from the river Ruhr. A rigorous and efficient treatment scheme was necessary to provide high-quality drinking water to the population (Figure 5.2 and discussion below).

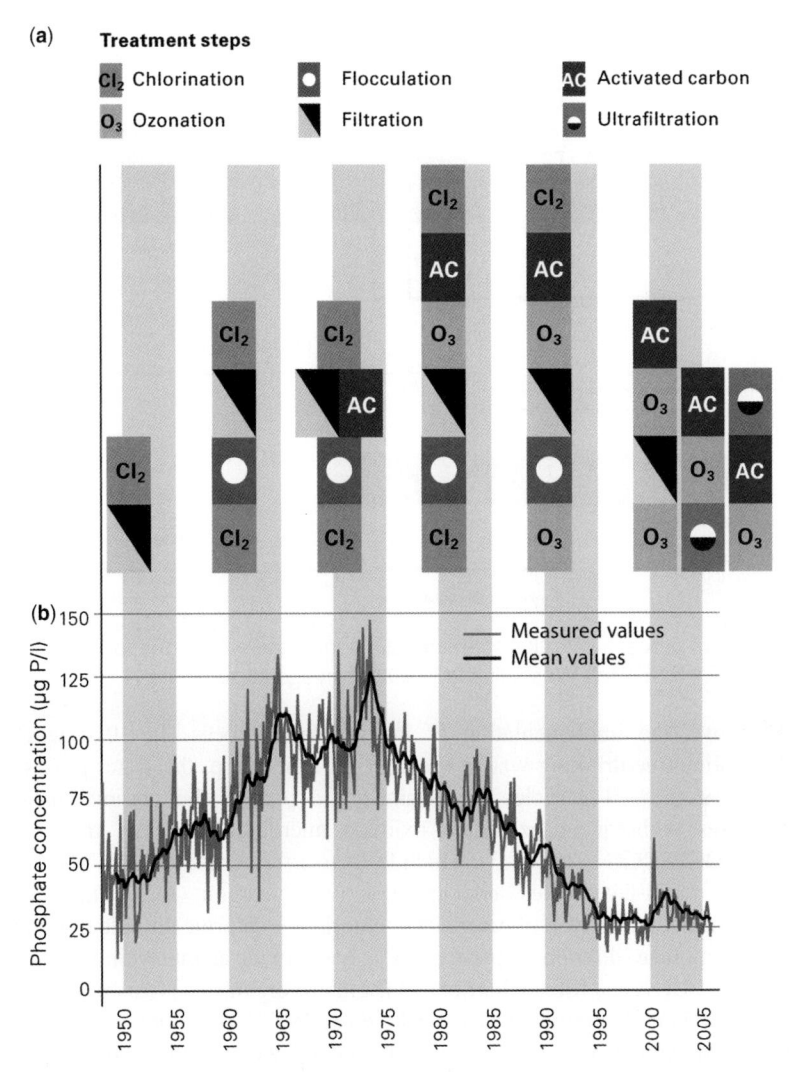

Figure 5.1 (a) Evolution of lake water treatment for Lake Zurich water from 1950 to 2005. (b) Evolution of the phosphate concentration as an indicator for the trophic state of the lake. According to von Gunten, 2008, with permission.

When lake water quality in Switzerland improved in the 2000s and membranes became affordable for drinking water treatment, treatment trains were simplified including a membrane filtration step (ultrafiltration, UF). Hence, treatment schemes could be reduced to three steps, always including the combination of ozone with biological activated carbon filtration and membranes. This combination guarantees a high drinking water quality with respect to hygiene, aesthetic properties and chemical contaminants and allows a distribution of drinking water without residual disinfectants such as chlorine, chloramine and chlorine dioxide. This has the advantage that no disinfection by-products are formed in the distribution system (Sedlak & von Gunten, 2011).

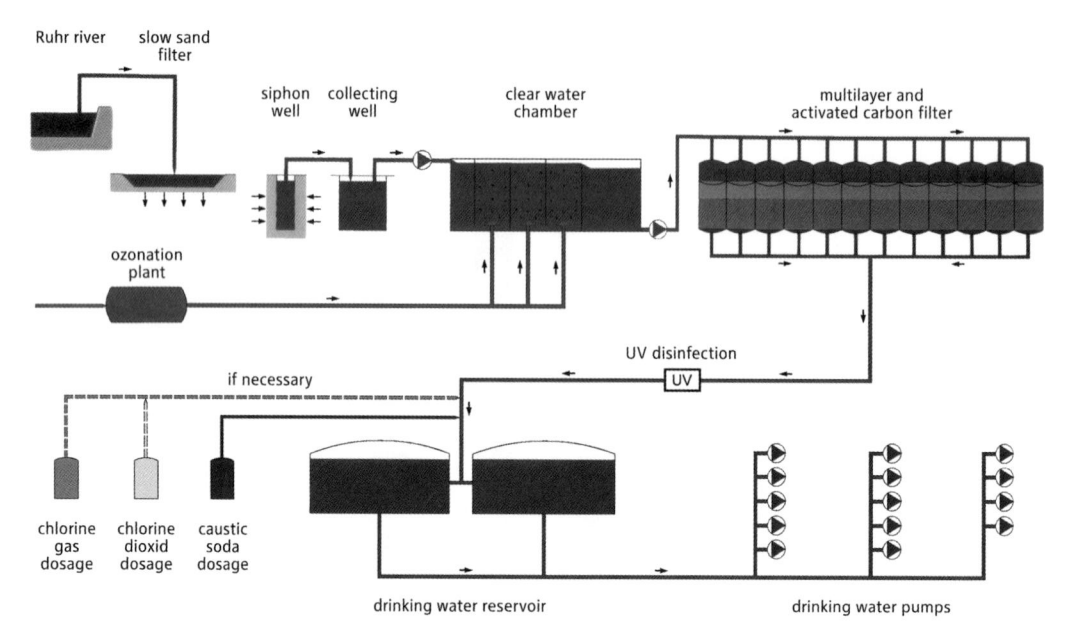

Figure 5.2 The Mülheim process with the characteristic combination of ozonation and biological filtration. With permission of RWW Rheinisch-Westfälische Wasserwerksgesellschaft mbH.

As mentioned above, the implementation of the Mülheim process in 1974 was a considerable break-through in chlorine-free drinking water treatment. In a first step, the water passes through a slow sand filter. Thereby, suspended particles are retained and part of the organic matter is consumed by microbial processes. Subsequent ozonation oxidises micropollutants and transforms part of the remaining DOM to AOC/BDOC (Chapter 3), which leads to a further reduction of DOC in the following biofiltration step with multi-layer filters containing activated carbon (AC). The water is then UV-disinfected prior to distribution. In case of emergency, chlorine or chlorine dioxide dosing is possible.

Since the 1990s, membrane filtration, in particular UF, has become an interesting alternative to deep bed sand filtration processes. UF is an efficient barrier against micro-organisms (viruses, bacteria and protozoa) but does not retain organic micropollutants (Jacangelo *et al.*, 1997). Therefore, a combination of UF with ozone oxidation and adsorption processes leads to a drinking water with good hygienic and chemical qualities.

Figure 5.3 shows a conventional process combination including ozonation and deep bed filtration processes and two possible process combinations including UF, ozonation and AC filtration (Pronk & Kaiser, 2008).

All three process combinations are currently used for the treatment of Lake Zurich water in Switzerland. Combination C may require a final disinfection with UV, because AC filters lose significant numbers of micro-organisms. In a pilot study with combination B, the total cell count determined by flow cytometry was $\approx 10^3$ cells/mL after ozonation and $>10^5$ cells/mL after AC filtration (cf. Figure 5.4) (Hammes *et al.*, 2008).

In the AC filter, bacteria can grow on AOC/BDOC, which leads to a significant increase in the total cell count (Figure 5.4). In combination B which is reflected in Figure 5.4, bacteria are removed by UF to below detection limit of flow cytometry, whereas in combination C, where AC filtration is the last treatment step, they would be released into the distribution system if not properly disinfected.

Combination A

Raw water → Pre-ozonation → [pH adjustment] → Rapid filtration → Intermediate ozonation → Activated carbon filtration → Slow sand filtration

Combination B

Raw water → Pre-filtration → Ozonation → Activated carbon filtration → Membrane filtration

Combination C

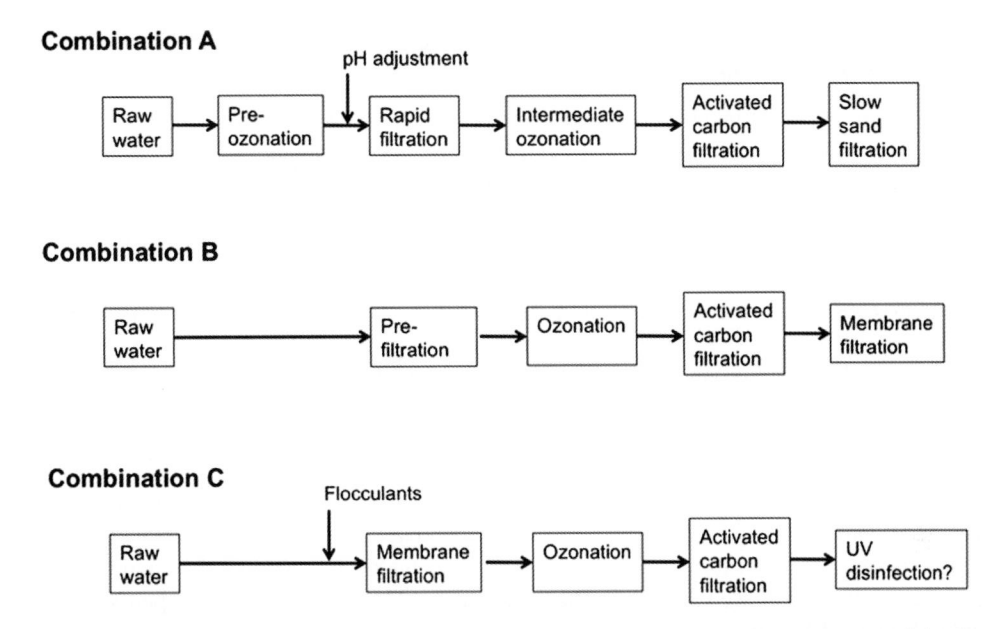

Raw water → [Flocculants] → Membrane filtration → Ozonation → Activated carbon filtration → UV disinfection?

Figure 5.3 Conventional multi-barrier treatment with ozonation (Combination A) and two possible alternative process combinations (B and C) including ozonation and ultrafiltration. Adapted from Pronk & Kaiser, 2008, with permission.

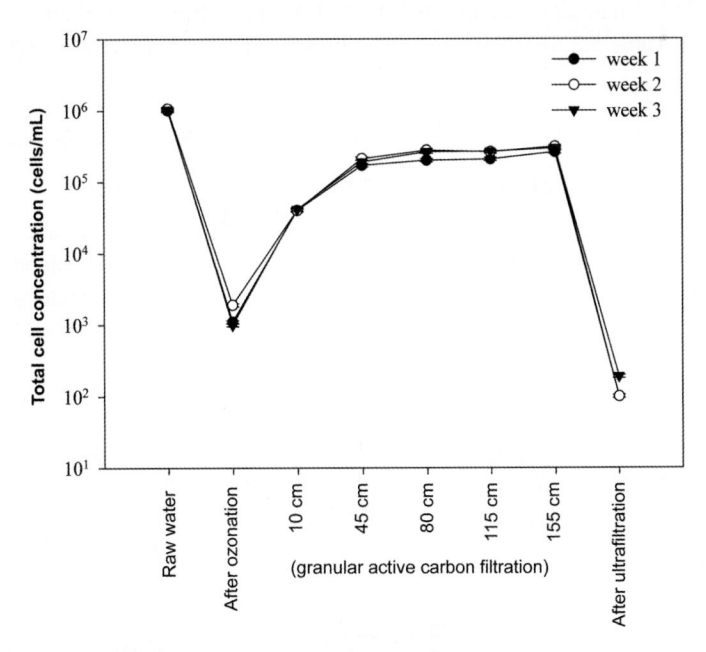

Figure 5.4 Total cell concentration determined with flow cytometry as a function of the treatment step for process combination B in Figure 5.3. According to Hammes *et al.*, 2008, with permission.

5.3 MICROPOLLUTANTS IN WATER RESOURCES, DRINKING WATER AND WASTEWATER

A large body of compounds found in waters and wastewater are considered as micropollutants (Fahlenkamp *et al.*, 2004). There is vast literature on the occurrence of micropollutants in water resources such as ground-water and surface waters and in urban water management systems. They have also been monitored in wastewater before and after biological treatment. These studies are too numerous to be dealt with in this book and the reader is referred to some review articles and books on this topic (Ternes, 1998; Kolpin *et al.*, 2002; Snyder *et al.*, 2003; Vanderford *et al.*, 2003; Schwarzenbach *et al.*, 2006, 2010; Kümmerer, 2010; Ternes & Joss, 2006; Benotti *et al.*, 2009; Kim *et al.*, 2007b; Richardson & Ternes, 2011).

Micropollutant abatement is one of the objectives of ozone in water treatment (Gerrity & Snyder, 2011). Mechanistic studies on the degradation of micropollutants by ozone are found in Chapters 6–13. Ozone rate constants of micropollutants as much as they are known thus far are also given in these chapters. Some micropollutants react too slowly with ozone to be eliminated by ozone under such conditions. They are typically called ozone-refractory, although this is not fully correct. Their reaction is just too slow to be of relevance. They may be eliminated by the much more reactive •OH radicals. For the formation of •OH in the reactions of ozone with the organic part of the water matrix (DOM) and the contribution of •OH to micropollutant abatement see Chapter 3. The chemistries of •OH and the ensuing peroxyl radicals are discussed in Chapter 14.

Combination B in Figure 5.3 was tested with regard to the potential for micropollutant removal from Lake Zurich water in pilot-scale (≈ 10 m^3/h). Three compounds with different physical chemical properties (rate constants for reactions with ozone and •OH, adsorption behaviour on AC, see legend of Table 5.3) were investigated, the fuel additive methyl-*t*-tbutylether (MTBE) and the taste and odour compounds 2-isopropyl-3-methoxypyrazine (IPMP) and β-ionone [for a compilation of typical taste and odour compounds see Table 5.2, for their occurrence in Swiss lakes see Peter *et al.* (2009)].

Table 5.2 Some abundant taste and odour compounds according to Peter & von Gunten (2007)

Compound	Odour	Odour threshold/ng L^{-1}	Source	Formula in Chapter
β-Cyclocitral	Fruity	19000	Cyanobacteria	6
Geosmin	Earthy	4	Cyanobacteria, Actinomycetes	14
cis-3-Hexen-1-ol	Grassy	70000	Algae	6
β-Ionone	Violets	7	Cyanobacteria, algae	6
2-Isopropyl-3-methoxypyrazine (IPMP)	Decaying vegetation	0.2	Actinomycetes	8
2-Methylisoborneol (MIB)	Musty	15	Cyanobacteria, Actinomycetes	14
trans,*cis*-2,6-Nonadienal	Cucumber	20	Algae	6
1-Penten-3-one	Fishy-rancid	1250	Cyanobacteria, algae	6

(*Continued*)

Table 5.2 Some abundant taste and odour compounds according to Peter & von Gunten (2007) (*Continued*)

Compound	Odour	Odour threshold/ng L^{-1}	Source	Formula in Chapter
2,6-Di-*t*-butyl-4-methylphenol	Plastic	Not available	Leaching from polyethylene pipes	7
2,4,6-Tribromoanisole (TBA)	Earthy–musty	0.03	Methylation of bromophenol by micro-organisms	7
2,4,6-Trichloroanisole (TCA)	Musty	0.03	Methylation of chlorophenol by micro-organisms	7

After ozonation, the relative residual concentrations of MTBE, IPMP and β-ionone are 80%, 50% and <5%, respectively (Table 5.3).

Table 5.3 Elimination of selected micropollutants in a multi-barrier system including ozonation, activated-carbon filtration and ultrafiltration (combination B in Figure 5.3). Relative residual concentrations are given in %. For MTBE and IPMP data are given for various activated-carbon running times (10, 150 and 200 days) MTBE: $k(O_3) = 0.15$ M^{-1} s^{-1}, $k(^{\bullet}OH) = 1.9 \times 10^9$ M^{-1} s^{-1}, log$K_{ow} = 0.94$; IPMP: $k(O_3) = 50$ M^{-1} s^{-1}, $k(^{\bullet}OH) = 5 \times 10^9$ M^{-1} s^{-1}, log$K_{ow} = 2.41$; β-ionone: $k(O_3) = 1.6 \times 10^5$ M^{-1} s^{-1}, $k(^{\bullet}OH) = 7.8 \times 10^9$ M^{-1} s^{-1}, log$K_{ow} = 3.84$. According to von Gunten, unpublished

Micropollutant	After ozonation	After activated-carbon filtration	After ultrafiltration
	Relative residual concentrations in %		
MTBE (10 days)	80	<d.l.	<d.l.
MTBE (150 days)	75	78	77
IPMP (10/200 days)	50	<d.l.	<d.l.
β-Ionone	1	<d.l.	<d.l.

MTBE methyl-*t*-butylether; IPMP 2-Isopropyl-3-methoxypyrazine; d.l. detection limit.

This can be explained on the basis of the rate constants for the reaction of these compounds with ozone and $^{\bullet}$OH. MTBE has the lowest reactivity towards ozone and $^{\bullet}$OH; IPMP is in an intermediate range, whereas β-ionone reacts rapidly with ozone and $^{\bullet}$OH (Table 5.3). For short running times, AC removes MTBE well, but at longer operation times (150 days) MTBE breaks through. IPMP is fully retained even after 200 days of operation of the granular activated carbon (GAC) filter. This behaviour of the two compounds can be explained by their octanol–water partitioning coefficients (K_{ow}) taken as a measure for the affinity to GAC. Based on the K_{ow} value, β-ionone would be even better adsorbed. But this has no consequences, since β-ionone is already fully removed by ozonation. For improving MTBE removal, the AOP ozone/H$_2$O$_2$ (Chapter 3 for details) was tested by injecting H$_2$O$_2$ in the third chamber of the four-chamber ozone reactor. An increase in the ozone dose in combination with H$_2$O$_2$ addition allowed a full elimination of MTBE (for the reaction of $^{\bullet}$OH with MTBE see Chapter 14).

5.4 ENHANCED WASTEWATER TREATMENT WITH OZONE

As discussed above, wastewater treatment with ozone has mainly been installed for disinfection purposes. In Europe and some other countries, the need for enhanced wastewater treatment for micropollutant removal to protect aquatic ecosystems is now considered (Ternes *et al.*, 2003; Joss *et al.*, 2008; Ort *et al.*, 2009). Currently, there are two options for enhanced treatment of secondary wastewater effluent, namely the addition of a powdered AC (Nowotny *et al.*, 2007) or ozonation (Joss *et al.*, 2008). For investigating ozonation of wastewater systems, a considerable number of pilot- and full-scale tests have been performed over recent years (Ternes *et al.*, 2003; Huber *et al.*, 2005; Wert *et al.*, 2009a; Hollender *et al.*, 2009; Reungoat *et al.*, 2010; Zimmermann *et al.*, 2011; Stalter *et al.*, 2010a, 2011; Macova *et al.*, 2010).

A typical treatment train for enhanced wastewater treatment with an ozonation step is shown in Figure 5.5. For minimising ozone consumption, ozone is placed after the activated sludge treatment, where DOC concentration is lowest.

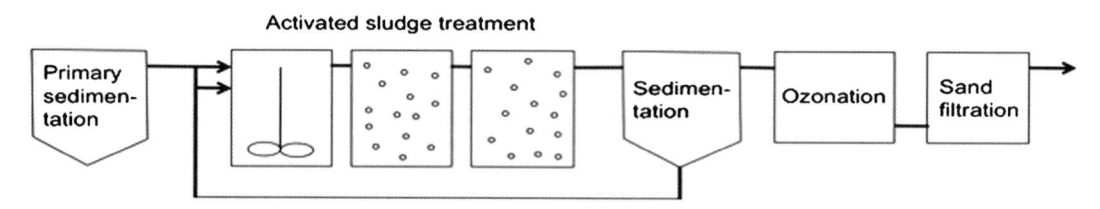

Figure 5.5 Treatment train for enhanced wastewater treatment with ozone.

Ozonation should be followed by biological sand filtration for degrading AOC/BDOC formed during ozonation (Zimmermann *et al.*, 2011) (Chapter 3). Ozonated wastewater leads to a developmental retardation of rainbow trout, but this effect disappears after sand filtration (Stalter *et al.*, 2010b). This might be an indication that easily biodegradable compounds such as aldehydes are responsible for these adverse health effects on fish (cf. Chapter 4). Aldehyde formation during ozonation of wastewater is quite significant (Wert *et al.*, 2007).

Formation and the degradation of AOC during ozonation and post-sand filtration is shown in Figure 5.6 for a full-scale wastewater treatment plant for a specific ozone dose of 1.24 g O_3/g DOC (Zimmermann *et al.*, 2011). The main increase in AOC is observed at the first sampling point in the ozone reactor (P1), where most of the ozone is consumed. Thereafter, AOC remains almost constant, mainly due to the small residual ozone concentration. After sand filtration, AOC is considerably reduced, indicating that a significant portion of low-molecular-weight compounds that are potentially toxic to fish can be removed by this process.

Figure 5.6 also shows the evolution of total cell counts (TCC) during ozonation and post-sand filtration. Depending on the ozone dose, TCC decreases by 2–4 orders of magnitude (Zimmermann *et al.*, 2011) reflecting the efficiency of ozone as a disinfectant. After sand filtration, total cell count increases again leading to an overall efficiency of the process of 1–2 logs reduction of TCC. This is in line with the observation of cell growth in activated carbon filters in drinking water treatment (Hammes *et al.*, 2008). A 0.5–3 log inactivation of *E. coli* was observed during ozonation of the same wastewater. In this case, however, sand filtration did not lead to an increase of bacterial counts for *E. coli* (Zimmermann *et al.*, 2011) suggesting that there is also no re-growth of pathogenic bacteria in the sand-filtered water after ozonation.

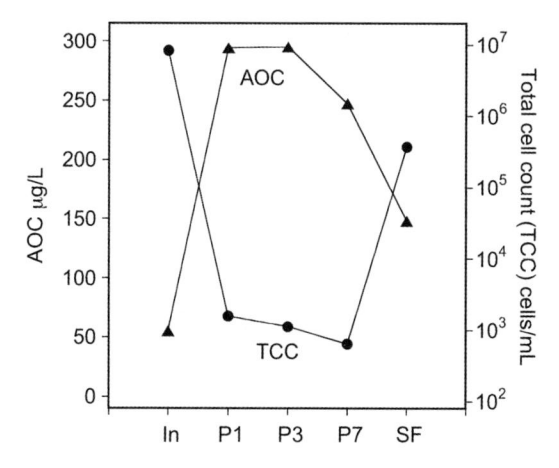

Figure 5.6 Assimilable organic carbon (AOC) formation and total cell counts (TCC, determined by flow cytometry) for ozonation followed by biological sand filtration for a full-scale wastewater treatment plant, Regensdorf, Switzerland (25000 population equivalent). Ozone dose 1.24 g O$_3$/g DOC. In: Secondary effluent, inlet to ozone reactor; P1, P3, P7 sampling points within the reactor; SF sand filtration. Adapted from Zimmermann et al., 2011, with permission.

5.5 ENERGY REQUIREMENTS FOR MICROPOLLUTANT TRANSFORMATION IN DRINKING WATER AND WASTEWATER

Energy requirements for micropollutant transformations during ozonation and advanced oxidation processes (AOPs) depend on the matrix of the water that consumes oxidants (Chapter 3, mainly type and concentration of DOM) and the rate constant for the reaction of a target compound with ozone and •OH radicals. Table 5.4 shows a comparison of the energy requirements for a 90% transformation of selected compounds during ozonation and the AOP UV/H$_2$O$_2$ for laboratory experiments. In general, an increase in energy of about 25% for O$_3$/H$_2$O$_2$ relative to the conventional ozonation is estimated based on production energy of 15 kWh/kg and 10 kWh/kg for ozone and H$_2$O$_2$, respectively (Katsoyiannis et al., 2011). For ozonation, Table 5.4 shows that for a given water quality, the required energy increases in the order SMX < pCBA < ATR < NDMA. This can be explained by a decrease in the second order rate constants for the reaction of these compounds with ozone and •OH from SMX to NDMA. Energy requirements also increase significantly from Lake Zurich water to Dübendorf wastewater, due to the higher concentrations of DOM (consumption of ozone and •OH, Chapter 3) and carbonate (consumption of •OH, Chapter 3). For a given water, energy requirements for UV/H$_2$O$_2$ are typically significantly higher and depend on the penetration depth of UV radiation. Only for NDMA, with low reactivity towards ozone and •OH (Lee et al., 2007b), does the energy requirement for UV/H$_2$O$_2$ become comparable to ozonation. This is due to the fact that NDMA undergoes mainly direct photolysis (Sharpless & Linden, 2003).

For a particular full-scale study (Hollender et al., 2009), the energy consumption for ozonation of secondary effluent, including all contributions (production of liquid oxygen, its transport, generation of ozone) was calculated. The energy requirement at the plant remained constant for process gas in the range of 100–170 g O$_3$ m^{-3} at 12 kWh/kg O$_3$. This translates into an energy requirement of 0.035 kWh m^{-3} for a specific ozone dose of 0.6 g O$_3$/g DOC, which corresponds to about 12% of the total energy consumption of a nutrient (C, N, P) removal plant (0.3 kWh m^{-3}). In addition, production of pure oxygen requires 0.01–0.015 kWh m^{-3}. Therefore, the overall energy requirement at this ozone dose (removal of SMX) is quite similar to that shown in Table 5.4 for laboratory systems. For a large range of wastewaters (10,000 to 500,000 person equivalents) and DOC contents (6 to 20 g DOC m^{-3}), total costs of ozonation (investment and operation including post-filtration step) were estimated to range between 0.05 and 0.15 € m^{-3}, depending on plant size and secondary effluent quality (Joss *et al.*, 2008).

Table 5.4 Energy requirements in kWh m^{-3} for 90% transformation of selected micropollutants by conventional ozonation in Lake Zurich water and Dübendorf wastewater and by using UV(254 nm)/H$_2$O$_2$ (0.2 mM) for varying optical path lengths (cm) in Lake Zurich water. Experimental conditions: target compound concentration = 0.5 µM, pH 8, T = 20 °C. According to Katsoyiannis *et al.*, (2011), with permission

Target compound	Lake Zurich Water Ozonation	Dübendorf wastewater Ozonation	Lake Zurich Water UV/H$_2$O$_2$ 1 cm	UV/H$_2$O$_2$ 5 cm	UV/H$_2$O$_2$ 10 cm
SMX	0.0015	0.045	0.39	0.15	0.11
pCBA	0.035	0.2	0.75	0.23	0.17
ATR	0.05	0.3	0.98	0.28	0.2
NDMA	0.5	0.9	1.62	0.44	0.3

SMX sulfamethoxazole; pCBA *p*-chlorobenzoic acid; ATR atrazine; NDMA *N*-nitrosodimethylamine; Lake Zurich water, DOC 1.3 mgC/L, carbonate alkalinity 2.6 mM; Dübendorf wastewater, DOC 3.9 mg C/L, carbonate alkalinity 6.5 mM.

5.6 SOURCE CONTROL

The removal of micropollutants from the wastewater stream by enhanced treatment of secondary wastewater effluent is an end-of-pipe solution. Other options include treatment of source-separated urine (Larsen & Gujer, 1996) or treatment of other point sources such as hospital wastewater. Urine separation and treatment with ozone has been demonstrated to be a feasible process for micropollutant removal. Even though it only accounts for about 1% of the wastewater stream, it requires more energy to remove micropollutants by ozonation than in wastewater (Dodd *et al.*, 2008). When the treatment of source-separated urine is combined with nutrient recovery (N, P), the overall energy requirement becomes even favourable for urine treatment compared to wastewater treatment (Dodd *et al.*, 2008). Nevertheless it has to be considered that some chemicals, among them high risk chemicals such as the antiarrhythmic compound amiodarone (cf. Chapter 8), are excreted via faeces (Escher *et al.*, 2011). Therefore, the full spectrum of compounds will not be removed by this approach. The contribution of hospital wastewater to the overall load of pharmaceuticals in municipal wastewater is typically quite low in the order of <15% (Ort *et al.*, 2010). Nevertheless, source control in hospitals has quite high acceptance among stakeholders, especially if the contribution of hospitals to the overall load is high (Lienert *et al.*, 2011). In this context it is noted that municipal wastewater receives an integrated load of chemicals used in households, which also include biocides, pesticides, personal care products, etc., which will also be removed by an ozone treatment of secondary effluents (Hollender *et al.*, 2009).

5.7 RECLAMATION OF WASTEWATER

Reclamation of wastewater as a resource for drinking water and for irrigation purposes (agriculture, golf courses, etc.) has become an important issue in arid and semi-arid areas due to population growth and climate change. Today, wastewater reuse is heavily based on membrane processes. Secondary effluent is typically treated by a combination of microfiltration/ultrafiltration with reverse osmosis (RO) (Figure 5.7) (Asano et al., 2007). Even in coastal areas this approach (<1 kWh/m^3) is more energy efficient than seawater desalination [3.5–4.5 kWh/m^3, (Sommariva, 2010)]. Water desalination has become increasingly important worldwide with large-scale plants (30,000–320,000 m^3/d) in Kuwait, Singapore, USA, Australia and China (Hemmi et al., 2010). The RO process is frequently followed by UV disinfection, and the water is then mostly used for replenishment of natural water bodies such as groundwaters or surface waters. In the treatment scheme outlined in Figure 5.7, ozonation is not applied. In principal, ozonation could be used to treat secondary wastewater effluent for removing NDMA precursors (Lee et al., 2007a). Such (unknown) precursors may lead to NDMA during chloramination which is routinely applied to hinder biofilm growth on the RO membrane. Typically, the rejection of micropollutants by RO is >90% (Busetti et al., 2009). Advanced oxidation of the RO permeate (UV/H$_2$O$_2$, UV/ozone or ozone/H$_2$O$_2$) is an option for an additional barrier for removing micropollutants such as NDMA, which are not fully retained by RO. AOPs in post-RO water are expected to be very efficient, because its •OH scavenging rate is very low due to its low DOC of <0.1 mg/L.

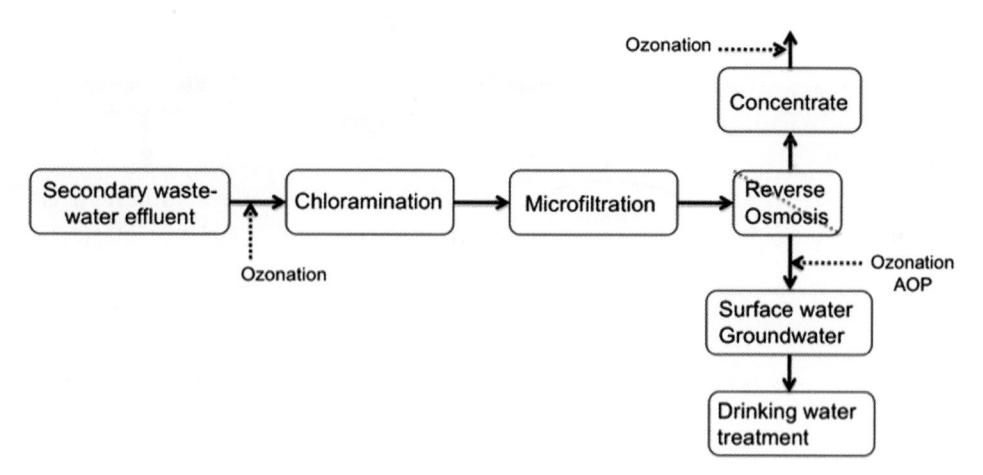

Figure 5.7 Water reuse scheme based on ultrafiltration/reverse osmosis. Points for potential ozonation steps are also indicated.

One of the problems in the RO-based water reuse scheme is the discharge of the RO concentrate. The concentration factor for micropollutants during RO treatment of wastewater is of the order of four. In a recent study, ozonation was investigated for the elimination of beta blockers from an RO concentrate with a DOC of 46 mg/L (Benner et al., 2008). For metoprolol (cf. Chapter 8) [$k(O_3) = 2 \times 10^3$ M^{-1}s^{-1} (pH 7), k(•OH) $= 7.3 \times 10^9$ M^{-1} s^{-1}] an ozone dose of >10 mg/L ($\approx$ 0.25g O$_3$/g DOC) was required for a removal of >90%. The bromate concentration for these conditions was of the order of <40 µg/L, which is relatively low considering the high Br$^-$ levels (1200 µg/L) in the RO concentrate. The ozonated water can then be released to the environment with less toxicological concern.

An alternative strategy for wastewater reuse consists of a combination of conventional processes such as ozonation, dissolved air flotation and active carbon filtration. Figure 5.8 shows a schematic representation of such a potential treatment train (Reungoat et al., 2010). Overall, this multi-barrier treatment removed, very efficiently, micropollutants and DOC. Fifty of fifty-four compounds detected in the secondary effluent were eliminated to below detection limits. The DOC concentration was reduced by 55–60%. Furthermore, toxicity determined by various biological endpoints (e.g. oestrogenicity, neurotoxicity, phytotoxicity), was significantly lower than in the influent, often very similar to the blank (Reungoat et al., 2010).

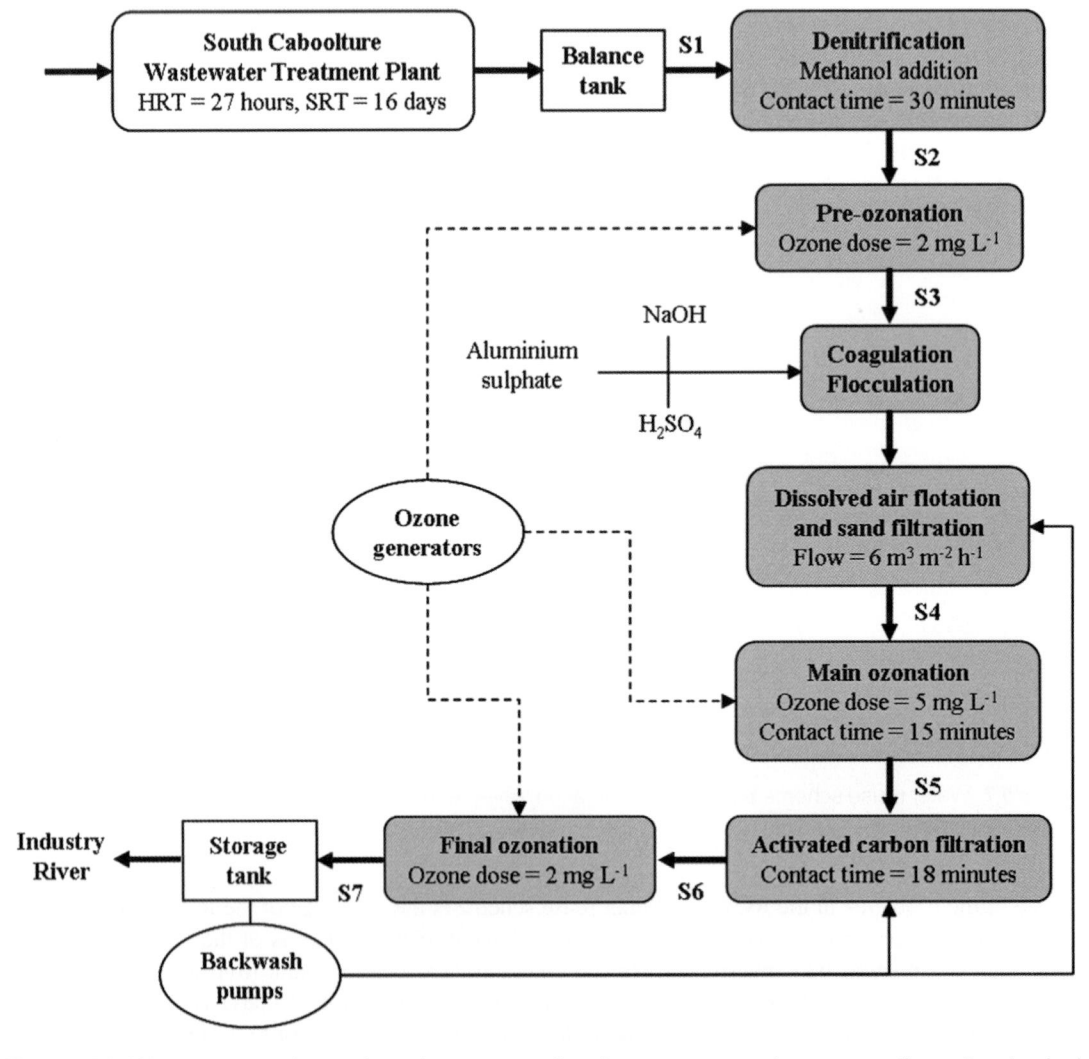

Figure 5.8 Water reuse scheme based on conventional processes such as ozonation, dissolved air flotation and activated carbon filtration. SRT: sludge retention time. According to Reungoat et al. (2010), with permission.

5.8 COMPARISON OF THE APPLICATION OF OZONE IN THE URBAN WATER CYCLE

Based on the discussion above, potential points of application of ozonation processes for micropollutant abatement in the urban water cycle are shown in Figure 5.9.

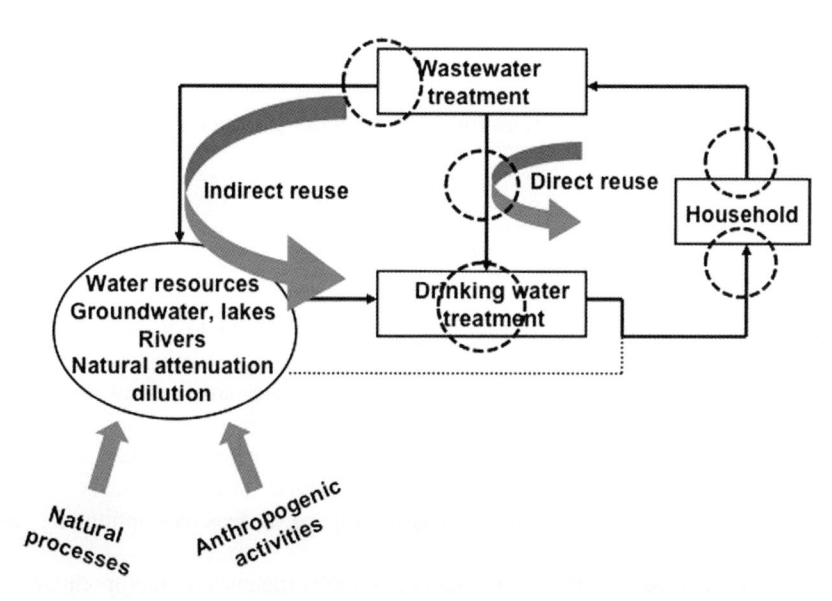

Figure 5.9 The urban water cycle with potential points of ozonation (marked with circles).

In principle, ozonation can be applied in households as point-of-entry or point-of-use treatment and to the wastewater, e.g. source separated urine (Dodd et al., 2008). Once the wastewater is collected, oxidative treatment may be carried out as post-treatment of secondary wastewater effluent (see above). Compared to treatment of source-separated urine, this also allows oxidative transformation of micropollutants which are derived from sources other than households. Both the treatment at the household level and the treatment in centralised wastewater treatment plants lead to a reduction of the micropollutant discharge to the receiving water bodies. As a consequence, ecosystems and water resources are protected from adverse impacts. When the urban water system is mainly driven by human toxicology, oxidative treatment (mainly ozonation or AOPs) may be placed within the drinking water treatment scheme (Westerhoff et al., 2005; Broséus et al., 2009; Vieno et al., 2007; Kruithof et al., 2007; Ternes et al., 2002).

This scenario has the advantage that micropollutants from diffuse sources such as agriculture, traffic and natural sources (e.g. cyanotoxins and taste and odour compounds) will also be removed (Acero et al., 2000, 2001; Benitez et al., 2007; Rodriguez et al., 2007; Peter & von Gunten, 2007; Onstad et al., 2007). For direct or indirect potable water reuse, an oxidation can be applied after a reverse osmosis treatment (Asano et al., 2007). The water quality for each treatment scenario is decisive for the efficiency of an ozonation process. The main parameter is the content of the dissolved organic matter, typically expressed as DOC concentration (Chapter 3). In addition, pH, alkalinity and ammonia also play an important role (Chapter 3). Table 5.5 summarises water qualities of hydrolysed and electrodialysed urine, municipal wastewater and water resources used for drinking water production.

Table 5.5 Water quality parameters relevant for oxidation processes, from Dodd *et al.* (2008); Pronk *et al.* (2006); Udert *et al.* (2003)

Water type	pH	DOC mg/L	Alkalinity mM	NH_3/NH_4^+ mg/L
Hydrolysed urine	9	≈2000	≈300	≈4000
Electrodialysed urine diluate*	8	≈400*	≈30	≈400
Wastewater effluent	7–8	≈5–20	2–4	≈20
Water resources for drinking water production				
Surface water	7–8	1–20	1–2	<0.005 to >1
Groundwater	7–8	<1 to 20	1–5	<0.005 to >1

*contained some methanol from dosing of micropollutants

A dramatic decrease of DOC is observed from hydrolysed urine to wastewater effluent. This is partially caused by dilution and partially by the DOC removal during activated sludge treatment. The DOC concentration in surface and groundwaters is typically much smaller and dominated by natural processes (soil weathering, algal growth, etc.). Because ozone demand is closely related to the DOM concentration, it is evident that ozone consumption gets smaller further away from the household source. However, it has to be taken into account, that human urine represents <1% of the total flow of municipal wastewater. Therefore, it might still be a feasible option for micropollutant elimination.

The water quality data (Table 5.5) have consequences for the efficiency of micropollutant elimination. A comparison of the required ozone doses for 90% micropollutant elimination is shown in Table 5.6 for hydrolysed urine, electrodialysed urine diluate, wastewater effluent and two surface waters for 17α-ethinyloestradiol (EE2) a synthetic steroid oestrogen and ibuprofen (IP) an antiphlogistic.

Table 5.6 Required ozone doses (mg/L) and corresponding O_3/DOC ratios (w/w) for a 90% elimination of 17α-ethinyloestradiol (EE2) and ibuprofen (IP) in various water sources (Lee & von Gunten, 2010; Dodd *et al.*, 2008; Huber *et al.*, 2003, 2005)

Water type	EE2 O_3 dose mg/L	EE2 O_3/DOC w/w	IP O_3 dose mg/L	IP O_3/DOC w/w
Hydrolysed urine	≈500	0.25	≈1000	0.5
Electrodialysed urine diluate*	≈150	0.375	≈600	1.5
Wastewater 1 (7.7 mg/L DOC)	>1	>0.13	n.d.	n.d.
Wastewater 2 (5 mg/L DOC)	0.5	0.1	≈4	0.8
Lake water (3.7 mg/L DOC)	0.1	0.03	n.d.	n.d.
River water (1.3 mg/L DOC)	<0.1	<0.08	>2	>2

n.d.: not determined;
*contained some methanol from dosing of micropollutants

While EE2 reacts quickly with ozone and •OH (pH 7: $k(O_3) \approx 3 \times 10^6 \, M^{-1} \, s^{-1}$, $k(•OH) = 9.8 \times 10^9 \, M^{-1} \, s^{-1}$), IP reacts mostly with •OH (pH 7: $k(O_3) = 9.6 \, M^{-1} \, s^{-1}$, $k(•OH) = 7.4 \times 10^9 \, M^{-1} \, s^{-1}$) (Huber *et al.*, 2003).

The efficiency of ozonation increases with decreasing DOC concentrations. For a 90% elimination of EE2, the ozone dose varies over more than three orders of magnitude, reflecting the difference in the DOC concentration between hydrolysed urine and the pre-treated river water. While this difference seems quite large, it has to be considered, that the volume of urine that needs to be treated is about two orders of magnitude smaller than that of wastewater. Furthermore, the ozone dose normalised to the DOC concentration for 90% abatement of the selected ozone-reactive and ozone-resistant compounds is similar for hydrolysed urine and wastewater. Therefore, urine treatment at the household level seems to be feasible, however as mentioned above, only part of the micropollutant load in the wastewater will be treated at this level. In addition, small-scale ozonation systems would have to be implemented at the household level which would require the appropriate maintenance. Table 5.6 also shows that the ozone doses and O_3/DOC ratios for drinking water sources are significantly smaller for compounds reacting rapidly with ozone and in a similar range as urine and wastewater for ozone-resistant compounds.

Chapter 6
Olefins

6.1 REACTIVITY OF OLEFINS

The rate constants for the reaction of ozone with olefins vary with the nature of the substituents at the C–C double bond by as much as eight orders of magnitude (Table 6.1).

Table 6.1 Compilation of rate constant of ozone with olefins. Published rate constants rounded to significant figures

Compound	pK_a	pH	k/M^{-1} s^{-1}	Reference
Acetamidoacrylic acid		6	9.4×10^5	Onstad et al., 2007
Aciclovir	3.4, 9.6		1.6×10^4	Prasse et al., 2012
anion			3.4×10^6	Prasse et al., 2012
protonated			250	Prasse et al., 2012
Acrylamide			1×10^5	Yao & Haag, 1991
Acrylic acid	4.25		2.8×10^4	Leitzke & von Sonntag, 2009
anion			1.6×10^5	Leitzke & von Sonntag, 2009
Acrylonitrile			670	Leitzke et al., 2003
			830	Pryor et al., 1983
Adenine	4.15		12	Theruvathu et al., 2001
protonated	9.8		5	Theruvathu et al., 2001
anion			1.3×10^5	Theruvathu et al., 2001
Adenosine	3.5		16	Theruvathu et al., 2001
protonated			5	Theruvathu et al., 2001
5'-Adenylic acid	4.4		200	Ishizaki et al., 1984
Anatoxin-a			2.8×10^4	Onstad et al., 2007
1,4-Benzoquinone			2.5×10^3	Mvula & von Sonntag, 2003
Buten-3-ol			7.9×10^4	Dowideit & von Sonntag, 1998

(Continued)

Table 6.1 Compilation of rate constant of ozone with olefins. Published rate constants rounded to significant figures (*Continued*)

Compound	pK_a	pH	k/M^{-1} s^{-1}	Reference
Carbamazepine			3×10^5	Huber *et al.*, 2003
Cephalexin		7	8.7×10^4	Dodd *et al.*, 2010
Chlordane			<0.04	Yao & Haag, 1991
5-Chlorouracil	8.0		4.3×10^3	Theruvathu *et al.*, 2001
anion			1.3×10^6	Theruvathu *et al.*, 2001
Cinnamic acid			5×10^4	Leitzke *et al.*, 2001
			1×10^5	Jans, 1996
anion			3.8×10^5	Leitzke *et al.*, 2001
			1.2×10^6	Jans, 1996
β-Cyclocitral			3.9×10^3	Peter & von Gunten, 2007
Cylindrospermopsin			~2.5×10^6	Onstad *et al.*, 2007
Cytidine	4.15		3.5×10^3	Theruvathu *et al.*, 2001
protonated			40	Theruvathu *et al.*, 2001
Cytosine	4.6, 12.2		1.4×10^3	Theruvathu *et al.*, 2001
			930	Ishizaki *et al.*, 1984
protonated			18	Theruvathu *et al.*, 2001
anion			1.5×10^6	Theruvathu *et al.*, 2001
				Ishizaki *et al.*, 1984
2'-Deoxyadenosine	3.8		14	Theruvathu *et al.*, 2001
protonated			5	Theruvathu *et al.*, 2001
2'-Deoxycytidine	4.3		3.5×10^3	Theruvathu *et al.*, 2001
protonated			44	Theruvathu *et al.*, 2001
5'-Deoxycytidylic acid	4.6		1.4×10^3	Ishizaki *et al.*, 1984
2'-Deoxyguanosine	2.5, 9.2		1.9×10^4	Theruvathu *et al.*, 2001
5'-Deoxyguanylic acid	2.9, 9.7		4×10^4	Ishizaki *et al.*, 1984
1,1-Dichloroethene			110	Dowideit & von Sonntag, 1998
cis-1,2-Dichloroethene			310	Yao & Haag, 1991
			540	Dowideit & von Sonntag, 1998
trans-1,2-Dichloroethene			6.5×10^3	Dowideit & von Sonntag, 1998
			6.5×10^3	Hoigné & Bader, 1983a
Dichloromaleic acid		3.3	10	Leitzke & von Sonntag, 2009
1,1-Dichloropropene			2.6×10^3	Dowideit & von Sonntag, 1998
Diethyl vinylphosphonate			3.3×10^3	Leitzke *et al.*, 2003
3,4-Dihydroxycinnamic acid			2×10^6	Jans, 1996
anion			1.2×10^7	Jans, 1996
1,3-Dimethyluracil			2.8×10^3	Theruvathu *et al.*, 2001
DNA			410	Theruvathu *et al.*, 2001
Endrin			<0.02	Yao & Haag, 1991

(*Continued*)

Table 6.1 Compilation of rate constant of ozone with olefins. Published rate constants rounded to significant figures (*Continued*)

Compound	pK_a	pH	k/M^{-1} s^{-1}	Reference
Ethene			1.8×10^5	Dowideit & von Sonntag, 1998
Flumequine	6.5		1.2	Dodd *et al.*, 2006a
anion			8.5×10^3	
Fumaric acid	3.0, 4.4		8.5×10^3	Leitzke & von Sonntag, 2009
		<3	1.5×10^5	Benbelkacem *et al.*, 2004
dianion			$\sim 6.5 \times 10^4$	Leitzke & von Sonntag, 2009
		2	6×10^3	Hoigné & Bader, 1983b
		5	1×10^5	Hoigné & Bader, 1983b
Guanosine	2.5, 9.2		1.6×10^4	Theruvathu *et al.*, 2001
protonated			<300	Theruvathu *et al.*, 2001
anion			4.0×10^6	Theruvathu *et al.*, 2001
Hexachlorocyclopentadiene			90	Yao & Haag, 1991
2-Hexenoic acid (dianion)	6.7		2.4×10^5	Pryor *et al.*, 1984
cis-3-Hexen-1-ol			5.4×10^5	Peter & von Gunten, 2007
Indigotrisulfonic acid			9.4×10^7	Muñoz & von Sonntag 2000a
β-Ionone			1.6×10^5	Peter & von Gunten, 2007
Isoorotic acid	4.2, 8.9		3.7×10^3	Theruvathu *et al.*, 2001
Maleic acid	1.6		1.4×10^3	Leitzke & von Sonntag, 2009
monoanion	6.1		4.2×10^3	Leitzke & von Sonntag, 2009
dianion			$\sim 7 \times 10^3$	Leitzke & von Sonntag, 2009
			2.4×10^4	Pryor *et al.*, 1984
Methacrylic acid	4.7		1.5×10^5	Leitzke & von Sonntag, 2009
anion			3.7×10^6	Leitzke & von Sonntag, 2009
4-Methoxycinnamic acid			1.3×10^5	Leitzke *et al.*, 2001
anion			6.8×10^5	Leitzke *et al.*, 2001
3-Methoxy-4-hydroxy cinnamic acid			1.1×10^6	Jans, 1996
anion			7.9×10^6	Jans, 1996
6-Methyluracil	9.8		140	Theruvathu *et al.*, 2001
anion			6×10^5	Theruvathu *et al.*, 2001
Microcystin-LR			4.1×10^5	Onstad *et al.*, 2007
cis,cis-Muconic acid		3.1	2.7×10^4	Leitzke & von Sonntag, 2009
trans,trans-Muconic acid	2.7, 4.7	4	1.6×10^4	Beltrán *et al.*, 2006
		3	1.5×10^4	Ramseier & von Gunten, 2009
		7	1.3×10^5	
cis,trans-Muconic acid	1.9, 4.8	3	1.6×10^4	Ramseier & von Gunten, 2009
		7	2.5×10^5	
4-Nitrocinnamic acid (anion)			1.2×10^5	Leitzke *et al.*, 2001
trans,cis-2,6-Nonadienal			8.7×10^5	Peter & von Gunten, 2007
Orotic acid	2.1, 9.5		5.9×10^3	Theruvathu *et al.*, 2001

(*Continued*)

Table 6.1 Compilation of rate constant of ozone with olefins. Published rate constants rounded to significant figures (*Continued*)

Compound	pK_a	pH	k/M^{-1} s^{-1}	Reference
Oseltamivir acid		7–8	1.7×10^5	Mestankova *et al.*, 2012
1-Penten-3-one			5.9×10^4	Peter & von Gunten, 2007
Progesterone			480	Barron *et al.*, 2006
Propene			8×10^5	Dowideit & von Sonntag, 1998
Sorbic acid	4.76	3	3.2×10^5	Onstad *et al.*, 2007
anion		8	9.6×10^5	
Tetrachloroethene			< 0.1	Hoigné & Bader, 1983a
Tetramethylethene			$> 1 \times 10^6$	Dowideit & von Sonntag, 1998
Thymidine	9.8		3×10^4	Theruvathu *et al.*, 2001
anion			1.2×10^6	Theruvathu *et al.*, 2001
5′-Thymidylic acid	10.0		1.6×10^4	Ishizaki *et al.*, 1984
Thymine	9.9		4.2×10^4	Theruvathu *et al.*, 2001
			2.3×10^4	Ishizaki *et al.*, 1984
anion			$\sim 3 \times 10^6$	Theruvathu *et al.*, 2001
Trichloroethene			17	Hoigné & Bader, 1983a
			15	Yao & Haag, 1991
			14	Dowideit & von Sonntag, 1998
Tylosin protonated	7.7		7.7×10^4	Dodd *et al.*, 2006a
Uracil anion	9.5		650	Theruvathu *et al.*, 2001
			9.2×10^5	Theruvathu *et al.*, 2001
Vinyl acetate			1.6×10^5	Leitzke *et al.*, 2003
Vinyl bromide			1×10^4	Leitzke *et al.*, 2003
Vinyl chloride			1.4×10^4	Dowideit & von Sonntag, 1998
Vinylene carbonate			2.6×10^4	Leitzke *et al.*, 2003
Vinyl phenyl sulfonate			~ 200	Leitzke *et al.*, 2003
Vinyl phosphonic acid			1×10^4	Leitzke *et al.*, 2003
monoanion			2.7×10^4	Leitzke *et al.*, 2003
dianion			1×10^5	Leitzke *et al.*, 2003
Vinyl sulfonate ion			8×10^3	Leitzke *et al.*, 2003

6.2 THE CRIEGEE MECHANISM

The 1,3-cyloadduct of ozone to olefins was named by *Carl Friedrich Harries* "ozonide" (Chapter 1). Mechanistic details were only later unravelled by *Rudolf Criegee* (Chapter 1), and the reaction of ozone with olefins bears his name. In water, olefins react essentially according to the Criegee mechanism [reactions (1)–(9)], originally studied in organic solvents (Criegee, 1975). In water, however, ionic intermediates are intercepted [reaction (7)], and the reaction sequence, that in organic solvents proceeds to the Criegee ozonide, is arrested at an early stage (Dowideit & von Sonntag, 1998). As here ozone reactions in aqueous solution are discussed, the Criegee ozonide [cf. reaction (9)] can be disregarded, and whenever an ozonide is mentioned, the first ozonide is referred to.

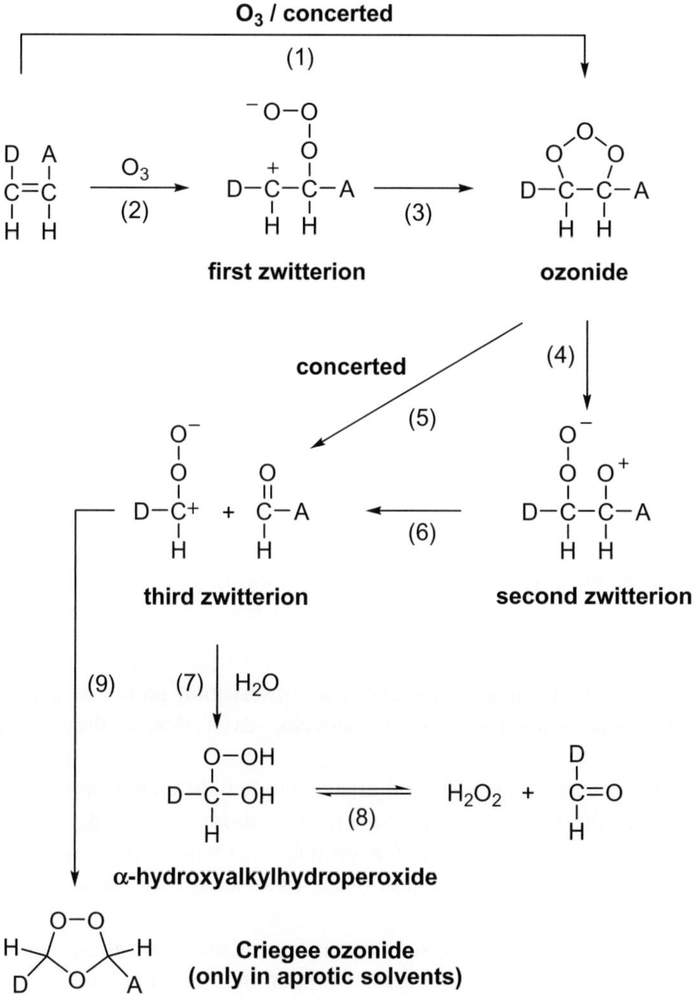

The Criegee mechanism has been challenged by suggesting that an electron transfer may occur instead [reaction (10)] (Zhang & Zhu, 1997).

$$\text{Olefin} + O_3 \rightarrow \text{Olefin}^{\bullet +} + O_3^{\bullet -} \qquad (10)$$

As will be seen from the following, this suggestion can be readily falsified on the basis of detailed product studies shown below. In fact, all data are in favour of the Criegee mechanism.

The reaction of ozone with olefins is usually described as a concerted, symmetry-allowed cycloaddition [reaction (1)] (Ponec et al., 1997; Anglada et al., 1999; Olzmann et al., 1997), that is, the reaction does not proceed via the first zwitterion as an intermediate [reactions (2) and (3)]. This even holds for polarised C–C double bonds such as in vinyl chloride.

Yet with tetramethylethene, a partial oxidation product is formed in 10% yield [reactions (11)–(13)] (Dowideit & von Sonntag, 1998).

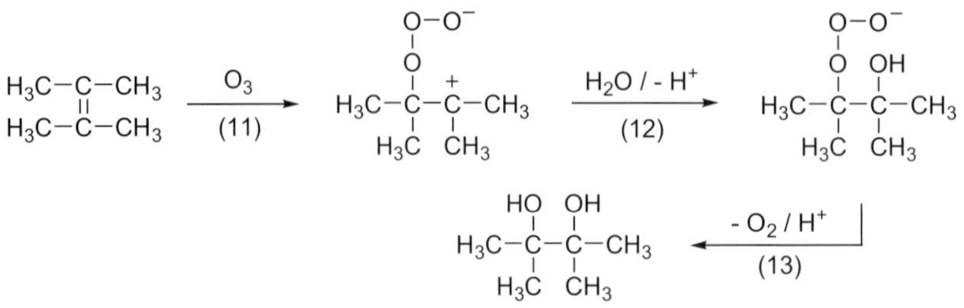

86   Chemistry of Ozone in Water and Wastewater Treatment

$$H_3C-C(CH_3)=C(CH_3)-CH_3 \xrightarrow{O_3} \underset{(11)}{} \underset{H_3C\ CH_3}{H_3C-C(O-O^-)(O)-C^+(CH_3)-CH_3} \xrightarrow{H_2O/-H^+} \underset{(12)}{} \underset{H_3C\ CH_3}{H_3C-C(O-O^-)(O-OH)-C(CH_3)-CH_3}$$

$$\underset{H_3C\ CH_3}{H_3C-C(HO)(OH)-C(CH_3)-CH_3} \xleftarrow[\ (13)\ ]{-O_2/H^+}$$

As cycloaddition proceeds, the C–C double bond turns into a C–C single bond and starts rotating to move the bulky methyl groups into positions of minimal steric hindrance (Figure 6.1).

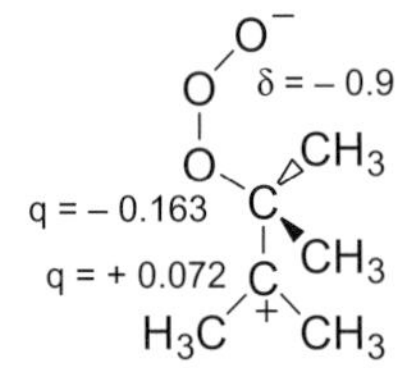

$$q = -0.163 \quad \delta = -0.9 \quad q = +0.072$$

Figure 6.1 Reaction of ozone with tetramethylethene. Rotation of the central C–C bond into positions of minimal steric hindrance of the substituents (interrupted calculation) Naumov & von Sonntag, unpublished.

Apparently, this developing zwitterion can be intercepted by water [reaction (12)] to some extent *en route* to the ozonide that must be the precursor of the major products. This observation points to a certain likelihood that in other olefins zwitterionic structures may also develop even though they are not noticeably intercepted by water.

The concerted reactions of ozone with fumaric and maleic acid give rise to two different stereoisomeric ozonides (see below). Their decay leads to the same products, but product ratios are markedly different (Leitzke & von Sonntag, 2009). This is an indication that if a primary zwitterion were formed, its lifetime would certainly be too short to allow a full equilibration of the two isomers by rotation around the central C–C bond.

In this context it is noted that in benzene and its derivatives, zwitterionic structures (ozone adducts) are sufficiently stable for calculating their standard Gibbs free energies of formation in water (Chapter 7). The reason for this marked difference to olefins may be that in aromatic compounds the developing positive charge is distributed over the entire aromatic ring, and a collapse to the ozonide is thus retarded.

Ozonides derived from asymmetrical olefins have different O–O and C–O bond lengths. An example is shown in Figure 6.2.

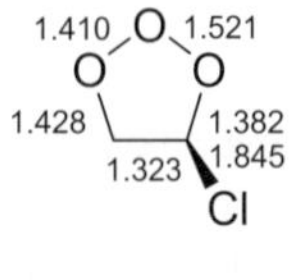

(bond lengths: 1.410, 1.521, 1.428, 1.382, 1.323, 1.845; substituent Cl)

Figure 6.2 Structure of the ozonide derived from vinyl chloride according to Naumov & von Sonntag, unpublished.

These differences in bond lengths indicate the direction the ozonide will take when opening the ring. The longer O–O bond will become even longer and eventually break, while the shorter C–O bond will further contract.

In the quantum-chemical calculations, the second zwitterion [reaction (4)] is not an intermediate, and the reaction continues by breaking the central C–C bond [concerted reaction (5)]. This predisposition in the ozonides for ensuing decay routes results in the observed high regioselectivity in product formation. The ozonides of vinyl chloride and propene may serve as examples. As expected, they follow opposite decay routes [reaction (15) vs. reaction (23)].

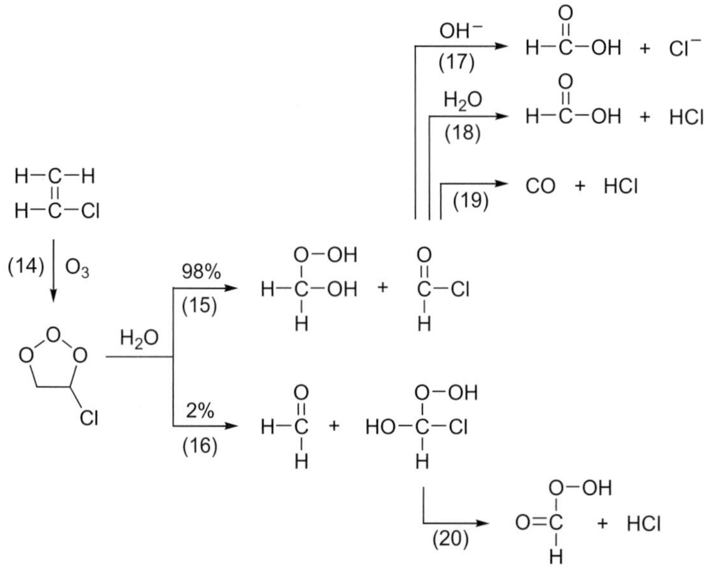

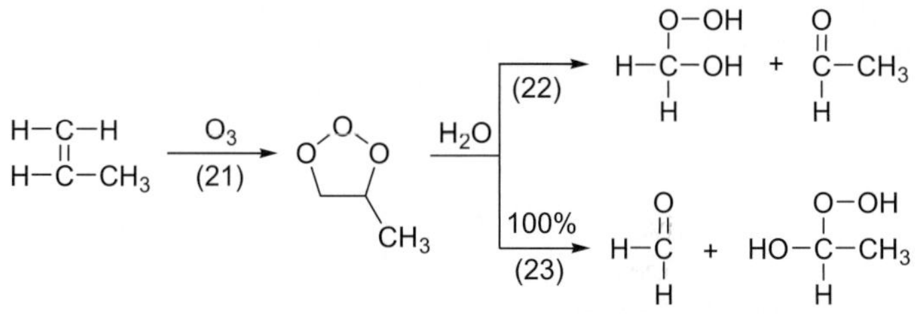

Even though the second zwitterion is neither a real intermediate nor a transition state in most ozone reactions, it is heuristically helpful. For differently substituted olefins, the route that leads to the zwitterion that can stabilise the positive charge at carbon the best is followed preferentially.

The third zwitterion is trapped by water giving rise to α-hydroxyalkylhydroperoxides [reaction (7)], and this reaction is much faster than a trapping by the carbonyl compound [reaction (9)] formed next to the third zwitterion within the solvent cage. Thus in the ozonolysis of olefins in water, there are typically two primary products, an α-hydroxyalkylhydroperoxide and a carbonyl compound. As will be shown below, these primary products may still be highly reactive and often decay very rapidly into the final products.

6.3 PARTIAL OXIDATION

Although the Criegee mechanism is the dominating process in the reactions of olefins with ozone, products may be observed in which the former C–C double bond is not fully cleaved, that is, the carbon skeleton remains intact. Such processes are called "partial oxidation". An example of this has been given above [reactions (11)–(13)]. Here, ring closure to the Criegee ozonide competes with a trapping of the carbocation site by water. In organic solvents, a methyl shift is observed instead [reactions (24) and (25)].

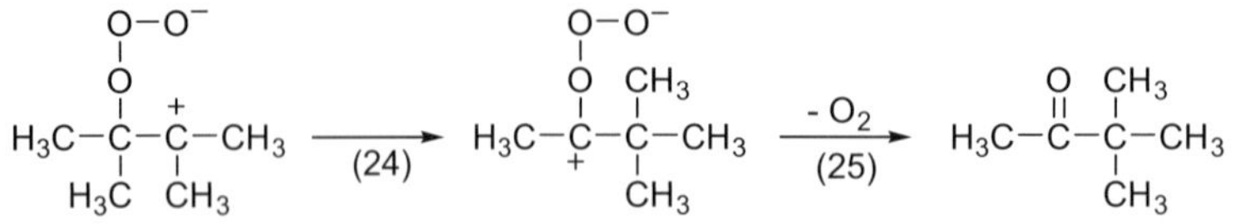

6.4 DECAY OF THE OZONIDE VIA FREE RADICALS

Although heterolytic cleavage of the ozonide is by far the most important process, there seem to be cases, where homolytic cleavage can occur. Strong experimental evidence for this has been obtained from the ozonolysis of dichloromaleic acid at pH 3 (largely the monoanion). The chloride yield is 3.9 mol per mol ozone and reduces to 1.7 in the presence of tertiary butanol (see paragraph 6.6.6) pointing to the formation of •Cl, which subsequently induces a short chain reaction. Quantum-chemical calculations (Naumov & von Sonntag, 2009, unpublished) indicate that, for example, reaction (26)/(27) is slightly exergonic ($\Delta G° = -2$ kJ mol^{-1}) as are subsequent decarboxylation plus $O_2^{•-}$ elimination [reaction (28), $\Delta G° = -108$ kJ mol^{-1}] and the hydrolysis of the ketene [reaction (29), $\Delta G° = -49$ kJ mol^{-1}]. There are other potential exergonic routes to the chain carrier •Cl, but at this stage it is premature to discuss other alternatives, as detailed product studies are missing.

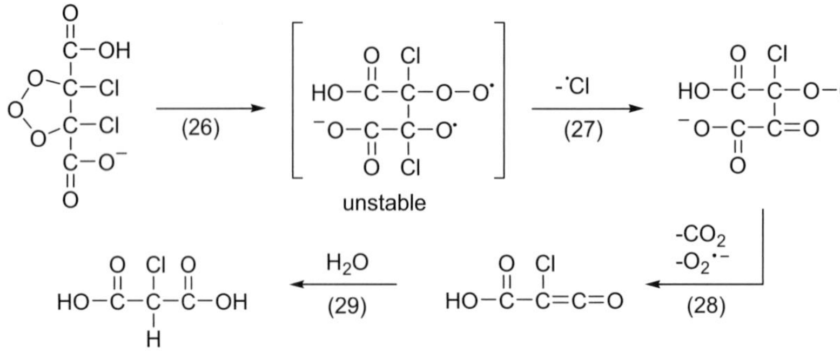

6.5 DETECTION OF α-HYDROXYALKYLHYDROPEROXIDES

α-Hydroxyalkylhydroperoxides are often quite stable, as equilibrium (8) is usually only slowly attained at pH 7 (Bothe & Schulte-Frohlinde, 1980). α-Hydroxyalkylhydroperoxides react, similarly to H_2O_2, with molybdate-activated iodide by forming I_3^-. The kinetics of such reactions are markedly different and can be used to characterise and quantify such intermediates (Dowideit & von Sonntag, 1998; Flyunt et al., 2003b; Stemmler & von Gunten, 2000). As the α-hydroxyalkylhydroperoxides react much more slowly than H_2O_2 (for a compilation of rate constants see Table 6.2), this approach can even be used to quantify H_2O_2 and α-hydroxyalkylhydroperoxides when formed side by side. The very reactive hydroperoxides, formic and acetic peracids, react even without molybdate catalysis. Formic peracid is reduced by $S(CH_2CH_2OH)_2$ to formic acid [reaction (30), $k = 220$ M^{-1} s^{-1} (Flyunt et al., 2003a)], and this reaction can be followed by changes in the conductance, as formic peracid ($pK_a = 7.1$) is barely dissociated at pH 6 while formic acid ($pK_a = 3.75$) is fully dissociated under such conditions.

$$S(CH_2CH_2OH)_2 + HC(O)OOH \rightarrow O=S(CH_2CH_2OH)_2 + H^+ + HC(O)O^- \qquad (30)$$

Table 6.2 Compilation of half-lives of the reactions of molybdate-activated iodide with some hydroperoxides

Hydroperoxide	$t_{1/2}$/s	Reference
HC(O)OOH[a]	0.0032	Dowideit & von Sonntag, 1998
CH$_3$C(O)OOH[b]	0.022	Flyunt et al., 2003b
CH$_3$C(O)C(O)NHC(O)N=C(OOH)C(O)H[b]	0.12, 0.16[c]	Flyunt et al., 2002
5-Hydroperoxy-5-methylhydantoin[b]	0.71, 0.9[c]	Flyunt et al., 2002
H$_2$O$_2$[d]	2.5	Dowideit & von Sonntag, 1998
H$_2$O$_2$[b]	1.8×10^{-3}	Flyunt et al., 2003a
HOCH(CH$_3$)OOH[d]	13.7	Dowideit & von Sonntag, 1998
HOCH$_2$OOH[d]	203	Dowideit & von Sonntag, 1998
HOOCH$_2$C(CH$_3$)$_2$OH	~260[c]	Schuchmann & von Sonntag, 1979
CH$_3$OOH	790	Flyunt et al., 2003b
HOOCH$_2$CH(CH$_3$)OH	420[c]	Flyunt et al., 2003b
HOC(CH$_3$)$_2$OOH[d]	~35 000	Dowideit & von Sonntag, 1998

[a]Hydroperoxide reacts equally rapidly without molybdate catalysis. [b]Reaction without molybdate catalysis. [c]Reagent A [0.4 M KI, 3.6×10^{-2} M KOH, 1.6×10^{-4} M (NH$_4$)$_6$Mo$_7$O$_{24}$], reagent B (0.1 M potassium hydrogen phthalate) and probe at 1 : 1 : 2. [d]Molybdate catalysis, [I$^-$] = 0.17 M.

6.6 OZONE REACTIONS OF OLEFINS – PRODUCTS AND REACTIONS OF REACTIVE INTERMEDIATES

In this paragraph, the ozonolysis products of a large number of olefins will be presented. This information will allow predictions on the ozonolyses of olefins for which no data are available. The high regioselectivity of ozone reactions with unequally substituted olefins assists these predictions.

6.6.1 Methyl- and halogen-substituted olefins

Final products and some of the longer-lived intermediates of some methyl- and chlorine-substituted olefins are compiled in Table 6.3.

The reactions of propene have been discussed above [reactions (21) and (23)]. Ethene is symmetrical, and from the above one expects equal amounts of formaldehyde and hydroxymethylhydroperoxide. This is indeed observed (Table 6.3). A more convenient source to produce hydroxymethylhydroperoxide in ozone reactions is buten-3-ol [reaction (31)].

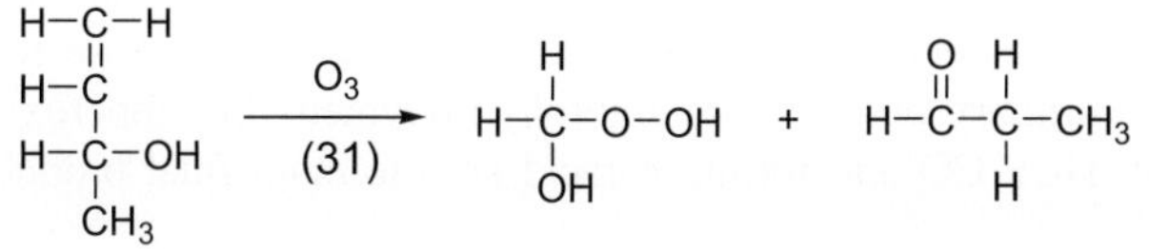

Table 6.3 Products and their yields [with respect to ozone consumption (mol/mol)] in the ozonolysis of ethene and some of its methyl- and chlorine-substituted derivatives in aqueous solution according to Dowideit & von Sonntag (1998). Empty fields indicate that a given product is not expected to be formed and has not been looked for

Product	Ethene	Propene	Me$_4$-Ethene	Cl-Ethene	1,1-Cl$_2$-Ethene	1,2-Cl$_2$-Ethene	Cl$_3$-Ethene	1,1-Cl$_2$-Propene
HCl				1.05	1.95	2.02	2.87	2.05
HC(O)OH				0.06	0.03	1.01	0.82	
HC(O)OOH				0.02[a]		0.98[a]	0.88[a]	
HOCH$_2$COOH					0.07[c]			
CO				1.01		1.08	0.04	
CO$_2$					0.90	0.02	0.95	1.01
CH$_2$O	2.04	1.03			0.96			
CH$_3$C(O)H		0.97						1.03
(CH$_3$)$_2$CO			1.74					
HOCH$_2$OOH	1.08[b]			1.06[b]	0.96[b]			
CH$_3$CH(OH)OOH		0.99[b]						0.98[b]
(CH$_3$)$_2$C(OH)OOH			[d]					
Cl$_2$CHC(O)H					<0.01[c]			
ClCH$_2$C(O)OH					0.08[c]			
Cl$_2$CHC(O)OH							0.04[c]	
[(CH$_3$)$_2$C(OH)]$_2$			~0.1					

[a]Precursor of formic acid. [b]Precursor of aldehydes. [c]Product of partial oxidation reaction. [d]The reaction with molybdate-activated iodide is too slow for its determination.

Its good solubility in water and its high rate constant (Table 6.1) makes buten-3-ol a very convenient competitor for the determination of ozone rate constants (Chapter 2).

With halogenated olefins, reactive halogen-containing intermediates are formed which show an interesting chemistry. With vinyl chloride the main reaction path leads to hydroxymethylhydroperoxide and formyl chloride [reaction (15)]. Formyl chloride rapidly decomposes into CO and HCl [reaction (19), $k = 600\,\text{s}^{-1}$; 94%] (Dowideit et al., 1996). Its hydrolysis into formic acid plus HCl [reaction (18), 6%] is slow in comparison but gains in importance at high pH [reaction (17), $k(\text{OH}^-) = 2.5 \times 10^4$ M^{-1} s^{-1}]. A minor route leads to chlorohydroxymethylhydroperoxide [reaction (16)]. The geminal chlorohydrine substructure in chlorohydroxymethylhydroperoxide is very unstable and rapidly loses HCl [on a low microsecond timescale (Köster & Asmus, 1971; Mertens et al., 1994)] giving rise to formic peracid [reaction (20)]. Formic peracid is generated in the reaction of 1,2-dichloroethene (see below), and this is a convenient method for its formation [for its reactions see (Flyunt et al., 2003a)].

Formyl bromide, the product of the analogous reaction of vinyl bromide with ozone also decomposes preferentially into CO plus HBr. With 3.6%, the formic acid yield is lower than the corresponding yield from formyl chloride (6%). Whether this is due to a faster decomposition or a slower hydrolysis is not yet known.

1,2-Dichloroethene is symmetrical, and the primary products are chlorohydroxymethylhydroperoxide and formyl chloride [reactions (32) and (33)]. HCl, CO and formic peracid are the main final products (see above).

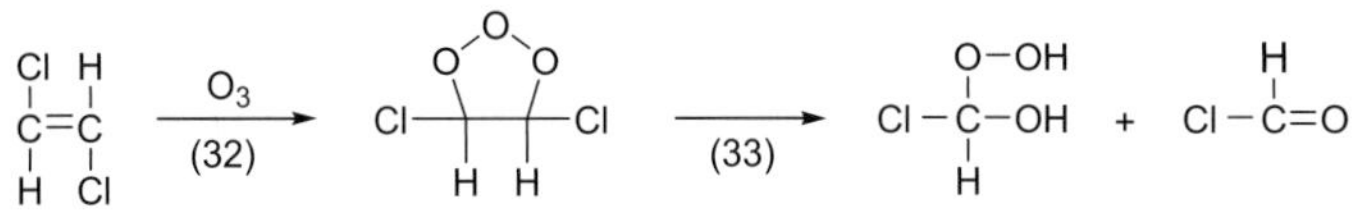

The reactions of 1,2-dibromoethene with ozone are analogous (Leitzke *et al.*, 2003).

6.6.2 Acrylonitrile, vinyl acetate, diethyl vinylphosphonate, vinyl phenyl sulfonate, vinylsulfonic acid and vinylene carbonate

Cyano, acetyl and diethyl phosphonate groups are electron-withdrawing substituents, and the ozonolysis of acrylonitrile, vinyl acetate and diethyl vinylphosphonate follow the same regioselectivity as observed for the ozone reaction with vinyl chloride discussed above, that is, the primary products are hydroxymethylhydroperoxide and the corresponding formyl derivatives, formyl cyanide, formyl acetate and diethyl formylphosphonate, respectively (Leitzke *et al.*, 2003). These mixed acid anhydrides are not stable in water. Stopped-flow with conductometric detection allowed the determination of the rate constant for their hydrolysis: formyl cyanide [$k = 3$ s^{-1}; $k(OH^-) = 3.8 \times 10^5$ M^{-1} s^{-1}]; formyl acetate ($k = 0.25$ s^{-1}); diethyl formylphosphonate [$k = 7 \times 10^{-3}$ s^{-1}; $k(OH^-) = 3.2 \times 10^4$ M^{-1} s^{-1}] (Leitzke *et al.*, 2003).

The ozonolysis of vinylphosphonic acid is somewhat more complex as the ozonide decays in two directions in a 1:3 ratio [reactions (35) and (36)] (Leitzke *et al.*, 2003). In the presence of excess H$_2$O$_2$, formic and phosphoric acids are the final products [reactions (37)–(39)]. In the presence of catalase, which destroys H$_2$O$_2$ but not α-hydroxyalkylhydroperoxides, the rate of formic acid formation is markedly slower and phosphonic acid is observed instead of phosphoric acid. Apparently, formylphosphonate hydrolyses only slowly [$k(OH^-) \approx 5$ M^{-1} s^{-1}]. The slowness of this reaction when compared to the formyl compounds discussed above may be due to the fact that here OH$^-$ has to react with a doubly negatively charged species. At pH 10.2 and in the presence of borate buffer, H$_2$O$_2$ reacts with formylphosphonate with an observed rate constant of $k_{obs} = 260$ M^{-1} s^{-1} (phosphonate is not oxidised by H$_2$O$_2$ at any pH). There seems to be also a pH dependence between reactions (−37) and (38). While at pH 7 H$_2$O$_2$ elimination is observed, decomposition is favoured at pH 10.2, where the hydroxyperoxy(hydroxy)methyl phosphonic acid is fully deprotonated.

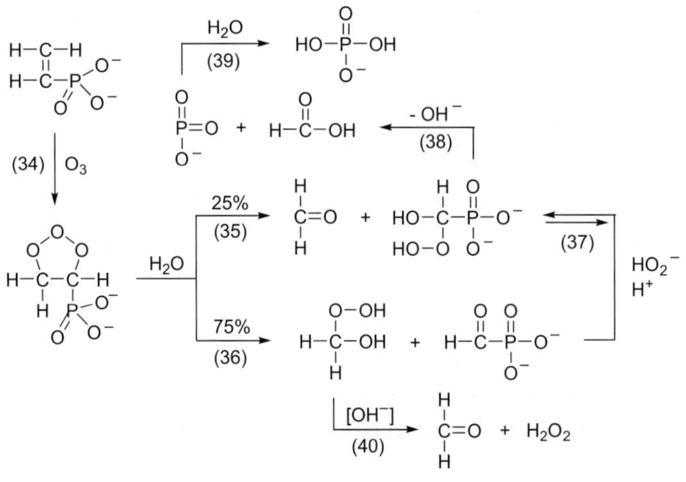

Ozonolysis of vinyl phenylsulfonate is complex as some of the products, notably SO_2 and phenol, react rapidly with ozone (Leitzke et al., 2003). Hydroxymethylhydroperoxide, formic acid and sulfuric acid are products, but the material balance is incomplete and reactions (41)–(45) must remain a tentative suggestion. Similar uncertainties prevail in the ozonolysis of vinylsulfonic acid (Leitzke et al., 2003).

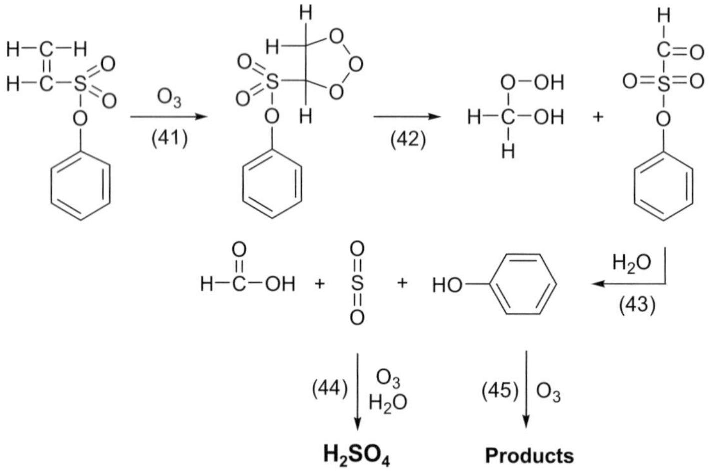

The ozonolysis of vinylene carbonate leads to the formation of formic acid, formic peracid, H_2O_2 and CO_2 (Leitzke et al., 2003). Potential decay reactions of this symmetrical ozonide are shown in reactions (46)–(52). The reaction that leads to H_2O_2 [reaction (50)] must be minor (4%).

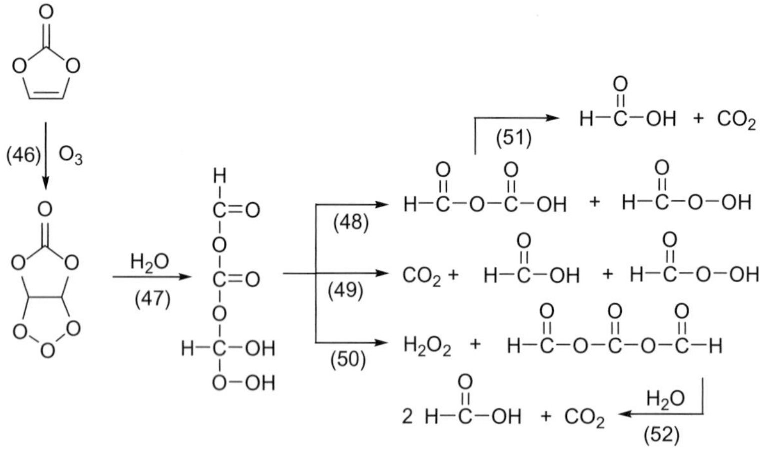

6.6.3 Acrylic, methacrylic, fumaric, maleic and muconic acids

For acrylic and methacrylic acids, the yields of the product based on ozone consumption are compiled in Table 6.4.

Table 6.4 Product yields (mol per mol O₃ consumed) in the reaction of O₃ with acrylic and methacrylic acids according to Leitzke & von Sonntag (2009)

Product	Acrylic acid Anion (free acid)	Methacrylic acid Anion (free acid)
Formaldehyde[a]	0.52 (0.72)	0.43 (0.62)
Hydroxymethylhydroperoxide	0.43 (ND)	see text (ND)
1-Hydroperoxypropan-2-one	–	0.57 (ND)
Glyoxylic acid	0.54 (ND)	–
2-Oxopropanoic acid	–	0.41 (ND)
Gylcol aldehyde	0.48 (ND)	–
Methylglyoxal	–	0.54 (ND)
Formic acid	absent (absent)	–
Acetic acid	–	absent (absent)
Total Peroxide	1.01 (1.0)	1.04 (1.0)
H_2O_2	0.58 (ND)	0.46 (ND)

[a]Includes the yield of hydroxymethylhydroperoxide that decomposes into formaldehyde during the assay, ND = not determined, – product cannot be formed.

Mechanistically, acrylic and methacrylic acids follow the same pattern [reactions (53)–(59) and reactions (63)–(67), respectively].

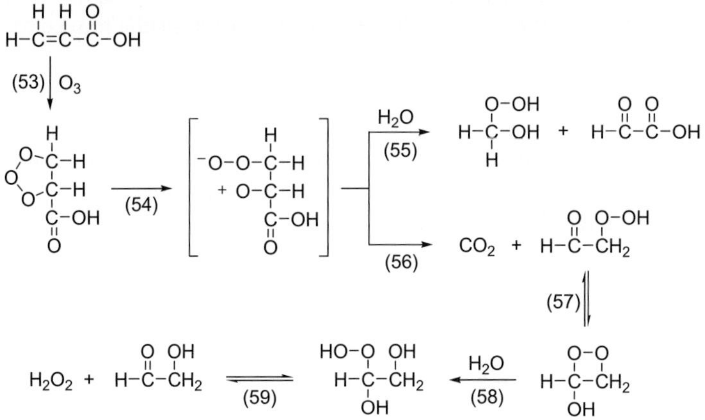

The second zwitterions are given in brackets as they may not be intermediates, but they indicate the directions the ozonides will take upon their decay. The absence of formic and acetic acids in the case of acrylic and methacrylic acids, respectively, shows that the other conceivable route, for example, reactions (60)–(62) is not taken in these acids.

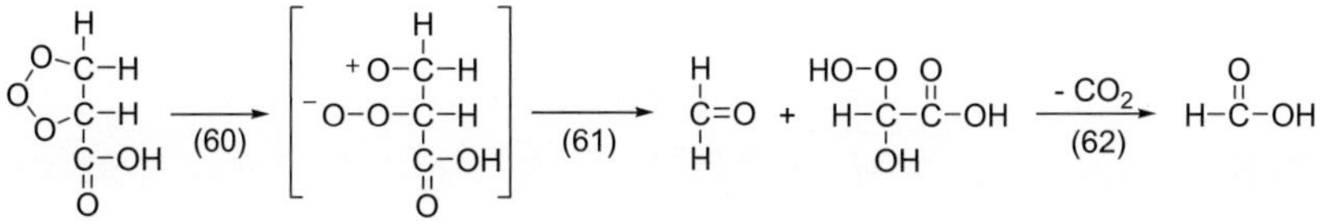

Moreover, cleavage along the O–O and C–C bonds of the envisaged dioxetane intermediate [cf. reaction (57)] does not take place.

Depending on the protonation state of the carboxylic group, there is some bias in the branching between reactions (55) and (56). As expected, the deprotonated carboxyl group decarboxylates more readily favouring reaction (56). Formaldehyde, which is produced from reaction (55), can be measured with some accuracy. A plot of the formaldehyde yield as a function of pH is shown in Figure 6.3.

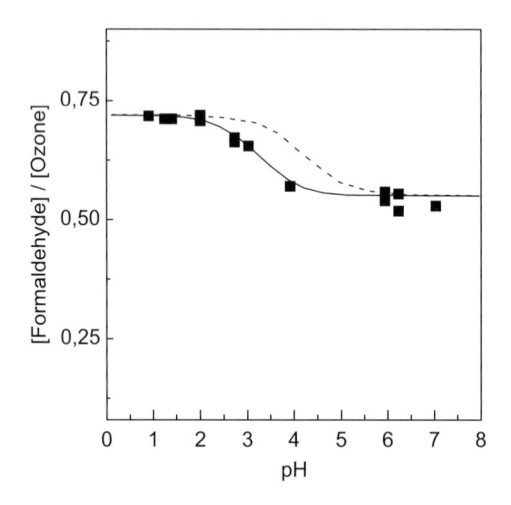

Figure 6.3 The pH dependence of the formaldehyde yield in the ozonolysis of acrylic acid. The dashed line indicates how this dependence would look if it were governed by the pK_a value of acrylic acid. From Leitzke & von Sonntag, 2009 with permission.

The inflection point is at pH 3, markedly distant from the pK_a value of acrylic acid, 4.25 (dashed line). It may reflect the pK_a value of the ozonide, as the zwitterion is most likely not an intermediate, but if it were it would be too short-lived for the pK_a equilibrium to become established during its lifetime.

The reactions of methacrylic acid ozonide are depicted in reactions (63)–(67).

Interestingly, hydroxymethylhydroperoxide, formed in reaction (64), cannot be detected by its slow reaction with molybdate-activated iodide. It is too short-lived here, as the buffering properties of methacrylic acid that decompose hydroxymethylhydroperoxide into formaldehyde and H_2O_2 [reaction (60)] are much more pronounced than in the case of acrylic acid. This has been shown by production of hydroxymethylhydroperoxide in the reaction of buten-3-ol with ozone [reaction (31)] and subsequent addition of methacrylic acid. The latter rapidly catalyses reaction (66) and hydroxymethylhydroperoxide is no longer detected [note that the somewhat lower value of hydroxymethylhydroperoxide as compared to that of formaldehyde in the case of acrylic acid (Table 6.4) may be due to the, albeit much lower, buffering effects of acrylic acid. Due to its higher pK_a value, methacrylic acid is a better buffer, and thus a more efficient catalyst, than acrylic acid in close to neutral solutions].

The hydroperoxide attributed to 1-hydroperoxypropan-2-one formed in reaction (65) reacts much faster with molybdate-activated iodide ($k = 0.15$ $M^{-1} s^{-1}$) than hydroxymethylhydroperoxide ($k = 3.4 \times 10^{-3}$ $M^{-1} s^{-1}$). Interestingly, the methyl substituent in this hydroperoxide has a marked effect on its decay. Compared to aldehydes, ketones form hydrates (and thus also hydroxyalkylhydroperoxides) less efficiently [in water, formaldehyde is practically fully hydrated, $CH_2(OH)_2$, acetaldehyde ~50% hydrated, acetone <0.1% hydrated (Bell, 1966)]. Thus, the reaction analogous to reaction (57) will be inefficient, and water elimination occurs instead [reaction (67)]. The buffer properties of methacrylic acid may speed up this reaction. There are other α-hydroxyalkylhydroperoxides, where water elimination rather than H_2O_2 elimination is observed. A case in point is the muconic compound formed in the ozonolysis of phenol (Mvula & von Sonntag, 2003). Reasons for this are as yet not known, but one may recall that H_2O_2 elimination is reversible while water elimination is not. A more detailed study of H_2O_2 with aldehydes would be required to elucidate this interesting aspect.

The reactions of O_3 with maleic and fumaric acids in aqueous solution has found some attention (Gilbert, 1977; Leitzke & von Sonntag, 2009; Ramseier & von Gunten, 2009), but mechanistic details were only understood recently (Leitzke & von Sonntag, 2009; Ramseier & von Gunten, 2009). These two isomeric acids are available in high purity. In aqueous solution, cis ⇌ trans isomerisation takes place. This may lead to an artefact, that is, the other isomer increases linearly with the ozone dose, when the work-up of the samples follows the sequence of ozone doses (Leitzke, 2003). It is now clear, that ozone does not induce cis ⇌ trans isomerisation. These two isomers give rise to the same products, but their yields differ remarkably (Table 6.5).

Table 6.5 Product yields (mol per mol O_3 consumed) in the reaction of O_3 with fumaric and maleic acids at pH 5.3 according to Leitzke & von Sonntag (2009)

Product	Fumaric acid	Maleic acid
Formic acid	0.77	0.92
Glyoxylic acid	0.85	1.05
Glyoxal	0.27	0.03
Total peroxide	0.31	ND

ND = not determined.

With fumaric acid, glyoxal is formed in 27% yield, while with maleic acid it is only 3%. This is a strong indication that these two isomers give rise to two ozonides that differ with respect to their preferred decay routes. This is in line with the concept that ozone addition to the double bond is a concerted reaction [reactions (68) and (69)]. In any case, a zwitterionic intermediate, if it is an intermediate, must be too short-lived to allow a rotation around the former C–C double bond.

Maleic acid + O_3 → (68) → [ozonide intermediate]

Fumaric acid + O_3 → (69) → [ozonide intermediate]

We are not yet in a position to present a mechanistic explanation for this difference, but the general mechanism suggested by reactions (70)–(74) accounts for the observed products.

(70) → [zwitterion intermediates]

(71) H_2O → hydroxy-hydroperoxide + glyoxylic acid derivatives

(72) H_2O → hydroxy-hydroperoxide with aldehyde + CO_2

(73) $-CO_2 / -H_2O$ → $H-\overset{O}{\underset{}{C}}-OH$ (formic acid)

(74) $-CO_2 / -H_2O$ → $H-\overset{O}{\underset{}{C}}-\overset{O}{\underset{}{C}}-H$ (glyoxal)

6.6.4 Muconic acids

The ozonolysis of *trans,trans*-muconic acid in aqueous solution has been studied under the conditions of extensive ozonation (Gilbert, 1980). No distinction between primary and secondary products has been made. The primary products and their yields in the ozonolysis of *cis,cis*-muconic acid (Leitzke et al., 2003) are compiled in Table 6.6, and additional information on primary as well as secondary products is also available (Ramseier & von Gunten, 2009) [reactions (75)–(79)].

Table 6.6 Product yields (mol per mol O₃ consumed) in the reaction of O₃ with *cis,cis*-muconic acid according to Leitzke *et al.* (2003)

Product	pH 2	pH 8
H_2O_2	0.98	0.9
Glyoxylic acid	0.98[a]	0.99[a]
4-Oxobut-2-enoic acid	not quantified[b]	not quantified[b]
Formic acid	absent	absent

[a]Catalase had been added immediately after ozonolysis; [b]detected by GC-MS.

There is no organic hydroperoxide detected, and the total peroxide yield is due to H_2O_2. The other product is glyoxylic acid. This aldehyde reacts readily with H_2O_2, resulting in the subsequent formation of formic acid [reactions (80) and (81)]. For suppressing this secondary reaction by destroying H_2O_2, catalase has been added immediately after ozonolysis. Under such conditions, glyoxylic acid matches the H_2O_2 yield, and formic acid is no longer formed. 4-Oxobut-3-enoic acid has been detected as a major product by GC-MS but was not quantified. Reactions (75)–(79) account for the observed products.

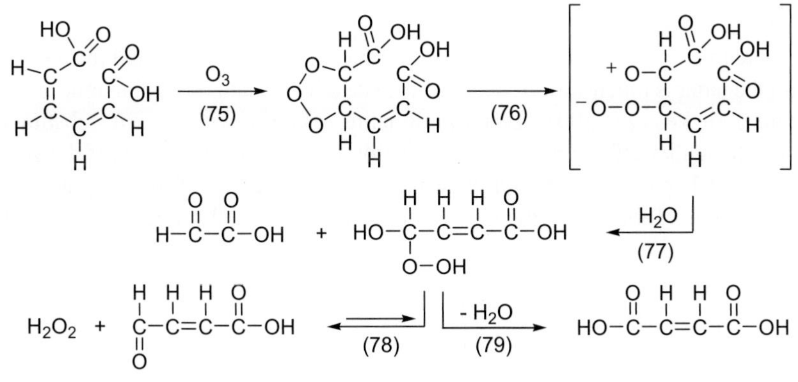

Besides the conversion of glyoxylic acid by hydrogen peroxide into CO_2 and formic acid [reactions (80) and (81)] (for the kinetics of these reactions see (Leitzke *et al.*, 2001), 4-oxobut-3-enoic acid is likely to undergo the analogous reactions (78) and (79). In the absence of catalase, the ensuing products, formic acid and maleic acid, are indeed formed (Ramseier & von Gunten, 2009).

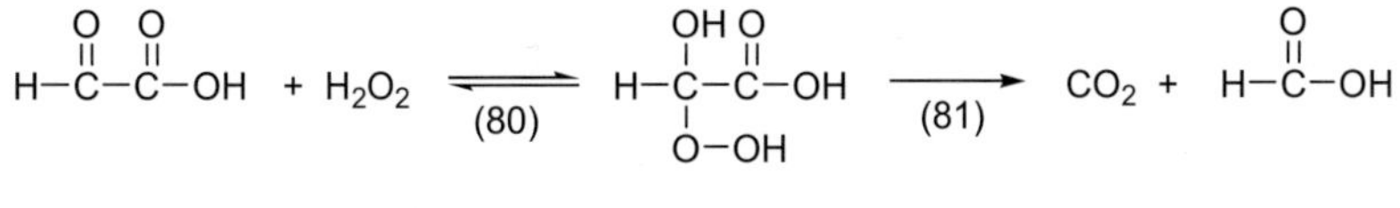

There is the interesting aspect that upon ozonolysis of *cis,cis*-muconic acid, 4% singlet oxygen is formed (Muñoz *et al.*, 2001). This points to an additional pathway for the reaction of muconic acid with ozone. As the formation of the ozonide is exergonic by ~43 kJ mol⁻¹ (Naumov & von Sonntag, 2009, unpublished) and seems not to decarboxylate readily, an additional pathway in competition to the ozonide formation seems to be of (minor) importance. Approach of ozone to muconic acid may develop a zwitterionic intermediate with the positive charge delocalised over the two conjugated C–C double bonds. This is not a stable structure, possibly not even a transition state, but decarboxylation may already occur [reaction (82)] as it tends to develop. Singlet oxygen is subsequently released [reaction (83)].

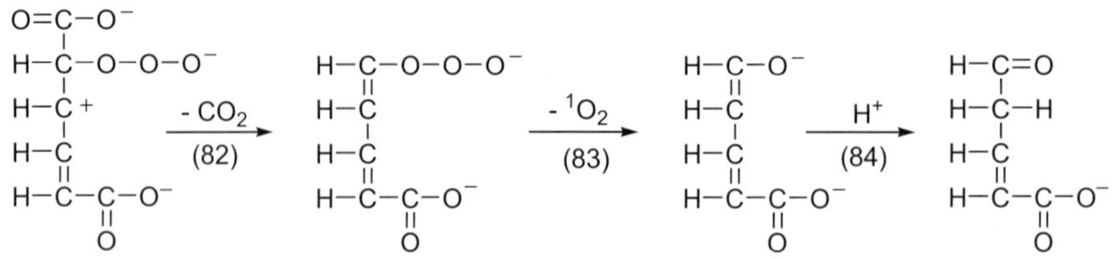

The standard Gibbs free energy including CO_2 elimination [reaction (82)] has been calculated at -77 kJ mol^{-1}. At the time when these measurements were made, the implications were not realised, and whether small amounts of 5-oxo-pent-2-enoic acid (or its decarboxylation products) are indeed formed has not been looked at. It seems worth mentioning that with aromatic compounds, where the positive charge of the zwitterionic ozone adduct can be distributed over one more double bond, such adducts are well-defined intermediates (Naumov & von Sonntag, 2010) (Chapter 7).

6.6.5 Cinnamic acids

Cinnamic acids contain two potential sites for ozone attack, the olefinic C–C double bond and the aromatic ring. The rate of reaction with the latter is much slower, and thus the site of ozone attack on cinnamic acids is the exocylic C–C double bond (Leitzke et al., 2001). Electron-donating/electron-withdrawing substituents at the aromatic ring have a much smaller effect on the rate of reaction than one would expect if the site of attack were the aromatic ring (cf. Chapter 7). Although there is only a small effect of methoxy and nitro substituents on the rate of reaction (Table 6.1), there is a marked effect on the decay route of the ozonides [reactions (85)–(88)].

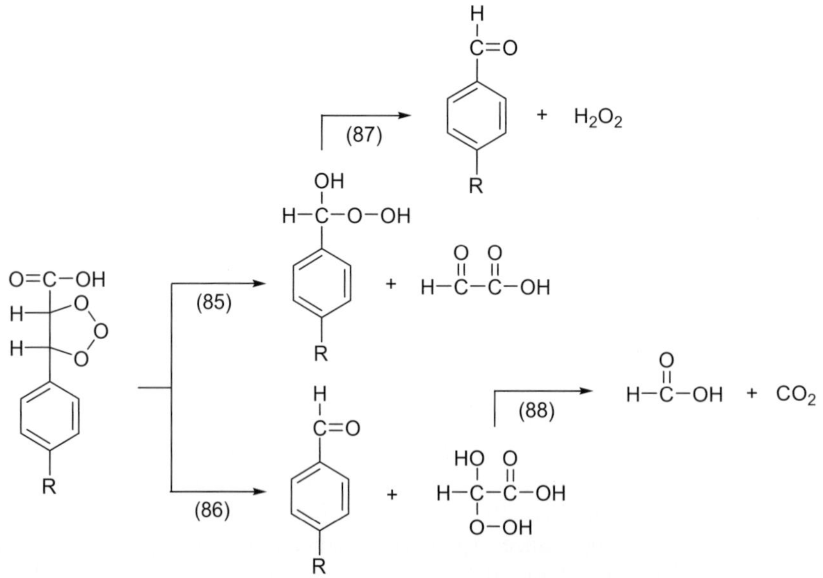

Olefins

The ozonides of cinnamic acid and its 4-methoxy derivative decay only according to reaction (85), whereas the electron-withdrawing nitro substituent makes it possible that reaction (86) proceeds to 30%. The hydroxyhydroperoxide only eliminates H_2O_2 [reaction (87)], and the potentially competing water elimination that would lead to benzoic acid is not observed (<2%).

6.6.6 Dichloromaleic acid

With dichloromaleic acid only some preliminary experiments have been carried out (Leitzke & von Sonntag, 2009). These few data are still sufficiently interesting to be reported as they may serve as a basis for future investigations. Ozonation has been carried out at pH 3.1. Under such conditions, the Cl^- yield was 3.9 with respect to ozone consumed, much higher than the maximum value of 2.0 for a non-chain reaction, and an acid is formed that has the same retention time as ketomalonic acid (not necessarily a correct assignment). When tertiary butanol (0.26 M), which scavenges reactive free radicals such as •Cl, is added in excess, the Cl^- yield drops to 1.7, and the unknown acid is no longer formed. This points to a free-radical-induced short chain reaction [reactions (26)–(29)] that runs in parallel to the normal non-radical reaction. The latter may be tentatively described by reactions (89) and (90) and ensuing HCl releasing reactions.

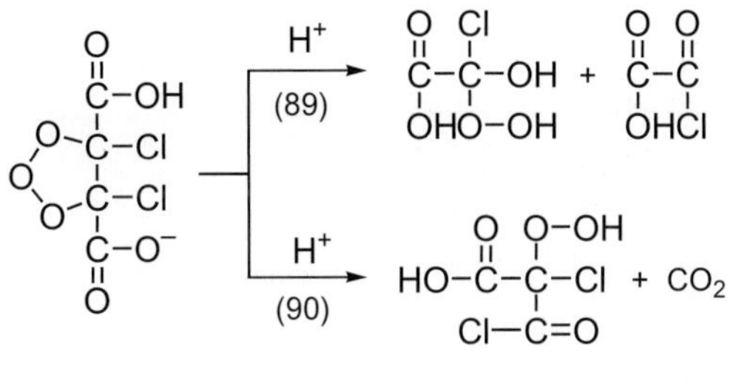

At this point, further suggestions would be mere speculation, as a detailed product study has not yet been carried out, but it is worth mentioning that short chain reactions are common in the peroxyl radical chemistry of chlorinated compounds (von Sonntag & Schuchmann, 1997) (cf. Chapter 14).

6.6.7 Pyrimidine nucleobases

Among the nucleic acid constituents, the pyrimidines react quite readily with ozone (Table 6.1 and Chapter 4). Among these, detailed product studies are only available on thymine and thymidine (Flyunt et al., 2002). Their reactions are governed by the addition of ozone to the C(5)–C(6) double bond. Some key reactions of thymine are shown in reactions (91)–(97). For further details and the reactions of thymidine that are similar but yet markedly different in several aspects the interested reader is referred to the original paper (Flyunt et al., 2002).

The ozonide formed in reaction (91) predominantly (75%) opens the ring in such a way that the positive charge in the virtual zwitterion resides at N(1) giving rise to the acidic [cf. equilibrium (94)] hydroperoxide 1-hydroperoxymethylen-3-(2-oxo-propanoyl)-urea [reaction (92), (34%)] and to its hydrate 1-(hydroperoxy (hydroxy)methyl)-3-(2-oxopropanoyl)urea [reaction (93), (41%)]. The potential intermediate involved in these reactions is not shown.

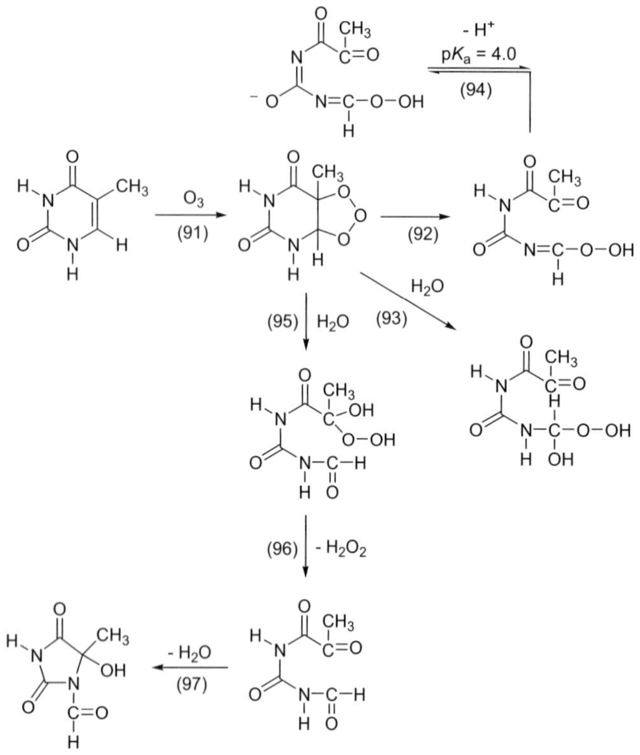

Besides the 75% organic hydroperoxides discussed above, there are 25% H_2O_2 formed in reactions (95)–(97). These hydroperoxides are short-lived and decay largely into 5-hydroperoxy-5-methylhydantoin, which can be reduced by a sulfide into 5-hydroxy-5-methylhydantoin and formic acid [reactions (98)–(102)], while the remaining 1-hydroperoxymethylen-3-(2-oxo-propanoyl)-urea is reduced to N-formyl-5-hydroxy-5-methylhydantoin (reaction not shown). Final treatment with base yields 5-hydroxy-5-methylhydantoin and formic acid in 100% yields.

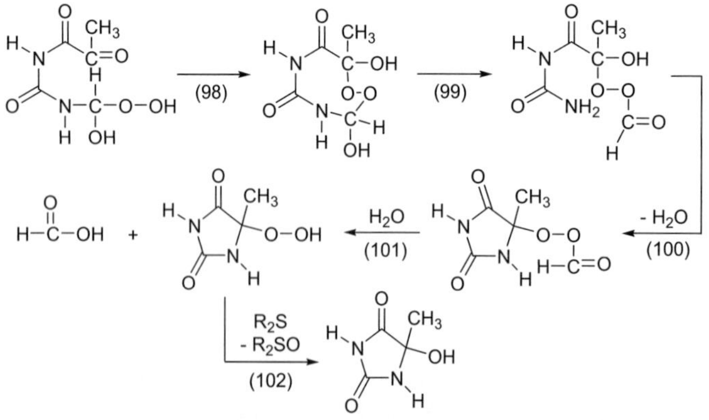

With Thy, singlet oxygen ($^1O_2$) formation, 4% at pH 7 and 8% at pH 8, is marginal but yet a very interesting process. No $^1O_2$ is detected with the nucleoside, Thyd (Muñoz et al., 2001). This is an indication that the hydrogen at N1 must be involved. Singlet oxygen formation is much more prominent (45%) with 5-chlorouracil (Table 6.7). For this nucleobase derivative some preliminary data are available (Muñoz et al., 2001) and shed some light on the mechanism. The hydrogen at N1 in the ozonide is most likely acidic [equilibrium (103)].

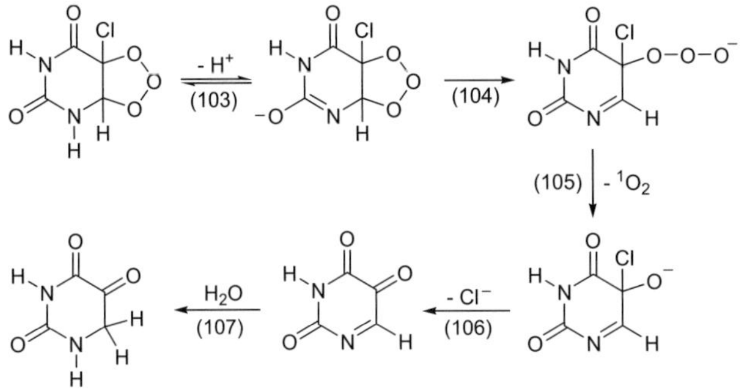

Subsequent cleavage of the ozonide yielding an isopyrimidine hydrotrioxide [reaction (104)] will be followed by the release of $^1O_2$ [reaction (105); for a compilation of $^1O_2$ yields in nucleic acid constituents see Table 6.7], Cl$^-$ release [reaction (106)] and water addition [reaction (107); for the formation and reactions of isopyrimidines see Al-Sheikhly et al. (1984); Schuchmann et al. (1984)]. At high pH isodialuric acid and HCl are indeed major products.

Table 6.7 Singlet oxygen yields (in % of mol O_3 consumed) in the reaction of ozone with nucleic acid constituents (Muñoz et al., 2001)

Substrate	pH	$^1O_2$ yield	Molar ratio substrate:ozone
Uracil (Ura)	3.5	No signal	9:1
	7	6	9:1
	11	7	9:1
1,3-Dimethyluracil	3.5	No signal	4:1
	11	No signal	4:1
6-Methyluracil	3.5	No signal	10:1
	7	12	10:1
	10	15	10:1
5-Chlorouracil	3.5	No signal	4:1
	7	45	4:1
	11	43	4:1
Thymine (Thy)	3.5	No signal	4:1
	7	4	4:1
	10	8	4:1
Thymidine (Thyd)	7	No signal	10:1
	10	No signal	4:1

(Continued)

Table 6.7 Singlet oxygen yields (in % of mol O₃ consumed) in the reaction of ozone with nucleic acid constituents (Muñoz et al., 2001) (*Continued*)

Substrate	pH	¹O₂ yield	Molar ratio substrate:ozone
Cytosine (Cyt)	7	No signal	16:1
	11	10	16:1
2'-Deoxycytidine (dCytd)	7	No signal	10:1
Cytidine (Cytd)	7	No signal	10:1
2'-Deoxyguanosine (dGuo)	7	40	10:1
Guanosine (Guo)	7	37	1:1
2'-Deoxyadenosine (dAdo)*	7	16	10:1
Adenosine (Ado)*	7	10	1:1
		21	10:1

*In the presence of tertiary butanol.

6.7 MICROPOLLUTANTS WITH OLEFINIC FUNCTIONS

Chlorinated olefins are abundant micropollutants. Trichloroethene and tetrachloroethene, react too slowly with ozone for an efficient elimination upon ozone treatment (Table 6.1). Their degradation is hence only achieved in DOM containing water by the •OH route (Chapter 3). Mechanistic details are discussed in Chapter 14. Bacterial degradation of the primary chlorinated olefins leads to *cis*-1,2-dichloroethene and to vinyl chloride. In certain ground waters, these secondary pollutants may be of major importance. While *cis*-1,2-dichloroethene also reacts slowly with ozone, vinyl chloride reacts rapidly (Table 6.1), and is degraded by ozone by the pathway shown above.

The high reactivity of the aromatic antiepileptic drug carbamazepine towards ozone (Table 6.1) is due to the reaction of its olefinic C–C double bond [reactions (108)–(111)] (McDowell et al., 2005). In this respect, it shares the cinnamic acids discussed above. The final product, 1-(2-benzaldehyde)-4-hydro(1*H*,3*H*)-quinazoline-2-one reacts only slowly with ozone ($k = 3$ M^{-1} s^{-1}), and upon ozonation of water and wastewater it is not readily further degraded [for carbamazepine degradation in other water treatment processes such as UV, chlorine dioxide, permanganate and ferrate see Kosiek et al. (2009); Hu et al. (2009)].

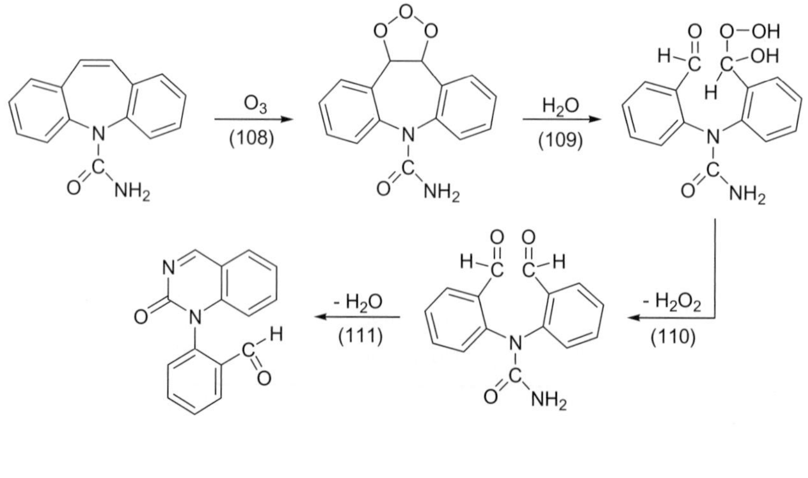

The β-lactam cephalosporin antibiotic cephalexin has two ozone-reactive sites, the sulfide function and a C–C double bond. Both contribute to its ozone-induced degradation. Details are discussed in Chapter 9. The sulfoxides that are formed upon the reaction of ozone continue to react readily with ozone at the C–C double bond, while the sulfoxide function is only of little reactivity (Chapter 9). The reaction of ozone with the cephalexin (*R*)-sulfoxide to the corresponding Criegee product [reaction (112)] takes place to only 30% (Dodd, 2008, 2010).

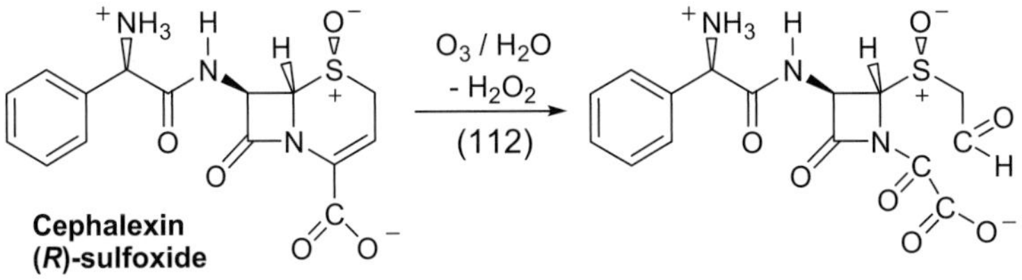

Cephalexin (*R*)-sulfoxide

Based on the reactions of acrylic and methacrylic acids discussed above, one would expect that a decarboxylation reaction should occur in competition. The product of this reaction has not yet been identified.

Tylosin, a macrolide antibiotic, has conjugated double bonds and a tertiary amine. The reactivity of the two conjugated C–C double bonds in conjugation with a carbonyl function must be an order of magnitude higher than a single C–C double bond in conjugation with a carbonyl group, as in progesterone, if a comparison of muconic acid with fumaric acid is a good guide (Table 6.1). The reactivity in the pH range relevant for water treatment is mostly controlled by the tertiary amine group (Dodd *et al.*, 2006) and will therefore be discussed in Chapter 8, where the formula of tylosin is also shown.

Tetracycline, another antibiotic compound has multiple sites for ozone attack, namely two olefins, a phenolic group and a tertiary amine. In the pH range relevant for water treatment, tetracycline will preferentially react with the phenolic moiety (Dodd *et al.*, 2006a) and will therefore be discussed in Chapter 7, where the formula of tetracycline is also given.

The antiviral drugs stavudine, zidovudine and lamivudine are thymine derivatives. They have been detected in WWTPs (Prasse *et al.*, 2010).

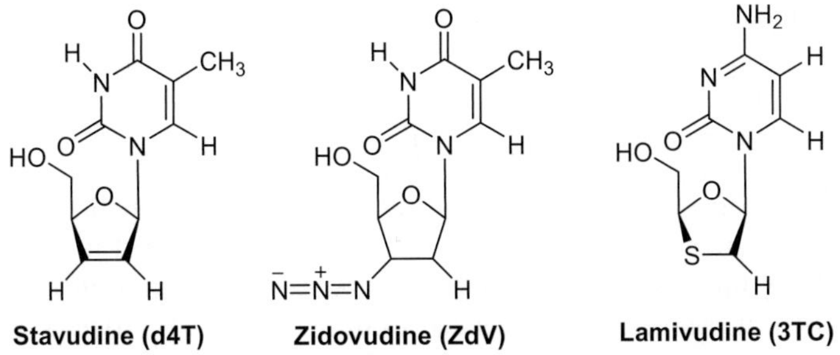

Stavudine (d4T)   **Zidovudine (ZdV)**   **Lamivudine (3TC)**

They should be readily eliminated upon ozonation (cf. the rate constant of thymidine in Table 6.1). Stavudine and lamivudine have additional sites for ozone attack, a further C–C double bond and a sulfur, respectively.

The antiviral drugs aciclovir (acyclovir) and penciclovir, also detected in wastewater, belong to the guanine family and with aciclovir ozone rate constants follow closely those of guanine derivatives (Table 6.1). Product formation is due to ozone attack at the C(4)–C(5) double bond (Prasse *et al.*, 2012).

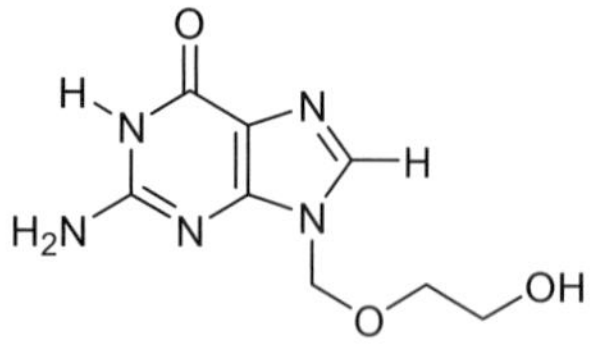

Aciclovir (AcV)

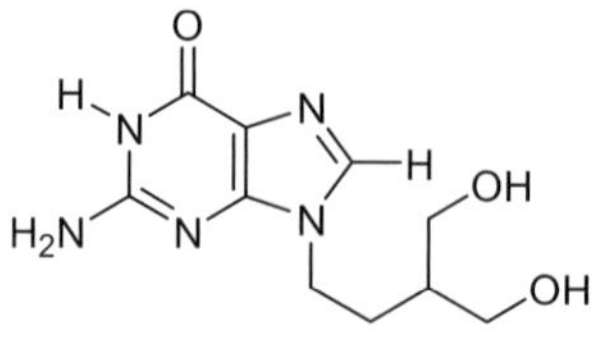

Penciclovir (PCV)

The antiviral drug abacavir, also detected in WWTPs (Prasse et al., 2010), belongs to the adenine family.

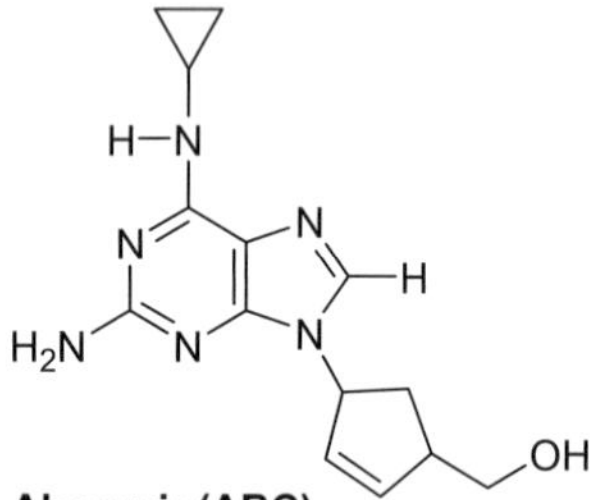

Abacavir (ABC)

Based on this, the ozone reactivity would be poor (cf. the rate constant of adenosine in Table 6.1). But there is also a highly ozone-reactive C–C double bond in the cyclopentene ring, and this guarantees an effective degradation upon ozonation (cf. the rate constant of propene in Table 6.1).

Oseltamivir acid is formed through the human metabolism when Tamiflu® (oseltamivir ethylester phosphate) is administered orally. A warning has been expressed that the widespread use of this antiviral drug during an influenza pandemic may lead to an increase in resistance (Singer et al., 2007).

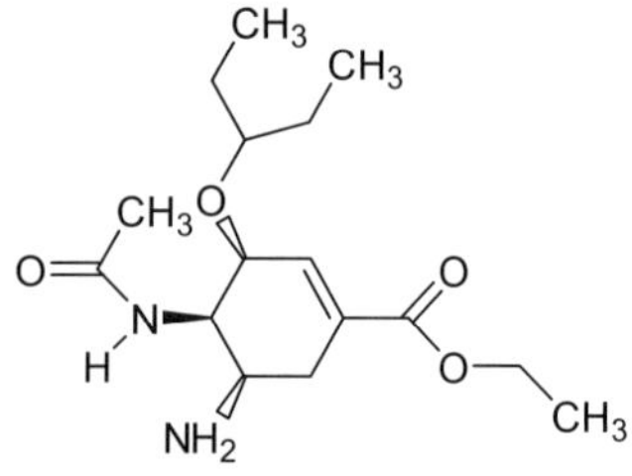

Oseltamivir

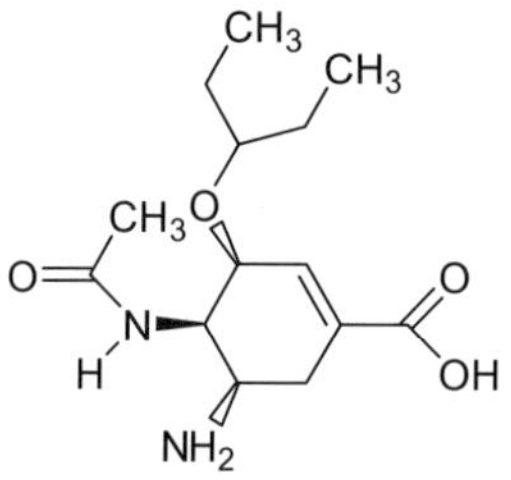

Oseltamivir acid

The activity of this drug is not markedly decreased in WWTPs (Fick et al., 2007). Its metabolite formed upon hydrolysis of the ester function, oseltamivir acid, is also active and has been detected in WWTP effluents and in rivers during the 2008–09 flu season in Japan (Ghosh et al., 2009). Yet, it is also found in German WWTPs (Prasse et al., 2010). It has been recommended to treat the WWTP effluents with ozone to minimise the risks (Singer et al., 2007). At a wastewater pH of 7–8, oseltamivir acid reacts with ozone mainly at the acrylate function due to its high ozone reactivity (Table 6.1). The nitrogen of the amino group is a second potential site, notably at higher pH. Upon ozonation, antiviral effects disappear proportionally with the degradation of the parent compound. A deviation from this behaviour was observed only for small degrees of transformation (Chapter 4) (Mestankova et al., 2012).

The reactivity of cyanotoxins such as microcystin-LR (MC-LR) is dominated by the ADDA group, a conjugated system of two double bonds (Onstad et al., 2007). When the ADDA group is bonded to the peptidic ring system, it is responsible for the hepatoxicity of the microcystins. The reaction of ozone with the conjugated double bonds of the ADDA group is fast (Table 6.1). It leads to a cleavage of the ADDA group and hence to a loss of toxicity. In the oxidation of MC-LR by permanganate, the degradation of the parent compound also leads to a loss of its toxicity (Rodríguez et al., 2008). Similar to ozone, it can be assumed from the reactivity of permanganate with olefins that the main site of attack for permanganate is the ADDA group (Waldemer & Tratnyek, 2006; Hu et al., 2009).

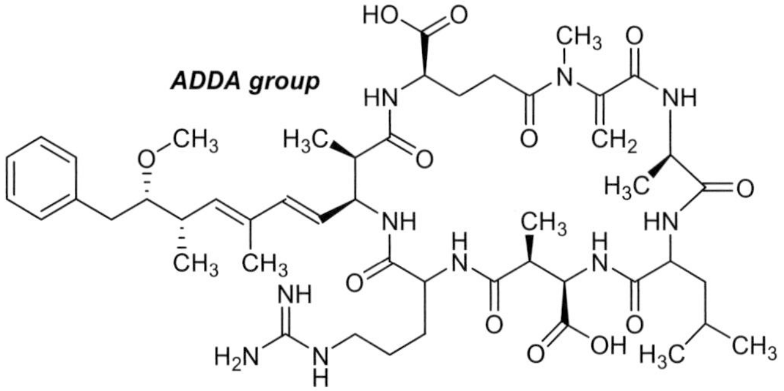

Microcystine-LR

Cylindrospermopsin, another cyanotoxin, contains a uracil derivative, which is the active moiety that inhibits protein translation to promote hepatoxicity or binds to DNA causing strand breakage and genotoxicity (Banker et al., 2001). The main site of attack is the uracil double bond (Onstad et al., 2007), and its ozone chemistry will be similar to that of thymine/thymidine discussed above. Most importantly, the uracil group will be altered significantly which should lead to a loss of the toxic properties of cylindrospermopsin.

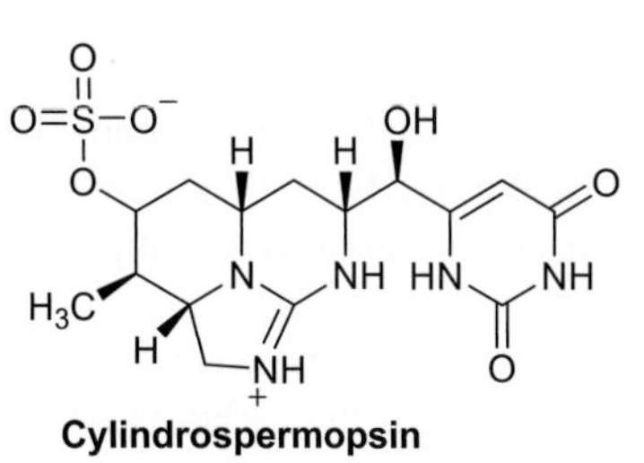

Cylindrospermopsin

Anatoxin-a is also a cyanotoxin which mimics acetylcholine and over-stimulates muscle cells and may cause paralysis (Carmichael et al., 1979).

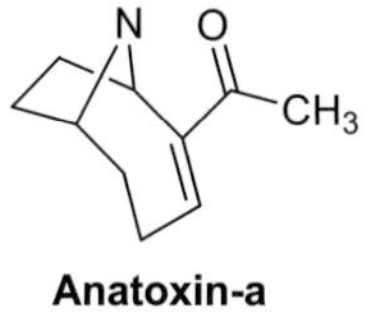

Anatoxin-a

Anatoxin has two functional groups, the amino group and an olefinic function. The reactivity pK_a of anatoxin-a is 7.88, which means that the olefinic group dominates the reactivity (Table 6.1) for most of the drinking water relevant pH range (Onstad *et al.*, 2007). The opening of the aliphatic ring caused by a Criegee-type reaction will most likely lead to a loss of anatoxin-a's toxicity.

The mycotoxin patulin is often found in apple juice, and its destruction by ozone has been suggested (Cataldo, 2008). Ozone rate constant and reaction products are as yet not known. The rate constant must be considerably lower than that of β-ionone (Table 6.1), as the carbonyl group in conjugation with the two C–C double bonds withdraws electron density from the sites of ozone attack.

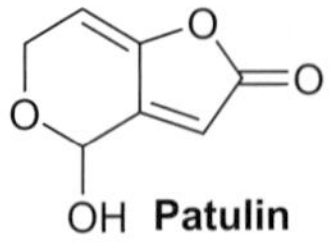

Patulin

Another class of natural olefin-containing compounds consists of algae derived taste and odour compounds, such as β-ionone, β-cyclocitral, 1-penten-3-one, cis-3-hexen-1-ol and *trans,cis*-2,6-nonadienal (for rate constants see Table 6.1; for an overview of taste and odour compounds and their properties see Chapter 5).

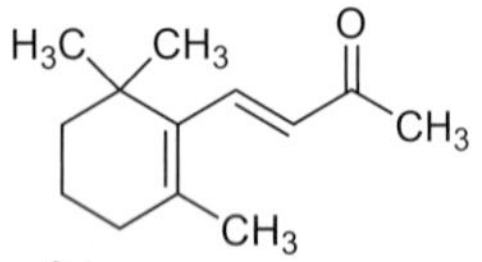

β-Ionone 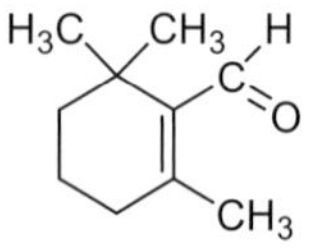   **β-Cyclocitral** 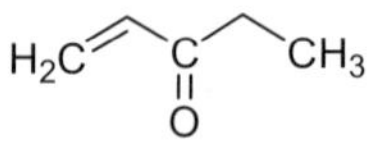   **1-Penten-3-one**

cis-3-Hexen-1-ol 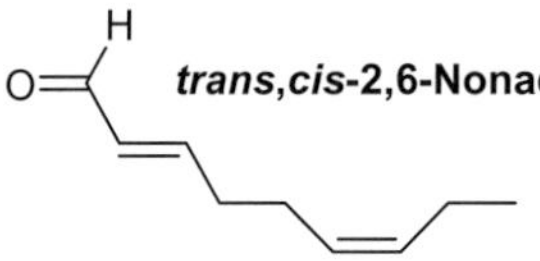   **trans,cis-2,6-Nonadienal**

These compounds will react with a Criegee-type reaction, leading to more hydrophilic compounds (Peter & von Gunten, 2007). Since evaporation of these compounds is an important property for their transfer into the cavity of the nose where the odour receptors are, oxidation by ozone is an efficient way to mitigate taste and odour problems caused by olefinic compounds in drinking waters.

The hormones progesterone and testosterone also contain an olefinic group. Progesterone reacts very moderately with ozone (Table 6.1), because of the electron-withdrawing effect of the carbonyl group in conjugation. The reactivity of testosterone is expected to be the same.

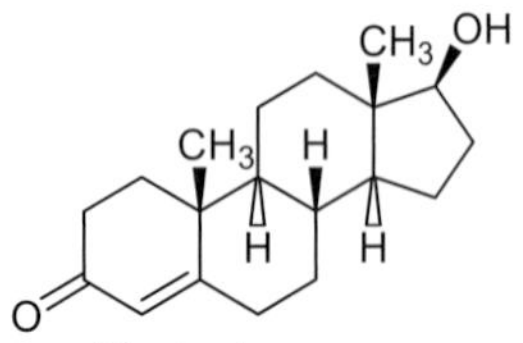

Testosterone 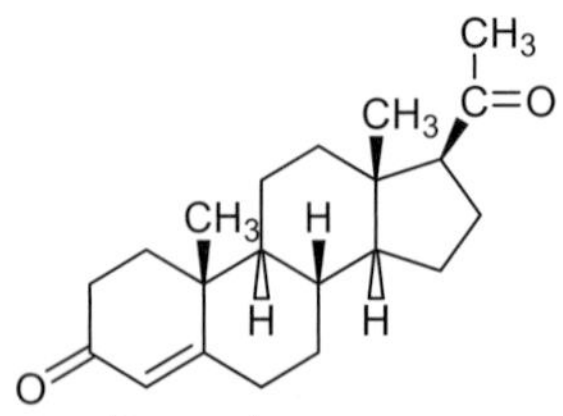   **Progesterone**

The reaction of ozone leads to a rupture of the 6-membered ring (Barron *et al.*, 2006). This must significantly reduce the biological activity of progesterone and testosterone. Product formation has been studied for progesterone and is shown in reactions (113)–(115) (Barron *et al.*, 2006).

The ozonide (not shown) cleaves the ring leading to aldehyde and hydroxyhydroperoxide functions. The H$_2$O$_2$ elimination [reaction (113)] is trivial and predominates under the given conditions, but a minor pathway leads to a (well-established) addition of the hydroperoxide to the carbonyl group [equilibrium (114)]. This adduct decomposes after deprotonation of the acidified hydroxyl group [reaction (115)].

The herbicides endrin and chlordane are particularly unreactive towards ozone with respect to its parent *cis*-1,2-dichloroethene (Table 6.1).

This low reactivity has been suggested to be due to a steric hindrance of the C–C double bond on both sides (Yao & Haag, 1991).

Chapter 7
Aromatic compounds

7.1 REACTIVITY OF AROMATIC COMPOUNDS

The rate constants of ozone with aromatic compounds vary strongly with the nature of substituents, with ten orders of magnitude between nitrobenzene and phenolate ion (Table 7.1).

Table 7.1 Compilation of rate constants for the reaction of ozone with aromatic compounds. For nitrogen heterocyclic aromatic compounds and aromatic compounds with nitrogen-containing groups in the side chain see Chapter 8. Published rate constants are rounded to significant figures

Compound	pK_a	pH	k/M^{-1} s^{-1}	Reference
Amidotrizoic acid			<0.8	Huber et al., 2003
			0.05	Real et al., 2009
Benzaldehyde			2.5	Hoigné & Bader, 1983b
Benzene			2	Hoigné & Bader, 1983a
Benzenesulfonate			0.23	Yao & Haag, 1991
Benzoate ion			1.2	Hoigné & Bader, 1983b
Bezafibrate			590	Huber et al., 2003
Bisphenol A	9.6, 10.2		1.7×10^4	Deborde et al., 2005
mono-anion			1.1×10^9	Deborde et al., 2005
di-anion			1.1×10^9	Deborde et al., 2005
Carbofuran			640	Yao & Haag, 1991
Catechol		7	5.2×10^5	Mvula & von Sonntag, 2003
			3.1×10^5	Gurol & Nekouinaini, 1984
Chlorobenzene			0.75	Hoigné & Bader, 1983a
2-Chlorophenol	8.3		1100	Hoigné & Bader, 1983b
anion			2×10^8	Hoigné & Bader, 1983b
4-Chlorophenol	9.2		600	Hoigné & Bader, 1983b
anion			6×10^8	Hoigné & Bader, 1983b

(Continued)

Table 7.1 Compilation of rate constants for the reaction of ozone with aromatic compounds. For nitrogen heterocyclic aromatic compounds and aromatic compounds with nitrogen-containing groups in the side chain see Chapter 8. Published rate constants are rounded to significant figures (*Continued*)

Compound	pK_a	pH	$k/M^{-1}s^{-1}$	Reference
2-(4-Chlorophenoxy)-2-methylpropionic acid (Clofibric acid)			<20	Huber et al., 2005
2-Cresol			1.2×10^4	Hoigné & Bader, 1983b
3-Cresol			1.3×10^4	Hoigné & Bader, 1983b
4-Cresol			3×10^4	Hoigné & Bader, 1983b
1,3-Dichlorobenzene			0.57	Yao & Haag, 1991
1,4-Dichlorobenzene			≪3	Hoigné & Bader, 1983b
2,3-Dichlorophenol	7.7		<2000	Hoigné & Bader, 1983b
2,4-Dichlorophenol	7.8		<1500	Hoigné & Bader, 1983b
anion			$\sim 8 \times 10^9$	Hoigné & Bader, 1983b
2,4-Dichlorophenoxyacetic acid (2,4-D)		2	5.3	Xiong & Graham, 1992
Diethyl-*o*-phthalate			0.1	Yao & Haag, 1991
1,4-Dimethoxybenzene			1.3×10^5	Muñoz & von Sonntag, 2000a
1,2-Dimethoxytoluene			8.9×10^4	Ragnar et al., 1999a
2,3-Dimethylphenol			2.47×10^4	Gurol & Nekouinaini, 1984
2,4-Dimethylphenol			1.95×10^4	Gurol & Nekouinaini, 1984
2,6-Dimethylphenol			9.88×10^4	Gurol & Nekouinaini, 1984
3,4-Dimethylphenol			9.88×10^4	Gurol & Nekouinaini, 1984
Dimethyl-*o*-phthalate			0.2	Yao & Haag, 1991
2,6-Di-*t*-butyl-4-methylphenol (BHT)			7.4×10^4	Peter & von Gunten, 2007
2,4-Dinitrotoluene		2	<14	Chen et al., 2008
2,6-Dinitrotoluene		2	<14	Chen et al., 2008
			5.7	Beltrán et al., 1998
17α-Ethinyloestradiol	10.4		1.8×10^5	Deborde et al., 2005
anion			3.7×10^9	Deborde et al., 2005
			$\sim 7 \times 10^9$	Huber et al., 2003
Ethylbenzene			14	Hoigné & Bader, 1983a
Galaxolide (HHCB)			140	Nöthe et al., 2007
Gemfibrozil			$\sim 2 \times 10^4$	Estimated, see section 7.4
2,2′,4,4′,5,5′-Hexachlorobiphenyl			<0.9	Yao & Haag, 1991
Hydroquinone		2.6	1.5×10^6	Gurol & Nekouinaini, 1984
		7	2.3×10^6	Mvula & von Sonntag, 2003
		3	1.8×10^6	Mvula & von Sonntag, 2003

(*Continued*)

Table 7.1 Compilation of rate constants for the reaction of ozone with aromatic compounds. For nitrogen heterocyclic aromatic compounds and aromatic compounds with nitrogen-containing groups in the side chain see Chapter 8. Published rate constants are rounded to significant figures (*Continued*)

Compound	pK_a	pH	$k/M^{-1} s^{-1}$	Reference
Ibuprofen [2-(4-Isobutylphenyl)propionic acid] deprotonated	4.9		7.2	Vel Leitner & Roshani, 2010
			9.6	Huber et al., 2003
Iomeprol			<0.8	Huber et al., 2003
Iopamidol			<0.8	Huber et al., 2003
Iopromide			<0.8	Huber et al., 2003
Isopropylbenzene			11	Hoigné & Bader, 1983b
Methoxybenzene			290	Hoigné & Bader, 1983a
Methoxychlor			270*	Yao & Haag, 1991
4-Methoxy-1-naphthalenesulfonic acid		7	3.6×10^3	Benner et al., 2008
Methylbenzoate			1.1	Hoigné & Bader, 1983b
2-Methyl-4-chlorophenoxyacetic acid (MCPA)		2	11.7	Xiong & Graham, 1992
2-Methyl-4-chlorophenoxypropionic acid (Mecoprop)		2	40	Beltrán et al., 1994
		7	111	Beltrán et al., 1994
		2	12.2	Xiong & Graham, 1992
1-Methylnaphthalene			1×10^3	Legube et al., 1986
2-Methylphenol	10.2		1.2×10^4	Hoigné & Bader, 1983b
3-Methylphenol	10.0		1.3×10^4	Hoigné & Bader, 1983b
Naphthalene			3000	Hoigné & Bader, 1983a
			~1500	Legube et al., 1986
Naproxen			~2×10^5	Huber et al., 2005
Nitrobenzene			9×10^{-2}	Hoigné & Bader, 1983a
			2.2	Beltrán et al., 1998
4-Nitrophenol	7.2		<50	Hoigné & Bader, 1983b
anion			1.7×10^7	Hoigné & Bader, 1983b
4-N-Nonylphenol	10.7		3.8×10^4	Deborde et al., 2005
			3.9×10^4	Ning et al., 2007b
anion			6.8×10^9	Deborde et al., 2005
4-Octylphenol			4.3×10^4	Ning et al., 2007b
17β-Oestradiol	10.4		2.2×10^5	Deborde et al., 2005
anion			3.7×10^9	Deborde et al., 2005
Oestriol	10.4		1.0×10^5	Deborde et al., 2005
anion			3.9×10^9	Deborde et al., 2005

(*Continued*)

Table 7.1 Compilation of rate constants for the reaction of ozone with aromatic compounds. For nitrogen heterocyclic aromatic compounds and aromatic compounds with nitrogen-containing groups in the side chain see Chapter 8. Published rate constants are rounded to significant figures (*Continued*)

Compound	pK_a	pH	$k/M^{-1}s^{-1}$	Reference
Oestrone	10.4		1.5×10^5	Deborde et al., 2005
anion			4.2×10^9	Deborde et al., 2005
Paracetamol		2.0	1.4×10^3	Andreozzi et al., 2003
		5.5	1.3×10^5	Andreozzi et al., 2003
Pentabromophenol anion			1.7×10^6	Mvula & von Sonntag, 2003
2,3,3',5,6-Pentachlorobiphenyl			<0.05	Yao & Haag, 1991
Pentachlorophenol	4.7			
anion		2	$>3 \times 10^5$	Hoigné & Bader, 1983b
			1.2×10^6	Mvula & von Sonntag, 2003
Phenol	9.9		1300	Hoigné & Bader, 1983b
anion			1.4×10^9	Hoigné & Bader, 1983b
1-Phenoxy-2-propanol			320	Benner et al., 2008
Resorcinol	9.8	2	$>3 \times 10^5$	Hoigné & Bader, 1983b
Salicylic acid	3.0,13.4		<500	Hoigné & Bader, 1983b
anion			2.8×10^4	Hoigné & Bader, 1983b
Tetracycline		7	1.9×10^6	Dodd, 2008
Toluene			14	Hoigné & Bader, 1983a
Tonalide (AHTN)			8	Nöthe et al., 2007
2,4,6-Tribromoanisole			0.02	Peter & von Gunten, 2007
2,4,6-Trichloroanisole			0.06	Peter & von Gunten, 2007
1,2,3-Trichlorobenzene			<0.06	Yao & Haag, 1991
2,4,5-Trichlorophenol	6.9		<3000	Hoigné & Bader, 1983b
anion			$>1 \times 10^9$	Hoigné & Bader, 1983b
2,4,6-Trichlorophenol	6.1		$<1 \times 10^4$	Hoigné & Bader, 1983b
anion			$>1 \times 10^8$	Hoigné & Bader, 1983b
Triclosan	8.1		1.3×10^3	Suarez et al., 2007
		7	3.8×10^7	Suarez et al., 2007
anion			5.1×10^8	Suarez et al., 2007
2,4,6-Triiodophenol anion			6.8×10^6	Mvula & von Sonntag, 2003
1,3,5-Trimethoxybenzene			9.4×10^5	Muñoz & von Sonntag, 2000
3,4,5-Trimethoxytoluene			2.8×10^5	Dodd et al., 2006a
1,2,4-Trimethylbenzene			400	Hoigné & Bader, 1983b
1,3,5-Trimethylbenzene			700	Hoigné & Bader, 1983b
Vancomycin		7	6.1×10^5	Dodd, 2008
m-Xylene			94	Hoigné & Bader, 1983b
o-Xylene			90	Hoigné & Bader, 1983b
p-Xylene			140	Hoigné & Bader, 1983b

*Consumption of substrate

In their pioneering work, Hoigné and Bader have shown that the logarithm of the rate constant for the reaction of ozone with substituted benzenes correlates linearly with the σ_p^+ values of the substituents (Hoigné & Bader, 1983a). As the ρ value (slope) of this Hammett-type plot was negative (ρ = −3.14), they concluded that in its reaction with benzene and its derivatives, ozone must act as an electrophilic agent. A similar plot with another set of σ_p^+ values also including the strongly electron-withdrawing NO$_2$ substituent is shown in Figure 7.1 (Naumov & von Sonntag, 2010). Here, the slope (ρ = −2.65) is slightly less negative, but such variations are common when using different sets of σ values.

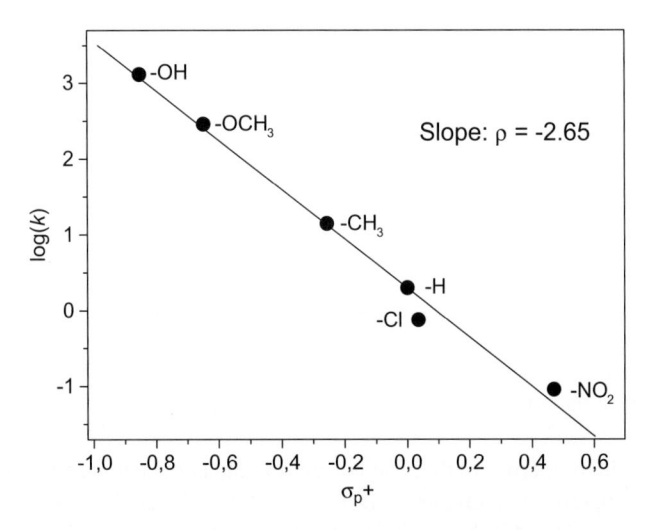

Figure 7.1 Hammett-type plot for ozone reactions with benzene and its derivatives according to Naumov & von Sonntag, 2010 with permission. Logarithm of the rate constant vs. the σ_p^+ values taken from Gordon & Ford, 1972.

The first step in the reaction of ozone with aromatic compounds is the formation of an adduct [reaction (1)].

Hammett type plots are based on free energy relationships ($\Delta G^0 = -\ln K \times RT$), and thus the standard Gibbs free energy of the formation of ozone adducts that can be calculated by quantum chemistry (Naumov & von Sonntag, 2010) must also correlate with the logarithm of the rate constant when the transition state is close to the adduct. This is apparently the case (Figure 7.2).

In the case of alkenes and alkynes, the ozone reaction seems to proceed via a concerted cycloaddition to the ozonide and ozone adducts cannot be calculated as intermediates (Chapter 6). In aromatic systems, however, the positive charge that develops upon formation of the ozone adduct [reaction (1)] is distributed over the whole aromatic ring, and a collapse to the ozonide [reaction (2)] is sufficiently retarded to allow an ozone adduct to become a distinct chemical entity (intermediate). This explains why in aromatic compounds there are further reaction pathways [reaction (3)], and the formation of the ozonide [reaction (2)] and its subsequent decay is only one of them.

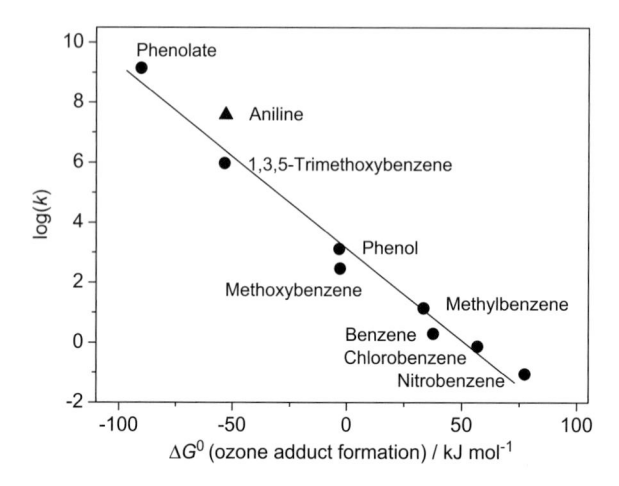

Figure 7.2 The logarithm of the rate constant for the reaction of ozone with benzene derivatives vs. the calculated standard Gibbs free energy of adduct formation. Adapted from Naumov & von Sonntag, 2010. The aniline rate constant, possibly somewhat too high, is taken from Tekle-Röttering et al., 2011, the corresponding standard Gibbs free energy was calculated by S. Naumov.

Several correlations with properties of the aromatic compound and the logarithm of the rate constant have been attempted, that is, properties that favour the formation of the adduct. The correlation with the energy of the highest occupied orbital (HOMO, Figure 7.3) comes out best.

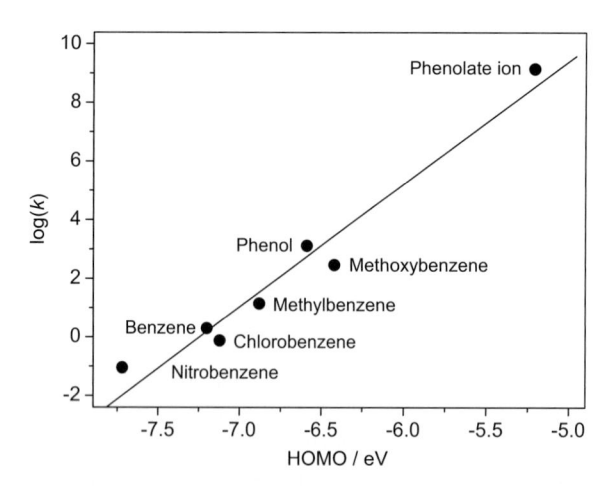

Figure 7.3 Correlation of the logarithm of the rate constant with the energy of the HOMO. Adapted from Naumov & von Sonntag, 2010, with permission.

According to the chemical reactivity theory of Klopman and Salem (Klopman, 1968; Salem, 1968; Fleming, 2002), the reaction between the reaction partners proceeds in a way to produce the most favourable interaction energy. This is controlled by two factors, an electrostatic interaction approximated by atomic charges and a frontier orbital interaction. The frontier orbital interaction is the interaction between occupied and unoccupied orbitals on the two interacting molecules. The frontier orbital theory contains the additional assumption that the reaction depends mostly on the interaction between the highest occupied molecular orbital (HOMO) of one reaction partner and the lowest unoccupied molecular orbital (LUMO) of the other (as shown in Figure 7.4 for the interaction of ozone with benzene). The smaller the energy separation ($\Delta\varepsilon$) between the HOMO of the donor and the LUMO of the acceptor the larger the interaction between reaction partners should be. With aromatic compounds, the most favourable interaction occurs between the HOMO of aromatic compounds (electron donor) and the LUMO of ozone as the electron acceptor.

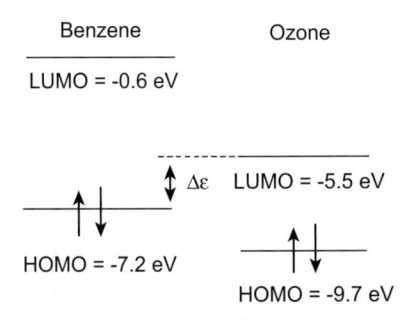

Figure 7.4 MO scheme [B3LYP/6-311G + (d)] of the frontier orbital interaction of benzene with ozone. Due to the small energy difference $\Delta\varepsilon$, the most favourable interaction takes place between the HOMO of benzene and the LUMO of ozone. This interaction has charge transfer character and leads to a strong shift of negative charge from benzene to ozone (about 0.8 e) according to Naumov & von Sonntag, 2010, with permission.

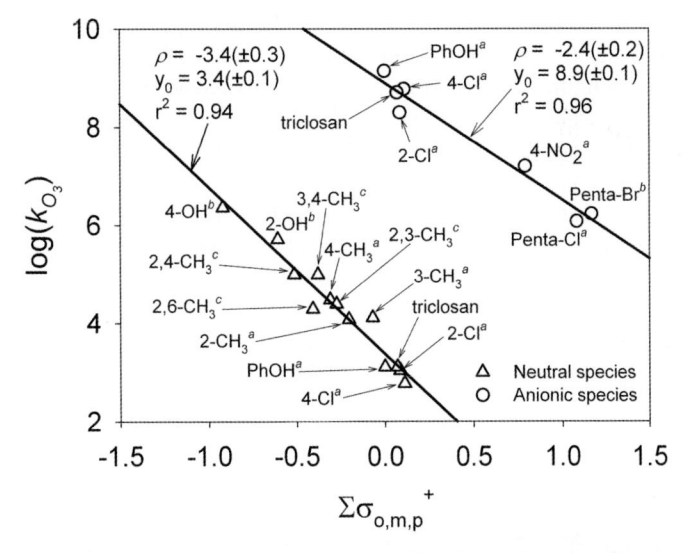

Figure 7.5 Correlation of the logarithm of the second order rate constants for the reaction of various phenols and triclosan with ozone with the Hammet coefficients. According to Suarez et al., 2007, with permission.

Aniline has also been included in Figure 7.1. Apparently, its ozone chemistry, as poorly as it is understood at present (Chapter 8), seems to be largely governed by an ozone addition to the ring. The yields of products that require an ozone addition to nitrogen are very minor.

Hammett plots are not only of academic interest; they are of high predictive value since σ values are additive and this allows the estimation of an unknown rate constant. Figure 7.5 demonstrates that the second order rate constant for triclosan (for structure see Paragraph 7.4) could be derived by a Hammet-type correlation with substituted phenols for both the neutral and anionic form. The measured values correspond nicely with the predictions.

In another study, it has been shown for two musk fragrances, with the same molecular weight but different structure, galaxolide and tonalide (for their structures see Paragraph 7.4), that this approach can be successful within a factor of two (Nöthe et al., 2007). While galaxolide has only electron-donating substituents, tonalide also has an electron-withdrawing one, the acetyl group. As a consequence, the ozone reactivity of the latter is much lower (Table 7.1).

7.2 DECAY OF OZONE ADDUCTS

The rate constants for the reaction of benzene and all its mono-substituted derivatives with ozone is $< 2 \times 10^3$ M^{-1} s^{-1}. Muconic products that arise from the breakdown of the ozonide [cf. reaction (6)] react with rate constants near 10^4 M^{-1} s^{-1} (Chapter 6). Thus, a detailed study of the primary products is not feasible. Although phenol has a rate constant of only 1.3×10^3 M^{-1} s^{-1}, the presence of low phenolate equilibrium concentrations [pK_a(phenol) = 9.9, k(phenolate + O$_3$) = 1.4×10^9 M^{-1} s^{-1}] results in the observed rate constant of $>1 \times 10^6$ M^{-1} s^{-1} at pH 7 (for the 'reactivity pK' see Chapter 2, Figure 2.3). Thus, phenols are good candidates for product studies. Upon substitution with two or more alkoxyl groups, the ozone rate constant is also enhanced to an extent, that a study of the primary products can be successful. Thus, our knowledge of the ozonolysis of aromatic compounds can be based only on electron-rich benzene derivatives, phenol (Mvula & von Sonntag, 2003; Ramseier & von Gunten, 2009), and di- and trimethoxybenzenes (Mvula et al., 2009). Here, it must be noted that in these compounds the high electron density in the benzene ring will allow reactions to occur that are not as likely to take place in the parent benzene and in benzene derivatives with electron-withdrawing substituents. With this *caveat* in mind, the ozone chemistry of aromatic compounds is discussed. The key reactions that were observed with such electron-rich aromatic compounds are presented here in general terms and will be substantiated below when discussing real systems.

A reaction that always occurs with aromatic compounds, and is well-documented in the reactions of olefins with ozone, is the formation of an ozonide and its breakdown [reactions (4)–(7)].

Aromatic compounds

This reaction is generally exergonic and will be given by all aromatic compounds. In competition, the ozone adduct may dissociate into a radical cation and an ozonide radical anion [reaction (8)]. This reaction may also proceed directly via an outer sphere electron transfer [reaction (9)].

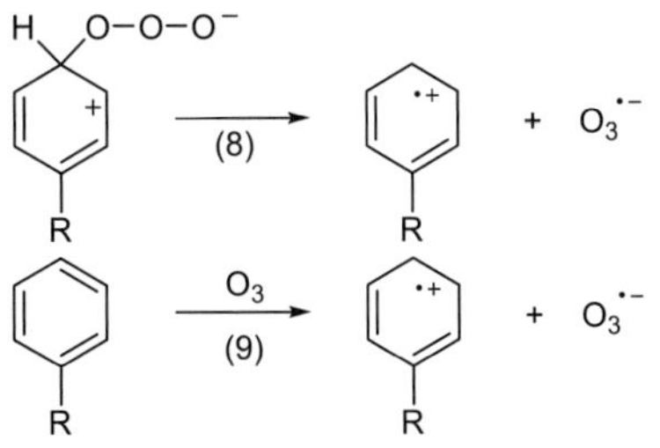

Which of the two processes will dominate depends on the influence of the substituents on the energetics of reactions (4), (8) and (9). For benzene, all these reactions are markedly endergonic (Table 7.2), and no $O_3^{\bullet-}$ is expected to be formed. With methoxybenzene, all reactions are exergonic, and the observed formation of •OH (via $O_3^{\bullet-}$) is possible by reaction (4) followed by (8) but also by direct electron transfer [reaction (9)]. With 1,3,5-trimethoxybenzene, reactions (4) and (9) are both exergonic but reaction (8) is endergonic. Here, direct electron transfer [reaction (9)] will be the process resulting in •OH formation.

Table 7.2 Standard Gibbs free energies (kJ mol^{-1}) for the formation of the ozone adduct, its decay into the radical cation and the ozonide radical anion, and the direct electron transfer for benzene, methoxybenzene and 1,3,5-trimethoxybenzene [Naumov and von Sonntag (2009), unpublished]

Starting material	Starting material → radical cation + $O_3^{\bullet-}$	Starting material → ozone adduct	Ozone adduct → radical cation + $O_3^{\bullet-}$
Benzene	+58 kJ mol^{-1}	+38 kJ mol^{-1}	+20 kJ mol^{-1}
Methoxybenzene	−12 kJ mol^{-1}	−3 kJ mol^{-1}	−9 kJ mol^{-1}
1,3,5-Trimethoxybenzene	−36 kJ mol^{-1}	−54 kJ mol^{-1}	+18 kJ mol^{-1}

Another process that always occurs is the hydroxylation of the starting material. It may be due to the release of oxygen from the ozone adduct followed by a hydrogen shift [reactions (10) and (11)] or by a re-aromatisation of the zwitterionic ozone adduct by proton loss [reaction (12)]. This aromatic hydrotrioxide anion would then have to lose oxygen [reaction (13)] followed by protonation of the phenolate ion [reaction (14)]. The spin conservation rule demands that oxygen is released in its excited singlet state, $O_2(^1\Delta_g)$, which lies 95.5 kJ mol^{-1} above the triplet ground state, $O_2(^3\Sigma_g^-)$. For a compilation of singlet oxygen yields in ozone reactions of aromatic compounds see Table 7.3.

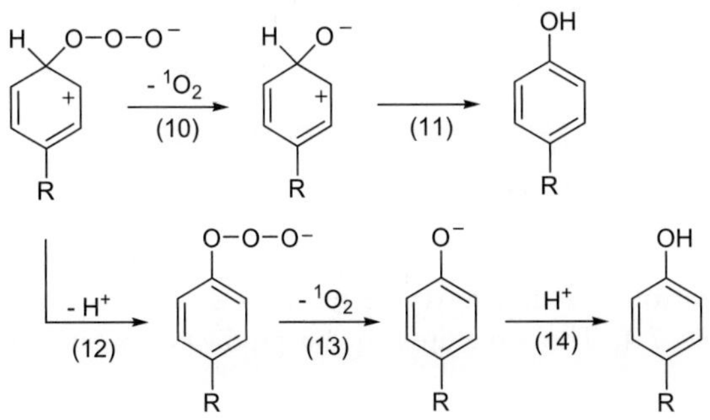

Table 7.3 Singlet oxygen yields in percentage of consumed ozone from the reaction of ozone with aromatic compounds according to (Muñoz et al., 2001). Aniline derivatives, discussed in Chapter 8, are included

Compound	Singlet oxygen yield/%
Phenol pH 1.8	Nil
Phenol pH 7	6
Phenol pH 9	9
Tyrosine pH 7	9
2,4,6-Trimethylphenol pH 9	17
Pentachlorophenol pH 8	58
Pentabromophenol pH 8	48
2,4,6-Triodophenol pH 9	19
1,4-Dimethoxybenzene	6
1,3,5-Trimethoxybenzene	30
N,N-Dimethylaniline pH 8.5	7
N,N,N',N'-Tetramethylphenylenediamine pH 3.5	9
N,N,N',N'-Tetramethylphenylenediamine pH 6	6

The decay of the ozone adduct into superoxide and phenoxyl radicals has also been suggested [reactions (15) and (16)] (Ragnar et al., 1999a, b).

$O_3^{\bullet-}$ and $O_2^{\bullet-}$ are the precursors of $^\bullet OH$, which may be formed in substantial yields in ozone reactions with aromatic compounds. Here, $O_2^{\bullet-}$ seems to be of a minor importance (see below). The $^\bullet OH$ radical yield (with respect to ozone consumption formed in the ozone treatment of wastewater will also arise mainly from such reactions (Chapter 3). Moreover, the hydroxylation process generates further electron-rich aromatic compounds, a requirement for a continuation of $^\bullet OH$ production at high ozone doses.

Phenols react very rapidly with ozone (Table 7.1), and in the reactions of ozone with aromatic compounds of relatively low reactivity (see above) they can build up to only very low steady-state concentrations and their formation may be difficult to detect.

7.3 OZONE REACTIONS OF AROMATIC COMPOUNDS – PRODUCTS AND REACTIONS OF REACTIVE INTERMEDIATES

7.3.1 Methoxylated benzenes

The products and yields formed upon ozonolysis of anisole, 1,2-dimethoxybenzene, 1,4-dimethoxybenzene and 1,3,5-trimethoxybenzene are compiled in Table 7.4. The structures of the products are shown in Figure 7.6.

Table 7.4 Ozonolysis of anisole, 1,2-dimethoxybenzene, 1,4-dimethoxybenzene and 1,3,5-trimethoxybenzene. Products and their yields in percentage of ozone consumed according to Mvula et al. (2009). Nitroform anion yields were determined in the presence of tetranitromethane and are an upper limit for $O_2^{•-}$ yields

Starting material/product	Yield/%
Anisole	
Methyl(2Z,4Z)-6-oxohexa-2,4-dienoate **1**	identified
2-Methoxyhydroquinone **2** and/or 2-methoxy-1,4-benzoquinone **3**	identified
Hydrogen peroxide	23
•OH (estimated)	~8
1,2-Dimethoxybenzene	
Dimethyl-(2Z,4Z)-hexa-2,4-diendioate **4**	25
Singlet oxygen (Muñoz et al., 2001)	9
Hydrogen peroxide	27
•OH (estimated)	~14
Nitroform anion	4
1,4-Dimethoxybenzene	
Methyl(2Z,4E)-4-methoxy-6-oxo-hexa-2,4-dienoate **5**	52
Hydroquinone **6**	2
1,4-Benzoquinone **7**	8
2,5-Dimethoxyhydroquinone **8**	identified
2,5-Dimethoxy-1,4-benzoquinone **9**	identified
Singlet oxygen (Muñoz et al., 2001)	6
Hydrogen peroxide	56
•OH (estimated)	~17
Nitroform anion	9
1,3,5-Trimethoxybenzene	
2,6-Dimethoxy-1,4-benzoquinone **10**	20
2,4,6-Trimethoxyphenol **11**	identified
(2E,4E)-methyl-3,5-dimethoxy-6-oxohexa-2,4-dieonate **12**	identified
Singlet oxygen (Muñoz et al., 2001)	22
Hydrogen peroxide	(6.5)*
•OH (estimated)	~15
Nitroform anion	10

*see text

In 1,3,5-trimethoxybenzene, the 2-, 4- and 6-positions are strongly activated, and hence one has to consider practically only one ozone adduct [reaction (18)].

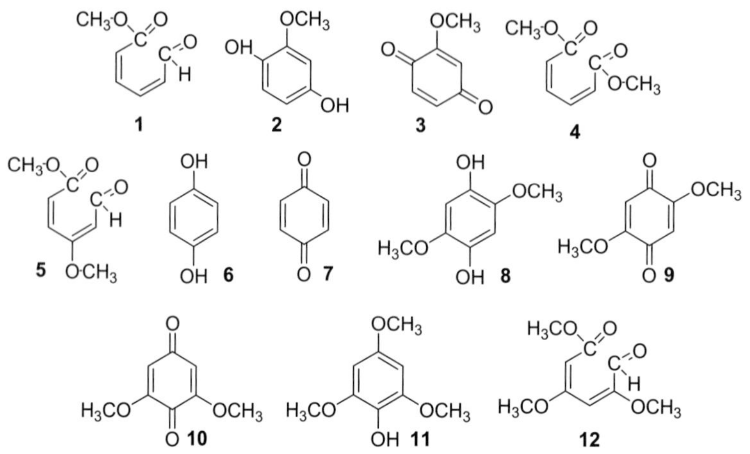

120 Chemistry of Ozone in Water and Wastewater Treatment

Figure 7.6 Structures identified in the ozonolysis of anisole (1–3), 1,2-dimethoxybenzene (4), 1,4-dimethoxybenzene (5–9) and 1,3,5-trimethoxybenzene (10–12).

Its ensuing decay routes are discussed in some detail in Mvula *et al.* (2009). Here it suffices to show only the essentials [reactions (19)–(25)], that is, neglecting acid/base equilibria. As in olefins, there is decay into the ozonide [reaction (19)]. This is the precursor of muconic compound **12** [reaction (20)]. In competition, the adduct may release $O_3^{•-}$ [reaction (21)], the origin of the relatively high $^{•}OH$ yield ($O_3^{•-} + H_2O \rightarrow {}^{•}OH + O_2 + OH^-$, details in Chapter 11). The release of singlet oxygen [reaction (22)], is connected with a hydroxylation of the aromatic ring [formation of **11**, reaction (23)]. Reaction (24) gives rise to $O_2^{•-}$ and reaction (25) to a hydrotrioxide.

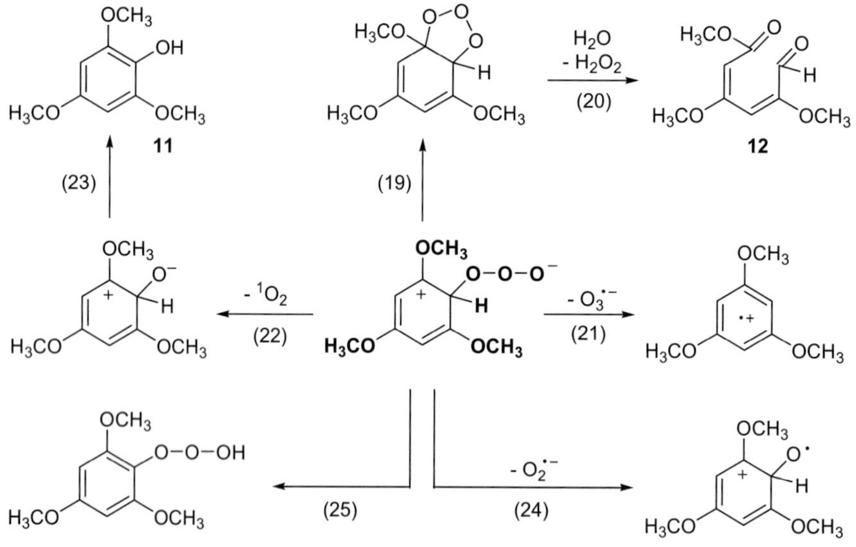

A major product is quinone **10** (Table 7.4). It has been suggested that it is mainly formed upon the decay of the hydrotrioxide [reactions (26)–(28)]. For the reaction of $HO_2^{•}/O_2^{•-}$ with phenoxyl radicals [cf. reaction (27)] see Jin *et al.* (1993).

7.3.2 Phenols

Phenols react moderately rapidly with ozone, but the reaction of their anions is close to diffusion controlled (Table 7.1). With pK_a(phenol) = 9.9, the rate of reaction observed at pH 7 is hence determined by the reaction of the phenolate anion in equilibrium (Chapter 2). Nevertheless, the product yields listed in Table 7.5 for pH 6–7 and for pH 10 are markedly different in many cases and must be due to different decay routes of the primary ozone adduct at these two different pH conditions.

Table 7.5 Ozonolysis of phenol. Products and their yields in percentage of ozone consumed. In several experiments, tertiary butanol (tBuOH) was added for scavenging •OH radicals. When more than one value is given, further runs at four or five different ozone concentrations were carried out Mvula & von Sonntag (2003)

Product (Scavenger)	≤pH 3	pH 6–7	pH 10
Hydroquinone	1.6/1.1[a]	13.3/16	0.8
Hydroquinone (tBuOH)	ND	<1	ND
Catechol	4.8/1.8[a]	13.6	20
Catechol (tBuOH)	ND	2	ND
1,4-Benzoquinone	9.6/10.4/6.1[a]	4.6/4.6	32
1,4-Benzoquinone (tBuOH)	ND	13	ND
cis,cis-Muconic acid	4.8/4.0/3[a]	2.8	1
cis,cis-Muconic acid (tBuOH)	ND	2.0	ND
4,4′-Dihydroxybiphenyl	ND	ND	~1
2,4′-Dihydroxybiphenyl	ND	ND	~1
Singlet dioxygen	absent	5.6	8
•OH (estimated)	20/22	28/26/24	22
Nitroform anion (TNM)	3	ND	ND
Organic (hydro)peroxide	ND	absent	ND
Organic (hydro)peroxide (tBuOH)	ND	2.6	ND
Hydrogen peroxide	8.5	4.8	2
Hydrogen peroxide (tBuOH)	ND	16	ND
Hydrogen peroxide (DMSO)	ND	13	ND
Phenol consumption	33[a]	48[a]	59[a]
Phenol consumption (tBuOH)	ND	42[a]	ND

[a][Phenol] = 2.5 × 10^{-4} M, ND = not determined, TNM = tetranitromethane, DMSO = dimethyl sulfoxide.

There is further study on phenol ozonolysis (Table 7.6) (Ramseier & von Gunten, 2009). There, product yields were given in percentage of phenol consumed, while in Table 7.5 product yields are based on ozone consumed.

Table 7.6 Ozonolysis of phenol (200 µM). Products and their yields in percentage of phenol consumed. Reactions of •OH were suppressed by tBuOH addition. According to Ramseier & von Gunten (2009)

Product	pH 3	pH 7
1,4-Benzoquinone	17/18	40/35
Hydroquinone	4/7	0/2
Catechol	0/1	27/33
cis,cis-Muconic acid	2/3	5/6

The values in Table 7.6 are necessarily higher than those of Table 7.5, since phenol consumption by ozone is substantially lower than unity (Table 7.5). This effect is also pH dependent. Moreover, for the presence of tBuOH, data are partially missing in Table 7.5 to allow a detailed comparison of the two data sets. The contribution of •OH in these ozone-induced reactions cannot be neglected as its yield is near 20–25% (Table 7.5). The formation of •OH (via $O_3^{•-}$) is balanced by an equivalent amount of phenoxyl radicals [reaction (29)].

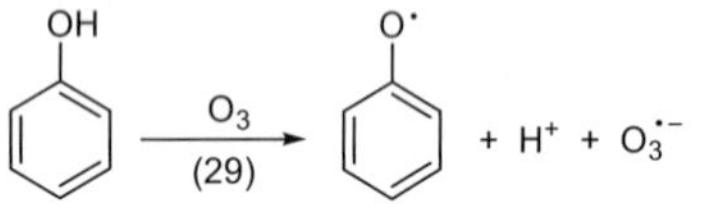

$$\text{PhOH} \xrightarrow{O_3} \text{PhO}^{\bullet} + H^+ + O_3^{\bullet-} \quad (29)$$

The reduction potential of phenol at pH 7 is 860 mV (Wardman, 1989). This may be compared with the corresponding reduction potentials of hydroquinone (549 mV) and catechol (530 mV). Hydroquinone and catechol are formed in hydroxylation reactions such as those depicted in reactions (10)–(14). The reduction potential of the peroxyl radical derived from tBuOH must be near 800 mV under such conditions (Chapter 14). Thus, notably hydroquinone is further oxidised in an ensuing electron transfer reaction [reactions (30) and (31)] and subsequent disproportionation [reaction (32)]. An electron transfer to O_2 is endergonic [reduction potential of 1,4-benzoquinone is 400 mV (Wardman, 1989), for $O_2^{•-}$ see Chapter 2]. Yet, when $O_2^{•-}$ is drawn out of equilibrium (34) the reaction may proceed nevertheless.

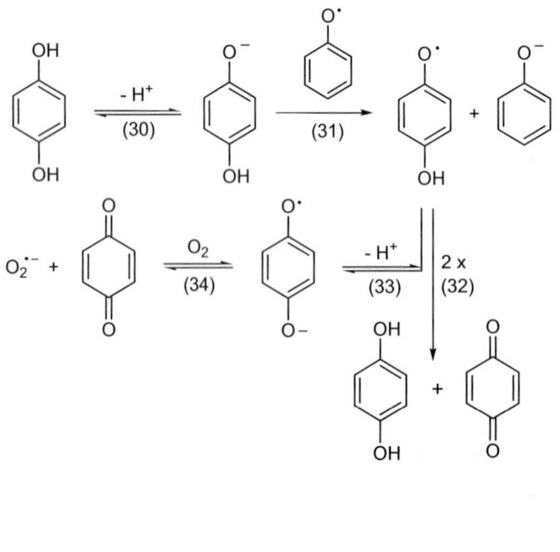

With hydroquinone as substrate, the yield of 1,4-benzoquinone is markedly enhanced in the presence of tBuOH and the data shown in Table 7.7 are supported by the study of Ramseier & von Gunten (2009), which gives a hydroquinone consumption related value of 83% at pH 7 and 52–61% at pH 3. The corresponding oxidation product of catechol, 1,2-benzoquinone, is unstable and would escape detection (cf. Table 7.8).

Table 7.7 Ozonolysis of hydroquinone (1 mM). Products and their yields in percentage of ozone consumed in absence and presence of tBuOH, added for scavenging •OH (Mvula & von Sonntag, 2003)

Product	Yield
1,4-Benzoquinone	13/12
1,4-Benzoquinone (t-BuOH)	36/32
1,4-Benzoquinone (DMSO)	30
2-Hydroxy-1,4-benzoquinone	11
2-Hydroxy-1,4-benzoquinone (tBuOH)	<1
2-Hydroxy-1,4-benzoquinone (DMSO)	<1
1,2,4-Trihydroxybenzene	absent
Singlet dioxygen	16
Hydrogen peroxide	5.6
Hydrogen peroxide (tBuOH)	14
Hydrogen peroxide (DMSO)	10.4
Organic (hydro)peroxides	absent
Organic (hydro)peroxides (tBuOH)	1.9
Formaldehyde	absent
Formaldehyde (tBuOH)	21/20
2-Hydroxy-2-methylpropionaldehyde	absent
2-Hydroxy-2-methylpropionaldehyde (tBuOH)	23
Methanesulfinic acid (DMSO)	6
Methanesulfonic acid (DMSO)	27
Hydroquinone consumption	47[a]
Hydroquinone consumption (tBuOH)	48[a]

[a] [Hydroquinone] $= 2.5 \times 10^{-4}$ M, TNM $=$ tetranitromethane, DMSO $=$ dimethyl sulfoxide.

cis,cis-Muconic acid is formed via the ozonide [cf. reaction (4)–(6)]. The hydroxyhydroperoxide that must be the intermediate typically loses H_2O_2 [reaction (35)], but here a water elimination is observed [reaction (36)]. This has been attributed to the fact that reaction (35) is reversible, while reaction (36) is not (Mvula & von Sonntag, 2003).

Table 7.8 Ozonolysis of catechol (1 mM). Products and their yields in % of ozone consumed (Mvula & von Sonntag, 2003)

Product	Yield
2-Hydroxy-1,4-benzoquinone	7.4
1,2,4-Trihydroxybenzene	absent
cis,cis-Muconic acid	7/10
Singlet dioxygen	14
Hydrogen peroxide	5.1
Hydrogen peroxide (DMSO)	7.2
Organic (hydro)peroxides	absent
Formaldehyde	absent
Formaldehyde (tBuOH)	11
Methanesulfinic acid (DMSO)	absent
Methanesulfonic acid (DMSO)	22
Catechol consumption	47[a]

[a] [Catechol] = 2.5×10^{-4} M, TNM = tetranitromethane, DMSO = dimethyl sulfoxide.

This implies that other α-hydroxyalkylhydroperoxides may also show water elimination, albeit on a much longer time scale. The efficiency of such a process will depend on the relative rates of such reactions.

7.4 MICROPOLLUTANTS WITH AROMATIC FUNCTIONS

Many micropollutants contain a benzene ring. This may be the site of ozone attack, but as most aromatic compounds react only slowly with ozone unless substituted by –OH and –OR functions, other ozone-reactive sites may dominate the reaction. Here, we deal only with those micropollutants whose reactivity is dominated by the aromatic ring. The aromatic micropollutant may contain an olefinic side chain, a nitrogen or a sulfur (even aniline derivatives) and these are discussed in Chapters, 6, 8 and 9, respectively.

Micropollutants that contain a phenolic function react very rapidly with ozone, because these phenols are in equilibrium with the phenolate which has a very high ozone rate constant, and it is the phenolate in equilibrium that determines their rate of degradation, notably in drinking water and wastewater which typically has a pH between 7 and 8.

Triclosan is widely applied in personal care products such as soaps and toothpastes due to its antibacterial properties. Paracetamol is a common painkiller.

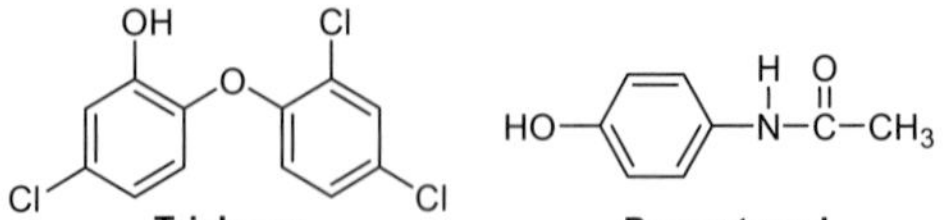

Triclosan
pK_a = 8.1

Paracetamol

Triclosan can be found in wastewater effluents in concentrations of up to 200 ng L^{-1} (Singer *et al.*, 2002). Its high second-order rate constant at pH 7 (Table 7.1) is due to the lower pK_a value of the phenol as compared to unsubstituted phenol. With the parent phenol (see above), triclosan shares a low consumption per ozone reacted (0.41) (Suarez *et al.*, 2007). Its reaction with ozone leads to a complete loss of its antibacterial properties (Chapter 4). For paracetamol, a series of degradation products are

Aromatic compounds

reported that are in line with the discussed mechanisms for the reaction of ozone with phenols (Andreozzi et al., 2003).

17α-Ethinyloestradiol is an artificial oestrogen, and, due to its abundant use as contraceptive, it can be found in the effluents of many WWTPs (Chapter 4).

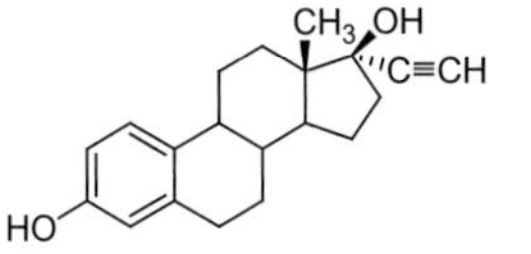

17α-Ethinyloestradiol (EE2)
pK_a = 10.4

Its ozonolysis has been investigated in some detail (Huber et al., 2004), and the observation has been made that some time after the ozone treatment about 3% of the 17α-ethinyloestradiol destroyed is regenerated. In ozone reactions, phenoxyl radicals and $O_2^{\bullet-}$ can be generated side by side (see above), and cage recombination can lead to a hydroperoxide (d'Alessandro et al., 2000) [reactions (37) and (38)].

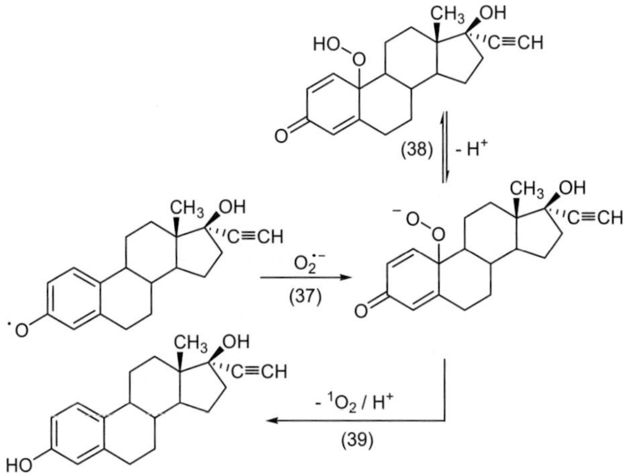

This must be far less ozone reactive than the phenol/phenolate system and escapes subsequent ozonolysis at moderate ozone doses. Out of equilibrium (38), $^1O_2$ is slowly eliminated [reaction (39)] regenerating the starting material. Complete mineralisation of 17α-ethinyloestradiol as discussed by Beltrán et al., (2009) is not required for elimination of its oestrogenic activity (Chapter 4).

Most common phenolic micropollutants are bisphenol A, t-butylphenol as well as octyl- and nonylphenols (cf. Chapter 4).

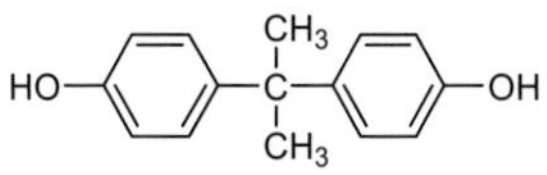

Bisphenol A 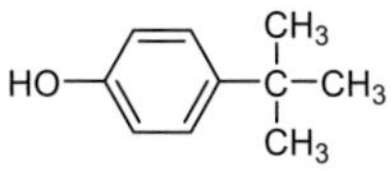   **t-Butylphenol**

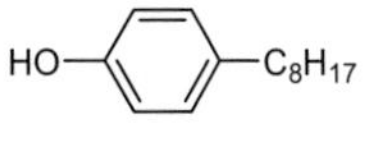

Octylphenol 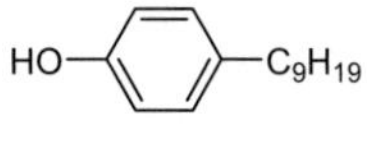   **Nonylphenol**

Most of the bisphenol A is used for the production of polycarbonate and epoxy resins, flame retardants and other speciality products (Staples et al., 1998). Bisphenol A is often detected in surface waters (Kolpin *et al.*, 2002). The transformation of bisphenol A by ozone leads to the expected products as described by the ozone-phenol reactions discussed above (Deborde *et al.*, 2008). Alkyl phenols are degradation products of the alkylphenolethoxylates. These tensides are produced worldwide at 650,000 t/a, 90% of which are nonylphenolethoxylates. The technical product is a mixture of many isomers the distribution of which depends on the manufacturing process (Günther, 2002) (see also Chapter 4).

2-*t*-Butyl-4-methoxyphenol ("butylated hydroxyanisol") (Lau *et al.*, 2007) is a commonly used antioxidant additive in food. It has some endocrine properties and has been found in river waters in Italy as the most abundant phenolic compound (Davi & Gnudi, 1999). Its degradation by ozone has found some attention (Lau *et al.*, 2007).

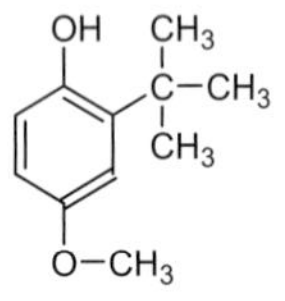

2-*t*-Butyl-4-methoxyphenol

The ozone chemistry of the antibiotic vancomycin may be largely governed by an attack at the phenolic moieties, but there is also a secondary amino group that can serve as an additional site of ozone attack (for a rate constant see Table 7.1). A single ozone attack destroys its biological activity (Dodd, 2008).

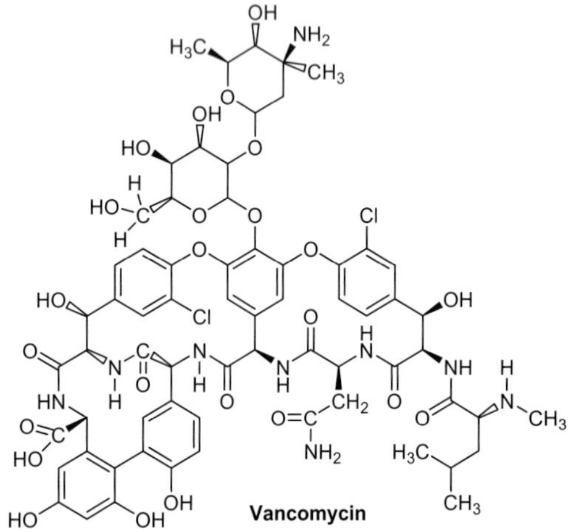

Vancomycin

There are further antibiotics containing a phenolic function, tetracycline and amoxillin (for rate constants see Table 7.1). Their biological activity is destroyed by a single ozone attack (Dodd, 2008) (Chapter 4).

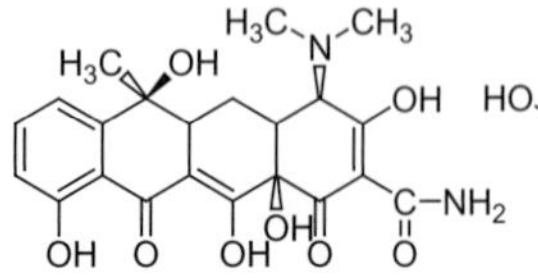

Tetracycline 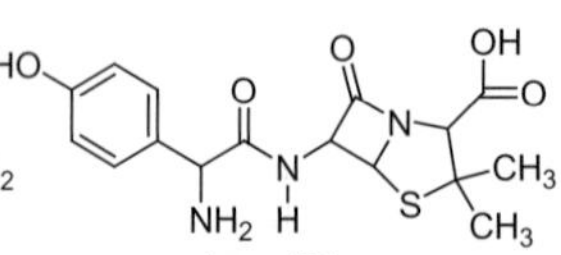   **Amoxillin**

Tetracycline has several ozone-reactive sites, namely a phenolic ring, enolic functions and an exocyclic tertiary amino group. The pK_a values for the tertiary amine and the phenol groups are at 7.7 and 9.7, respectively. Details of the preferred sites of attack as a function of pH are not fully established, but these functions and the enolic groups are obvious candidates.

Amoxillin has three ozone-reactive sites, the phenol, amine and sulfide groups. It has pK_a values at 2.68 (carboxylic acid), 7.49 (amine) and 9.68 (phenol). Rate constants as a function of pH have not been reported, but it is most likely that at the pH values relevant for water ozonation, amoxillin degradation by ozone is largely governed by reactions with the deprotonated phenol site in equilibrium (Andreozzi et al., 2005). The amine (Chapter 8) and sulfide (Chapter 9) sites would react more slowly.

Aflatoxins belong to the group of mycotoxins which are produced by a number of fungi such as *Aspergillus*, *Penicillum* and *Fuscarum*. The structures of the most abundant ones are shown below. Their aromatic rings carry strongly electron-donating substituents, that is, their rate constants toward ozone must be high, most likely close to that of 1,3,5-trimethoyxbenzene (Table 7.1). As a consequence, they are readily degraded (detoxified) by ozone (McKenzie et al., 1997). Their olefinic functions are all deactivated by electron-withdrawing substituents (Chapter 6), and thus their contributions to detoxifications by ozone are expected to be minor.

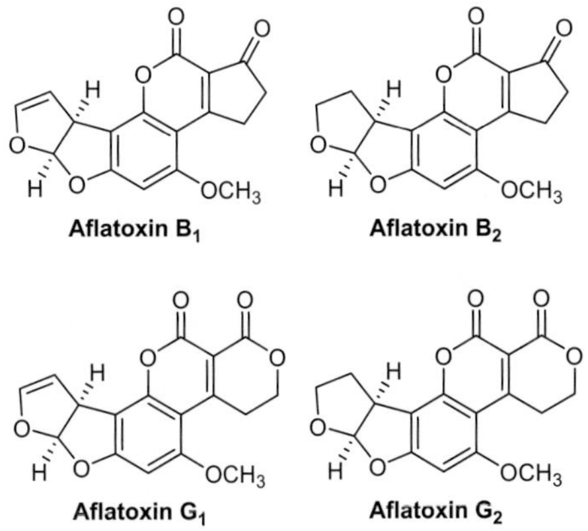

Aflatoxin B₁

Aflatoxin B₂

Aflatoxin G₁

Aflatoxin G₂

Naphthalene sulfonic acid derivatives are used as textile auxiliaries. Some aspects of their degradation by ozone have been addressed (Babuna et al., 2009). The reaction of ozone with such compounds is moderately fast, near $10^3 \, M^{-1} \, s^{-1}$, if the parent, naphthalene, is a good guide (Table 7.1). Naxopren is an antiphlogistic pharmaceutical with a naphthalene moiety, additionally activated by a methoxy and an alkyl group. This leads to a marked enhancement of the rate constant (Table 7.1) (Huber et al., 2005).

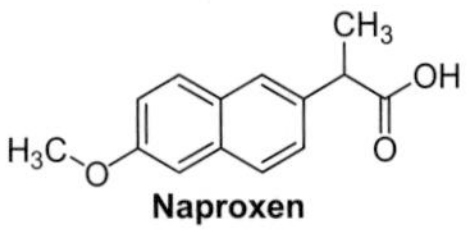

Naproxen

The lipid-lowering drug gemfibrozil contains a benzene ring that is activated by two alkyl and one alkoxyl groups. Its ozone rate constant has been estimated to $\sim 2 \times 10^4 \, M^{-1} \, s^{-1}$, by quantitative structure–activity relationship (QSAR) (Lee & von Gunten, 2012), which lies between naxopren and bezafibrate (Table 7.1). The herbicide methoxychlor also belongs to the group of aromatic compounds activated by alkoxyl groups.

Gemfibrozil **Bezafibrate**

Methoxychlor

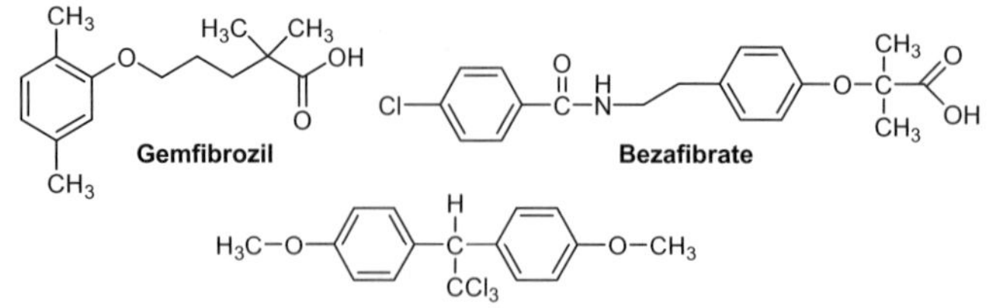

Bezafibrate has two aromatic rings. One carries one alkyl and one alkoxyl group, while the other is deactivated by the chlorine and amide substituents. The ozone attack is expected to occur preferentially on the alkyl-/alkoxyl-activated benzene ring. There is a host of products; muconic and hydroxylated ones seem to dominate (Dantas *et al.*, 2007).

When the number of activating substituents is decreased and/or deactivating substituents such as chlorine are present, the rate of reactions drops substantially (Paragraph 7.1). The analgesic ibuprofen reacts very slowly indeed as do the herbicides MCPA, 2,4-D, mecoprop and clofibric acid (Table 7.1) [for reviews on the degradation of herbicides by ozone and AOPs see Ikehata & El-Din (2005a); Ikehata & El-Din (2005b)].

2-(4-Isobutylphenyl)propionic acid (Ibuprofen, p*K*a = 4.9)

2-Methyl-4-chlorophenoxy-acetic acid (MCPA) **2,4-Dichlorophenoxy-acetic acid (2,4-D)**

2-Methyl-4-chlorophenoxy-propionic acid (Mecoprop) **2-(4-Chlorophenoxy)-2-methyl-propionic acid (Clofibric acid)**

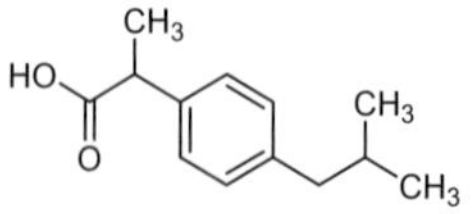

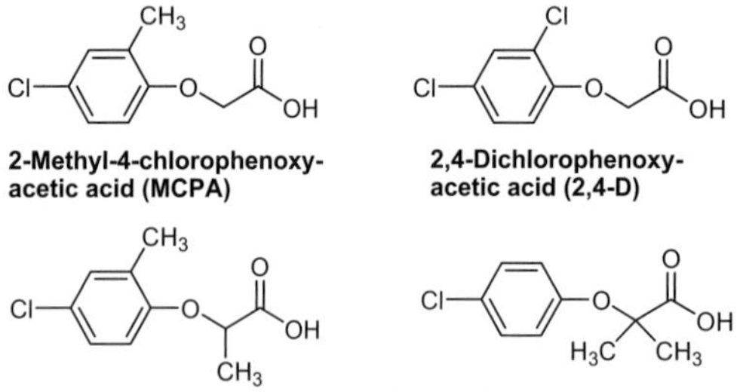

With clofibrate, the chloride yield is near 40% of ozone consumption [Sein *et al.*, (2006), unpublished results]. Ozone-refractory chlorinated products do not seem to be uncommon, as with 4-chlorophenol incomplete chloride release is also observed (Andreozzi & Marotta, 1999).

The musk fragrances galaxolide and tonalide have the same molecular weight but differ markedly in their ozone rate constants (Table 7.1).

Tonalide (AHTN) **Galaxolide (HHCB)**

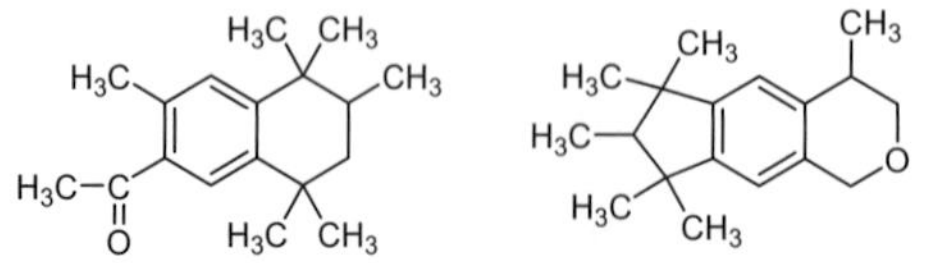

The reason for this has been discussed above. In a bubble column, the disappearance of the less reactive one, tonalide, is partly due to stripping effects (Nöthe *et al.*, 2007).

Aromatic compounds

X-ray contrast media such as amidotrizoic acid, iomeprol, iopamidol and iopromide are abundant micropollutants (Schulz et al., 2008). All contain a benzene ring with three iodine substituents to provide the x-ray response and further hydrophilic substituents to guarantee high water solubility.

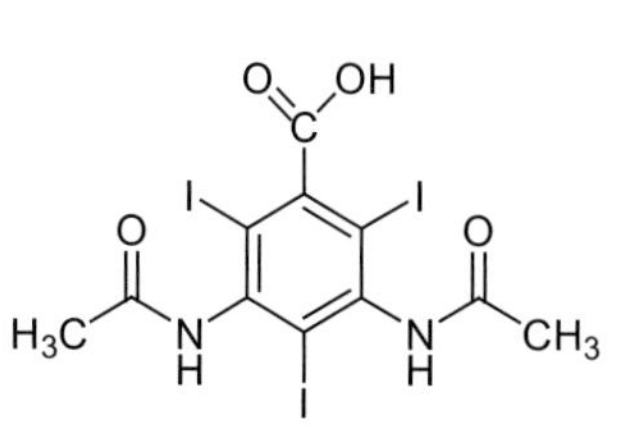

Amidotrizoic acid

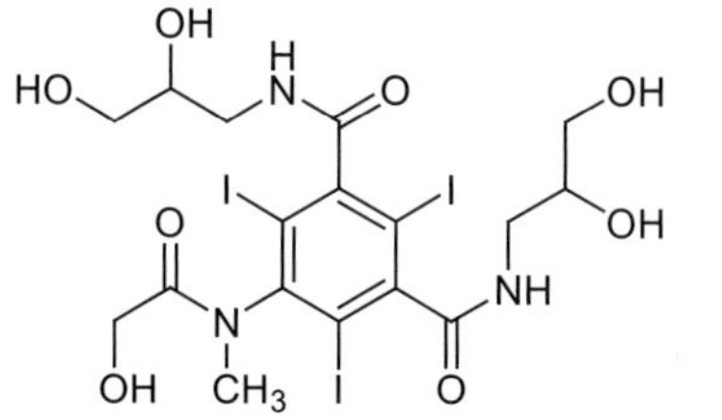

Iomeprol

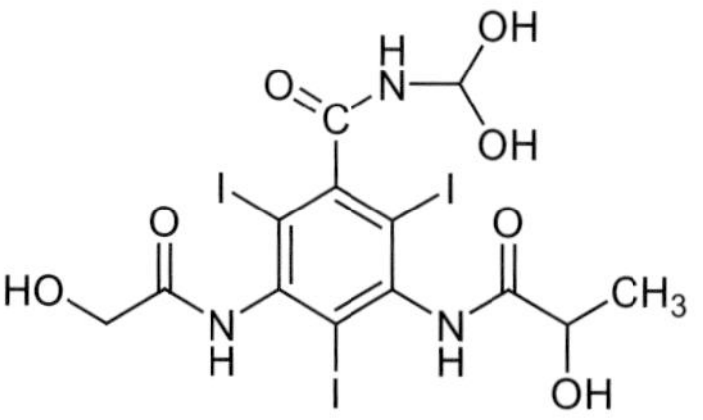

Iopamidol

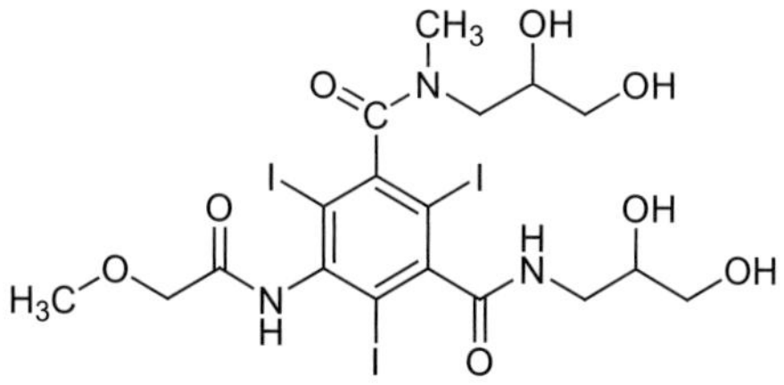

Iopromide

All substituents are electron-withdrawing, and thus their ozone rate constants are all <0.8 M^{-1} s^{-1} (Table 7.1). This agrees with a more recent iomeprol study (Seitz et al., 2006). Upon ozone treatment of wastewater, they are only eliminated via the •OH route. Therefore high doses of ozone are required for their elimination (Chapter 3). They share this problem with activated carbon. Here, a high dosage of powdered activated carbon is required for the adsorption of these highly water-soluble compounds (Nowotny et al., 2007) (for GAC data see (Seitz et al., 2006).

Benzophenone is used for protection against UV radiation in perfumes, soaps, colours and plastic packaging material. Its ozone rate constant must be similarly low as that of the structurally related analgesic ketoprofen [$k = 0.4$ M^{-1} s^{-1}, Real et al. (2009)].

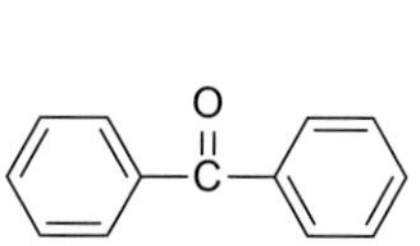

Benzophenone

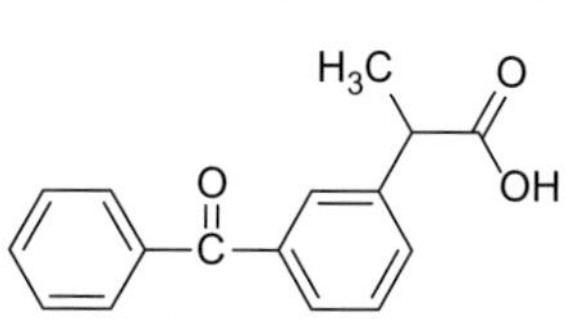

Ketoprofen

Polycyclic aromatic compounds are poorly soluble in water (Schwarzenbach et al., 2010) and are not expected to play a major role as contaminants in water. Typical examples are naphthalene (Legube et al., 1983, 1986), phenanthrene, pyrene, benzo[a]pyrene and the chlorinated biphenyls.

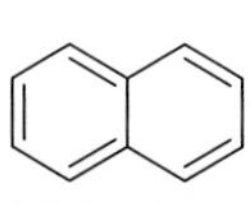

Naphthalene 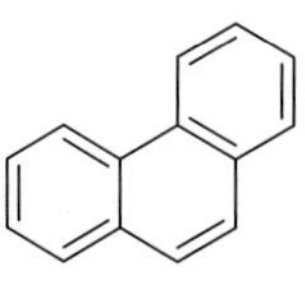   **Phenanthrene** 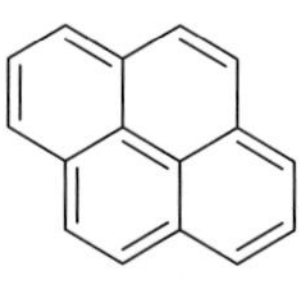   **Pyrene** 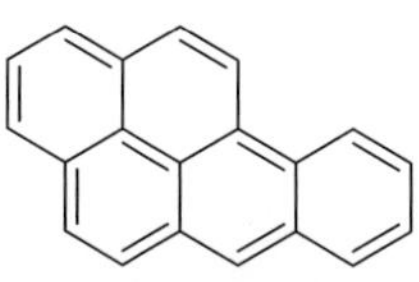   **Benzo[a]pyrene** 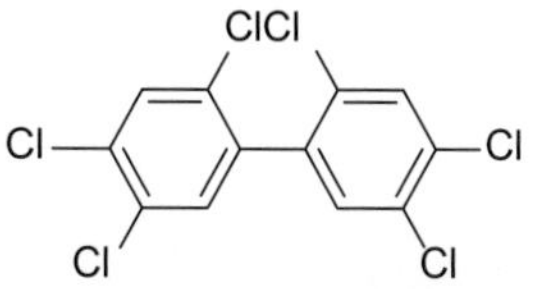   **2,2',4,4',5,5'-Hexachloro-biphenyl**

Ozone rate constants with polycyclic aromatic compounds are high [for naphthalene see Table 7.1; missing in this table are the rate constants for phenanthrene ($k = 1.5 \times 10^4$ M^{-1}s^{-1}), pyrene ($k = 4 \times 10^4$ M^{-1}s^{-1}) and benzo[a]pyrene ($k = 6 \times 10^3$ M^{-1}s^{-1}) reported by Butković et al. (1983)].

Of the aromatic taste and odour compounds (Peter & von Gunten, 2007), only 2,6-di-*t*-butyl-4-methylphenol (BHT), leached from polyethylene pipes, reacts rapidly with ozone, while 2,4,6-tribromoanisole and 2,4,6-trichloroanisole, derived by methylation from the corresponding phenols by micro-organisms, are practically ozone-refractory (Table 7.1).

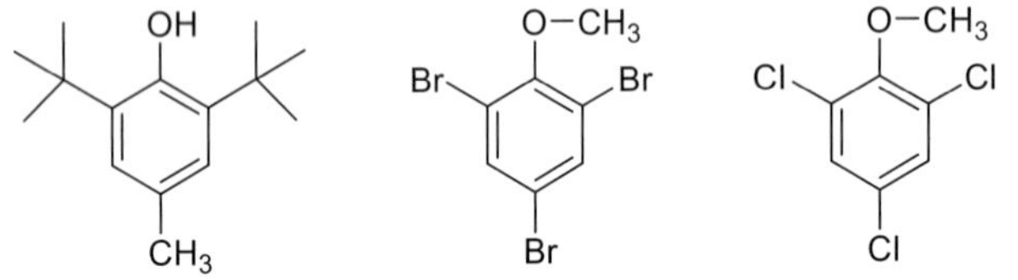

2,6-Di-*t*-butyl-4-methylphenol (BHT)   **2,4,6-Tribromo-anisole (TBA)**   **2,4,6-Trichloro-anisole (TCA)**

For their taste and odour properties see Chapter 5.

Chapter 8
Nitrogen-containing compounds

8.1 REACTIVITY OF NITROGEN-CONTAINING COMPOUNDS

For aliphatic amines to be ozone-reactive, the lone electron pair at nitrogen has to be accessible. Protonation of aliphatic amines or complexation of the nitrogens as in Fe(III)–EDTA or Fe(III)–DTPA eliminates the ozone reactivity or at least strongly reduces it (Table 8.1).

Table 8.1 Compilation of ozone rate constants of nitrogen-containing compounds in aqueous solution. Rate constants for sulfur-containing amino acids are given in Chapter 9. Published rate constants are rounded to significant figures

Compound	pK_a	pH	k/M^{-1} s^{-1}	Reference
Acebutolol	9.2		2.9×10^5	Benner et al., 2008
		7	1.9×10^3	Benner et al., 2008
protonated			60	Benner et al., 2008
N-Acetylglycine		3.7	0.3	Pryor et al., 1984
N-α-Acetylhistidine	7.3		8.4×10^5	Pryor et al., 1984
N-α-Acetyllysine	10.5		1.0×10^5	Pryor et al., 1984
N-ε-Acetyllysine	9.46		2.4×10^4	Pryor et al., 1984
protonated			no reaction	Pryor et al., 1984
N-(4)-Acetylsulfamethoxazole	5.5		260	Dodd et al., 2006a
protonated			20	Dodd et al., 2006a
Alachlor			3.4	de Laat et al., 1996
Alanine	9.87		6.4×10^4	Hoigné & Bader, 1983b
			7.6×10^4	Pryor et al., 1984
			2.8×10^4	Muñoz & von Sonntag, 2000b
protonated			no reaction	Pryor et al., 1984
			~3×10^{-3}	Hoigné & Bader, 1983b
β-Alanine			6.2×10^4	Hoigné & Bader, 1983b

(Continued)

Table 8.1 Compilation of ozone rate constants of nitrogen-containing compounds in aqueous solution. Rate constants for sulfur-containing amino acids are given in Chapter 9. Published rate constants are rounded to significant figures (*Continued*)

Compound	pK_a	pH	k/M^{-1} s^{-1}	Reference
Amikacin	6.7, 8.4 8.4, 9.7	7	1.8×10^3	Dodd, 2008
4-Aminophenylmethyl sulfone			4.7×10^4	Dodd *et al.*, 2006a
Ammonia	9.24		20	Hoigné & Bader, 1983b
			44	Garland *et al.*, 1980
protonated			no reaction	Hoigné & Bader, 1983b
Aniline	4.63		9.0×10^7	Hoigné & Bader, 1983
			3.8×10^7	Tekle-Röttering *et al.*, 2011
		6.5	1.4×10^7	Pierpoint *et al.*, 2001
		1.5	5.9×10^4	Pierpoint *et al.*, 2001
Arginine	8.99		5.7×10^4	Pryor *et al.*, 1984
Asparagine monoanion	2.0, 8.8		4.2×10^5	Pryor *et al.*, 1984
Aspartate ion	9.82		4.1×10^4	Pryor *et al.*, 1984
protonated			1.0	Pryor *et al.*, 1984
Atenolol	9.6		6.3×10^5	Benner *et al.*, 2008
protonated			110	Benner *et al.*, 2008
Atrazine		2	24	Yao & Haag, 1991
			2.3	Xiong & Graham, 1992
			5.65	de Laat *et al.*, 1996
			6.0	Acero *et al.*, 2000
Azithromycin	8.7, 9.5	7	1.1×10^5	Dodd *et al.*, 2006a
Azobenzene		2	220	Yao & Haag, 1991
1-(2-Benzaldehyde)-4-hydro(1*H*,3*H*)-quinazoline-2-one			3.0	Kosjek *et al.*, 2009
Benzotriazole	1.6, 8.2		35	Lutze *et al.*, 2011a
			36	Vel Leitner & Roshani, 2010
anion			2650	Lutze *et al.*, 2011a
Benzylamine	9.33		6.3×10^4	Pryor *et al.*, 1984
Butylamine	10.77		1.2×10^5	Pryor *et al.*, 1984
			1.7×10^5	Hoigné & Bader, 1983b
protonated			≤ 0.1	Pryor *et al.*, 1984
s-Butylamine	10.83		5.2×10^4	Pryor *et al.*, 1984
protonated			no reaction	Pryor *et al.*, 1984
3-Chloroaniline		6	7.84×10^6	Pierpoint *et al.*, 2001
4-Chloroaniline		6	1.04×10^7	Pierpoint *et al.*, 2001
5-Chlorobenzotriazole	−0.04, 7.5		13	Lutze *et al.*, 2011a
anion			630	Lutze *et al.*, 2011a

(*Continued*)

Nitrogen-containing compounds

Table 8.1 Compilation of ozone rate constants of nitrogen-containing compounds in aqueous solution. Rate constants for sulfur-containing amino acids are given in Chapter 9. Published rate constants are rounded to significant figures (*Continued*)

Compound	pK_a	pH	$k/M^{-1} s^{-1}$	Reference
Chlorotoluron			50.5	de Laat *et al.*, 1996
			394	Benitez *et al.*, 2007
Ciprofloxacin	6.2, 8.8		9×10^5	Dodd *et al.*, 2006a
monoprotonated			7.5×10^3	Dodd *et al.*, 2006a
Clarithromycin		7	4×10^4	Lange *et al.*, 2006
Creatine		2	~0.5	Hoigné & Bader, 1983b
Creatinine		6	~2	Hoigné & Bader, 1983b
Cyclohexanemethylamine	10.3		7.1×10^4	Dodd *et al.*, 2006a
protonated			<1	Dodd *et al.*, 2006a
Cyclohexylamine	10.6		4.9×10^4	Dodd *et al.*, 2006a
protonated			<1	Dodd *et al.*, 2006a
Cyclophosphamide			2.27	Fernández *et al.*, 2010
			3.0	Garcia-Ac *et al.*, 2010
Deethylatrazine		2	0.18	Acero *et al.*, 2000
Deethyldeisopropylatrazine		2	<0.1	Acero *et al.*, 2000
Deisopropylatrazine		2	3.1	Acero *et al.*, 2000
2,4-Diamino-5-methylpyrimidine	3.2, 7.1		1.3×10^6	Dodd *et al.*, 2006a
monoprotonated			2.9×10^3	Dodd *et al.*, 2006a
diprotonated			5.0×10^2	Dodd *et al.*, 2006a
1,4-Diazabicyclo[2.2.2]octane (DABCO)	3.0, 8.2		3.2×10^6	Muñoz & von Sonntag, 2000b
monoprotonated			3.5×10^3	Muñoz & von Sonntag, 2000b
Diazepam			0.75	Huber *et al.*, 2003
Dichloramine			1.3	Haag & Hoigné, 1983b
Diclofenac			1.8×10^4	Vogna *et al.*, 2004
			6.8×10^5	Sein *et al.*, 2008
			~10^6	Huber *et al.*, 2003
Diethylamine	10.49		6.2×10^5	Pryor *et al.*, 1984
			9.1×10^5	Muñoz & von Sonntag, 2000b
protonated			11 ± 6	Pryor *et al.*, 1984
Diethylenetriaminepentaacetic acid (DTPA)	−0.1, 0.7 1.6, 2.0 2.6, 4.3 8.6, 10.5			
CaDTPA^{3-}			6200	Stemmler *et al.*, 2001
Fe(III)DTPA^{2-}			<10	Stemmler *et al.*, 2001
Fe(III)(OH)DTPA^{3-}			2.4×10^5	Stemmler *et al.*, 2001

(*Continued*)

Table 8.1 Compilation of ozone rate constants of nitrogen-containing compounds in aqueous solution. Rate constants for sulfur-containing amino acids are given in Chapter 9. Published rate constants are rounded to significant figures (*Continued*)

Compound	pK_a	pH	k/M^{-1} s^{-1}	Reference
ZnDTPA^{3-}			3500	Stemmler et al., 2001
N,N-Dimethylacetamide		7	0.7	Pryor et al., 1984
Dimethylamine	11		1.2×10^7	Hoigné & Bader, 1983b
protonated			<0.1	Hoigné & Bader, 1983b
4-Dimethylaminoantipyrine	5.1		1.7×10^8	Lee et al., 2007a
protonated			1.7×10^5	Lee et al., 2007a
3-(Dimethylaminomethyl) indole	10.0		2×10^9	Lee et al., 2007a
protonated			2.4×10^6	Lee et al., 2007a
Dimethylaniline	5.1		2×10^9	Lee et al., 2007a
5,6-Dimethylbenzotriazole	~2		880	Lutze et al., 2011a
anion	~9.28		3.3×10^4	Lutze et al., 2011a
Dimethylchloramine			1.9×10^3	Haag & Hoigné, 1983b
N,N-Dimethylcyclohexylamine	10.7		3.7×10^6	Dodd et al., 2006a
protonated			<1	Dodd et al., 2006a
Dimethylethanolamine	9.2		1.1×10^7	Lee et al., 2007a
protonated			<0.2	Lee et al., 2007a
Dimethylformamide			0.24	Lee et al., 2007a
3,5-Dimethylisoxazole			5.4×10^1	Dodd et al., 2006a
Dimethylsulfamide (DMS)		7	~20	von Gunten et al., 2010
Diuron			14.7	de Laat et al., 1996
			16.5	Benitez et al., 2007
		2.4	13.3	Chen et al., 2008
Enrofloxacin	6.1, 7.7	7	1.5×10^5	Dodd et al., 2006a
Ethylenediaminetetraacetic acid (EDTA)	0.26, 0.96, 2.0, 2.7, 6.2, 10.2		3.3×10^6	Muñoz & von Sonntag, 2000b
monoprotonated (HEDTA)			1.6×10^5	Muñoz & von Sonntag, 2000b
CaEDTA^{4-}		6	~1×10^5	Muñoz & von Sonntag, 2000b
Fe(III)EDTA^{3-}		6	320	Muñoz & von Sonntag, 2000b
Ethyl N-piperazine-carboxylate	8.3		1.1×10^6	Dodd et al., 2006a
protonated			<1	Dodd et al., 2006a
2-Fluoroaniline		6	7.3×10^6	Pierpoint et al., 2001
3-Fluoroaniline		6	8.1×10^6	Pierpoint et al., 2001
4-Fluoroaniline		6	1.2×10^7	Pierpoint et al., 2001
Glutamate ion	9.47		2.6×10^4	Pryor et al., 1984
protonated			0.2	Pryor et al., 1984
Glycine	9.78		1.3×10^5	Hoigné & Bader, 1983b

(*Continued*)

Table 8.1 Compilation of ozone rate constants of nitrogen-containing compounds in aqueous solution. Rate constants for sulfur-containing amino acids are given in Chapter 9. Published rate constants are rounded to significant figures (*Continued*)

Compound	pK_a	pH	k/M^{-1} s^{-1}	Reference
			2.1×10^5	Pryor et al., 1984
			2.1×10^5	Muñoz & von Sonntag, 2000b
protonated			0.05	Hoigné & Bader, 1983b
Hexahydro-1,3,5-trinitro-1,3,5-triazin (RDX)		2.4	0.97	Chen et al., 2008
Histidine	6.0		2.1×10^5	Pryor et al., 1984
Hydroxylamine	6.0		2×10^4	Hoigné et al., 1985
Imidazole	6.95		2.4×10^5	Pryor et al., 1984
			4×10^5	Hoigné & Bader, 1983b
protonated			22	Hoigné & Bader, 1983b
Iminodiacetic acid	9.12		2.8×10^6	Muñoz & von Sonntag, 2000b
3-Iodoaniline		6	7.4×10^6	Pierpoint et al., 2001
4-Iodoaniline		6	9.2×10^6	Pierpoint et al., 2001
Isoleucine	9.76		5.6×10^4	Pryor et al., 1984
2-Isopropyl-3-methoxypyrazine (IPMP)			50	Peter & von Gunten, 2007
Isoproturon			141	de Laat et al., 1996
			2218	Benitez et al., 2007
Leucine	9.74		5.3×10^4	Pryor et al., 1984
Lincomycin	7.79		2.8×10^6	Qiang et al., 2004
protonated			3.3×10^5	Qiang et al., 2004
Linuron	2.4		3	de Laat et al., 1996
			1.9	Benitez et al., 2007
			2.6	Chen et al., 2008
protonated			no reaction	Pryor et al., 1984
Lysine	9.18		3.1×10^4	Pryor et al., 1984
Methotrexate			>10^4	Garcia-Ac et al., 2010
2-Methoxyaniline		6	1.7×10^7	Pierpoint et al., 2001
N-Methylacetamide			0.6	Pryor et al., 1984
Methylamine	10.6		<1×10^5	Hoigné & Bader, 1983b
3-Methylaniline		6	1.9×10^7	Pierpoint et al., 2001
5-Methylbenzotriazole	~1.65, ~8.5		164	Lutze et al., 2011a
anion			1.0×10^4	Lutze et al., 2011a
Methylchloramine			810	Haag & Hoigné, 1983a
Methyldichloramine			<1×10^{-2}	Haag & Hoigné, 1983a
4-Methylimidazole	7.52		3.1×10^6	Pryor et al., 1984

(*Continued*)

Table 8.1 Compilation of ozone rate constants of nitrogen-containing compounds in aqueous solution. Rate constants for sulfur-containing amino acids are given in Chapter 9. Published rate constants are rounded to significant figures (*Continued*)

Compound	pK_a	pH	k/M^{-1} s^{-1}	Reference
1-Methylpyrrolidine	10.2		2.0×10^6	Dodd et al., 2006a
protonated			<1	Dodd et al., 2006a
4-Methylsulfonylaniline		6	4.7×10^4	Pierpoint et al., 2001
Metolachlor			3	de Laat et al., 1996
Metoprolol	9.7		8.6×10^5	Benner et al., 2008
protonated			330	Benner et al., 2008
		7	2.0×10^3	Benner et al., 2008
Monobromamine			40	Haag et al., 1984
Monochloramine			26*	Haag & Hoigné, 1983
Morpholine	8.36		1.8×10^5	Tekle-Röttering et al., 2011
Nitrilotriacetic acid (NTA)	1.65, 2.94, 10.33		9.8×10^5	Muñoz & von Sonntag, 2000b
		2	830	Games & Staubach, 1980
3-Nitroaniline		6	2.4×10^6	Pierpoint et al., 2001
4-Nitroaniline		6	1.4×10^5	Pierpoint et al., 2001
Nitroform anion			1.4×10^4	Flyunt et al., 2003a
N-Nitrosodimethylamine			~10	Hoigné & Bader, 1983a
			0.052	Lee et al., 2007b
1-Phenoxy-3-propanol		7	320	Benner et al., 2008
Phenylalanine	9.24		3.8×10^5	Pryor et al., 1984
Proline	10.6		4.3×10^6	Pryor et al., 1984
Prometon	4.3	3	~2	Chen et al., 2008
		7.5	<12	Chen et al., 2008
Propachlor			0.94	de Laat et al., 1996
Propranolol	9.5		~1×10^5	Benner et al., 2008
protonated			~1×10^5	Benner et al., 2008
Propylamine		2	$\leq 1 \times 10^{-2}$	Hoigné & Bader, 1983b
Pyridine	5.14		~3	Hoigné & Bader, 1983b
			2.0	Andreozzi et al., 1991
protonated			0.01	Hoigné & Bader, 1983b
Quinoline			51	Wang et al., 2004
Roxithromycin	9.2	7	6.3×10^4	Dodd et al., 2006a
Serine	9.21		1.29×10^5	Pryor et al., 1984
protonated			7±4	Pryor et al., 1984
Simazine		4.3	4.8	Yao & Haag, 1991
		2	3.3	Xiong & Graham, 1992
Spectinomycin	6.8, 8.8		1.3×10^6	Qiang et al., 2004

(*Continued*)

Table 8.1 Compilation of ozone rate constants of nitrogen-containing compounds in aqueous solution. Rate constants for sulfur-containing amino acids are given in Chapter 9. Published rate constants are rounded to significant figures (*Continued*)

Compound	pK_a	pH	k/M^{-1} s^{-1}	Reference
protonated			3.3×10^5	Qiang et al., 2004
Sulfamethoxazole	5.6		4.7×10^4	Dodd et al., 2006a
conjugate base			5.7×10^5	Dodd et al., 2006a
Tetranitromethane			10	Flyunt et al., 2003a
Threonine	9.11		4.6×10^4	Pryor et al., 1984
Tramadol	9.4		1.0×10^6	Zimmermann et al., 2012
protonated			7.7×10^4	Zimmermann et al., 2012
Triethylamine	11.01		2.1×10^6	Pryor et al., 1984
			4.1×10^6	Muñoz & von Sonntag, 2000b
protonated			5 ± 4	Pryor et al., 1984
Trimethoprim	3.2, 7.1	7	2.7×10^5	Dodd et al., 2006a
Trimethylamine	9.76		4.1×10^6	Hoigné & Bader, 1983b
			5.1×10^6	Muñoz & von Sonntag, 2000b
Tryptophan	9.38		7×10^6	Pryor et al., 1984
Tylosin	7.7	7	5.1×10^5	Dodd et al., 2006a
Valine	9.72		6.8×10^4	Pryor et al., 1984
protonated			≤ 3	Pryor et al., 1984

*a rate constant of 7 M^{-1}s^{-1} was determined for substrate decrease

For aromatic amines (cf. Figure 8.1), the aromatic ring is a second potential site of ozone attack.

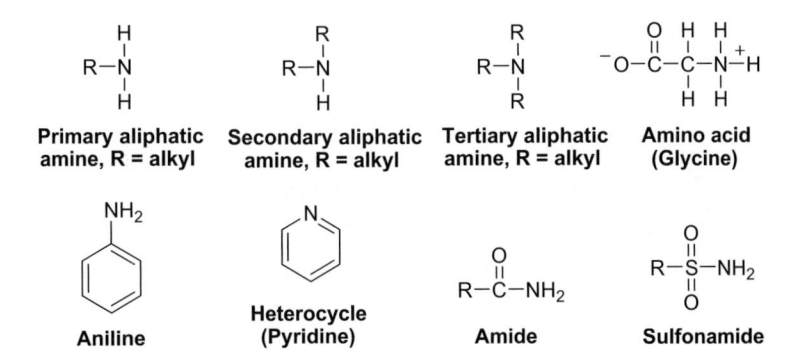

Figure 8.1 Nitrogen-containing compounds.

The reactivity of the nitrogen in aromatic heterocyles is low (cf. pyridine). In amides, the free electron pair is drawn towards the electron-withdrawing carbonyl function with the effect that amides show extremely low rate constants, e.g. *N*-methylacetamide (Table 8.1). In sulfonamides, the electron-withdrawing effect of the neighbouring SO$_2$ group is so strong that they become weak acids [e.g. pK_a(sulfamethoxazole) = 5.7 (Huber *et al.*, 2005)]. With compounds that carry two amino groups such as EDTA (for structure see below) and DABCO, the rate constants are high, when both amino groups are deprotonated, but drops by three orders of magnitude when one of them is protonated (cf. Table 8.1 and Figure 8.2). Apparently, protonation of one of the nitrogens also affects the electron density of the other, thereby reducing its reactivity towards ozone.

DABCO
(1,4-Diazabicyclo[2,2,2]octane)

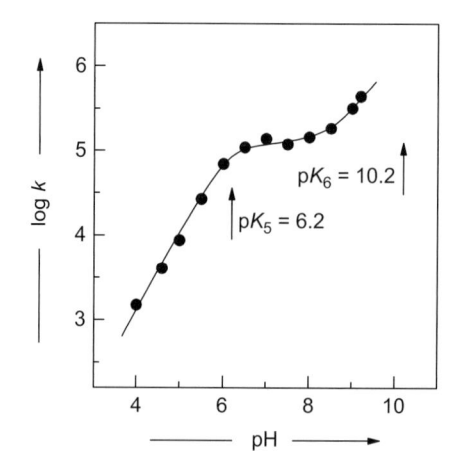

Figure 8.2 Logarithm of the ozone rate constant of ethylenediaminetetraacetic acid (EDTA) as a function of pH according to Muñoz & von Sonntag, 2000b, with permission. For structure of EDTA see Section 8.3.

8.2 GENERAL MECHANISTIC CONSIDERATIONS
8.2.1 Aliphatic amines

Ozone reacts with aliphatic amines and amino acids by adding to the lone pair at nitrogen [reaction (1)].

Ammonia, the parent of the amines, reacts only very slowly with ozone (Table 8.1; for a catalysis by Br$^-$ see Chapter 11). Alkylamines react several orders of magnitude faster. Electron-donating alkyl groups enhance the electron density at nitrogen allowing the electrophilic ozone to undergo the addition reaction (1) more readily. This increase in electron density can be seen from the dependence of the HOMO as a function of the number of alkyl substituents (Figure 8.3).

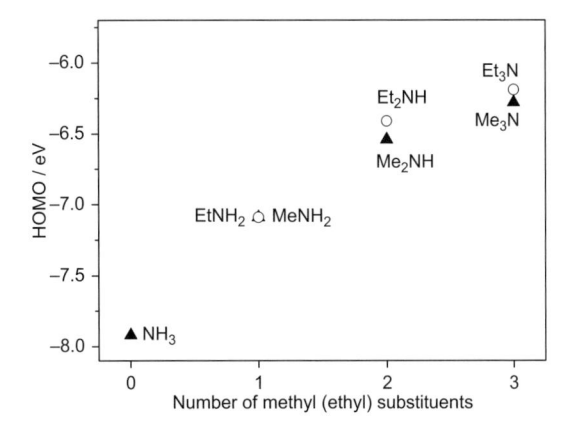

Figure 8.3 Energy of the HOMO of ammonia and its methyl and ethyl derivatives vs. the number of methyl (ethyl) substituents (Naumov & von Sonntag, 2009, unpublished).

An increase in the electron density at nitrogen should correlate with a decrease of the standard Gibbs free energy of reaction (1), and it would be tempting to correlate the rate constants of the reaction of ozone with the standard Gibbs free energy of adduct formation, reaction (1), as has been successfully done in the case of the addition of ozone to aromatic compounds (Chapter 7). A major problem in arriving at reliable data is, however, the fact that a water molecule hydrogen bonded to the nitrogen is replaced by ozone and the loss of this hydrogen bond (ca. 25 kJ mol^{-1}) may have to be taken into account (note that this problem does not arise with aromatic compounds, Chapter 7). Moreover, steric effects seem to play an important role. This is apparent from the drop in the pK_a value on going from the (protonated) dimethylamine (pK_a = 10.64) to trimethylamine (pK_a = 9.8) despite the fact that the nitrogen in trimethylamine has a higher electron density (Figure 8.3).

The ozone adducts to amines may lose O_2 [reaction (2)].

$$R_2N^+(R)-O-O-O^- \xrightarrow{(2)} R_2N^+(R)-O^- + {}^1O_2$$

Since the ozone adduct in its ground state is in its singlet state, the overall spin multiplicity of the products must also be a singlet (spin conservation rule). As a consequence, O_2 is released in its (excited) singlet state [$^1O_2$, $O_2(^1\Delta_g)$], which lies 95.5 kJ mol^{-1} above the (triplet) ground state of O_2 [$^3O_2$, $O_2(^3\Sigma_g^-)$]. The singlet oxygen yields in ozone reactions are typically high (Table 8.2).

With tertiary amines, this reaction results in the formation of N-oxides [reaction (2)] (Muñoz & von Sonntag, 2000b; Lange et al., 2006; Zimmermann et al., 2012) (Table 8.3). For trimethylamine, quantum-chemical calculations indicate that reactions (1) and (2) are both exergonic [$\Delta G^0 = -48$ kJ mol^{-1} and -63 kJ mol^{-1}, respectively; Naumov & von Sonntag (2010, unpublished)].

Table 8.2 Singlet oxygen ($^1O_2$) yields (in mol % of mol ozone consumed) in the reaction of ozone with amines at different pH values. The ratio of substrate concentration to ozone concentration is given in parentheses. According to Muñoz et al. (2001)

Substrate	pH	$^1O_2$ yield/% ([substrate] : [ozone])
Trimethylamine	7	63 (10:1)
	9.5	60 (10:1), 69 (100:1)
Triethylamine	7	No signal (1:1), 42 (10:1), 72 (10:1)
	8	No signal (1:1), 63 (10:1), 80 (100:1)
	9.5	52 (1:1), 73 (10:1), 85 (100:1)
	10.5	59 (1:1)
	11.5	78 (100:1)
DABCO	7	70 (1:1), 80 (10:1)
	9	90 (10:1)
Ethylenediaminetetraacetic acid (EDTA)	4.5	No signal (1:1), 19 (10:1)
	5.5	No signal (1:1), 40 (10:1), 39 (100:1)
	7	39 (1:1), 59 (10:1), 43 (100:1)
	9.5	31 (1:1), 33 (10:1), 37 (100:1)
EDTA/Ca^{2+}	~3	15 (10:1)
EDTA/Fe^{3+}	~3	No signal (1:1), No signal (10:1)
Nitrilotriacetic acid (NTA)	7	No signal (1:1), 18 (10:1)
	9.5	22 (1:1), 21 (10:1)
Diethylamine	9	20 (1:1)
	10.5	24 (1:1), 20 (10:1)
Iminodiacetic acid (IDA)	9	17 (1:1)
	10.5	20 (1:1), 18 (10:1)
Ethylamine	9	No signal (1:1)
	10.5	11 (1:1), 17 (10:1)
Glycine	7	No signal (1:1), (10:1)
	10.5	4 (10:1)
N,N-Diethylaniline	8.5	7 (17:1)
N,N,N',N'-Tetramethyl-phenylenediamine	3.5	9 (3:1)
	6	4 (3:1)

Detailed data on primary and secondary amines are not yet available, but for dimethylamine quantum-chemical calculations indicate that reactions (1) and (2) are also both exergonic ($\Delta G^0 = -41$ kJ mol^{-1} and -74 kJ mol^{-1}, respectively) and that the corresponding N-oxide is only a short-lived intermediate and rearranges into the isomeric hydroxylamine [reaction (3); $\Delta G^0 = -20$ kJ mol^{-1}; Naumov & von Sonntag (2010, unpublished)].

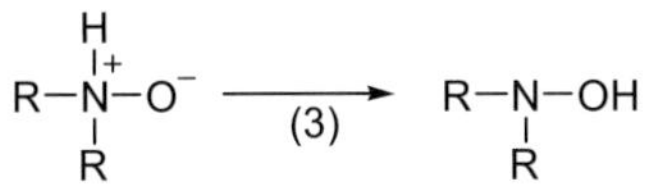

$$\begin{array}{c} H \\ | + \\ R-N-O^- \\ | \\ R \end{array} \xrightarrow{(3)} \begin{array}{c} \\ R-N-OH \\ | \\ R \end{array}$$

Table 8.3 Products of the reaction of ozone with trimethylamine, triethylamine and 1,4-diazabicyclo[2.2.2]octane (DABCO) in % of ozone consumed (Muñoz & von Sonntag, 2000b)

Product	Trimethylamine	Triethylamine	DABCO
Aminoxide	93	85	90
Singlet oxygen	69	85	90
Secondary amine	Not determined	15 (7)[a]	Not observed
Aldehyde	9	19	Not observed

[a]in the presence of tertiary butanol

In the case of propranolol, the hydroxylamine has been identified (Benner & Ternes, 2009b).

In competition with $^1O_2$ elimination [reaction (2)], the ozone adduct may also cleave heterolytically, thereby forming an amine radical cation and an ozonide radical anion, $O_3^{\bullet-}$ [reaction (4)]. The latter reacts with water giving rise to $^\bullet$OH [reaction (5), for details see Chapter 11].

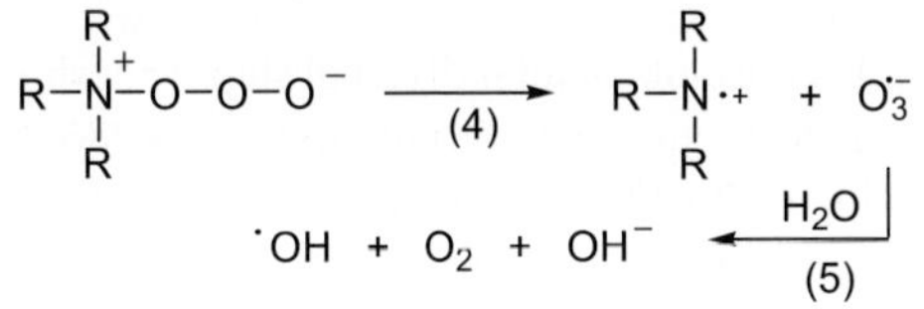

$$\begin{array}{c} R \\ | + \\ R-N-O-O-O^- \\ | \\ R \end{array} \xrightarrow{(4)} \begin{array}{c} R \\ | \\ R-N^{\bullet+} \\ | \\ R \end{array} + O_3^{\bullet-}$$

$$^\bullet OH + O_2 + OH^- \xleftarrow{H_2O}{(5)}$$

In principle, this reaction could also proceed via a direct electron transfer without an adduct as intermediate [reaction (6)].

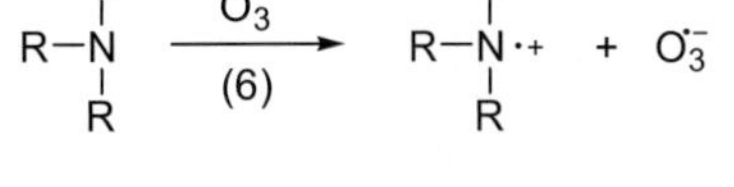

$$\begin{array}{c} R \\ | \\ R-N \\ | \\ R \end{array} \xrightarrow[(6)]{O_3} \begin{array}{c} R \\ | \\ R-N^{\bullet+} \\ | \\ R \end{array} + O_3^{\bullet-}$$

In reactions (4) and (6), two radicals are formed. Whenever two radicals are formed side by side, they are held together by the solvent for a short period of time. If they react with one another at diffusion-controlled rates there is a given chance that they will react within the solvent cage in competition with diffusing out of the cage. The importance of such cage reactions was first addressed by Rabinowitch (Rabinowitch & Wood, 1936; Rabinowitch, 1937). The kinetics for diffusion-controlled reactions were discussed by Noyes (Noyes, 1954). A typical example is the photodecomposition of H_2O_2. In the gas phase, two $^\bullet$OH are formed per photodissociation event. In water, however, there is a 50% probability of the two $^\bullet$OH formed recombining within the solvent cage to H_2O_2. Thus in water, the yield of free $^\bullet$OH is only 50% with respect to the primary $^\bullet$OH yield. Viscosity is one parameter that has a substantial effect on the yield of the cage product and so is temperature [for examples of cage effects in free-radical reactions see (Barrett *et al.*, 1968; Chuang *et al.*, 1974; Mark *et al.*, 1996; Goldstein *et al.*, 2005)]. In the given trialkylamine case, this may lead to the formation of the *N*-oxide [reaction (7)].

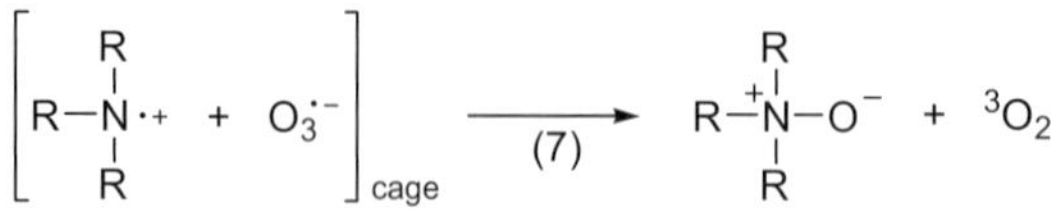

$$[R-\overset{R}{\underset{R}{N}}\overset{\bullet}{^+} + O_3^{\bullet-}]_{cage} \xrightarrow{(7)} R-\overset{R}{\underset{R}{\overset{+}{N}}}-O^- + {}^3O_2$$

In contrast to reaction (2), triplet (ground state) O_2 can be formed, as radicals have a doublet spin multiplicity, and their disproportionation can lead to one molecule in the singlet state (here: the *N*-oxide) and one molecule in the triplet state (here: $^3O_2$). This can explain why in the trialkylamines singlet oxygen yields are below *N*-oxide yields. Such a gap between the yields of singlet oxygen and O-transfer products is not observed with sulfides, but sulfides do not give rise to free-radicals in their reactions with ozone (Chapter 9).

Besides reactions (4) and (5), •OH radicals can also be formed via $O_2^{\bullet-}$ ($O_2^{\bullet-} + O_3 \rightarrow O_2 + O_3^{\bullet-}$, cf. Chapter 13). This ozone-reactive intermediate is formed in the course of the decay of aliphatic amine radical cations [reactions (8)–(11)].

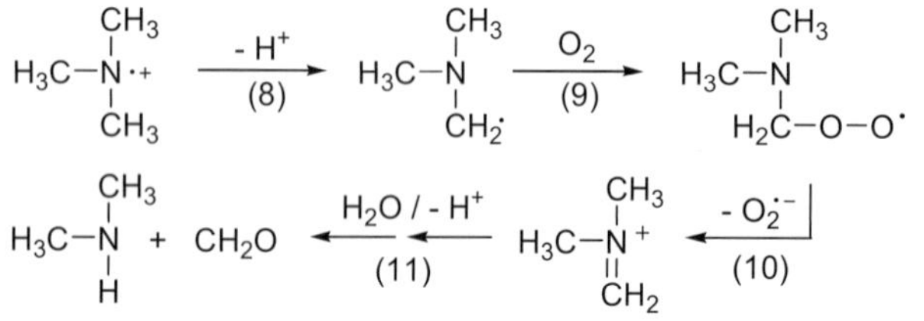

$$H_3C-\overset{CH_3}{\underset{CH_3}{N}}\overset{\bullet}{^+} \xrightarrow[(8)]{-H^+} H_3C-\overset{CH_3}{\underset{CH_2^\bullet}{N}} \xrightarrow[(9)]{O_2} H_3C-\overset{CH_3}{\underset{H_2C-O-O^\bullet}{N}}$$

$$H_3C-\overset{CH_3}{\underset{H}{N}} + CH_2O \xleftarrow[(11)]{H_2O/-H^+} H_3C-\overset{CH_3}{\underset{CH_2}{\overset{+}{N}}}\| \xleftarrow[(10)]{-O_2^{\bullet-}}$$

Details of the kinetics of reactions (8)–(11), including the rate constants of all individual steps (not shown here), have been elucidated with the help of the pulse radiolysis technique (Das & von Sonntag, 1986; Das *et al.*, 1987) (for a potential role of these reactions in contributing to •OH formation in the reaction of ozone with DOM see Chapter 3). The •OH yields in the reaction of ozone with some amines are compiled in Table 8.4.

Table 8.4 Compilation of •OH yields in % of ozone consumed (mol •OH per mol ozone) in the reaction of ozone with amines

Amine	•OH yield/%	Reference
Adenosine	43	Flyunt *et al.*, 2003a
Aniline	27	Tekle-Röttering *et al.*, 2011
Benzotriazole anion	~32	Lutze *et al.*, 2011a
Diclofenac	30	Sein *et al.*, 2008
N,N-Diethylaniline	28	Flyunt *et al.*, 2003a
N,N-Diethyl-*p*-phenylenediamine[a]	23	Jarocki, 2011, unpublished
5,6-Dimethylbenzotriazole anion	~20	Lutze *et al.*, 2011a
5-Methylbenzotriazole anion	~28	Lutze *et al.*, 2011a
Morpholine	33	Tekle-Röttering *et al.*, 2011

(*Continued*)

Table 8.4 Compilation of •OH yields in % of ozone consumed (mol •OH per mol ozone) in the reaction of ozone with amines (*Continued*)

Amine	•OH yield/%	Reference
o-Phenylenediamine	30	Flyunt et al., 2003a
N,N,N,N-Tetramethylphenylenediamine	68[b]	Flyunt et al., 2003a
Trimethylamine	15	Flyunt et al., 2003a

[a]see Chapter 2, [b]for a *caveat* see Flyunt et al. (2003)

In the reaction of ozone with activated carbon, •OH radicals are formed (Jans & Hoigné, 1998). This reaction has been attributed to the presence of nitrogen-containing sites in the activated carbon (Sánchez-Polo et al., 2005), and when these sites are oxidised by ozone, •OH production ceases (Chapter 3).

8.2.2 Aromatic amines (anilines)

The reaction of ozone with aniline is very fast (Table 8.1). The rate constant determined by competition with buten-3-ol ($k = 3.8 \times 10^7$ M^{-1}s^{-1}) may still be somewhat too high, because <0.5 mol aniline are destroyed per mol ozone consumed. Aniline shares this phenomenon with diclofenac, and a potential explanation of this has been made below. A rate constant just above 10^7 M^{-1} s^{-1} would be compatible with an addition to the strongly activated aromatic ring [Chapter 7, e.g. reaction (12)], but an addition to the nitrogen must also be envisaged [reactions (13)–(15)].

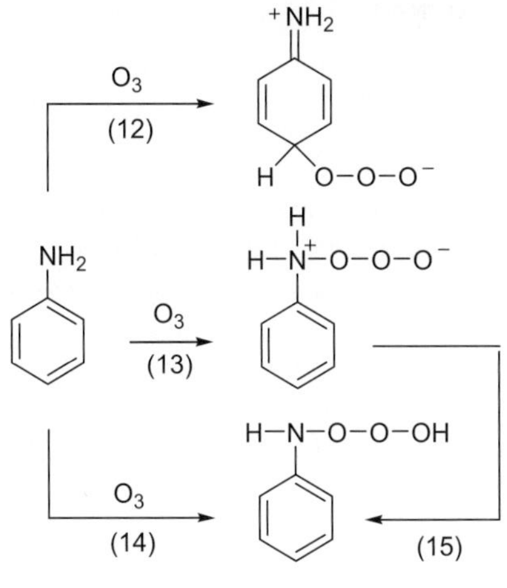

Quantum-chemical calculations indicate that addition to the ring is markedly exergonic [reaction (12), $\Delta G^0 = -53$ kJ mol^{-1}, Naumov & von Sonntag (2009, unpublished)]. This value and the above rate constant fall reasonably well on the plot of calculated standard Gibbs free energies vs. the logarithm of the rate constant such as shown in Figure 7.2 (Chapter 7) (Tekle-Röttering et al., 2011). An addition to the nitrogen is slightly endergonic [reaction (13), $\Delta G^0 = +15$ kJ mol^{-1}]. However, it cannot be excluded that an addition to nitrogen is accompanied with a concomitant proton transfer (in analogy to the well-documented proton transfer coupled electron transfer) [reaction (14), $\Delta G^0 = -77$ kJ mol^{-1}]. Thus, a reaction at nitrogen, possibly competing with an addition to the ring, is not unlikely.

Despite the considerable number of papers that deal with the ozone chemistry of aniline (Gilbert & Zinecker, 1980; Dore et al., 1980; Turhan & Uzman, 2007; Caprio & Insola, 1985), a material balance is still missing. Products that are evidently due to an attack at nitrogen are nitrosobenzene, nitrobenzene and azobenzene (after completion of the nitrosobenzene plus aniline reaction). With respect to ozone consumption, their yields are very low (~0.5%, ~0.7% and ~0.02%, respectively; Tekle-Röttering, private communication). This indicates that ozone attack at nitrogen is a very minor reaction.

Concerning the mechanism of their formation, one has to take into account that nitrobenzene is a primary product despite the fact that it requires two oxidation equivalents.

The NO–OO⁻ bond length in the species formed in reaction (13) is long (1.812 Å), and this intermediate would readily release $O_2^{\bullet-}$. Once protonated [cf. reactions (14) and (15)], the intermediate becomes so unstable that upon optimisation of the structure it releases $HO_2^{\bullet}$ [reaction (16)]. A bimolecular decay of the nitroxyl radical, assisted by water [reaction (17)] giving rise to aniline and nitrobenzene is energetically feasible [reaction (18), $\Delta G^0 = -83$ kJ mol^{-1}].

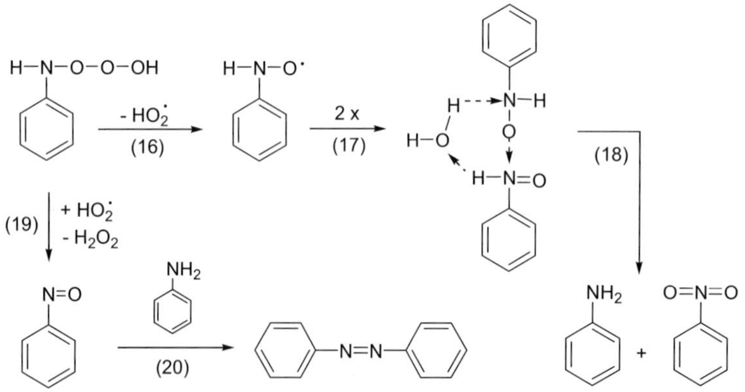

Nitrosobenzene may be formed in the disproportionation of the nitroxyl radical with $HO_2^{\bullet}$ [reaction (19), e.g. in the cage]. The reaction of nitrosobenzene with aniline to form azobenzene [reaction (20)] is a slow postozonation process. The kinetics of this reaction is well documented (Yunes et al., 1975; Dalmagro et al., 1994).

The reactions at nitrogen, although quite interesting as such, are of little importance in comparison to an addition at the ring. Here most product escaped detection, except pyridine-2-carboxylic acid which seems to be derived from the ozonide formed upon closing of the *ortho* ozone adduct (not shown, cf. Chapter 7) [reactions (21)–(24)].

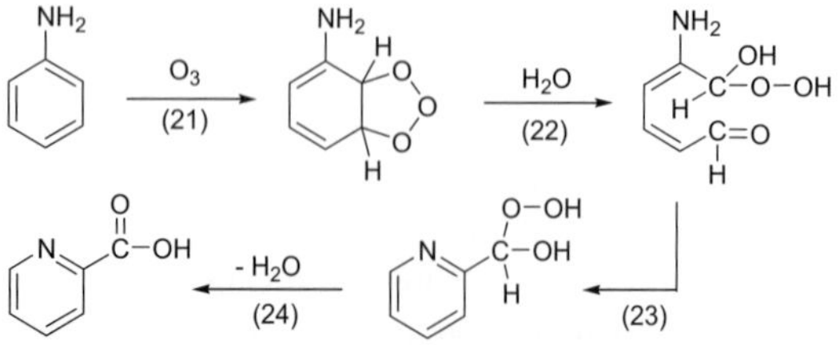

Ozone addition to the ring may be followed by a release of $O_3^{\bullet-}$ [e.g. reaction (25)].

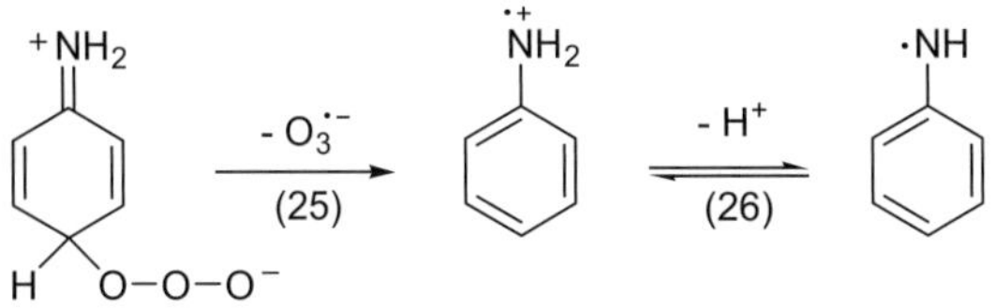

The high •OH yield (precursor: $O_3^{•-}$) given in Table 8.4 must be due to this reaction. Aromatic radical cations are in equilibrium with the corresponding radicals. The pK_a value of the aniline radical cation is 7.05. As expected, substituents experience a strong effect of the pK_a values expressed as a free-energy relationship (27) (Jonsson et al., 1994, 1995).

$$pKa = 7.09 - 3.17 \sum_{i=2}^{6} \sigma_i^+ \qquad (27)$$

Aniline radical cations are also formed upon •OH attack on aniline (Qin et al., 1985), and products observed there may also be of some relevance for aniline ozonolysis. Beyond those mentioned above, there are no further products at the level of low-molecular-weight compounds, and, most likely, the major products are hidden in the oligomer fraction (condensation products with aniline), which is not yet fully elucidated.

Deficits in the material balance between ozone and amine consumption have been reported for diclofenac, and it has been suggested that there may be a chain reaction that destroys ozone in competition with its reaction with amines (Sein et al., 2008) [reactions (28) and (29)] (see also Chapter 13).

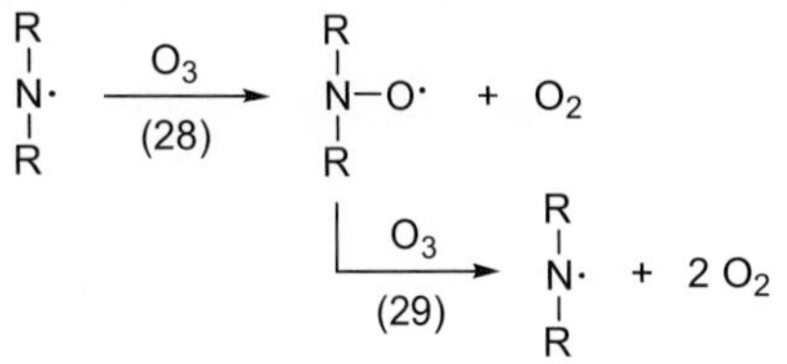

Quantum chemical calculations were carried out with morpholine as a model system. They show that both reactions are exergonic ($\Delta G° = -77$ kcal mol^{-1} and -23 kcal mol^{-1}, respectively, Naumov & von Sonntag, 2011, unpublished). A stable nitroxyl radical, TEMPO, reacts indeed very fast with ozone ($k = 1.3 \times 10^7$ M^{-1} s^{-1}) (Muñoz & von Sonntag, 2000a).

8.2.3 Nitrogen-containing heterocyclic compounds

Pyridine reacts with ozone (Table 8.1) about as fast as benzene ($k = 2$ M^{-1} s^{-1}, Table 7.1), many orders of magnitude slower than most other nitrogen-containing compounds. In the presence of tBuOH as •OH scavenger, which substantially protects pyridine from getting additionally degraded, the major product (ca. 80% of pyridine degraded) is the N-oxide (Andreozzi et al., 1991) [reactions (30) and (31)]. This indicates that the major primary site of ozone attack is the nitrogen.

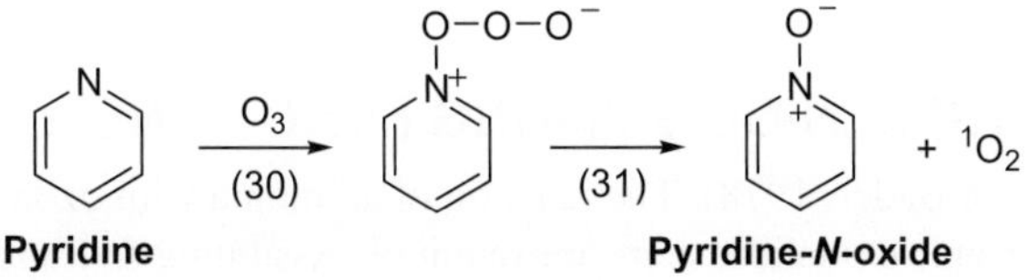

According to quantum-chemical calculations, N-adduct formation [reaction (30)] is exergonic ($\Delta G^0 = -49$ kJ mol^{-1}) and the NO–OO bond in the adduct is very long (2.085 Å, Naumov & von Sonntag 2011, unpublished). Thus, the subsequent release of $^1O_2$ is already preformed in the adduct [reaction (31), $\Delta G^0 = -81$ kJ mol^{-1}]. The low rate constant despite the high driving force for this reaction is analogous to the reaction of ozone with dimethyl sulfoxide (Chapter 9), but is not yet fully understood and may be due to a coulombic repulsion of ozone and the pyridine nitrogen resulting in substantial activation energy of this reaction. A reaction by addition to the ring seems unlikely as the standard Gibbs free energy for addition is too positive ($\Delta G^0 > +158$ kJ mol^{-1}) to account for the observed rate constant (cf. Chapter 7).

Quinoline reacts much faster than pyridine (Table 8.1). Oxidation of intermediates with H_2O_2 after ozonation leads to quinolinic acid as the major product (Andreozzi et al., 1992).

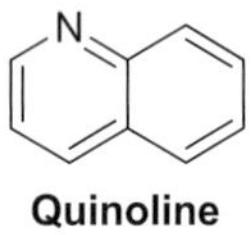

Quinoline

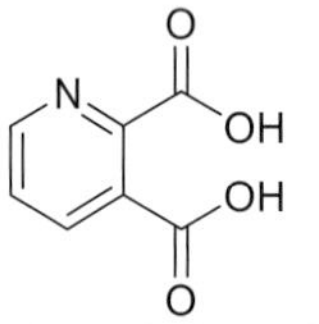

Quinolinic acid

This indicates a change in mechanism. It is now the benzene ring that is mainly attacked (note the high ozone rate constant of naphthalene (Chapter 7, Table 7.1). It seems that the formation of quinoline-N-oxide has not been looked for. Thus, to what extent an attack at nitrogen also takes place is as yet not known.

With atrazine, the ozone rate constant is low (Table 8.1). Ozone attack at the ring can be excluded (cf. pyridine) and ozone attack must be at the nitrogens.

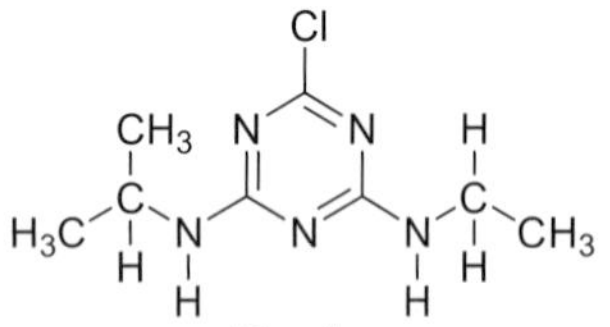

Atrazine

There are two types of nitrogens, and data are not yet available for deciding the preferred site.

2-Isopropyl-3-methoxypyrazine (IPMP) is a potent taste and odour compound (Chapter 5). Similarly to atrazine, ozone attack will probably occur at the nitrogens. An activation by the methoxy group leads to a higher rate constant.

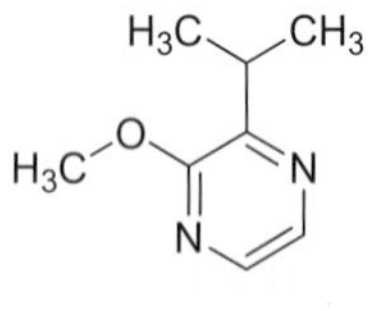

2-Isopropyl-3-methoxy-pyrazine (IPMP)

8.3 MICROPOLLUTANTS WITH NITROGEN-CONTAINING FUNCTIONS

Ammonia is oxidised by ozone to nitrate (Hoigné & Bader, 1978). The reaction of ammonia with ozone is very slow (Table 8.1), and *en route* to nitrate several oxidation steps are required. Assuming O-transfer reactions, these must be hydroxylamine (H$_2$NOH), hyponitrous acid (HNO and its dimer) and nitrite (NO$_2^-$). Of these intermediates, it is known that hydroxylamine (Table 8.1) and nitrite (Chapter 11) react many orders of magnitude faster with ozone than ammonia itself, and thus these products can only attain very low steady states in the course of the reaction. Moreover, HNO is unstable. It decays bimolecularly to N$_2$O, and to what extent this reaction can be intercepted by ozone and the corresponding ozone rate constant are unknown. In general, mechanistic details of systems that react only slowly with ozone are not readily accessible.

Many pharmaceuticals have amino groups as essential functional groups. Cases in point are the macrolide antibiotics such as erythromycin, clarithromycin, roxithromycin and tylosin. They are not fully degraded upon biological wastewater treatment. For the occurrence of macrolide antibiotics in the aquatic environment see Chapter 4.

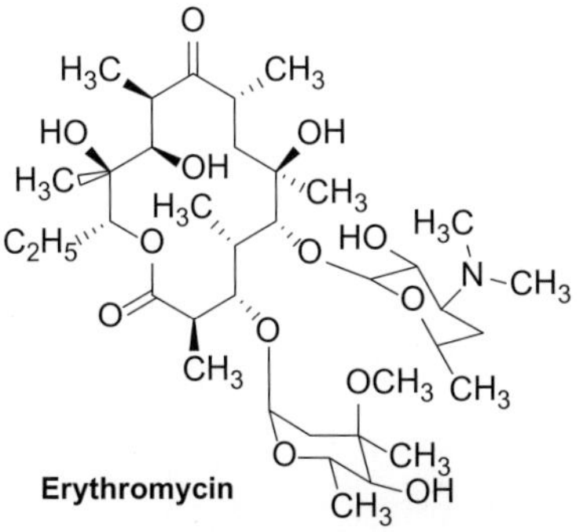

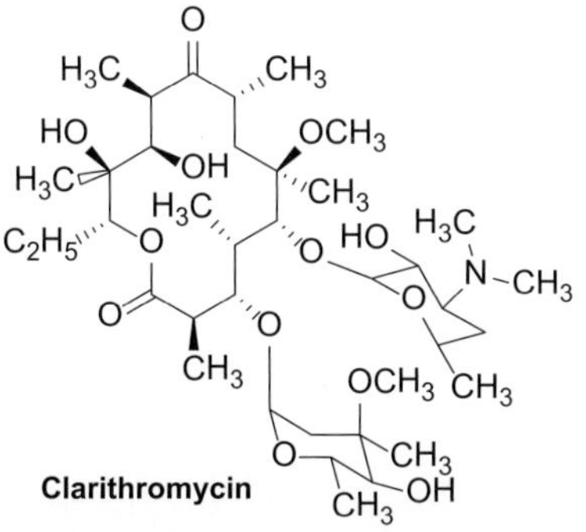

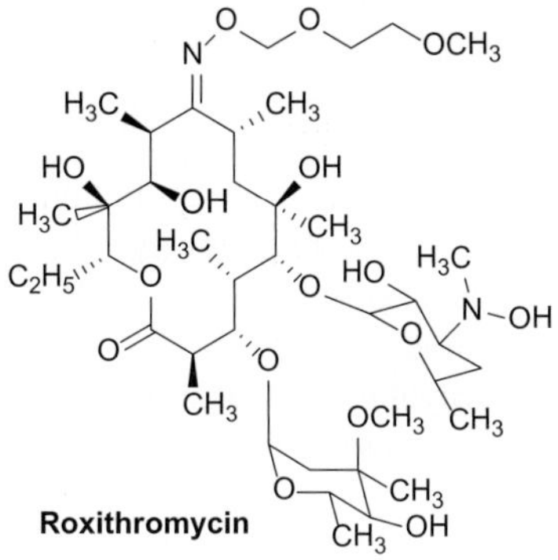

Their site of ozone attack is the tertiary amino group. A product study has been carried out with clarithromycin, and it has been found that the major product is the *N*-oxide [> 90%, cf. reaction (2)], and as a minor route, demethylation and formaldehyde formation occurs [cf. reactions (8)–(11)] (Lange *et al.*, 2006). Since the tertiary amino group is essential for the antibacterial activity of these drugs (Goldman *et al.*, 1990; Schlünzen *et al.*, 2001), its conversion to the *N*-oxide or demethylation eliminates the biological activity (Chapter 4) (Lange *et al.*, 2006).

Demethylation has also been reported for roxithromycin, but the expected *N*-oxide formation escaped detection (Radjenovic *et al.*, 2009). Further degradation requires the action of •OH (Radjenovic *et al.*, 2009). Among others, the elimination of a sugar moiety was observed. Scission of the glycosidic linkage induced by •OH is common in carbohydrate chemistry (von Sonntag, 1980).

The macrolide drug tylosin has two ozone-reactive sites, the tertiary amino group and the dienone function. Their contribution to degradation of tylosin by ozone is pH dependent. The degradation above pH ~5 is controlled by the tertiary amino group (Dodd et al., 2006a).

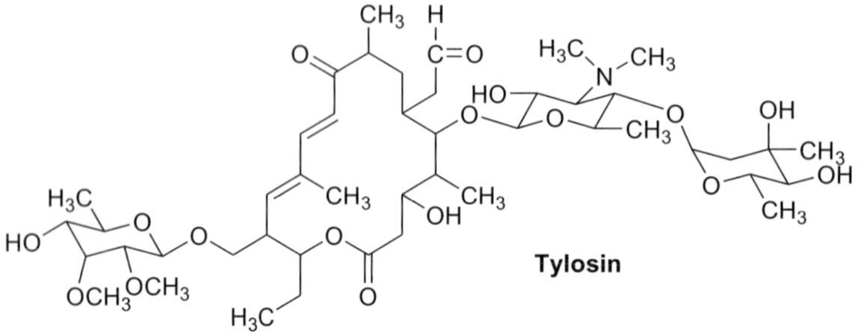

Tylosin

Tramadol is an opioid which is frequently prescribed in case of moderate to severe acute and chronic pain.

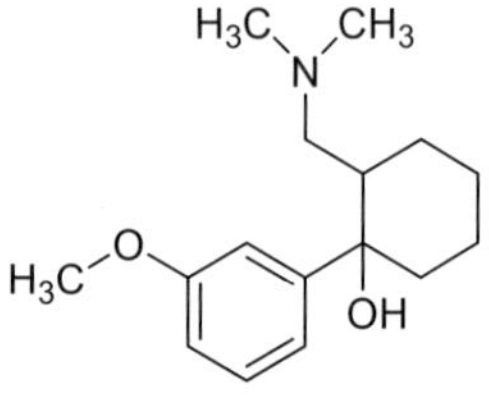

Tramadol

About 10–30% of the tramadol is excreted unchanged via urine and concentrations of up to 97 µg/L and 6 µg/L were reported in secondary effluent and surface waters, respectively (Kasprzyk-Hordern et al., 2009). As expected from the above, the main ozone product of tramadol is its N-oxide (ca. 90%) and a minor one, N-dealkylated tramadol (<10%) (Zimmermann et al., 2012).

Amiodarone, a commonly used drug against cardiac arrhythmia, also contains a tertiary amine function, and its ozone chemistry, as yet not studied, must be analogous to that of the other tertiary amines.

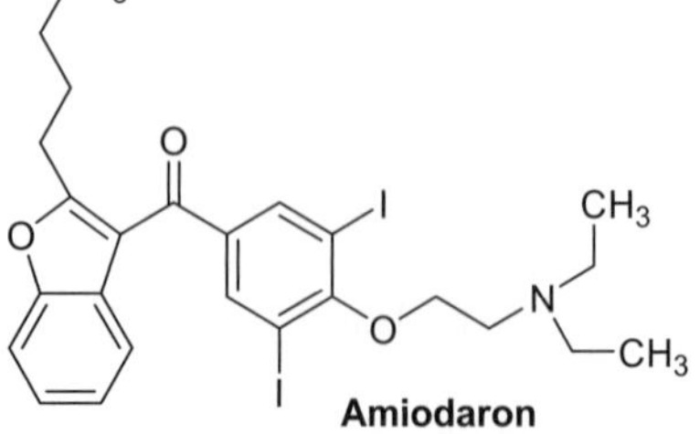

Amiodaron

Beta blockers, for example, atenolol, metoprolol, bisoprolol or propranolol, which are commonly observed as micropollutants in wastewater (Huggett et al., 2003), [for their fate in wastewater see

Wick et al. (2009)] contain secondary aliphatic amino groups, and this side group seems to be required for the biological response.

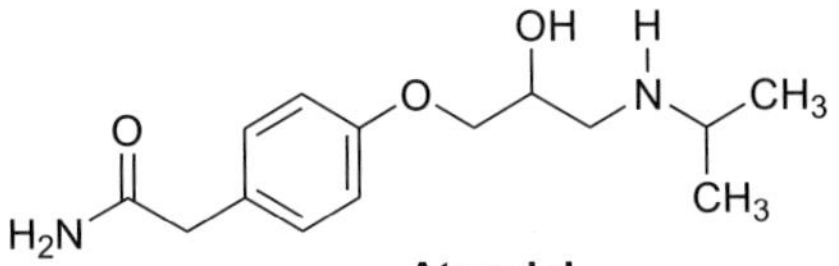

Atenolol

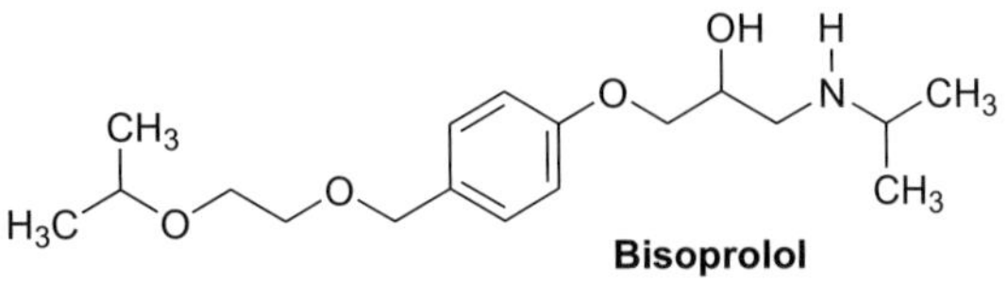

Bisoprolol

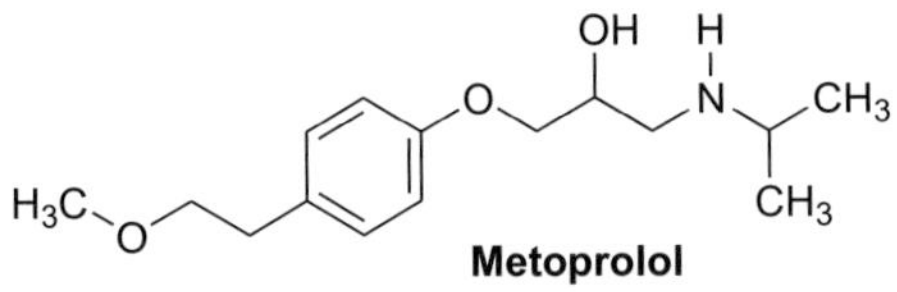

Metoprolol

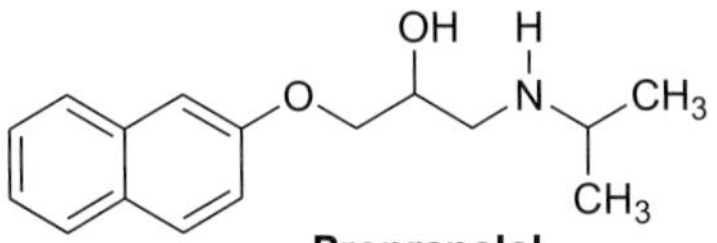

Propranolol

At a wastewater pH of 7–8, the reaction of ozone with these beta blockers is largely determined by the reaction of the free amine in equilibrium (cf. Chapter 2) with the exception of propranolol, in which ozone attack at the more reactive naphthalene ring also takes place (Benner & Ternes, 2009b).

This was confirmed in a product study on metoprolol (Benner & Ternes, 2009a) in which the product from amine oxidation (hydroxylamine) has been detected. The chemistry is complex as substantial amounts of $^{\bullet}OH$ are also formed (Sein et al., 2009).

The ozone-reactive site of the psychotropic drug fluoxetine is also the secondary amine in the side group. The aromatic rings are of comparatively little ozone reactivity.

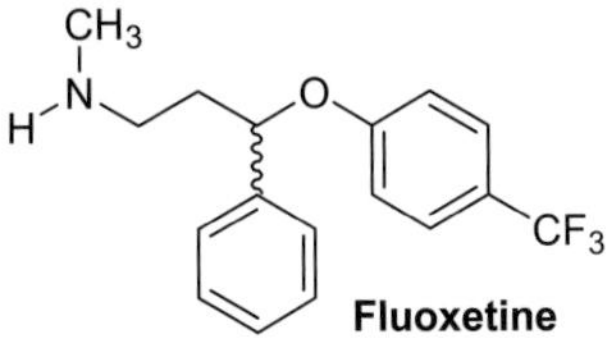

Fluoxetine

For the antibiotics ciprofloxin and enrofloxacin (for rate constants see Table 8.1), sites of ozone attack are the nitrogens at N1 and N4. The reason for the higher reactivity of enrofloxacin at pH 7 compared to that of ciprofloxin is most likely due to the differences of the pK_a values at N4, reported as 7.7 and 8.8, respectively (Dodd et al., 2006a).

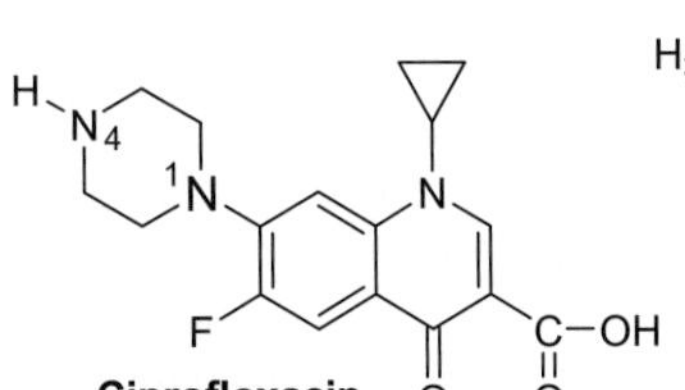

Ciprofloxacin

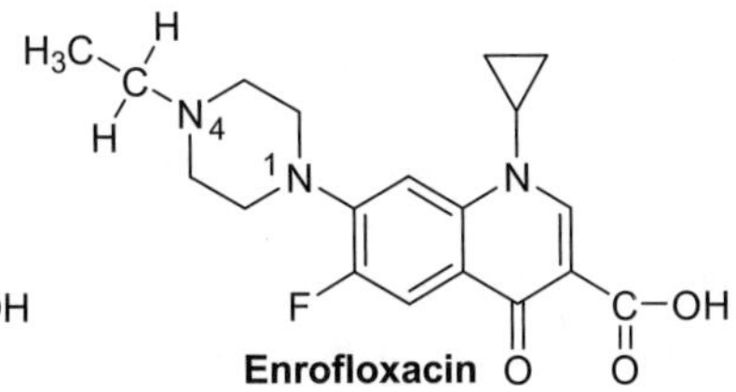

Enrofloxacin

The sites of ozone attack at the antibiotic spectinomycin are the secondary nitrogens (Qiang et al., 2004).

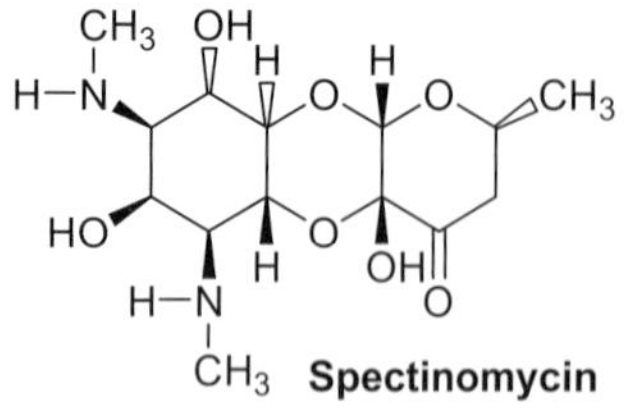

[Structures: Spectinomycin; Lincomycin]

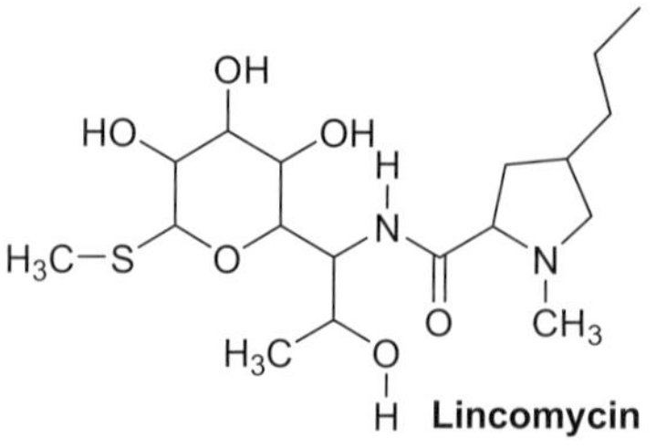

In lincomycin, the tertiary nitrogen in the five-membered ring and the sulfur in the sulfide group are competing sites for ozone attack (Qiang et al., 2004; Dodd, 2008). The pH dependence of the ozone rate constant showed that at pH < 5 the sulfur group is the reactive moiety, whereas at higher pH values the tertiary amine becomes dominant.

The antibiotic amikacin has four primary amino groups, which are the ozone-reactive sites (Dodd, 2008).

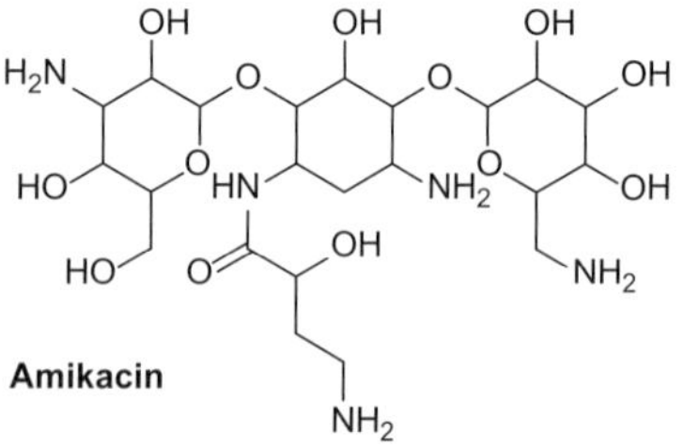

[Structure: Amikacin]

The sulfonamide sulfamethoxazole has two pK_a values; the one at 1.7 is due to protonation at the aniline group (note that the model compound 4-aminophenyl methylsulfone has a pK_a value of 1.5). The strong electron-withdrawing sulfonamide function is the reason for this low value.

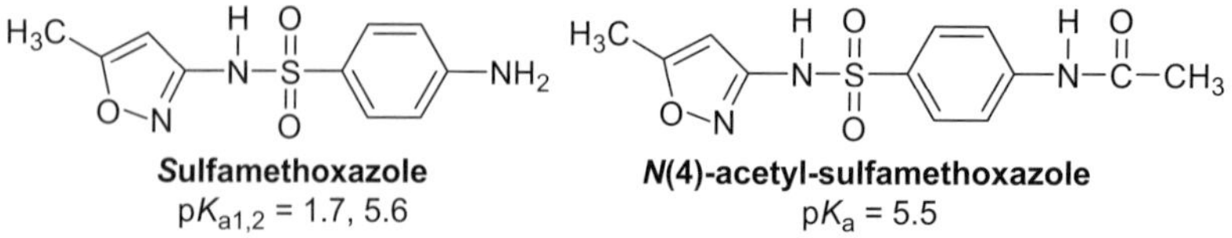

[Structures: Sulfamethoxazole, $pK_{a1,2}$ = 1.7, 5.6; N(4)-acetyl-sulfamethoxazole, pK_a = 5.5]

N-Acetylation, as in N(4)-acetyl-sulfamethoxazole, shows that the pK_a at 5.5 must be due to a deprotonation at the sulfonamide nitrogen. At a pH where the sulfonamide function is deprotonated, the sulfamethoxazole rate constant is an order of magnitude higher than with the neutral molecule (Table 8.1) (Dodd et al., 2006a). The model compound 4-aminophenyl methyl sulfone has a rate constant of 4.7×10^4 M^{-1} s^{-1}, while N(4)-acetyl-sulfamethoxazole displays rate constants of 260 M^{-1} s^{-1} (deprotonated) and 20 M^{-1} s^{-1} (neutral). The latter observation excludes the sulfonamide nitrogen as the ozone-reactive site, although its deprotonation increases the rate of reaction by adding further electron density into the ring.

These data point to the nitrogen at the anilino group and/or the now activated aromatic ring in sulfamethoxazole as the major site of attack. The slow reaction found for neutral N(4)-acetyl-sulfamethoxazole may be attributed to an addition to the C–C double bond on the oxazole ring. This reaction would become faster upon deprotonation of the sulfonamide function, which channels electron density into the oxazole ring.

The psychotropic drug diazepam has two nitrogens. One is an amide nitrogen and certainly of negligible ozone reactivity. The other is an enamine nitrogen with the double bond conjugated into the aromatic ring.

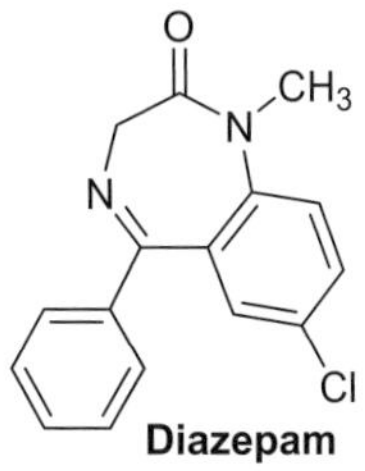

Diazepam

The rate constant is so low (Table 8.1, below that of benzene, Chapter 7, Table 7.1) that ozone addition may take place at the aromatic rings (one deactivated by chlorine, the other deactivated by the conjugated enamine that could act like a carbonyl function).

The metabolite of the antiviral drug Tamiflu®, oseltamivir carboxylate has been discussed in Chapter 6, because at the pH of wastewaters its ozone reactivity will be largely dominated by the C–C double bond and not by its amino group. Further nitrogen-containing antiviral drugs found in WWTP effluents are ribavirin, and nevirapine (Prasse *et al.*, 2010). For other antiviral drugs, besides Tamiflu®, see Chapter 6.

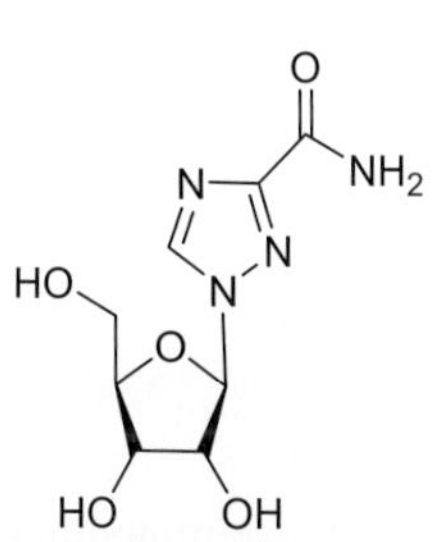

Ribavirin (RBV) 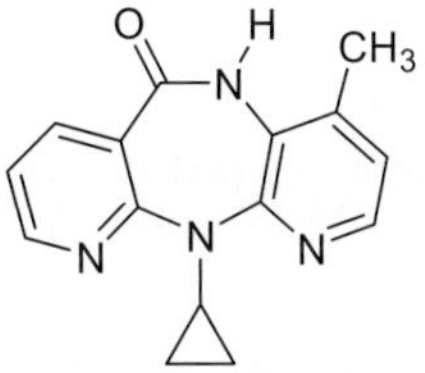   **Nevirapine (NVP)**

An abundant micropollutant is the nonsteroidal anti-inflammatory drug diclofenac. It is highly toxic (Henschel *et al.*, 1997) to aquatic life such as fish (Schwaiger *et al.*, 2004), but also to birds, and the dramatic decline in the vulture population on the Indian subcontinent has been attributed to diclofenac poisoning (Oaks *et al.*, 2004; Shultz *et al.*, 2004).

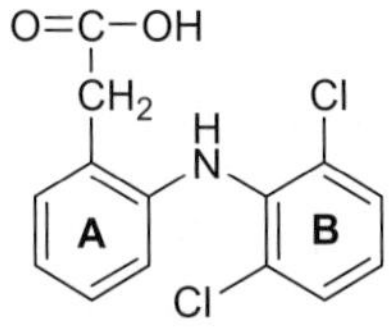

Diclofenac

It reacts fast with ozone (Table 8.1) and is thus readily eliminated from drinking water and wastewater upon ozonation. Its ozone chemistry has found considerable attention (Vogna *et al.*, 2004; Sein *et al.*, 2008, 2009). Potential sites of ozone attack are the nitrogen but also the aromatic ring **A**. Both rings are activated by the imino group (cf. aniline), but ring **B** is deactivated by the two chlorines (Chapter 7). Amines and electron-rich aromatic compounds give rise to $O_3^{\bullet-}$ and thus to $^{\bullet}OH$ radicals, and this is also observed with diclofenac (Sein et al., 2008) (Table 8.4). In the original paper, it has been suggested that it arises from the ozone adduct at nitrogen, but ozone attack at ring **A** would give the same result [cf. reaction (25)].

In absence of other •OH scavengers, •OH radicals will react with diclofenac. These reactions have been discussed (Sein et al., 2008), but under wastewater or drinking water conditions they are scavenged by the water matrix (Chapter 3), and these reactions must then largely fall away.

The diclofenac radical cation is in equilibrium with the corresponding aminyl radicals [cf. equilibrium (26); for pK_a values of such intermediates see above], and these may interact with other radicals, notably peroxyl radicals, present.

In competition with $O_3^{\bullet-}$ elimination, the adduct can undergo $^1O_2$ elimination [reaction (32)], and this leads to the formation of hydroxylated products [reactions (33) and (34)].

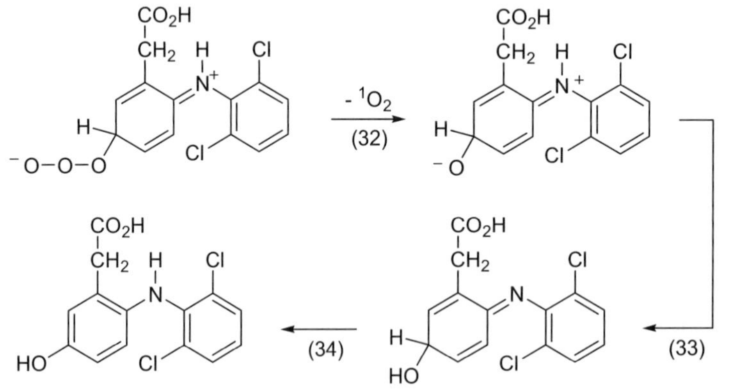

The reaction of the aminyl radical with peroxyl radicals yields 1,5-diclofenaciminoquinone [reactions (35)–(37)].

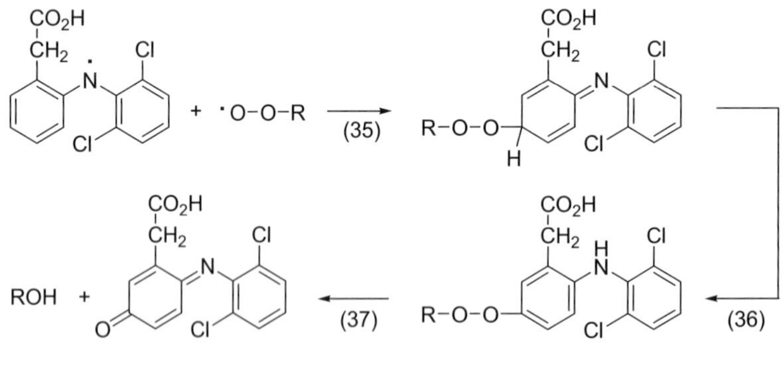

The consumption of diclofenac by ozone is rather low, and it has been suggested that this might be due to an ozone-consuming chain reaction that gives rise to no other products but O_2 [cf. reactions (28) and (29)] (Sein et al., 2009) (for the reactions of radicals with ozone see Chapter 13).

The cytostatic drug methotrexate also reacts rapidly with ozone (Table 8.1) and poses no problem in detoxifying waters with ozone (Garcia-Ac et al., 2010).

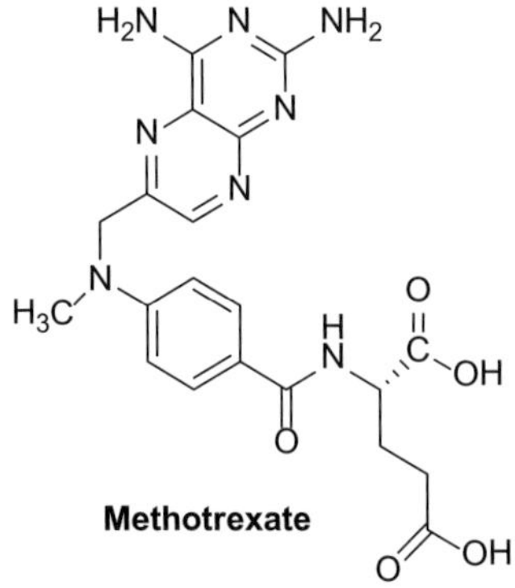

Methotrexate

Benzotriazoles are widespread micropollutants (Cancilla *et al.*, 2003; Janna *et al.*, 2011). In Switzerland, they are, after EDTA, the second most abundant individual contaminants in many natural waters (Cancilla *et al.*, 2003; Voutsa *et al.*, 2006). Among other applications, benzotriazoles are used as corrosion inhibitors in dishwasher detergents and for de-icing of aircraft (Hart *et al.*, 2004; Weiss *et al.*, 2006). Benzotriazole acts as an inhibitor of the nitrification process (Callender & Davis, 2002). There are some studies on its reaction with ozone (Legube *et al.*, 1987; Vel Leitner & Roshani, 2010; Lutze *et al.*, 2011a; Müller *et al.* 2012). Its protonated form dominates below pH 1.6 and its deprotonated form above pH 8.2. The ozone rate constants strongly depend on its protonation state (Table 8.1).

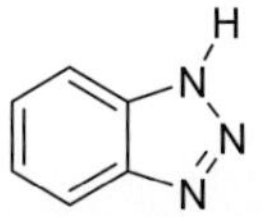

Benzotriazole
pK_a = 1.6, 8.2

Similarly to the more simple aromatic compounds (Chapter 7), there is a good correlation of the standard Gibbs free energy of formation of ozone adducts to the ring with the logarithm of the rate constants on benzotriazole, its derivatives and their anions (Lutze *et al.*, 2011). The electron rich anions give rise to •OH (Table 8.4).

The herbicide glyphosate is an amine containing three acidic groups (one carboxylate and two phosphonate groups).

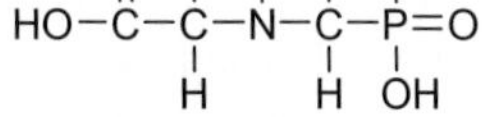

Glyphosate 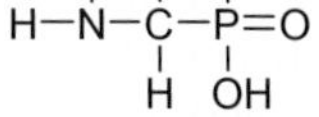   **AMPA**

Thus, it is largely present as a zwitterion. Protonation at nitrogen must prevent a rapid reaction at pH 6.5, where it is degraded only slowly (Assalin *et al.*, 2010). AMPA is one of the ensuing products. At pH 10, degradation is more efficient and AMPA is also degraded in the course of reaction.

The triazine herbicide atrazine has been mentioned above. Due to its low ozone rate constant, atrazine is mainly degraded in DOM-containing water via the •OH route. These •OH-induced reactions are discussed in some detail in Chapter 14. For reviews on the degradation of herbicides by ozone and AOPs see (Ikehata & El-Din, 2005a, b).

Azobenzene is a building block of azo dyes. It has a moderate reactivity with ozone (Table 8.1).

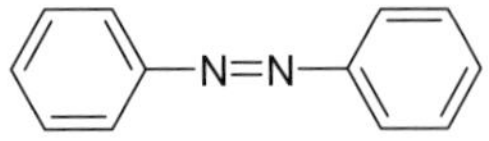

Azobenzene

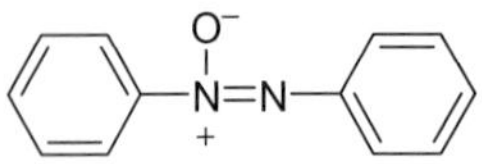

Azoxybenzene

In its reactions with ozone, it gives rise to azoxybenzene (von Sonntag, unpublished). Azoxybenzene still retains some absorption in the visible region, and this may have to be taken into account when the discoloration of azo dyes by ozone is attempted (López-López et al., 2007). For the oxidation of azo dyes by ozone in industrial wastewaters see Khadhraoui et al. (2009); Konsowa et al. (2010); Tabrizi et al. (2011) and references cited therein.

Abundant nitrogen-containing micropollutants are the complexing agents EDTA, NTA and DTPA.

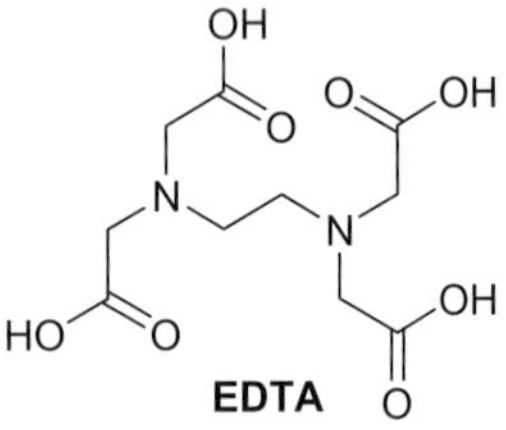

EDTA

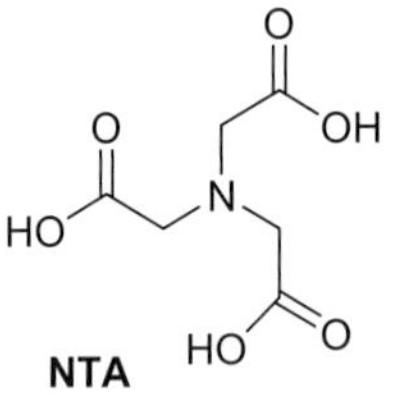

NTA

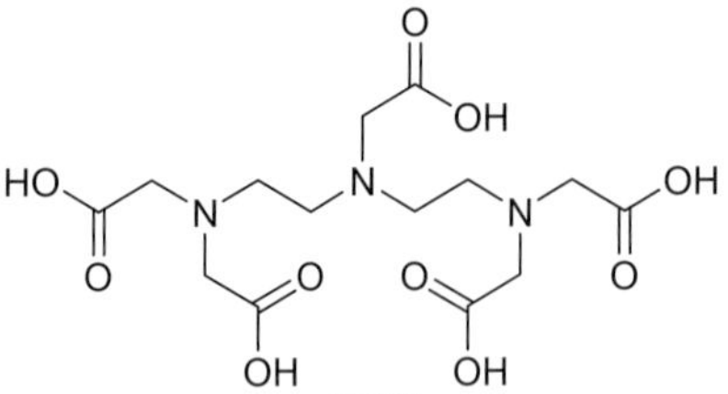

DTPA

Whenever the lone electron pair at nitrogen is not available for ozone attack by protonation or by complexation to a transition metal ion, the reactivity drops dramatically. A case in point is EDTA. It has pK_a values at 0.26, 0.96, 2.0, 2.7, 6.2 and 10.2. The latter two are relevant to protonation at nitrogen and this is reflected in the rate of ozone reaction (Figure 8.2). When both amino groups are free, the rate constant is 3.2×10^6 M^{-1} s^{-1} and it drops to 1.6×10^5 M^{-1} s^{-1} when EDTA is monoprotonated. Further lowering the pH results in a drop of a factor of ten in the rate constant per pH unit. That is, the remaining ozone reactivity is only due to the concentration of free amine in equilibrium and fully protonated amines are practically non-reactive towards ozone. Complexation of EDTA by Ca^{2+} has only a small effect ($k = 10^5$ M^{-1} s^{-1} at pH 6), as Ca^{2+} is largely complexed by the carboxyl groups of EDTA [for a review on stability constants see Smith & Martell (1987); for the mechanism of the degradation of EDTA by •OH see Höbel & von Sonntag (1998)]. In contrast, complexation by Fe(III), which involves also the lone electron pairs at the nitrogens, lowers the rate constant to 330 M^{-1} s^{-1} at the same pH 6 (Muñoz & von Sonntag, 2000b). A similar effect was observed for the kinetics of the oxidation of DTPA. Complexation of DTPA with Fe(III) leads to a dramatic decrease of the ozone reaction rate constant relative to the free DTPA (Stemmler et al., 2001). However, when the Fe(III)-hydroxo-DTPA complex is formed at pH > 6 there is an increase in the rate of DTPA oxidation (Table 8.1). Therefore, DTPA can be quite readily degraded in the pH-range 7–8 of drinking and wastewaters.

It has been mentioned above that the nitrogen in amides is not a target for ozone attack. Hence urethane-based micropollutants such as the sedative meprobamate and the cytostatic drug cyclophosphamide are practically ozone-refractory (for cyclophosphamide see Table 8.1) and can only be eliminated by the •OH route (cf. Chapters 3 and 14).

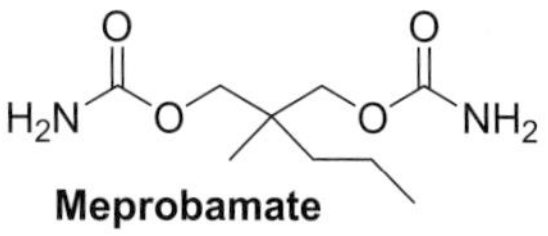

Meprobamate

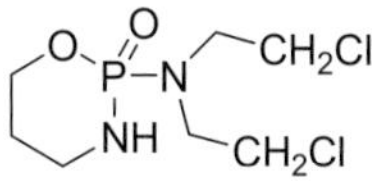

Cyclophosphamide

The ozone chemistry of the antibiotics tetracycline and vancomycin may be largely governed by an attack at the phenolic moieties and is discussed in Chapter 7, but there is also a secondary amino group that can serve as an additional site of ozone attack. Anatoxin-A and cylindrospermopsin contain a C–C double bond besides a secondary amine function and have been discussed in Chapter 6.

The phenylurea-derived herbicides diuron, chlortoluron, linuron and isoproturon react from slowly to moderately fast with ozone (Table 8.1; note the poor agreement among the reported rate constants). The nitrogens at the urea functions are non-reactive and, ozone addition to the aromatic ring (Chapter 7) is slow due to the chlorine functions at the ring.

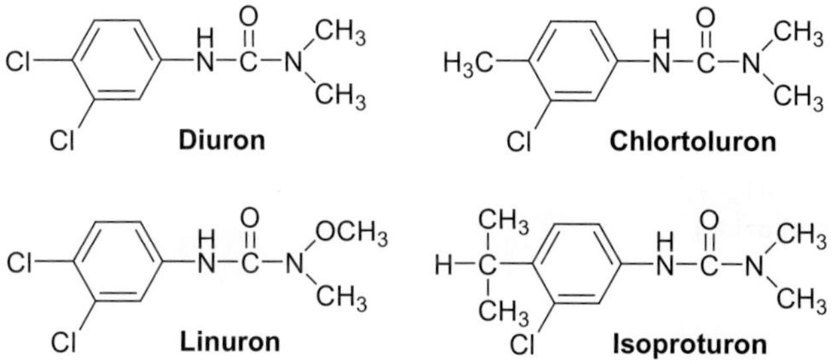

Diuron

Chlortoluron

Linuron

Isoproturon

Although in wastewater diuron will be mainly degraded by the •OH route, direct action of ozone also eliminates its toxicity. Some products have been determined, and their toxicity towards algae has been assessed (Mestankova et al., 2011). The toxicity of diuron towards algae is completely eliminated by ozone and •OH (Mestankova et al., 2011).

In the acetamide-derived herbicides, alachlor, metochlor and propachlor, the same principle governs their ozone reactivity (Table 8.1) as discussed for the phenylureas. For relative efficiencies of the degradation of herbicides by ozone see Ormad et al. (2010).

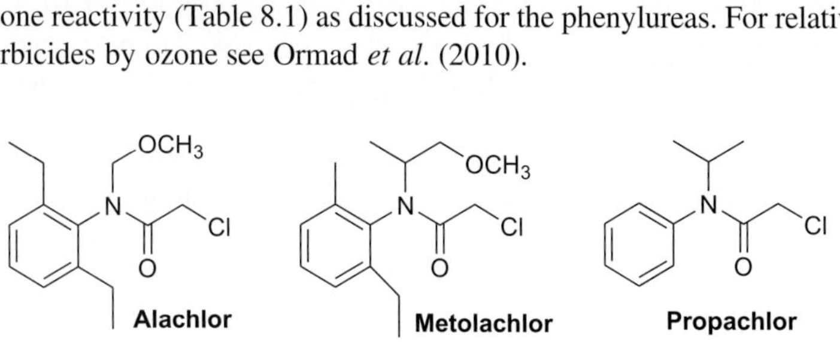

Alachlor

Metolachlor

Propachlor

Amoxillin ozonolysis is determined by its phenol group and discussed in Chapter 7.

Further abundant micropollutants in wastewater are the antiepileptic drug phenytoin, the anticonvulsant primidone and the analgesic phenazone.

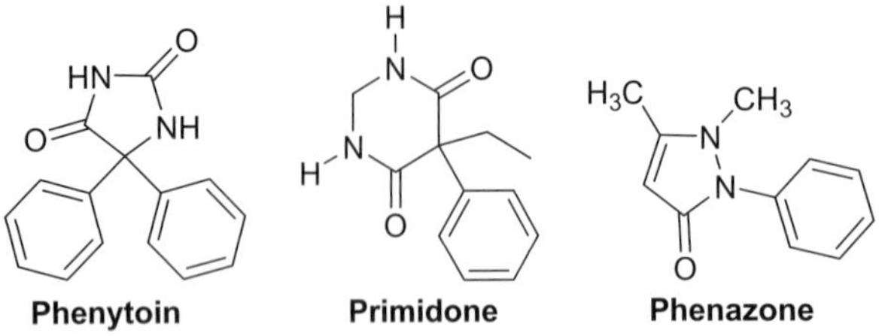

Primidone [$k = 1.0 \, M^{-1} s^{-1}$, Real et al. (2009)] and phenytoin have very low ozone reactivity (Lee & von Gunten, 2012). This has been demonstrated in the relative resistance of primidone to ozone degradation during full-scale wastewater treatment (Hollender et al., 2009). For phenazone, a high reactivity with ozone can be expected due to the structural similarity to 4-dimethylaminoantipyrine (Table 8.1).

8.3.1 The *N*-Nitrosodimethylamine (NDMA) puzzle

N-Nitrosodimethylamine is a disinfection by-product, notably observed after chloramine disinfection (Mitch et al., 2003; Choi & Valentine, 2002; Gerecke & Sedlak, 2003). In wastewater ozonation, the antiyellowing agents HDMS and TMDS have been recognised as NDMA precursors (Kosaka et al., 2009).

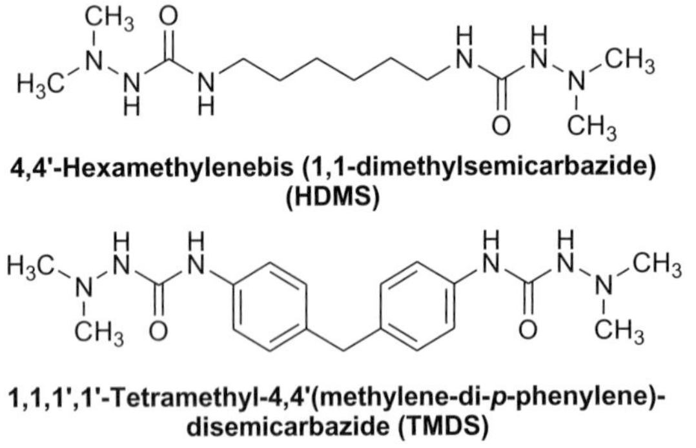

For this suspected human carcinogen, a notification level of 10 ng/L has been published by the California Department of Public Health (NDMA, 2009). The standard for an excess lifetime cancer risk of 10^{-5} for drinking waters was calculated to be 100 ng/L (WHO, 2004). For the same risk, bromate, another known ozonation by-product was calculated to be 3000 ng/L and bromoform, a known chlorination by-product 100,000 ng/L (WHO, 2004).

Also with ozone disinfection, NDMA can be a disinfection by-product. The fungicide tolylfluanide is hydrolytically and biologically degraded at pH 7 with a half-life of two days into dimethylsulfamide (DMS).

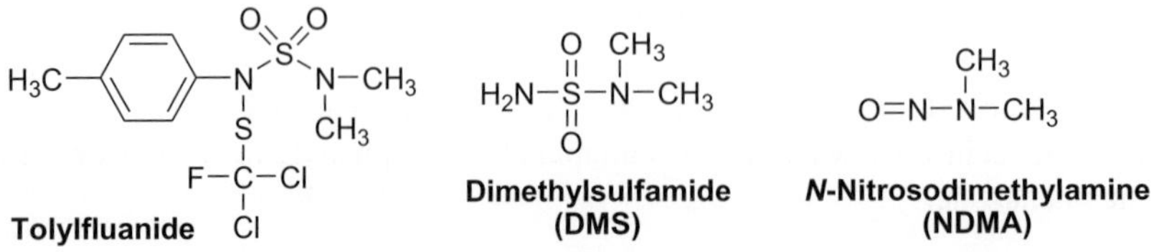

Due to its short half-life, tolylfluanide as such poses no problem, but its degradation product DMS has a much longer half-live, 50–70 days. In Germany for example, DMS concentrations of 100–1000 ng L^{-1} and 50–90 ng L^{-1} were detected in ground waters and surface waters, respectively. Its elimination by typical water treatment processes such as riverbank filtration, activated carbon filtration, treatment with permanganate, chlorine dioxide or UV-radiation proved to be ineffective (Schmidt & Brauch, 2008). DMS as such does not pose a problem. Ozonation of waters taken from areas where tolylfluanide had been used as fungicide in, e.g. strawberry cultivation led, however, to high NDMA levels (Schmidt & Brauch, 2008). This shows that NDMA, even though typically formed in small yields only, has to be monitored carefully. It is noteworthy that in most European countries the use of tolylfluanide is no longer allowed.

The reaction of ozone with DMS in distilled water does not yield NDMA. Br$^-$, present in trace amounts in practically all waters, catalyses the formation of NDMA from DMS during ozonation (von Gunten et al., 2010). Below, the present mechanistic concept of NDMA formation is discussed.

In the reaction of ozone with Br$^-$, HOBr is formed (Chapter 11). This reacts with DMS according to reaction (38).

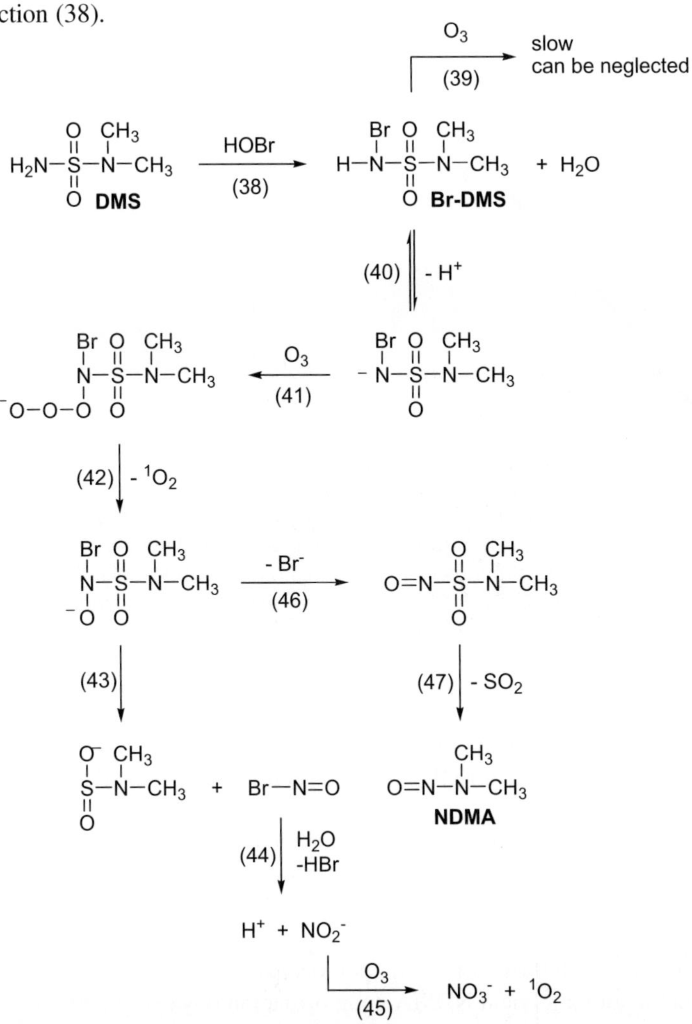

Brominated DMS (Br–DMS) reacts in its deprotonated form with ozone to an adduct [reactions (40) and (41)], and this loses singlet oxygen [reaction (42)]. The ensuing unstable intermediate can either lose nitroxyl bromide [reaction (43)] or, in competition, give rise to Br$^-$ plus nitrosodimethylsulfamide [reaction (46)]. Nitroxyl bromide hydrolyses to nitric acid which is quickly oxidised to nitrate [reactions (44) and (45)]. Nitrosodimethylsulfamide undergoes an intramolecular rearrangement forming NDMA by SO$_2$ extrusion [reaction (47)].

HOBr formation is the rate-limiting step in NDMA production (von Gunten *et al.*, 2010). Already Br$^-$ concentrations in the order of 10 µg/L are sufficient to catalyse this reaction. The formation of NDMA from DMS as a function of ozone exposure during ozonation of Br$^-$-containing waters is shown in Figure 8.4.

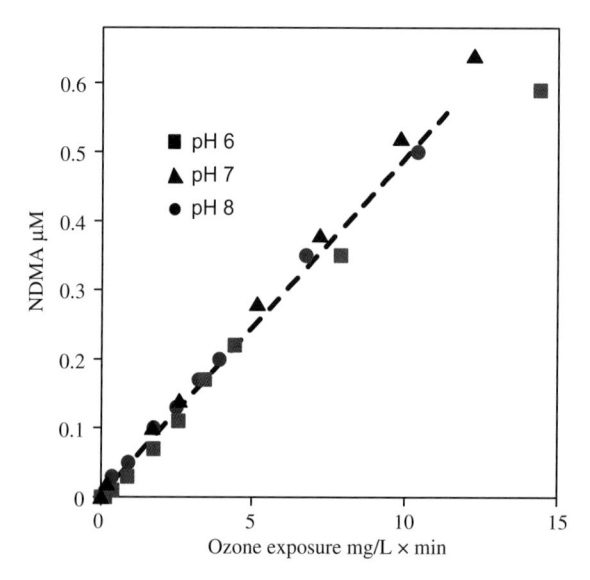

Figure 8.4 NDMA formation upon ozonation of DMS- and Br$^-$-containing ultrapure water in the presence of 5 mM tBuOH as a function of the ozone exposure at pH 6 (squares), 7 (triangles), and 8 (circles). The ozone dose was 1 mg/L (20 µM), initial DMS and Br$^-$ concentrations were 2.5 and 0.5 µM, respectively. The dashed line is a kinetic simulation for the cumulative HOBr formation during ozonation at pH 8. Reprinted with permission from von Gunten *et al.* (2010). Copyright (2010) American Chemical Society.

In the pH range 6–8, NDMA formation is independent of pH in agreement with the ozone/Br$^-$ reaction. The dashed line in Figure 8.4 shows the calculated rate of HOBr formation confirming that oxidation of Br$^-$ is the rate-limiting step.

NDMA is also formed during ozonation of DMS in presence of HOCl with a lower yield of 20–30% (von Gunten *et al.*, 2010). This is, however, not a catalytic process, because Cl$^-$, which is released in the reaction of Cl-DMS with ozone, is only re-oxidised to a non-measurable extent to HOCl upon ozonation (Chapter 11).

DMS is not the only potential precursor of NDMA formation upon ozonation. In fact, some dyes that contain a dimethylamino function were reported to yield NDMA upon ozonation (Oya *et al.*, 2008).

Nitrogen-containing compounds

The chemical yield of NDMA (in percentage of the target molecule consumed at excessive ozonation as it seems) were low, although the NDMA yields are sufficiently high to be of potential relevance (Table 8.5).

Table 8.5 NDMA formation in the ozonation of some N-dimethyl groups containing compounds (0.5 mM) (Oya et al., 2008)

Target compound	NDMA/ngL^{-1}	Chemical yield/%
Methylene blue	310	8.3×10^{-3}
Methyl orange	270	7.2×10^{-3}
Methyl violet B	460	1.2×10^{-2}
Auramine	480	1.3×10^{-2}
Brilliant green	<5	$<1 \times 10^{-4}$
N,N-Dimethylformamide	13	3.5×10^{-4}
N,N-Dimethylaniline	240	6.4×10^{-3}
N,N-Dimethylphenylenediamine	1600	4.3×10^{-2}

Dimethylamine can also yield NDMA upon ozonation (Yang et al., 2009; Andrzejewski et al., 2008). Also in this case, NDMA yields are extremely low (<0.01% at pH 7.5, ozone dose 66 mg/L). For such low chemical yields, mechanistic suggestions cannot be made, especially since NDMA cannot be a primary product. This raises the question as to how relevant these data are for ozone treatment of wastewaters containing dyes that have N-dimethylamino groups, but they are a most useful *caveat* to make one aware of the potential formation of NDMA.

Chapter 9

Reactions of sulfur-containing compounds

9.1 REACTIVITY OF SULFUR-CONTAINING COMPOUNDS

Ozone reacts with sulfur compounds in their low oxidation states such as bisulfide/dihydrogen sulfide (HS^-/H_2S) (Chapter 11), thiols (RSH) sulfides (RSR), disulfides (RSSR), sulfoxides (R_2SO), sulfinic acids [RS(O)OH] and sulfite [HOS(O)O$^-$] (Chapter 11). All sulfur(VI) compounds (e.g. sulfate ion, organic sulfates and sulfonic acids) do not react with ozone. Reported rate constants are compiled in Table 9.1.

Table 9.1 Compilation of ozone rate constants and singlet oxygen yields in percentage of ozone consumption (Muñoz et al., 2001) of some organic sulfur-containing compounds. For inorganic sulfur-containing compounds see Chapter 11. For cephalexin see Chapter 6, for lincomycin see Chapter 8. Published rate constants are rounded to significant figures

Compound	pH	Rate constant $M^{-1} s^{-1}$	Singlet oxygen yield	Reference
Aldicarb	2.1	4.4×10^4		Yao & Haag, 1991
	7.0	4.3×10^5		
Cysteine		4.2×10^4		Pryor et al., 1984
		3.0×10^4		Hoigné & Bader, 1983b
		3.0×10^4		Hoigné & Bader, 1983b
anion		2.4×10^6		Pryor et al., 1984
Cystine	1	555		Hoigné & Bader, 1983b
	3.1	1×10^3		
Dimethyl sulfoxide		8		Pryor et al., 1984
		1.8 at 20 °C[a]		Reisz[a]
trans-1,2-Dithiane-4,5-diol		2.1×10^5	105	Muñoz et al., 2001
1,4-Dithiothreitol	4.8		46	Muñoz et al., 2001
	9		17	

(Continued)

Table 9.1 Compilation of ozone rate constants and singlet oxygen yields in percentage of ozone consumption (Muñoz et al., 2001) of some organic sulfur-containing compounds. For inorganic sulfur-containing compounds see Chapter 11. For cephalexin see Chapter 6, for lincomycin see Chapter 8. Published rate constants are rounded to significant figures (*Continued*)

Compound	pH	Rate constant $M^{-1} s^{-1}$	Singlet oxygen yield	Reference
EPTC (S-ethyl-N,N-dipropylthiocarbamate)		500[b]		Chen et al., 2008
Glutathione		4×10^6		Pryor et al., 1984
protonated, $pK_a = 8.75$		2×10^4		Pryor et al., 1984
bis(2-Hydroxyethyl) disulfide		1.7×10^5		Muñoz et al., 2001
Methanesulfinate		2×10^6	96	Flyunt et al., 2001a
Methionine		1.8×10^6	104	Pryor et al., 1984
Molinate		500[b]		Chen et al., 2008
Penicillin G	7	4.8×10^3		Dodd et al., 2010
Vydate (Oxamyl)	2–7	620		Yao & Haag, 1991

[a] $\log A = 11.5$, $E_a = 63$ kJ mol^{-1} [Reisz & von Sonntag (2011), unpublished results], [b] Estimated.

For thiols, rate constants increase upon deprotonation (cf. cysteine), and this is expected, as the higher electron density in the anion favours the attack by the electrophilic ozone. This effect is even more pronounced in the HS^-/H_2S system (Chapter 11). The reaction of sulfides is also very fast (cf. methionine). The rate constants for the reactions of uncharged disulfides such as *trans*-1,2-dithiane-4,5-diol bis(2-hydroxyethyl) disulfide are about 2×10^5 M^{-1} s^{-1}. Positively charged neighbouring groups such as in cystine at pH 3.1 and pH 1 dramatically reduce the rate of reaction. The rate constants of sulfoxides (cf. dimethyl sulfoxide) are low and this is of major importance for the elimination of the biological activity of penicillin G upon ozonation of wastewater (see below). In contrast to dimethyl sulfoxide, methylsulfinate reacts very fast with ozone.

Only some ozone adducts that are likely intermediates in the (dominating) $^1O_2$ elimination reactions (for yields see Table 9.1) are sufficiently stable to calculate the Gibbs energy of adduct formation (see below). This prevents us from presenting a similar correlation of Gibbs energies of adduct formation with the logarithm of the rate constant as has been found for aromatic compounds (Chapter 7).

9.2 THIOLS

There is very little information as to the ozone reactions of thiols. In contrast to other sulfur-containing compounds that show singlet oxygen ($^1O_2$) yields near 100% (Table 9.1), it is only near 50% for thiols and even less for thiolate ions (cf. 1,4-dithiothreitol, Table 9.1). In this respect, there is an analogy to HS^-/H_2S (Chapter 11). The understanding of the mechanism of the HS^- reaction increased considerably by quantum-chemical calculations, and it became apparent, that the ozone adduct can rearrange readily by transferring an O-atom in competition with the release of $^1O_2$ (Mark et al., 2011). If such a reaction were to proceed in thiols, sulfinic peracids would be the primary reaction products. Thus far, this is only speculation, and detailed studies would be required for substantiating this hypothesis.

9.3 SULFIDES, DISULFIDES AND SULFINIC ACIDS

For sulfides, disulfides and sulfinic acids the singlet oxygen ($^1O_2$) yield is unity (Table 9.1). This points to O-transfer as the only reaction, as shown in reactions (1) and (2) for a sulfide as an example.

$$\underset{R}{\overset{R}{S}} \xrightarrow{O_3 \atop (1)} \underset{R}{\overset{R}{^+S-O-O-O^-}} \xrightarrow{(2)} \underset{R}{\overset{R}{S=O}} + {}^1O_2$$

With sulfides, the sulfoxide yield equals that of singlet oxygen, as has been shown for methionine (Muñoz et al., 2001). With the disulfides trans-1,2-dithiane-4,5-diol disulfide and bis(2-hydroxyethyl) disulfide, the S-alkylsulfinates are the corresponding products (Mvula, 2002), and with methanesulfinic acid, methanesulfonic acid is also formed with a yield of unity (Flyunt et al., 2001a). In the case of an electron transfer occurring as well, a chain reaction would set in (Flyunt et al., 2001a). This is not observed.

In DFT calculations [Naumov & von Sonntag (2010), unpublished results], ozone adducts were stable only for H_2S, methanethiol, the sulfides and dimethyl disulfide.

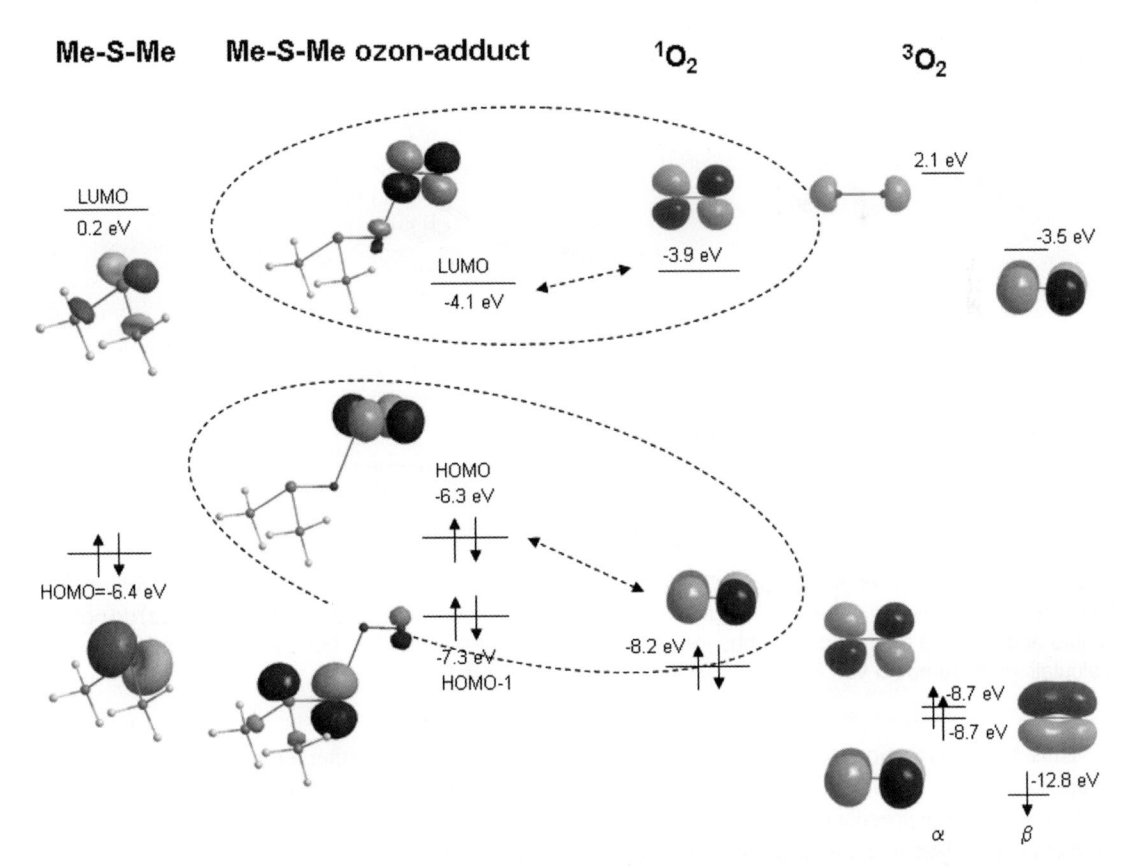

Figure 9.1 Molecular orbitals of dimethyl sulfide, its ozone adduct, of singlet oxygen ($^1O_2$) and triplet (ground state) oxygen ($^3O_2$) according to Naumov & von Sonntag (2010), unpublished.

Some data for dimethyl sulfide are shown in Figure 9.1. The orbital of highest energy (HOMO, Highest Occupied Molecular Orbital) is at sulfur. This is the position where ozone adds to the molecule. In this adduct, the SO–OO bond is already very long (2.18 Å), and $^1O_2$ is practically preformed. This can be seen from orbitals of the two outer oxygens (HOMO and LUMO, Lowest Unoccupied Molecular Orbital) that already have the symmetry and spin distribution of $^1O_2$. This differs substantially from that of $^3O_2$ given for comparison. Moreover, the O–O bond length in $^1O_2$ (1.205 Å) is very close to the SOO–O bond length in the adduct (1.233 Å), while the S–OOO bond length in the adduct is only marginally longer (1.558 Å) than that of the final product dimethyl sulfoxide (1.549 Å).

With dimethyl sulfoxide and methanesulfinate ion, the ozone adducts even decomposed on the way to optimising their structures, but shortly before this occurred single-point values indicated that their standard Gibbs free energies of adduct formation must be near –116 kJ mol^{-1} and –178 kJ mol^{-1}, respectively. With methyl methanesulfinate the adduct was also practically decomposed but the standard Gibbs free energy of adduct formation could still be calculated. There seems to be a correlation between the standard Gibbs free energy of the overall reaction and the standard Gibbs free energy of adduct formation (Figure 9.2). The value for the ozone adduct to HS$^-$ has not been included, as the adduct is generally unstable and the value obtained was only a local minimum. In fact, O-transfer is only of minor importance, and the reaction proceeds in several directions, which are discussed in Chapter 11.

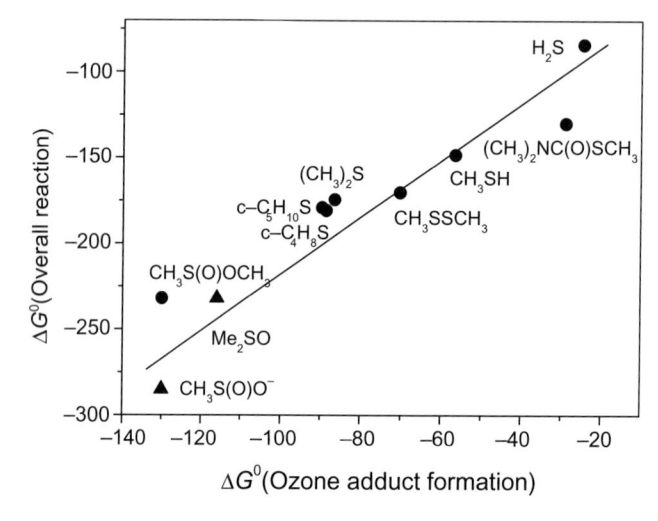

Figure 9.2 Correlation of the standard Gibbs free energy of ozone adduct formation with the standard Gibbs free energy of the overall O-transfer reaction. Dimethyl sulfoxide and methylsulfinate ion (triangles) do not form stable adducts and the standard Gibbs free energy of adduct formation has been estimated by single-point calculations [Naumov & von Sonntag (2011), unpublished results].

Thus, in these reactions the adduct is often not a real intermediate but rather a transition state. This may happen, when the standard Gibbs free energy of the overall reaction exceeds about 200 kJ mol^{-1}.

It is tempting to speculate that in benzene and its derivatives the transition state in the formation of the adduct is very close to the adduct, and thus we observe an excellent correlation of the standard Gibbs free energy of adduct formation with the logarithm of the rate constant (Chapter 7). In the sulfur compounds, in contrast, we may have an early transition state, and the overall energetics as expressed by the standard Gibbs free energy does not correlate with the rate of reaction. This is not without precedence. In the H-abstraction of alkyl radicals from thiols (R$^•$ + R'SH → RH + R'S$^•$), the rate constant of reaction *increases with decreasing* energy gain of product formation (Reid *et al.,* 2002). This intriguing observation has been rationalised by a detailed quantum mechanical analysis of the reaction path, and it has been concluded that there must be an early transition state and that charge polarisation effects dominate the rate of reaction.

Dimethyl sulfoxide reacts only very slowly with ozone (Table 9.1) despite the high exergonicity of the overall reaction (Figure 9.2). An adduct sufficiently stable to account as an intermediate cannot be calculated (see above). It may be speculated that the oxygen withdraws much of the electron density at sulfur and the zwitterionic adduct with a positive charge at sulfur cannot be established.

9.4 SULFOXIDES

Sulfoxides react very slowly with ozone (Table 9.1). Thus as far as micropollutant abatement is concerned, these typical products of the reactions of sulfides must be considered as ozone-refractory. With dimethyl sulfoxide (DMSO), dimethyl sulfone is the only product detectable by ^{1}H NMR analysis [Jarocki & von Sonntag (2011), unpublished results]. Quantum-chemical calculations indicate that there is indeed a high likelihood for this reaction [Naumov & von Sonntag (2010), unpublished results].

Due to its low ozone rate constant and its high $^•$OH rate constant, dimethyl sulfoxide (DMSO) can be used to detect and quantify $^•$OH yields in ozone reactions (Chapter 14).

Because, for the determination of $^•$OH in ozone reaction, the rate constant of ozone with DMSO is crucial for setting up adequate competition conditions, its ozone rate constant has been re-determined [Reisz & von Sonntag (2011), unpublished results]. According to the new data, the earlier value (8 M^{-1}s^{-1}, Table 9.1) seems to be on the high side. In Figure 9.3, an Arrhenius plot for the reaction of ozone with DMSO is given.

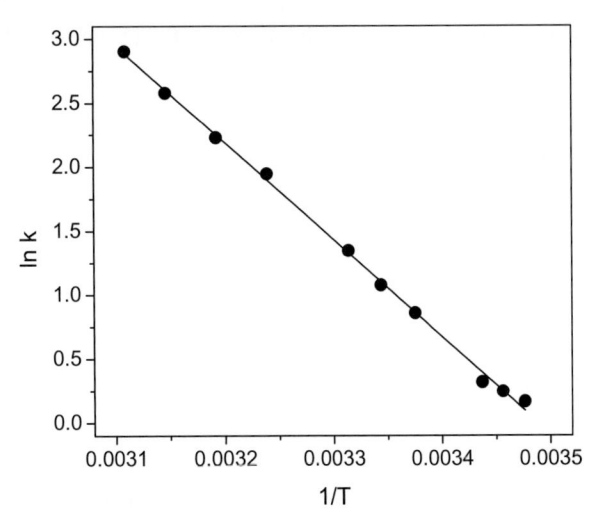

Figure 9.3 Arrhenius plot for the reaction of ozone with dimethyl sulfoxide according to Reisz & von Sonntag (2011), unpublished results.

This yields $k = 1.8$ at 20 °C (log $A = 11.5$, $E_a = 62$ kJ mol^{-1}). The Hammond–Leffler postulate assumes early transition states for strongly exergonic reactions, and data shown in Figure 9.2 indicate that this situation may be met in the DMSO/ozone reaction. The fact that the reaction is nevertheless slow is due to the high activation energy. Similarly high activation energies are observed in ozone reactions with C–H functions (Chapter 10).

9.5 MICROPOLLUTANTS CONTAINING AN OZONE-REACTIVE SULFUR

The ozone-reactive site in the antibacterial agent penicillin G is its sulfide function. Adequately substituted sulfoxides are chiral, and with penicillin G two enantiomeric sulfoxides are formed, the *R*-isomer dominating over the *S*-isomer in a 55/45 ratio [reactions (3) and (4)]. The *R*-isomer is the energetically preferred product (Naumov & von Sonntag, 2012, unpublished quantum-chemical calculations). Practically no other products are formed (Dodd *et al.*, 2010).

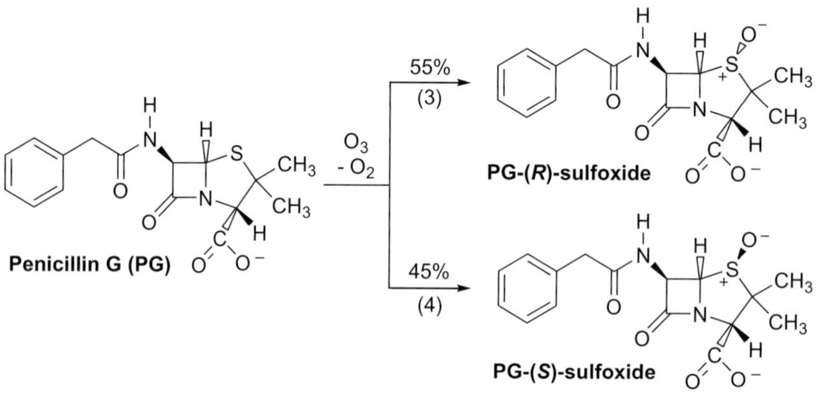

Similarly, the main products of cephalexin, which belongs to the group of cephalosporin antibiotics are the *R*- and *S*-sulfoxides.

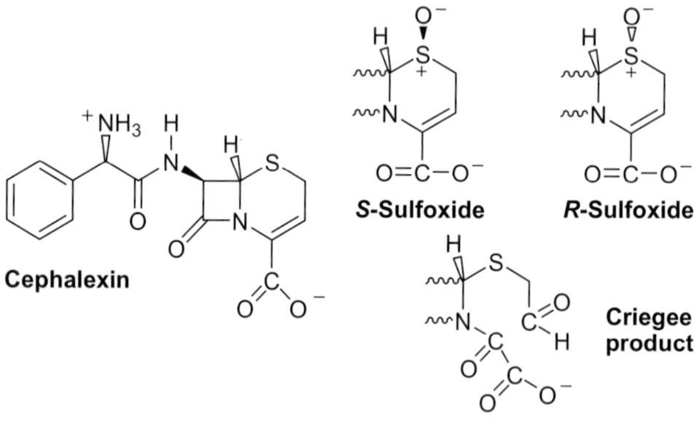

However, the C–C double bond is another site of attack, and thus the Criegee product (Chapter 6) is also of importance here. In principle, the amino group could also be a third site of ozone attack, but at pH 7, where these experiments were carried out, it is largely protonated. At a higher pH, the amino group must be another site of attack (for the pH dependence of different ozone-reactive groups see Chapter 2). At pH 7, k_{obs} is reported as 8.7×10^4 $M^{-1}s^{-1}$, and at this pH amines would react an order of magnitude more slowly (Chapter 8). A material balance of the initially formed products shows that the *R*-sulfoxide dominates (~34%) over the *S*-sulfoxide (~18%). The Criegee product (≥28%) plus a hydrolysed Criegee product

($\sim$5%) are formed in a somewhat lower but comparable yield. This indicates that the observed rate constant must be due to about equal contributions of the reactivity of the sulfide function and the C–C double bond.

With penicillin G, k_{obs} is only 4.8×10^3 M^{-1}s^{-1} at pH 7. One would thus have to conclude that the sulfide function in the five-membered ring is less reactive than in the six-membered ring.

The insecticides aldicarb and vydate and the antifouling agent (algicide) irgarol have sulfide groups which are the primary sites of ozone attack (Table 9.1). It is expected that mainly the corresponding sulfoxides are formed in these reactions.

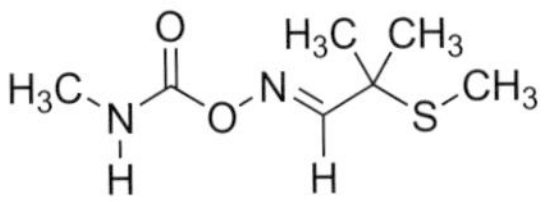

Aldicarb

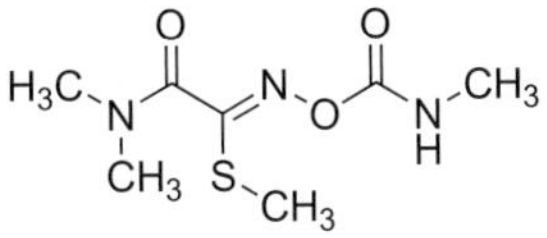

Vydate (Oxamyl)

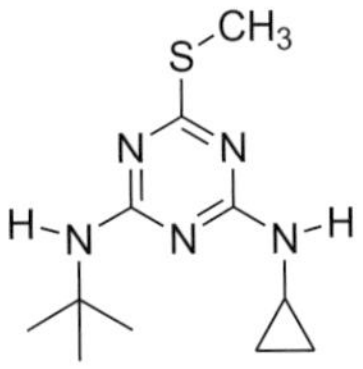

Irgarol (Cybutryn)

The antibiotic lincomycin has two potential sites of ozone attack, a sulfide group and the tertiary nitrogen in the five-membered ring. Because the reaction rate is largely controlled by the tertiary amine at near-neutral pH, lincomycin ozonolysis is discussed in Chapter 8. The antibiotic amoxillin contains a sulfide function, but also amine and phenolic groups. It is the ozone chemistry of the latter that dominates its degradation (Chapter 7).

The thiocarbamate herbicides molinate and EPTC have been estimated to react moderately fast with ozone (Table 9.1).

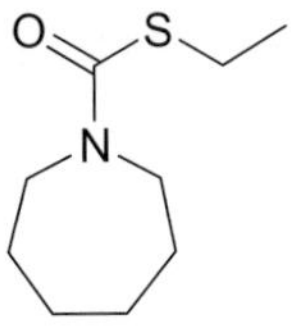

Molinate

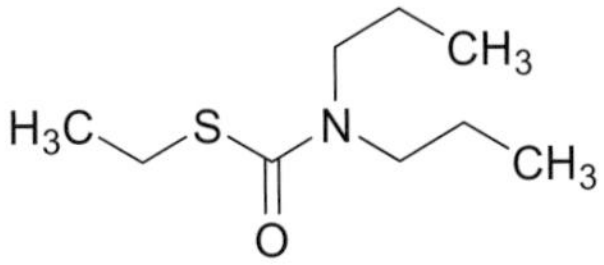

S-Ethyl-N,N-dipropylthiocarbamate (EPTC)

The reaction may lead by O-transfer to the corresponding sulfinylamides. Details are as yet not known.

Chapter 10

Compounds with C–H functions as ozone-reactive sites

10.1 REACTIVITY OF COMPOUNDS WITH C–H FUNCTIONS AS OZONE-REACTIVE SITES

The title compounds react generally so slowly with ozone that in pollutant abatement they are typically called "ozone refractory". Yet, within this generally low reactivity there is a spread in their ozone rate constants of seven orders of magnitude (Table 10.1).

Table 10.1 Compilation of ozone rate constants of compounds containing only C–H functions as reactive sites. Published rate constants are rounded to significant figures

Compound	pKa	pH	$k/M^{-1}\,s^{-1}$	Reference
Acetate ion			$<3 \times 10^{-5}$	Hoigné & Bader, 1983b
Acetic acid	4.75		$\leq 3 \times 10^{-5}$	Hoigné & Bader, 1983b
			$<1 \times 10^{-6}$	Sehested et al., 1992
Acetone			3.2×10^{-2}	Hoigné & Bader, 1983a
Bromoform			$\leq 2 \times 10^{-2}$	Hoigné & Bader, 1983a
1-Butanol		2	0.6	Pryor et al., 1984
2-Butanone		2	0.7	Hoigné & Bader, 1983a
Butyrate ion			$\leq 6 \times 10^{-3}$	Hoigné & Bader, 1983b
Butyric acid	4.8		$\leq 6 \times 10^{-3}$	Hoigné & Bader, 1983b
Carbon tetrachloride			$<5 \times 10^{-3}$	Hoigné & Bader, 1983a
Chloroform			≤ 0.1	Hoigné & Bader, 1983a
Cyclopentanol		2	2.0	Hoigné & Bader, 1983a
Dichloromethane			≤ 0.1	Hoigné & Bader, 1983a
Diethyl ether		2	1.1	Hoigné & Bader, 1983a
Diethyl malonate		2	0.06	Hoigné & Bader, 1983a
1,4-Dioxane			0.32	Hoigné & Bader, 1983a
Ethanol			0.45	Pryor et al., 1984
			0.37	Hoigné & Bader, 1983a

(Continued)

Table 10.1 Compilation of ozone rate constants of compounds containing only C–H functions as reactive sites. Published rate constants are rounded to significant figures (*Continued*)

Compound	pKa	pH	k/M^{-1} s^{-1}	Reference
Formaldehyde			0.1	Hoigné & Bader, 1983a
Formate ion[a]			100 ± 20	Hoigné & Bader, 1983b
no additive[b]			82	Reisz & von Sonntag, 2011
with tBuOH[c]			46	Reisz & von Sonntag, 2011
Formic acid	3.75		very low	Hoigné & Bader, 1983b
Geosmin			0.1	Peter & von Gunten, 2007
Glucose		2	0.45	Hoigné & Bader, 1983a
Glutarate ion			(8 ± 2) × 10^{-3}	Hoigné & Bader, 1983b
Glutaric acid	4.3, 5.4		<8 × 10^{-3}	Hoigné & Bader, 1983b
Glyoxylate ion			1.9 ± 0.2	Hoigné & Bader, 1983b
Glyoxylic acid	3.2		0.17 ± 0.4	Hoigné & Bader, 1983b
Malonate ion			7 ± 2	Hoigné & Bader, 1983b
Malonic acid	2.8, 5.7		<4	Hoigné & Bader, 1983b
Methanol			~0.02	Hoigné & Bader, 1983a
2-Methylisoborneol			0.35	Peter & von Gunten, 2007
2-Methyl-2-propanol (tertiary Butanol)			~3 × 10^{-3}	Hoigné & Bader, 1983a
			1.1 × 10^{-3}[d]	Reisz & von Sonntag, 2011
Octanal			8 ± 0.8	Hoigné & Bader, 1983a
1-Octanol			≤0.8	Hoigné & Bader, 1983a
Oxalate dianion (Oxalic acid)	1.2, 4.2		≤4 × 10^{-2}	Hoigné & Bader, 1983b
2-Pentanone			~0.02	Hoigné & Bader, 1983a
Pivalate ion			2 × 10^{-3}	Pryor et al., 1984
Pivalic acid	5.03		<1 × 10^{-3}	Pryor et al., 1984
Propanal			2.5	Hoigné & Bader, 1983a
1-Propanol			0.37	Hoigné & Bader, 1983a
2-Propanol			1.9	Hoigné & Bader, 1983a
			1.9	Pryor et al., 1984
			0.83[e]	Reisz & von Sonntag, 2011
Propionate ion			(1 ± 0.5) × 10^{-3}	Hoigné & Bader, 1983b
Propionic acid	4.9		<4 × 10^{-4}	Hoigné & Bader, 1983b
Propyl acetate			0.03	Hoigné & Bader, 1983a
Succinate ion			(3 ± 1) × 10^{-2}	Hoigné & Bader, 1983b
Succinic acid	4.2, 5.6		<3	Hoigné & Bader, 1983b
Sucrose			0.012	Hoigné & Bader, 1983a
			0.05	Pryor et al., 1984
Tetrahydrofuran			6.1	Pryor et al., 1984

[a]In the presence of 2-propanol as •OH scavenger, [b]at 20°C, log$A$ = 10.88, E_a = 50.3 kJ mol^{-1}, [c]tBuOH as •OH scavenger, at 20°C, log$A$ = 11.36, E_a = 54.6 kJ mol^{-1}, [d]at 20°C, log$A$ = 9.30, E_a = 68.7 kJ mol^{-1}, [e]at 20°C, log$A$ = 12.51, E_a = 70.7 kJ mol^{-1}.

For the ozone reactions of formate ion, 2-propanol and tertiary butanol (2-methyl-2-propanol, tBuOH) *A* factors and activation energies have been determined (Table 10.1). For formate ion, data in the presence of tBuOH show that scavenging •OH radicals not only lowers the observed rate constant, but also enhances the activation energy (Section 10.3). The importance of •OH radicals in such systems is discussed below, but similar effects are expected for other compounds such as alcohols and ethers, but information is more difficult to obtain for these because of their lower rate constants (unfavourable competition conditions).

10.2 GENERAL MECHANISTIC CONSIDERATIONS

For the reaction of ozone with compounds that have only C–H functions as reactive sites, such as methane, three primary reactions have been considered: (i) H-abstraction [reaction (1)], (ii) hydride transfer [reaction (2)], and (iii) insertion [reaction (3)] (Nangia & Benson, 1980; Giamalva et al., 1986). For aqueous solutions, one may even add electron transfer [reaction (4)].

$$CH_4 + O_3 \rightarrow {}^\bullet CH_3 + HO_3^\bullet \tag{1}$$

$$CH_4 + O_3 \rightarrow HO_3^- + {}^+CH_3 \tag{2}$$

$$CH_4 + O_3 \rightarrow CH_3OOOH \tag{3}$$

$$CH_4 + O_3 \rightarrow O_3^{\bullet -} + {}^{+\bullet}CH_4 \tag{4}$$

With methane as a model substrate that must react very slowly with ozone (rate constant not available, but may be similar to or lower than tertiary butanol, Table 11.1), quantum chemical reactions have been carried out [Naumov & von Sonntag, (2011), unpublished]. H-abstraction [reaction (1)], is mildly endergonic ($\Delta G^0 = +20$ kJ mol^{-1}). Hydride transfer [reaction (2)] is markedly endergonic ($\Delta G^0 = +150$ kJ mol^{-1}) as is electron transfer [reaction (4), $\Delta G^0 = +254$ kJ mol^{-1}]. While reactions (1), (2) and (4) are mechanistically straightforward, insertion, reaction (3), is somewhat problematic. In the transition state (cf. Figure 10.1), the carbon has to become pentavalent, and the transfer of the hydrogen is related to the H-abstraction reaction (1). Although reaction (3) is the most exergonic of these reactions, $\Delta G^0 = -168$ kJ mol^{-1}, there may be a considerable activation energy that would make this reaction slow. This may generally hold for insertion reactions which are all highly exergonic,

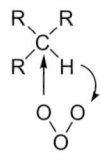

Figure 10.1 Likely transition state in the insertion reaction.

The formate ion reacts fastest within this group of compounds, and here one may write reactions (5)–(8).

$$HC(O)O^- + O_3 \rightarrow CO_2^{\bullet -} + HO_3^\bullet \tag{5}$$

$$HC(O)O^- + O_3 \rightarrow CO_2 + HO_3^- \tag{6}$$

$$HC(O)O^- + O_3 \rightarrow HOOOC(O)O^- \tag{7}$$

$$HC(O)O^- + O_3 \rightarrow HC(O)O^\bullet + O_3^{\bullet -} \tag{8}$$

Compared to methane, the calculated energetics are now very different. For the H-abstraction reaction (5), a standard Gibbs free energy of -32 kJ mol^{-1} has been calculated. Hydride transfer is now highly exergonic [reaction (6), $\Delta G^0 = -249$ kJ mol^{-1}]. The insertion reaction (7) continues to be exergonic [$\Delta G^0 = -239$ kJ mol^{-1}]. An electron transfer remains endergonic [reaction (8), $\Delta G^0 = +43$ kJ mol^{-1}], but is no longer dramatically so when compared to methane [reaction (4)]. Interestingly, the insertion product is sufficiently stable to account as an intermediate. Its decay into the final products [reaction (9), $\Delta G^0 = -61$ kJ mol^{-1}] is exergonic, as expected.

$$HOOOC(O)O^- \rightarrow CO_2 + HO_3^- \quad (9)$$

These calculations indicate that it will not be possible for this group of compounds to come up with a uniform reaction mechanism for the primary step as in the case of olefins and, at least to a large extent, for aromatic compounds (Chapters 6 and 7). H-abstraction and electron transfer [cf. reactions (1), (4), (5) and (8) give rise to $HO_3^\bullet/O_3^{\bullet-}$ and hence to $^\bullet OH$ (Chapter 11)]. There are ways of quantifying $^\bullet OH$ in ozone reactions (Chapter 14), and use has been made of this for addressing the importance of these radical-forming processes.

Hydride transfer [cf. reactions (2) and (5)] has been favoured by Benson (Nangia & Benson, 1980), while the Plesnicar and Pryor groups (Plesnicar et al., 1998; Giamalva et al., 1986) suggest that H-abstraction [cf. reactions (1) and (5)] dominates. It will be shown below that both processes seem to contribute to varying extents based on the few preliminary data that are available at present (formate ion, 2-propanol and tertiary butanol). As long as H-abstraction contributes to a significant extent, the observed rate constant must correlate with the C–H bond dissociation energy (BDE). Such a correlation can inherently not be perfect, as the contribution of H-abstraction varies from compound to compound, but at least the order of magnitude will be approximately correct.

A plot of the logarithm of the rate constant vs. the standard Gibbs free energies (ΔG^0) for H-transfer should then give a straight line. For this plot, formate ion, 2-propanol, ethanol, methanol and tertiary butanol (tBuOH) have been selected (Table 10.2 and Figure 10.2).

Table 10.2 Rate constants of ozone with some saturated alcohols and the formate ion (Table 10.1) and calculated standard Gibbs free energies for H-abstraction by ozone [Naumov & von Sonntag (2011), unpublished] as well as some available BDEs (von Sonntag, 2006) of the relevant C–H bonds

Compound	k/M^{-1} s^{-1}	ΔG^0 (H-abstraction)/kJ mol^{-1}	BDE kJ mol^{-1}
H–C(O)O$^-$	46	-32	
H–C(CH$_3$)$_2$OH	0.83	-23	393
H–C(CH$_3$,H)OH	0.4	-14	396
H–CH$_2$OH	0.02	-10	402
H–CH$_2$C(CH$_3$)$_2$OH	0.001	$+17$	423[a]

[a]Taken from the value for ethane.

If it is correct that H-abstraction and hydride transfer are the main competing ozone reactions for C–H compounds in aqueous solution, their rate constants must be similar and, as a consequence, kinetic parameters that determine H-abstraction seem to also determine hydride transfer to a similar extent. These ideas are presented here to elicit further research. Quite obviously, the approach outlined above is an oversimplification. For example, the acetate ion reacts much slower with ozone than tBuOH (Table 10.1) although the C–H BDE (calculated at $\Delta G^0 = +14$ kJ mol^{-1}) is lower (consider mesomeric stabilisation of the ensuing radical) than that of tBuOH. Yet, in this respect, ozone shares this surprising effect with the much more reactive $^{\bullet}$OH radical [$k(^{\bullet}\text{OH} + \text{acetate}) = 8 \times 10^7$ M^{-1}s^{-1}, $k(^{\bullet}\text{OH} + \text{tBuOH}) = 6 \times 10^8$ M^{-1}s^{-1} (Buxton et al., 1988)]. Thus, factors beyond mere BDEs play a role in H-abstraction reactions.

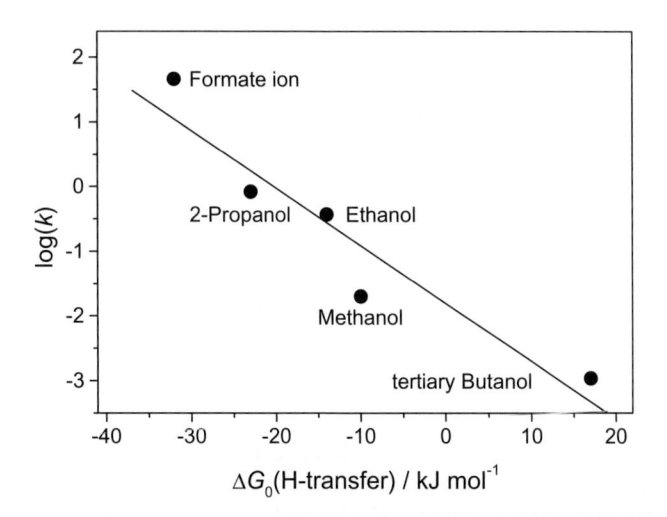

Figure 10.2 Plot of the logarithm of the rate constants for the reactions of ozone with formate ion, 2-propanol, ethanol, methanol and tertiary butanol vs. the calculated standard Gibbs free energy for H-transfer from the substrate to ozone.

There is a host of data regarding the products in the reactions of ozone with neat substrates or in carbon tetrachloride solutions (Bailey, 1982; Plesnicar et al., 1976, 1991, 1998, 2000; Cerkovnik & Plesnicar, 1983, 1993, 2005; Koller et al., 1990; Koller & Plesnicar, 1996; Kovac & Plesnicar, 1979; Murray et al., 1970), but the information regarding their reactions in aqueous solution is meagre. Presently, information on the ozone chemistry of compounds containing only C–H functions as ozone reactive sites (carboxylic acids, alcohols, carbonyl compounds) is very limited, and data using excessive ozonation (Niki et al., 1983) cannot be used for elucidating mechanistic aspects. One may be tempted to transfer mechanistic concepts derived in non-polar environments to aqueous solutions, but, as will be shown, the presence of water has a dramatic effect on the rates of some reactions. To this point, one of us (CvS) together with two younger colleagues (Erika Reisz and Alexandra Jarocki) has carried out some experiments on the formate ion, 2-propanol and tertiary butanol and these, as yet unpublished, data now allow us to put the ozone chemistry of C–H reactive sites into perspective.

10.3 FORMATE ION

Of all C–H reactive compounds, ozone reacts fastest with the formate ion (Table 10.1). Upon addition of the $^{\bullet}$OH scavenger tBuOH, k_{obs} is lowered (Table 10.1), and the observed activation energy is higher (54.6 kJ mol^{-1}) than in the absence of tBuOH (50.3 kJ mol^{-1}) (Table 10.1 and Figure 10.3).

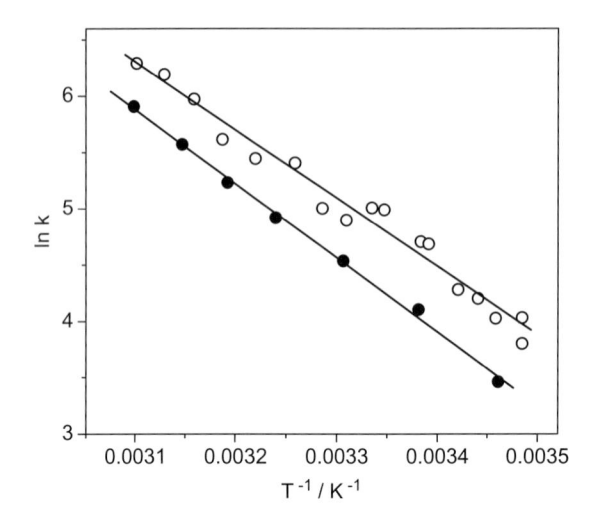

Figure 10.3 Arrhenius plot for the reaction of ozone with the formate ion (open circles) and in the presence of the •OH scavenger tBuOH (closed circles, [formate ion] = 10–100 μM, [tBuOH] = 0.2 M) [Reisz & von Sonntag (2011) unpublished results].

The yield of free •OH is near 40% (Reisz et al., 2012a). There are two conceivable reactions that could lead to •OH radicals, that is, to their precursors $HO_3^•/O_3^{•-}$ [reactions (5) and (8)]. For the H-abstraction reaction (5), a standard Gibbs free energy of -32 kJ mol^{-1} has been calculated, while electron transfer (8) is endergonic, $\Delta G^0 = +43$ kJ mol^{-1}. Of these two conceivable reactions, the exergonic reaction is favoured here. H-abstraction reactions may be connected with considerable activation energies, and it is not surprising that reaction (5) is slow despite the exergonicity of the reaction.

Cage recombination (reaction of $CO_2^{•-}$ with $HO_3^•$) will lower the free •OH yield. The efficiency of cage recombination depends on the recombination rate constants of the radicals involved. With •OH, for example, it is high ($k = 5.5 \times 10^9$ M^{-1} s^{-1}), while for $SO_4^{•-}$ it is much lower ($k = 8 \times 10^8$ M^{-1} s^{-1}). As a consequence, the free •OH yield in the photolysis of H_2O_2 is 1.0 at room temperature (Legrini et al., 1993), while in the photolysis of peroxodisulfate the free $SO_4^{•-}$ yield is 1.4 (Mark et al., 1990). Upon increasing the temperature, the viscosity of water is lowered, the $SO_4^{•-}$ radicals diffuse even more freely into the bulk solution, and at 50°C the quantum yield of free $SO_4^{•-}$ approaches the maximum value of 2.0. With respect to the radicals formed in reactions (5) and (8), one has to consider that the competing reactions (10) and (11) occur on the 10^{-12} s timescale [cf. $pK_a(HO_3^•) = -2.0$, Chapter 13], while the timescale of cage reactions is in the order of 10^{-9} s.

$$HO_3^• \longrightarrow H^+ + O_3^{•-} \tag{10}$$
$$HO_3^• \longrightarrow {}^•OH + O_2 \tag{11}$$

Reactions $O_3^{•-} + O_3^{•-}$ ($k = 9 \times 10^8$ M^{-1} s^{-1}) and $CO_2^{•-} + CO_2^{•-}$ ($k = 5 \times 10^8$ M^{-1} s^{-1}) are relatively slow, and $k(O_3^{•-}+CO_2^{•-})$ may be similar, while the reaction •OH + $CO_2^{•-}$ may be as fast as •OH + •OH. Thus, cage recombination efficiency may be close to or even less than that of $S_2O_8^{2-}$ photolysis. Based on this, the free •OH yield is not far from 57% for total radical formation if one assumes the same cage escape efficiency as in $S_2O_8^{2-}$ photolysis (Reisz et al., 2012a).

With this, the free-radical generating reactions (5) [plus (8)] are important processes, but certainly not the only ones. There must be non-radical processes such as (6) and (7) that contribute about as much to the primary reaction as the free-radical reactions. Neglecting the insertion reaction (7), leaves hydride transfer (6) as the competing reaction. The hydride transfer is highly exergonic ($\Delta G^0 = -249$ kJ mol^{-1}), yet the observed reaction is slow. This can be reconciled considering that in the transition state the strong solvation of the formate ion disappears while that of HO$_3^-$ has to build up. As far as products are concerned, hydride transfer and insertion are indistinguishable. With tertiary butanol, however, these two processes give rise to very different products, and hydride transfer has been shown to be by far the dominating process (see below). In analogy, it is also favoured here.

10.4 2-METHYL-2-PROPANOL (TERTIARY BUTANOL)

Tertiary butanol (tBuOH) reacts with ozone nearly five orders of magnitudes more slowly than the formate ion (Table 10.1). From the data in Figure 10.4, an activation energy of $E_a = 68.7$ kJ mol^{-1} is calculated.

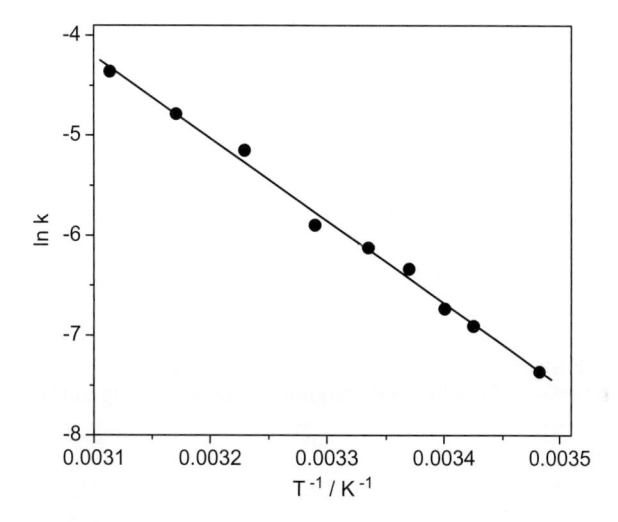

Figure 10.4 Arrhenius plot for the reaction of ozone with tertiary butanol (1 M) (Reisz et al., 2012a).

At a tBuOH concentration of 1 M, formaldehyde (7%), acetone (9%) and methyl ethyl ketone (92%) have been quantified (Reisz et al., 2012a). While formaldehyde and acetone are typical •OH-induced products (Chapter 14), methyl ethyl ketone is not formed via free-radical reactions.

Upon trying to calculate the standard Gibbs free energy for hydride transfer, the quantum-chemical calculations proceeded automatically to protonated methyl ethyl ketone by a methyl shift [reaction (12)/(13)]. The carbocation depicted in reaction (12) is not an intermediate in this reaction. It is a transition state, at the best, *en route* to the products [reaction (13); $\Delta G^0(12)/(13) = -114$ kJ mol^{-1}].

Protonated ketones are very strong acids, the pK_a value of protonated acetone is given as -3.06 (Bagno et al., 1995), and protonated methyl ethyl ketone is also only a very short-lived intermediate. Insertion of ozone into tBuOH yielding HOOOCH$_2$C(CH$_3$)$_2$OH is strongly exergonic ($\Delta G^0 = -145$ kJ mol^{-1}). This potential intermediate would most likely decay into $HO_2^\bullet$ and $^\bullet OCH_2C(CH_3)_2OH$ ($\Delta G^0 = -17$ kJ mol^{-1}) rather than release of HO_3^- ($\Delta G^0 = +31$ kJ mol^{-1}). This renders insertion as a route to methyl ethyl ketone unlikely.

Compared to hydride transfer, H-abstraction is minor [reaction (14), $\Delta G^0 = +17$ kJ mol^{-1}].

$$(CH_3)_3OH + O_3 \rightarrow {}^\bullet CH_2C(CH_3)_2OH + HO_3^\bullet \tag{14}$$

As $HO_3^\bullet$ is converted to $^\bullet OH$ that again abstracts a hydrogen from tBuOH, two $^\bullet CH_2C(CH_3)_2OH$ radicals are generated by one primary event (for details of the ensuing chemistry see Chapter 14). This closes the product balance (ozone vs. products) shown above, which is only seemingly above 100%.

10.5 2-PROPANOL

Detailed information on the ozonolysis of 2-propanol in organic solvents is available, and data in aqueous solution have recently been obtained [Reisz et al. (2011), unpublished results]. The difference between product yields in organic solvents and in water is remarkable, and first the results of studies in an organic solvent are discussed.

With 2-propanol in acetone-d_6, H_2O_3 and the α-hydroxyalkylhydrotrioxide were detected by low-temperature ^{17}O-NMR (Table 10.3) in a 1:2 ratio (Plesnicar et al., 1998). The final products (1 M in methyl-t-butyl ether) are compiled in Table 10.4.

Table 10.3 ^{17}O-NMR shifts (δ) of $(CH_3)_2C(OH)OOOH$, H_2O_3 and H_2O_2 in acetone-d_6 at -10°C. δ Values in ppm downfield of the internal standard $H_2^{17}O$. According to Plesnicar et al. (1998)

Compound	δO^1	δO^2	δO^3
$(CH_3)_2C(OH)O^1$–O^2–O^3–H	368	445	305
H–O^1–O^2–O^3–H	305	421	305
H–O^1–O^2–H	187	187	

Table 10.4 Ozonolysis of 2-propanol (1 M in methyl-t-butyl ether) at -78°C. Product determination after warming. Material balance as by GC/MS and NMR (sum of all products = 100%) (Plesnicar et al., 1998) and in aqueous solution [Jarocki et al. (2011), preliminary unpublished results]. Material balance based on ozone consumed

Product	Organic solvent	Water
Acetone	37±5	88
Acetic acid	39±5	3.6
Acetic peracid	11±2	
Formic acid	7±2	
Formaldehyde (as hemiacetal)	5±1	
Formaldehyde		3.9
H_2O_2	11±3	n.d.

n.d.: not determined

In the organic solvent, there is, however, a major material balance deficit of one-carbon compounds (formaldehydeisopropylacetal plus formic acid, 12%) vs. two-carbon products (acetic acid plus acetic peracid, 50%). This could be due to the fact that GC without derivatisation is not a reliable method for determining formaldehyde. Assuming that the missing one-carbon compound is formaldehyde, one may follow the proposed mechanistic suggestions (Plesnicar et al., 2000) with some additions.

It has been suggested that ozone reacts as an H-abstractor. This would create a radical pair that is held together for a short time by the solvent cage, where it can react by combination or disproportionation. Such reactions are shown here for the 2-propanol system [reactions (15)–(18)].

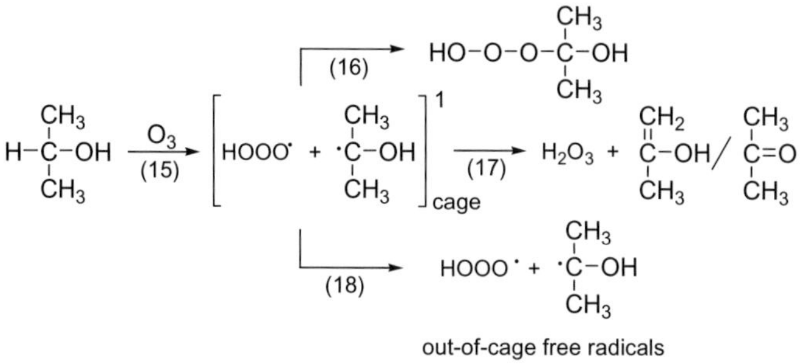

$$\begin{array}{c}
CH_3 \\
H-C-OH \\
CH_3
\end{array} \xrightarrow[(15)]{O_3} \left[HOOO^\cdot + \begin{array}{c} CH_3 \\ \cdot C-OH \\ CH_3 \end{array} \right]_{cage} \begin{array}{c} \xrightarrow{(16)} HO-O-O-\underset{CH_3}{\overset{CH_3}{C}}-OH \\ \xrightarrow{(17)} H_2O_3 + \underset{CH_3}{\overset{CH_2}{C}}-OH \Big/ \underset{CH_3}{\overset{CH_3}{C}}=O \\ \xrightarrow{(18)} HOOO^\cdot + \underset{CH_3}{\overset{CH_3}{\cdot C}}-OH \end{array}$$

out-of-cage free radicals

Evidence for the formation of the enol of acetone (2-hydroxypropene) is obtained from experiments with 2-propanol deuterated at methyl, where deuterium is incorporated into the H_2O_3 formed.

Preferred H-transfer from the methyl groups as compared to the OH group is typical for the disproportionation of the 2-hydroxymethyl radical (von Sonntag, 1969; Blank et al., 1975). This is an amusing reaction in so far as it follows, as so often in free-radical chemistry, Ostwald's step rule, that is, the thermodynamically less favoured product, here the enol of acetone, is formed preferentially [ΔG^0(2-hydroxypropene → acetone) = −69 kJ mol^{-1}, Naumov, private communication]. In the photolysis of acetone in the presence of 2-popanol, two 2-hydroxyprop-2-yl radicals are formed within the solvent cage [reaction (19)].

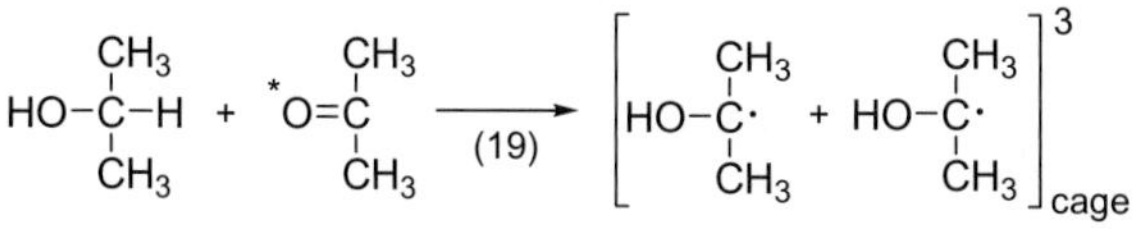

$$\begin{array}{c} CH_3 \\ HO-C-H \\ CH_3 \end{array} + \begin{array}{c} CH_3 \\ {}^*O=C \\ CH_3 \end{array} \xrightarrow{(19)} \left[\begin{array}{c} CH_3 \\ HO-C\cdot \\ CH_3 \end{array} + \begin{array}{c} CH_3 \\ HO-C\cdot \\ CH_3 \end{array} \right]^3_{cage}$$

As this reaction is caused by triplet acetone, the multiplicity of the ensuing radical pair is also triplet. This prevents a ready recombination/disproportionation in the cage, as activation energy is required for their reaction as the crossing over point triplet → singlet is at higher energies. Thus most radicals will diffuse out of the cage, and it is mainly the reaction of the free radicals that accounts for the products [reactions (20)–(22)].

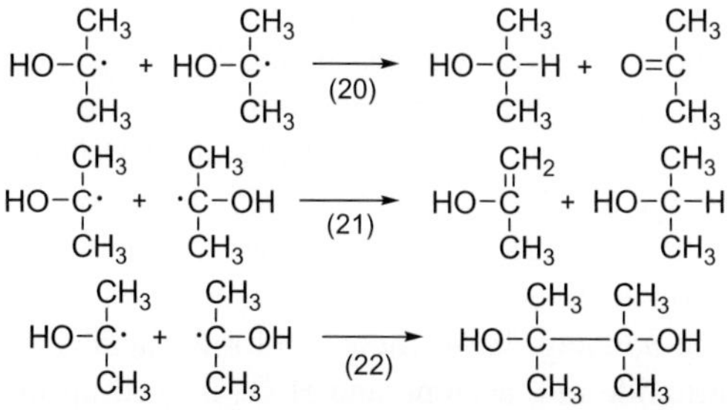

$$\begin{array}{c} CH_3 \\ HO-C\cdot \\ CH_3 \end{array} + \begin{array}{c} CH_3 \\ HO-C\cdot \\ CH_3 \end{array} \xrightarrow{(20)} \begin{array}{c} CH_3 \\ HO-C-H \\ CH_3 \end{array} + \begin{array}{c} CH_3 \\ O=C \\ CH_3 \end{array}$$

$$\begin{array}{c} CH_3 \\ HO-C\cdot \\ CH_3 \end{array} + \begin{array}{c} CH_3 \\ \cdot C-OH \\ CH_3 \end{array} \xrightarrow{(21)} \begin{array}{c} CH_2 \\ HO-C \\ CH_3 \end{array} + \begin{array}{c} CH_3 \\ HO-C-H \\ CH_3 \end{array}$$

$$\begin{array}{c} CH_3 \\ HO-C\cdot \\ CH_3 \end{array} + \begin{array}{c} CH_3 \\ \cdot C-OH \\ CH_3 \end{array} \xrightarrow{(22)} \begin{array}{c} CH_3 \; CH_3 \\ HO-C—C-OH \\ CH_3 \; CH_3 \end{array}$$

An analysis of the data using 2-propanol-d_6 as a reactant indicated that $k_{CH}/k_{OH} = 3.3$ and $k_{CH}/k_{recomb} = 3.4$ (Blank et al., 1975).

In the above ozone plus 2-propanol reaction (15), the radical pair has singlet multiplicity, and compared to the excited acetone plus 2-propanol reaction (19), there is no barrier for the cage reaction. Thus, enol (2-hydroxypropene) formation is most likely a cage reaction [cf. reaction (17)]. In the ozonolysis of 2-propanol in an organic solvent, ozonolysis of 2-hydroxypropene is believed to compete with the transformation of enol–acetone (2-hydroxypropene) to acetone (Plesnicar et al., 1998), and it has been suggested that acetic peracid and formic acid may result from this reaction (Plesnicar et al., 2000). Based on what has been reported in Chapter 6 for aqueous solutions, one may suggest reactions (23)–(25).

$$\begin{array}{c}CH_2\\||\\C-OH\\|\\CH_3\end{array} \xrightarrow{O_3 \atop (23)} \begin{array}{c}H\\|\\O\diagdown C-H\\|\quad|\\O\diagup C-OH\\|\\CH_3\end{array} \begin{array}{c}\xrightarrow{(24)} CH_2O + CH_3-\overset{O}{\underset{||}{C}}-O-OH\\[4pt] \xrightarrow{(25)} H-\underset{\underset{OH}{|}}{\overset{\overset{H}{|}}{C}}-O-OH + CH_3-\overset{O}{\underset{||}{C}}-OH\end{array}$$

The electron-withdrawing OH substituent at the C–C double bond will drive the reaction preferentially toward reaction (25) if vinyl chloride (Chapter 6) is a good guide, that is, acetic acid will be favoured over acetic peracid, as observed (Table 10.4).

Formic acid is not accounted for in this scheme. Yet, formic acid may result from the reaction of acetic peracid with formaldehyde considering the reaction of acetic peracid with acetaldehyde (Schuchmann & von Sonntag, 1988).

In competition, the radicals may diffuse out of the cage [reaction (18)]. Here, the viscosity of the solvent is of major importance, and at low temperatures, where the solvent is more viscous, cage reactions will be strongly favoured. In these reactions, the above-mentioned products would be formed. The α-hydroxyalkylhydrotrioxide, in analogy to hemicacetals and the α-hydroxyalkylhydroperoxides mentioned in Chapter 6, may decompose into H_2O_3 plus acetone [reaction (26)].

$$\begin{array}{c}CH_3\\|\\HO-O-O-C-OH\\|\\CH_3\end{array} \xrightarrow{(27)} HO_2^\bullet + {}^\bullet O-\underset{\underset{CH_3}{|}}{\overset{\overset{CH_3}{|}}{C}}-OH$$

$$(26)\Big\downarrow \qquad\qquad {}^\bullet CH_3 + O=\underset{\underset{CH_3}{|}}{C}-OH \xleftarrow{(28)}$$

$$H_2O_3 + \underset{\underset{CH_3}{|}}{\overset{\overset{CH_3}{|}}{C}}=O$$

For this reaction, there is low-temperature NMR evidence (Plesnicar et al., 1998). In competition, in analogy to tetroxides (von Sonntag & Schuchmann, 1997; von Sonntag, 2006), the α-hydroxyalkylhydrotrioxide may undergo homolytic cleavage into $HO_2^\bullet/O_2^{\bullet-} + H^+$ and an oxyl radical [reaction (27)]. In accordance with this, a standard Gibbs free energy of $\Delta G^0 = +9$ kJ mol^{-1} has been calculated for the decay of CH_3OOOH into $CH_3O^\bullet + HO_2^\bullet$ (O–O bond length 1.438Å), while the competing decay into $CH_3OO^\bullet + {}^\bullet OH$ (O–O bond length 1.427Å) is considerably more endergonic ($\Delta G^0 = +48$ kJ mol^{-1} (Naumov & von Sonntag 2011, unpublished). The ensuing β-fragmentation [reaction (28)] will be very fast, cf. Chapter 14.

This reaction would also account for acetic acid and formaldehyde. The latter is a major product in the reaction of the methyl radical with O_2 (Schuchmann & von Sonntag, 1984) (note that formic acid is also a minor product). The decay of the α-hydroxyalkylhydrotrioxide into acetone and H_2O_3 is sped-up in the presence of water, and it is conceivable that oxyl radical formation is minor in water compared to 2-propanol in an organic solvent, as H_2O_3 elimination [reaction (26)] and oxyl radical formation followed by β-fragmentation [reactions (27) and (28)] are competing reactions (note that β-fragmentation products are not observed with D-glucose, see below).

According to this mechanistic concept, one of the primary products is the $HO_3^{\bullet}$ radical [reaction (18)]. It readily decomposes into $^{\bullet}OH$ plus O_2 [reaction (11)] (for details see Chapter 13).

Thus, any $HO_3^{\bullet}$ that escapes the cage will induce $^{\bullet}OH$ reactions. For 2-propanol, they are well documented. Reaction (29) is preferred over reaction (30) in a 4:1 ratio (Asmus et al., 1973).

Both radicals thus formed react rapidly with O_2 [e.g. reaction (31)].

The α-hydroxyalkylperoxyl radical eliminates (in water) $HO_2^{\bullet}$ [reaction (32), $k = 650$ s^{-1} (Bothe et al., 1977)]. The OH^--induced $O_2^{\bullet-}$ elimination is diffusion-controlled ($k = 10^{10}$ M^{-1}s^{-1}). The other 2-propanol-derived peroxyl radical decays bimolecularly (for details see Chapter 14).

The question now arises as to what extent the above reactions also dominate the ozonolysis of 2-propanol in water.

The reaction of ozone with 2-propanol in water requires an activation energy of $E_a = 70.7$ kJ mol^{-1} ($\log A = 12.51$, for an Arrhenius plot see Figure 10.5).

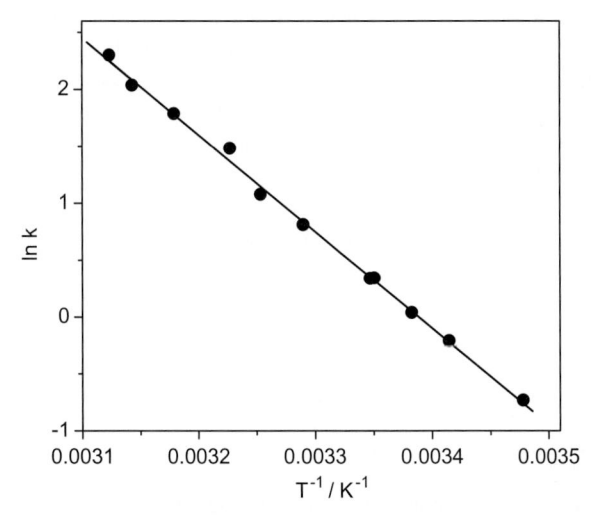

Figure 10.5 Arrhenius plot for the reaction of ozone with 2-propanol in water [Reisz & von Sonntag, 2011, unpublished results].

The formaldehyde yield is low, 3.9% with respect to ozone consumption, and is enhanced to 5.4 in the presence of an excess of tBuOH, that is, when •OH radicals are scavenged by tBuOH ([2-propanol] = 0.1 M, [tBuOH] = 1 M; efficiency of •OH scavenging by tBuOH = 75%; efficiency of ozone scavenging by tBuOH = 1%). The difference of 1.7% points to a free •OH yield of 3.4% (for a discussion of the use of tBuOH in the determination of •OH in ozone reactions see Chapter 14). This is the minimum value of H-abstraction in this system. It could double if the cage recombination product [cf. reaction (16)] would decompose homolytically [reaction (27)] rather than eliminate H_2O_3 according to reaction (26). Note, that the liberation of $HO_2^{\bullet}$ would give rise to $O_2^{\bullet-}$, which would react with ozone yielding •OH (for details see Chapters 11 and 13). Thus in aqueous solution, H-abstraction can be ~7% at the most. This is now in agreement with the formate and tBuOH systems where other, non-radical processes compete (formate, tBuOH) or even dominate (tBuOH). With tBuOH, insertion could be excluded, and all data would be compatible with a hydride transfer. For 2-propanol, one would then write reactions (33)–(35).

$$\underset{\substack{|\\CH_3}}{\overset{\substack{CH_3\\|}}{H-C-OH}} \xrightarrow[(33)]{O_3} HOOO^- + \underset{\substack{|\\CH_3}}{\overset{\substack{CH_3\\|}}{C=\overset{+}{O}H}} \xrightarrow[(34)]{} \underset{\substack{|\\CH_3}}{\overset{\substack{CH_3\\|}}{HOOOC-OH}}$$

$$(35) \downarrow H_2O$$

$$\underset{\substack{|\\CH_3}}{\overset{\substack{CH_3\\|}}{C=O}} + H_3O^+$$

In reaction (33) an intimate ion pair is formed that, in principle, could recombine [reaction (34)]. Protonated acetone formed in reaction (33) is, however, a very strong acid [$pK_a = -3.06$ (Bagno et al., 1995)], and deprotonation (35) must compete very effectively. Reaction (33) is very exergonic ($\Delta G^0 = -160$ kJ mol^{-1}). The energy difference between acetone and its enol is -69 kJ mol^{-1}. This leaves ample room for the enol being formed as well in analogy with what has been discussed for H-abstraction [cf. reaction (17)]. The formaldehyde yield is matched by that of acetic acid. As discussed above, their formation could be due to an ozonolysis of the enol [reaction (23)], or a decomposition of the hydrotrioxide [reaction (27) and subsequent reactions]. In the presence of OH$^-$, the enol lifetime is reduced and thus its ozonolysis suppressed. Indeed, the acetate yield drops to 10% of its former value when ozonolysis is carried out at pH 10. As OH$^-$ would also enhance the rate of hydrolysis of the hydrotrioxide into H_2O_3 and acetone [reaction (25)], a distinction between these two mechanistic possibilities cannot be made. Assuming that additional ozone is required for the ozonolysis of the acetone enol, the material balance is fair (~96%).

10.6 CARBOHYDRATES

In the reaction with D-glucose, ozone attacks all six C–H sites leading to the corresponding two-electron oxidised products (Schuchmann & von Sonntag, 1989). Their structures are shown in Figure 10.6, and their yields are compiled in Table 10.5.

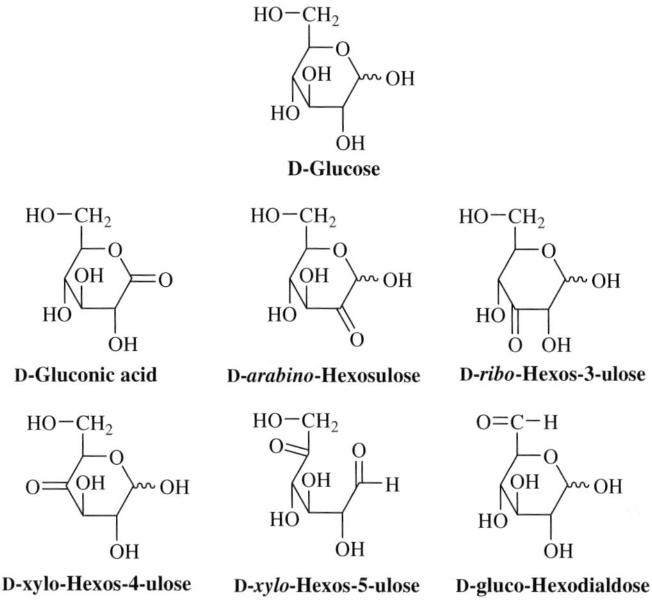

Compounds with C–H functions as ozone-reactive sites

Figure 10.6 Glucose and its ozone products.

Table 10.5 Products and their yields (in percent age of ozone/•OH radicals) of the reaction of D-glucose (4 × 10^{-3} M) with ozone and •OH radicals (γ-radiolysis in N_2O/O_2-saturated solution) at pH 6.5. Tertiary butanol (tBuOH, 0.1 M) has been added in one set of experiments for scavenging 90% of the •OH radicals (Schuchmann & von Sonntag, 1989)

Product	Ozone	Ozone/tBuOH	•OH Radicals
D-Gluconic acid	30	67	21
D-*arabino*-Hexosulose	17	12	17
D-*ribo*-Hexos-3-ulose	9	7	11
D-*xylo*-Hexos-4-ulose	9	5	10
D-*xylo*-Hexos-5-ulose	5	2	8
D-*gluco*-Hexodialdose	30	6	32
Sum	93	65	90

Addition of tBuOH, which scavenges 90% of the •OH radicals formed in these reactions, lowers the total yield of carbohydrate products from 93% to 65%. The effect of •OH can be read from the data obtained by γ-radiolysis in N_2O/O_2-saturated solution. One notices that these radicals are not very selective as far as the C–H sites are concerned, and the relatively high yield of D-gluco-hexodialdose is largely due to the fact that C6 carries two hydrogens. In contrast, the distribution in the •OH-scavenged system shows a marked preponderance with respect to an attack at C1, which carries the weakest bound hydrogen. D-Arabinose is not among the products, and other carbohydrates containing five carbon atoms are also not formed. This indicates that oxyl radicals and ensuing β-fragmentations [cf. reactions (40)–(42)] do not occur.

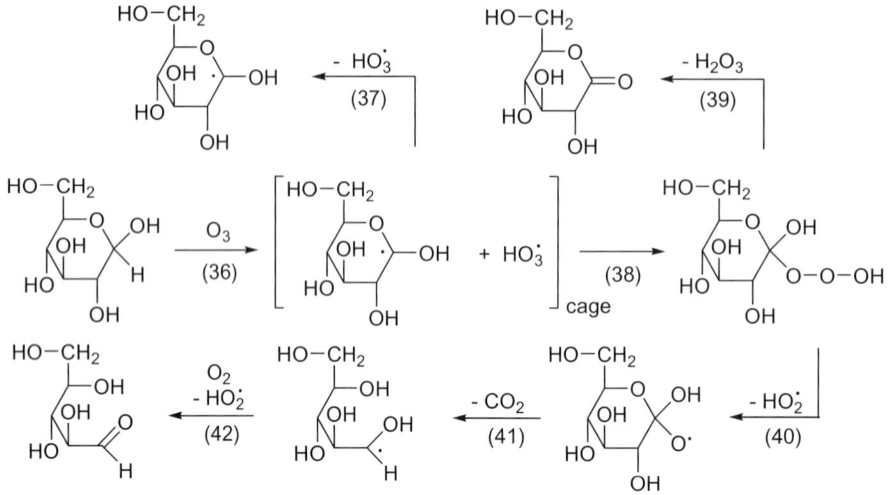

The release of H_2O_3 [cf. reaction (39)] terminates the reaction as H_2O_3 decomposes into H_2O and O_2 without giving rise to free radicals (see below). Reaction (37), however, yields an •OH radical, and this reacts rapidly with glucose. The glucose radicals react with O_2 and the peroxyl radicals thus formed release $HO_2^•/O_2^{•-}$ [for the kinetics see Bothe *et al.* (1978)]. As $O_2^{•-}$ reacts rapidly with O_3 yielding further radicals, a chain reaction is induced (note the similarities of product distribution of ozone without a scavenger and γ-radiolysis). The scavenging data do not lead to a firm conclusion as to the primary •OH yield for reactions such as reaction (37), but 15% may be a good guess. Based on more recent information, the original interpretation of the glucose data (Schuchmann & von Sonntag, 1989) had to be somewhat modified as shown above. Based on the above, competing reactions could be analogous to those discussed above for formate ion and for 2-propanol.

Alginic acids are polysaccharides consisting mainly of 1,4-linked β-D-manuronic acid (**A**) and varying amounts of 1,4-linked α-L-guluronic acid (**B**).

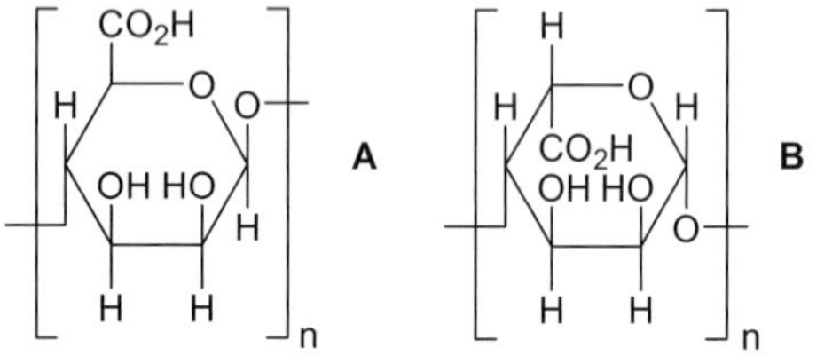

Upon reacting with ozone, chain breaks are formed with an efficiency of 18% (Akhlaq *et al.*, 1990). This is mainly due to the formation of •OH (see above). With alginic acid, radiolytically generated •OH causes chain breakage with an efficiency of 22% and $O_2^{•-}$ release with an efficiency of 73%. The reactions of carbohydrates with •OH have been reviewed (von Sonntag, 1980; von Sonntag & Schuchmann, 2001).

10.7 DIHYDROGEN TRIOXIDE – PROPERTIES OF A SHORT-LIVED INTERMEDIATE

In the above reactions, dihydrogen trioxide, H_2O_3, is an important, albeit short-lived, product. It has also been quoted as a reactive intermediate in other contexts (Wentworth Jr *et al.*, 2001, 2003; Czapski &

Bielski, 1963, 1970; Xu & Goddard III, 2002; Fujii et al., 1997; Engdahl & Nelander, 2002; Lesko et al., 2004), hence it seems worthwhile to report what is known thus far about this interesting intermediate, notably its fate in aqueous solution.

In water ["slightly acidic", as acids are formed (Plesnicar et al., 2000)], the lifetime of H_2O_3 is much shorter (estimated at 20 ms) than in organic solvents (16 min) (Plesnicar, 2005). More detailed information for the reactions of H_2O_3 in aqueous solution, is available from pulse radiolysis experiments (Czapski & Bielski, 1963; Bielski, 1970). H_2O_3 has been generated in acidic, O_2-saturated solutions. Under such conditions $HO_2^{\bullet}$ and $^{\bullet}OH$ are generated in a 1 : 0.8 ratio. Their bimolecular reaction gives rise to H_2O_3 [reaction (43)].

$$HO_2^{\bullet} + {}^{\bullet}OH \rightarrow H_2O_3 \tag{43}$$

This short-lived intermediate decays H^+-catalysed [reaction (44), $k = 6\ M^{-1}s^{-1}$] and as anion, HO_3^-, in equilibrium [reaction (45)].

$$H_2O_3 + H^+ \rightarrow H_3O^+ + O_2 \tag{44}$$
$$H_2O_3 \rightleftarrows HO_3^- + H^+ \tag{45}$$
$$HO_3^- \rightarrow OH^- + {}^1O_2 \tag{46}$$

The pK_a value of H_2O_3 [reaction (45)] can be estimated from relationship (47) (Bielski & Schwarz, 1968).

$$pK_a(H_xO_y) = 19 - 7([O]/[H]) \tag{47}$$

The decay of HO_3^- according to reaction (46), that is, even with the release of $^1O_2$ is exergonic [$\Delta G^0 = -51\ kJ\ mol^{-1}$, HO–OO$^-$ distance 1.482 Å, HOO–O$^-$ distance 1.402 Å, Naumov and von Sonntag (2011), unpublished results]. The much higher solvation energy of OH^- as compared to HO_3^- will contribute substantially to the exergonicity of this reaction.

From relationship (47), the pK_a value of H_2O_3 is calculated as 8.5. A plot of this relationship is shown in Figure 10.7.

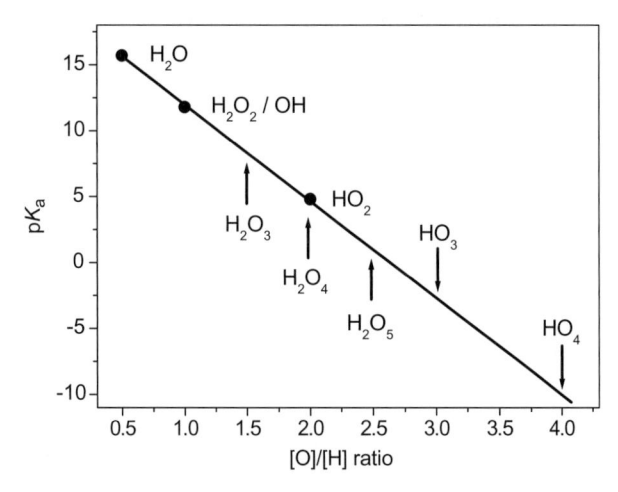

Figure 10.7 Plot of pK_a values of H_xO_y compounds as a function of the H/O ratio according to relationship (47). The pK_a values of H_2O, $^{\bullet}OH$, H_2O_2 and $HO_2^{\bullet}$ are experimental values. From Merényi et al., 2010b. Copyright Wiley-VCH Verlag. Reproduced with permission.

It allows one to also estimate the pK_a values of other H_xO_y intermediates such as $HO_3^\bullet$, H_2O_4, and so on that are of interest in the chemistry of ozone in aqueous solution. The validity of this plot has been confirmed by calculating the pK_a of $HO_3^\bullet$ using a thermodynamic cycle (Naumov & von Sonntag, 2011b).

The ratio of k_{-45}/k_{45} has been given as 1720 M^{-1} (Czapski & Bielski, 1963), and taking $k_{-45} = 10^{10}$ $M^{-1}s^{-1}$ (Eigen, 1963), the intrinsic decay rate of HO_3^- must be near 6×10^6 s^{-1}. This rather long lifetime has been taken as a hint that reaction (46) may even give rise to triplet (ground state) O_2, that is, a spin conversion without a heavy atom effect may take place (Sein et al., 2007). This suggestion was based on the report that in the ozonolysis of 2-propanol in water (note that 2-propanol is likely to also give rise to H_2O_3 in water) no singlet oxygen signal could be recorded (Muñoz et al., 2001). Inspection of these data shows that the 2-propanol concentration chosen, although 500 times the ozone concentration, was possibly too low for arriving at a firm conclusion. At an ozone concentration of 1×10^{-4} M, the 2-propanol concentration would have been 5×10^{-2} M. With k(2-propanol + O_3) = 0.83 $M^{-1}s^{-1}$, this would have resulted in a half-life of 14 seconds. Strong signals with half-lives of 1 s were successfully detected, but it may well be possible that a weaker 2-propanol signal might have come out too broad and thus too low for adequate detection. Thus, this reference is not a proof for the absence of singlet oxygen in the aqueous 2-propanol system, and the option that reaction (46) proceeds in the spin-allowed mode, that is, under the release of singlet oxygen, is not ruled out.

The chemical reactivity of H_2O_3 is low. It seems that its reaction is somewhat similar to that of H_2O_2. It does not react with alcohols or I^-, and 1.6 M Fe^{2+} were required for its reduction within its short lifetime in aqueous solution (Czapski & Bielski, 1963).

10.8 SATURATED MICROPOLLUTANTS LACKING OZONE-REACTIVE HETEROATOMS

All saturated compounds lacking ozone-reactive heteroatoms react so slowly with ozone (Table 10.1) that the direct reaction of ozone in removing such micropollutants is not efficient. They mostly react, however, rapidly with •OH. Their reactions with •OH are discussed in Chapter 14. In DOM-containing waters such as wastewater and drinking water, the reaction of ozone with the water matrix leads to •OH formation, and by this route ozone-refractory compounds are eliminated (Chapter 3).

Chapter 11

Inorganic anions and the peroxone process

11.1 INTRODUCTORY REMARKS

Inorganic anions are negatively charged, and thus react much faster than their conjugate acids, for example, HS^-/H_2S (Table 11.1).

Table 11.1 Compilation of ozone rate constants of inorganic anions (for metal-derived anions see Chapter 12). Published rate constants are rounded to significant figures; pK_a values of conjugate acids

Compound	pK_a	k / M^{-1} s^{-1}	Reference
Azide ion	9.4	4×10^6	Hoigné et al., 1985
protonated		$<4 \times 10^3$	Hoigné et al., 1985
Borate ion protonated		$<6 \times 10^{-2}$	Hoigné et al., 1985
diprotonated		$<4 \times 10^{-3}$	Hoigné et al., 1985
Bromate ion		1×10^{-3}	Hoigné et al., 1985
Bromide ion		160 at 20 °C	Haag & Hoigné, 1983a
		258 at 25 °C	Liu et al., 2001
Bromite ion		$>1 \times 10^5$	Hoigné et al., 1985
		8.9×10^4	Nicoson et al., 2002
Carbonate ion	10.3, 6.3	<0.1	Hoigné et al., 1985
Bicarbonate ion		$\ll 0.01$	Hoigné et al., 1985
Chlorate ion		$\ll 1 \times 10^{-4}$	Hoigné et al., 1985
Chloride ion		$<3 \times 10^{-3}$	Hoigné et al., 1985
		6×10^{-4} at 9.5 °C, $E_a = 74$ kJ mol^{-1}	Yeatts & Taube, 1949
Chlorite ion		4×10^6	Kläning et al., 1985
		8.2×10^6	Nicoson et al., 2002
Cyanate ion	3.9	$\leq 10^{-2}$	Hoigné et al., 1985

(Continued)

Table 11.1 Compilation of ozone rate constants of inorganic anions (for metal-derived anions see Chapter 12). Published rate constants are rounded to significant figures; pK_a values of conjugate acids (*Continued*)

Compound	pK_a	k / M^{-1} s^{-1}	Reference
Cyanide ion	9.1	1×10^3	Hoigné et al., 1985
		2.6×10^3	Gurol & Bremen, 1985
protonated		$\leq 10^{-3}$	Hoigné et al., 1985
Hydroperoxide ion	using 11.6	5.5×10^6	Staehelin & Hoigné, 1982
	using 11.8	9.6×10^6	Sein et al., 2007
protonated		<0.01	Staehelin & Hoigné, 1982
		0.065**	Sehested et al., 1992
Hydroxide ion		48	Forni et al., 1982
		70	Staehelin & Hoigné, 1982
protonated (water)		$<1 \times 10^{-7}$	Hoigné et al., 1985
Hypobromite ion	8.8	430 at 25 °C	Haag & Hoigné, 1983a
protonated		$\leq 1 \times 10^{-2}$	Haag & Hoigné, 1983a
Hypochlorite ion	7.5	170*	Haag & Hoigné, 1983b
Hypoiodite ion	10.4	1.6×10^6	Bichsel & von Gunten,
protonated		3.6×10^4	1999b
Iodate ion		$<10^{-4}$	Hoigné et al., 1985
Iodide ion		2×10^9	Garland et al., 1980
		1.2×10^9	Liu et al., 2001
Nitrate ion		$<10^{-4}$	Hoigné et al., 1985
Nitrite ion	3.29	5.83×10^5	Liu et al., 2001
		1.8×10^5	Hoigné et al., 1985
protonated		<500	Hoigné et al., 1985
Perchlorate ion		$\ll 2 \times 10^{-5}$	Hoigné et al., 1985
Periodate ion		<0.01	Hoigné et al., 1985
Phosphate ion protonated	2.1, 7.2, 12.3	$<2 \times 10^{-4}$	Hoigné et al., 1985
diprotonated		<0.02	Hoigné et al., 1985
Sulfate ion protonated	2	$<10^{-4}$	Hoigné et al., 1985
Sulfide ion protonated	6.9	3×10^9	Hoigné et al., 1985
diprotonated		$\sim 3 \times 10^4$	Hoigné et al., 1985
Sulfite ion	7	1.5×10^9	Hoffmann, 1986
		1.3×10^9	Liu et al., 2001
protonated		3.7×10^5	Hoffmann, 1986
		8×10^5	Liu et al., 2001
		8.6×10^5	Erickson et al., 1977
		2×10^6	Nahir & Dawson 1987
Sulfur dioxide		2.4×10^4	Neta et al., 1988 based on Hoffmann, 1986
		$<10^4$	Erickson et al., 1977
Thiosulfate ion		2.2×10^8	Utter et al., 1992
		$(0.7–2.2) \times 10^8$	Kanofsky & Sima, 1995

* For the decay of ozone; for the decay of ClO$^-$, the rate constant is 140 M^{-1} s^{-1}.
** At 31 °C.

Water does not react with ozone, but its conjugate base, OH⁻, reacts with ozone, albeit slowly (Table 11.1). In all these reactions, O-transfer reactions dominate, and an ozone adduct must be the first intermediate or is at least a transition state as in the reaction of NO_2^- with ozone (see below). In other cases, the standard Gibbs free energies of adduct formation can be calculated by quantum chemistry. When this reaction is slightly endergonic, adduct formation is not only slow but the reaction is also reversible. A case in point is the reaction with Br⁻. There may be also other faster decay routes of the adduct, and then no reversibility is observed (as in OH⁻). Since the substrate anion (A⁻) and ozone are in their singlet ground states, the spin conservation rule demands that the multiplicities of the adducts and their products must be singlets as well [reactions (1) and (2)]. Thus, the oxygen that is released in reaction (2) must be in its excited singlet state, $^1O_2$ [$O_2(^1\Delta_g)$]. The excited state of oxygen ($^1O_2$) lies 95.5 kJ mol⁻¹ above its ground state [$^3O_2$, $O_2(^3\Sigma_g^-)$]. This is of major importance when the energetics of such O-transfer reactions is assessed.

$$A^- + O_3 \rightleftarrows AOOO^- \longrightarrow AO^- + {}^1O_2 \qquad (1)/(2)$$

The heavy atom effect can enhance the rate of singlet → triplet conversion by spin-orbit coupling, and this is the reason why the $^1O_2$ yield in the reaction of ozone with Br⁻ and I⁻ is lowered to 54% and 14%, respectively (see below). These low $^1O_2$ yields are a further indication for a certain lifetime of these adducts.

Nitrite is sometimes used as a competitor for the determination of ozone rate constants because nitrate, the main product, can be readily quantified. The authors of the higher and later value, state that their value has been measured with special care (Liu *et al.*, 2001), and this value may be used for such competition studies.

11.2 HYDROXIDE ION

The reaction of ozone with OH⁻ gives rise to •OH radicals (Staehelin & Hoigné, 1982). As far as the mechanism is concerned, it has been suggested that an O-transfer may occur [reaction (3)] and HO_2^- thus formed may induce the peroxone process.

$$OH^- + O_3 \longrightarrow HO_2^- + O_2 \qquad (3)$$

A recent thermokinetic analysis comes to the conclusion that in the first step an adduct is formed [reaction (4), $\Delta G^\circ = 3.5$ kJ mol⁻¹] (Merényi *et al.*, 2010b).

$$OH^- + O_3 \longrightarrow HO_4^- \qquad (4)$$

This decays into $HO_2^\bullet$ plus $O_2^{\bullet-}$ [equilibrium (5), $\Delta G^\circ = 14.8$ kJ mol⁻¹].

$$HO_4^- \rightleftarrows HO_2^\bullet + O_2^{\bullet-} \qquad (5)$$

In the absence of O_3, these radicals eventually decay by H-transfer [reaction (6), $k = 9.7 \times 10^7$ M⁻¹ s⁻¹ (Bielski *et al.*, 1985)], about 50 times more slowly than the reverse of reaction (5).

$$HO_2^\bullet + O_2^{\bullet-} \longrightarrow O_2 + HO_2^- \qquad (6)$$

$HO_2^\bullet$ is in equilibrium with $O_2^{\bullet-}$ [equilibrium (7), $pK_a(HO_2^\bullet) = 4.8$ (Bielski *et al.*, 1985)].

$$HO_2^\bullet \rightleftarrows O_2^{\bullet-} + H^+ \qquad (7)$$

With ozone, $O_2^{•-}$ gives rise to $O_3^{•-}$ [reaction (8), $k = 1.5 \times 10^9$ M^{-1} s^{-1} (Sehested et al., 1983), 1.6×10^9 M^{-1} s^{-1} (Bühler et al., 1984)] that is the precursor of •OH (for details see Chapter 13).

$$O_2^{•-} + O_3 \rightarrow O_2 + O_3^{•-} \tag{8}$$

In the literature, the reaction of ozone with OH$^-$ is often wrongly cited as the source for •OH in a given system. This reaction has a very low rate constant (Table 11.1) and can only play a role at very high pH. An example may show this. Even at pH 8, the upper end for pH values for drinking waters and wastewaters, the reaction is so slow ($k_{obs} = $ [OH$^-$] $\times k_4 = 7 \times 10^{-5}$ s^{-1}; $t_{½} = 10.000$ s $= 2.8$ h) that it must be neglected as other ozone reactions that occur in competition are typically much faster [for example, the reaction of ozone with DOM (Chapter 3)].

11.3 HYDROPEROXIDE ION – PEROXONE PROCESS

The combination of ozone with H_2O_2 for water treatment is often called the "peroxone process". It gives rise to •OH and is one of the most common AOPs. It was first described in a brilliant paper by (Staehelin & Hoigné, 1982). There, it has been shown that the rate constant for the reaction of ozone with H_2O_2 is very low (Table 11.1) and that the observed reaction is due to its anion, HO$_2^-$, present in equilibrium [equilibrium (9)].

$$H_2O_2 \rightleftarrows HO_2^- + H^+ \tag{9}$$

The rate of reaction at a pH below the pK_a of H_2O_2, k_{obs}, is thus given by equation (10). In the original paper, a $pK_a(H_2O_2) = 11.6$ was taken (Staehelin & Hoigné, 1982). A more recent paper uses $pK_a(H_2O_2) = 11.8$ (Sein et al., 2008), and subsequent discussions are based on this higher pK_a value. This has an effect on k but not on the experimentally observed k_{obs}.

$$k_{obs} = k(HO_2^- + O_3) \times 10^{(pH - pK_a)} \tag{10}$$

It has been suggested (Staehelin & Hoigné, 1982) that the reaction proceeds by electron transfer [reaction (11)].

$$HO_2^- + O_3 \rightarrow HO_2^• + O_3^{•-} \tag{11}$$

It has been proposed that the $O_3^{•-}$ radical becomes protonated [reaction (12)] and the ensuing HO$_3^•$ radical decomposes into •OH and O_2 [reaction (13), $pK_a(HO_3^•) = 8.2$ (Bühler et al., 1984)].

$$O_3^{•-} + H^+ \rightleftarrows HO_3^• \tag{12}$$
$$HO_3^• \rightleftarrows {}^•OH + O_2 \tag{13}$$

The HO$_2^•$ is in equilibrium with $O_2^{•-}$ [equilibrium (14)], and this undergoes electron transfer to ozone [reaction (15)].

$$HO_2^• \rightleftarrows O_2^{•-} + H^+ \tag{14}$$
$$O_2^{•-} + O_3 \rightarrow O_2 + O_3^{•-} \tag{15}$$

The overall stoichiometry has been concluded to be $H_2O_2 + 2O_3 \rightarrow 2\,{}^•OH + 3O_2$. In fact, the efficiency of •OH formation is only half of this (von Sonntag, 2008; Jarocki et al., 2012), and this calls for a revision of this mechanism. Before discussing details, a completely different view of the peroxone process must be mentioned. Based on the observed kinetic isotope effect, it has been suggested that in the primary step a hydride is transferred (HO$_2^-$ + O$_3$ → O$_2$ + HO$_3^-$), and the ensuing chemistry has been attributed to the reaction of H_2O_3 (Lesko et al., 2004) [for formation and decay of H_2O_3 see Chapter 10]. Since it has been shown that this observed kinetic isotope effect was due to a kind of experimental artefact (Sein et al., 2007), this mechanism must be disregarded, and the earlier concept of •OH formation continues to be valid, although it requires modifications (Merényi et al., 2010a).

As the first step, one must assume the formation of an adduct [reaction (16), $\Delta G° = -39.8$ kJ mol^{-1} (Merényi et al., 2010a)].

$$HO_2^- + O_3 \rightarrow HO_5^- \tag{16}$$

This reaction is slightly more exergonic than the electron transfer [reaction (11), $\Delta G° = -26.6$ kJ mol^{-1}]. The adduct may decompose in two directions [reactions (17), $\Delta G° = 13.2$ kJ mol^{-1}, and (18), $\Delta G° = -197$ kJ mol^{-1}].

$$HO_5^- \rightarrow HO_2^\bullet + O_3^{\bullet-} \tag{17}$$

$$HO_5^- \rightarrow 2\,^3O_2 + OH^- \tag{18}$$

The experimental observation that the efficiency of •OH formation is only 50% can be rationalised if these reactions occur to about the same extent. At first sight, it may be surprising that k_{18} is not significantly larger than k_{17}. However, a reaction involving making or breaking of a single bond is inherently much faster than a reaction in which several bonds are simultaneously rearranged. Thus, the closeness of the two rate constants makes sense.

The ozonide radical anion, $O_3^{\bullet-}$, is in rapid equilibrium with O_2 and $O^{\bullet-}$ [equilibrium (19); $k_{19} = 1.4 \times 10^3$ s^{-1}, $k_{-19} = 3.5 \times 10^9$ M^{-1} s^{-1} (Elliot & McCracken, 1989)], and $O^{\bullet-}$ is in equilibrium with its conjugate acid, •OH [equilibrium (20); pK_a(•OH) = 11.8].

$$O_3^{\bullet-} \rightleftarrows O^{\bullet-} + O_2 \tag{19}$$

$$O^{\bullet-} + H_2O \rightleftarrows \,^\bullet OH + OH^- \tag{20}$$

Since only $O^{\bullet-}$ but not •OH reacts with O_2, these equilibria drive $O_3^{\bullet-}$ into •OH and O_2 (except at very high pH, where these equilibria stabilise $O_3^{\bullet-}$). This mechanistic concept replaces the current one, where $O_3^{\bullet-}$ is protonated and $HO_3^\bullet$ decomposes into •OH and O_2 [reaction (13)]. The reported $pK_a(HO_3^\bullet) = 6.15$ (revised at 8.2, Bühler, private communication) must be a kinetic artefact (so-called "reactivity pK", see Chapter 2) (Merényi et al., 2010a). It is now clear that it must be rather near –2 (Merényi et al., 2010a; Naumov & von Sonntag, 2011b). In this context, it may be mentioned that the decay of $HO_3^\bullet$ in the gas phase [reaction (13)] has been calculated at a high level of theory [CCSD(T)/6-311++G**] and found to be slightly endothermic ($\Delta H = 21$ kJ mol^{-1} for the *trans* form and 20 kJ mol^{-1} for the *cis* form) (Plesnicar et al., 2000). In water, the decay of $HO_3^\bullet$ is markedly exergonic, $\Delta G° = -47$ kJ mol^{-1} (Naumov & von Sonntag, 2011b).

11.4 FLUORIDE

There are no reports on the reaction of the F$^-$ with ozone. It seems not to react with ozone, and this is fairly well understood considering that both adduct formation and subsequent oxygen release as $^1O_2$ as required by the spin conservation rule are endergonic (Table 11.2).

On top of the endergonicities of these reactions, there must be substantial activation energies, and this must be the reason for the low reactivity, if any, of the fluoride ion.

Table 11.2 Reactions of ozone with halide ions. Standard Gibbs free energies (ΔG^0/kJ mol^{-1}) for the formation of ozone adducts and their decay by oxygen release according to Naumov & von Sonntag (2010) unpublished

Halide ion	ΔG^0 (Adduct formation)	ΔG^0 (Oxygen release)	State of O_2
Fluoride	+25	+45	$^1O_2$, spin allowed
		−50	$^3O_2$, spin forbidden
Chloride	+21	+4	$^1O_2$, spin allowed
		−92	$^3O_2$, spin forbidden
Bromide	−8	+8	$^1O_2$, spin allowed
		−86	$^3O_2$, spin forbidden
Iodide	−20	−3	$^1O_2$, spin allowed
		−98	$^3O_2$, spin forbidden

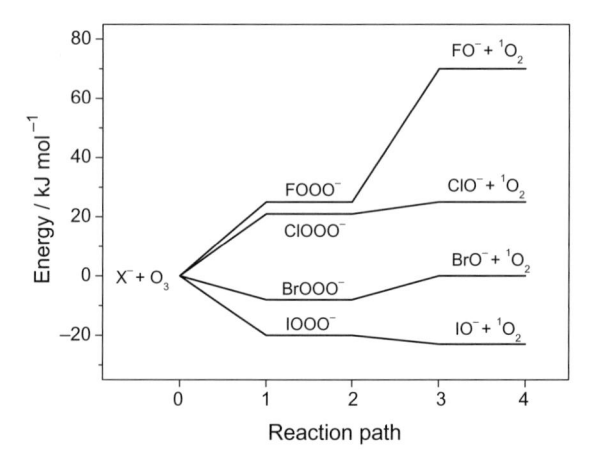

Figure 11.1 Energetics of the reaction of ozone with halide ions along the reaction path according to Naumov & von Sonntag, 2010, unpublished. For Br$^-$ and I$^-$, there is also a competing release of ground state O_2 ($^3O_2$). The latter makes the last step much more exergonic (Table 11.2).

In Figure 11.1, the data in Table 11.2 are visualised. With F$^-$ and Cl$^-$, all steps along the reaction path are endergonic, with Br$^-$ there is reversibility of the adduct, and with I$^-$ reactions are exergonic throughout, as will be discussed below.

11.5 CHLORIDE

For Cl$^-$, the situation is similar in so far as the addition reaction is also noticeably endergonic, and a substantial activation energy of 74 kJ mol^{-1} has been reported (Yeatts & Taube, 1949). The ensuing release of $^1O_2$ from an adduct is weakly endergonic (Table 11.2), and there may be an additional barrier for the release of $^1O_2$ potentially favouring the reverse of adduct formation. Altogether, this may result in a very slow reaction (Table 11.1). It has been suggested that hypochlorite and oxygen are formed

(Yeatts & Taube, 1949). Since the potential product hypochlorite [pK_a(HOCl) = 7.5] has a comparatively high rate constant (Table 11.1), product interference with the reaction of ozone plus Cl⁻ is likely.

The reaction of ozone with Cl⁻ can be typically neglected. Even in seawater ([Cl⁻] ≈ 0.5 M in the Atlantic) the reaction of ozone with chloride will be largely overrun by competing reactions (e.g. by a reaction with Br⁻, [Br⁻] ≈ 1 mM).

11.6 HYPOCHLORITE

In aqueous solution, chlorine, Cl_2, is in equilibrium with HOCl, OCl⁻ and Cl_3^- [reactions (21)–(23)].

$$Cl_2 + H_2O \rightleftarrows HOCl + H^+ + Cl^- \quad (21)$$

$$HOCl \rightleftarrows ClO^- + H^+ \quad (22)$$

$$Cl_2 + Cl^- \rightleftarrows Cl_3^- \quad (23)$$

UV–Vis absorption maxima and equilibrium constants are compiled in Table 11.3.

Hypochlorous acid (HOCl) does not react with ozone which slows down the reaction towards chlorate at circum-neutral pH. Hypochlorite reacts moderately rapidly with ozone (Table 11.1). It gives rise to Cl⁻ and ClO_2^- in a 3.4:1 ratio (Haag & Hoigné, 1983b). This is an indication that there are two competing reactions [reactions (24) k_{24} = 30 M⁻¹ s⁻¹ and (25) k_{25} = 110 M⁻¹ s⁻¹].

$$ClO^- + O_3 \longrightarrow ClO_2^- + O_2 \quad (24)$$

$$ClO^- + O_3 \longrightarrow Cl^- + 2O_2 \quad (25)$$

Table 11.3 UV–Vis absorption maxima and equilibrium constants of halogen-derived species in water (20–25°C)

Compound	λ_{max}/nm	ε/M⁻¹ cm⁻¹	K	Reference
Cl_2	323	68	$K_{Hydrolysis}$ = 4 × 10⁻⁴ M²	Soulard et al., 1981
HOCl	233	97	K_a = 4 × 10⁻⁸ M⁻¹	Soulard et al., 1981
			pK_a = 7.5	Morris, 1966
ClO⁻	292	350		Soulard et al., 1981
Cl_3^-	233	18 500		Soulard et al., 1981
	325	85		Soulard et al., 1981
			$K(Cl^- + Cl_2)$ = 0.18 M⁻¹	Eigen & Kustin, 1962
Br_2	392	175	$K_{Hydrolysis}$ = 7.2 × 10⁻⁹ M²	Soulard et al., 1981
			$K_{Hydrolysis}$ = 3.5-6 × 10⁻⁹ M²	Beckwith et al., 1996
HOBr	260	160	K_a = 1.8 × 10⁻⁹ M⁻¹	Soulard et al., 1981
			pK_a = 8.74	Haag & Hoigné, 1983a
			pK_a = 8.8	
BrO⁻	329	345		Soulard et al., 1981
Br_3^-	266	35 000		Soulard et al., 1981
		40 900		Beckwith et al., 1996

(Continued)

Table 11.3 UV–Vis absorption maxima and equilibrium constants of halogen-derived species in water (20–25°C) (*Continued*)

Compound	λ_{max}/nm	$\varepsilon/M^{-1}\,cm^{-1}$	K	Reference
I_2			$K(Br^- + Br_2) = 17\ M^{-1}$	Eigen & Kustin, 1962
			$K_{Hydrolysis} = 5.44 \times 10^{-13}\ M^2$	Burger & Liebhafsky, 1973
HOI			$K_a = 4 \times 10^{-11}\ M^{-1}$	Bichsel & von Gunten, 2000b
			$pK_a = 10.4 \pm 0.1$	
			$pK_a = 10.6 \pm 0.8$	Chia, 1958
I_3^-	350	25 500		Allen *et al.*, 1952
	351	25 700		Bichsel & von Gunten, 1999a
			$K(I^- + I_2) = 725\ M^{-1}$	Burger & Liebhafsky, 1973
			$830\ M^{-1}$	Eigen & Kustin, 1962

The mechanisms of these reactions must be analogous to the reactions of BrO$^-$ with ozone, discussed in more detail below, in that adducts are the most likely intermediates. Reaction (25) is one of the examples, where ozone acts as a reducing agent.

11.7 CHLORITE

Chlorite, ClO_2^-, reacts rapidly with ozone (Table 11.1). Since chlorate was found to be the final product, it has been suggested that ozone reacts with ClO_2^- by O-transfer (Haag & Hoigné, 1983b). However, more recently it has been shown that this reaction proceeds by electron transfer (Kläning *et al.*, 1985) [reaction (26), $\Delta G^0 = -43$ kJ mol^{-1} (Naumov & von Sonntag, 2011b, SI)].

$$ClO_2^- + O_3 \longrightarrow ClO_2^\bullet + O_3^{\bullet -} \qquad (26)$$

The subsequent reactions of ozone with $ClO_2^\bullet$ are discussed in Chapter 13. Although it has been stated that perchlorate is not a product in the reaction of ClO_2^- with ozone (Haag & Hoigné, 1983b), traces of perchlorate (in the order of 1% of chlorate) are also formed (Rao *et al.*, 2010). Cl_xO_y radicals are potential intermediates (Rao *et al.*, 2010).

11.8 BROMIDE

The reaction of ozone with Br$^-$ is of major interest, as the final oxidation product is bromate, BrO_3^-, which is classified as a potential human carcinogen with a drinking water standard of 10 µg/l (WHO, 2004). This system has been widely studied (Haag *et al.*, 1982; Haag & Hoigné, 1983a; von Gunten & Hoigné, 1992, 1994, 1996 von Gunten *et al.*, 1995; Liu *et al.*, 2001; Naumov & von Sonntag, 2008b). The oxidation of Br$^-$ to bromate is a multi-stage process. Hypobromite and bromite are intermediates. In the first step, an adduct is formed [reactions (27) and (28)].

$$Br^- + O_3 \rightleftarrows BrOOO^- \longrightarrow BrO^- + O_2 \qquad (27)/(28)$$

This reaction is slow (Table 11.1) and reversible (Liu *et al.*, 2001). Quantum-chemical calculations support this experimental observation [$\Delta G^0 = +6$ kJ mol^{-1} (Naumov & von Sonntag, 2008b), cf. Table 11.4].

Table 11.4 Calculated Gibbs energies (ΔG) taking the solvent water into account (B3LYP/6-311+G(d)/SCRF=COSMO) according to Naumov & von Sonntag (2008b)

Reaction	ΔG/kJ mol^{-1}
$Br^- + O_3 \rightarrow BrOOO^-$	+6
$BrOOO^- \rightarrow BrO^- + O_2(^1\Delta_g)$ spin allowed	+13
$BrOOO^- \rightarrow BrO^- + O_2(^3\Sigma_g^-)$ spin forbidden	−81
$Br^- + O_3 \rightarrow BrOOO^- \rightarrow BrO^- + O_2(^1\Delta_g)$ spin allowed	+19
$Br^- + O_3 \rightarrow BrOOO^- \rightarrow BrO^- + O_2(^3\Sigma_g^-)$ spin forbidden	−75
$BrO^- + O_3 \rightarrow$ Adduct $\rightarrow BrO_2^- + O_2(^1\Delta_g)$ spin allowed	−36
$BrO^- + O_3 \rightarrow$ Adduct $\rightarrow BrO_2^- + O_2(^3\Delta_g)$ spin forbidden	−130
$BrO^- + O_3 \rightarrow$ Adduct $\rightarrow Br^- + 2\,O_2(^3\Sigma_g^-)$ spin allowed	−347
$BrO_2^- + O_3 \rightarrow$ Adduct $\rightarrow BrO_3^- + O_2(^3\Sigma_g^-)$ spin allowed	−93
$BrO_2^- + O_3 \rightarrow$ Adduct $\rightarrow BrO_3^- + O_2(^3\Sigma_g^-)$ spin forbidden	−187

In competition with the reverse of reaction (1), the adduct loses O_2 [reaction (2)]. The spin conservation rule demands that O_2 must be released in its excited singlet state, $^1O_2$ [$O_2(^1\Delta_g)$], because ozone, Br^- and $BrOOO^-$ are all in their singlet ground states. The adduct is, however, sufficiently long-lived that the heavy atom effect exerted by the bromine in the adduct also allows, by spin orbit coupling, the release of (triplet) ground-state O_2 [$^3O_2$, $O_2(^3\Sigma_g^-)$]. This route is energetically more favourable as the excited state ($^1O_2$) lies 95.5 kJ mol^{-1} above the ground state ($^3O_2$). The $^1O_2$ yield in this system is only 54% (Muñoz et al., 2001).

11.9 HYPOBROMITE

The reaction of ozone with BrO^- gives rise to two products, BrO_2^- plus O_2, the trivial and *oxidative* pathway [reactions (29) and (30); $k_{obs} = 100$ M^{-1} s^{-1} (Haag & Hoigné, 1983a); Δ$G^0 = -36$ kJ mol^{-1} for the spin-allowed reaction, formation of $^1O_2$ (Naumov & von Sonntag, 2008b)], but there is also a *reductive* pathway leading to Br^- plus 2 O_2 [reactions (31) and (32) or reactions (29), (33) and (34); $k_{obs} = 330$ M^{-1} s^{-1} (Haag & Hoigné, 1983a); Δ$G^0 = -347$ kJ mol^{-1} (Naumov & von Sonntag, 2008b)].

For the two pathways, one may assume two different adducts [reactions (29) vs. (31)]. The alternative, one common adduct [reaction (29)] followed by reactions (33) and (34) for the reductive pathway has, however, been favoured as BrO^- is more negative at oxygen (86% of excess electron density) than at bromine (14%) (Naumov & von Sonntag, 2008b). The electronic structure of O_3 is largely bipolar with negative charges at the outer oxygens. An addition to oxygen as compared to bromine may be disfavoured by charge repulsion despite the fact that O_3 is electrophilic.

The interesting reductive pathway finds its analogy in the reactions of ozone with ClO^- (see above) and Mn^{2+} *en route* to MnO_4^- (Reisz et al., 2008) (Chapter 12).

11.10 BROMITE

The fast reaction of BrO⁻ with ozone was originally suggested to be an O-transfer [reactions (35) and (36)] (Hoigné et al., 1985).

$$BrO_2^- + O_3 \longrightarrow O_2BrOOO^- \longrightarrow BrO_3^- + O_2 \qquad (35)/(36)$$

This reaction is energetically feasible [$\Delta G^0 = -93$ kJ mol^{-1} for the spin allowed reaction (Naumov & von Sonntag, 2008b)]. Yet, it has been shown that the reaction is dominated by an electron transfer [reaction (37); $\Delta G^0 = -64$ kJ mol^{-1} (Naumov & von Sonntag, 2011a, SI)] (Nicoson et al., 2002).

$$BrO_2^- + O_3 \longrightarrow BrO_2^\bullet + O_3^{\bullet-} \qquad (37)$$

In the reaction of ozone with Br⁻ en route to bromate, •OH radicals (precursor: $O_3^{\bullet-}$) have indeed been detected [Jarocki & von Sonntag (2011), unpublished results]. The ensuing rapid bimolecular decay [equilibrium (38)] and reaction of the dimer with water and a faster reaction with OH⁻ [cf. reaction (39)] give rise to bromate (Buxton & Dainton, 1968; Field & Försterling, 1986).

$$2\,BrO_2^\bullet \rightleftarrows Br_2O_4 \qquad (38)$$
$$Br_2O_4 + H_2O \longrightarrow BrO_2^- + BrO_3^- + 2H^+ \qquad (39)$$

The forward rate constant of $k_{38} = 1.4 \times 10^9$ M^{-1}s^{-1} and also the equilibrium constant K_{38} given by Buxton & Dainton (1968) are somewhat in doubt (Buxton & von Sonntag, 2010, unpublished). Yet, the error in k_{38} may not be high, as radical–radical recombination reactions are typically close to diffusion controlled. When the value of $k_{38} = 1.4 \times 10^9$ M^{-1}s^{-1} is assumed, $k_{-38} = 7.4 \times 10^4$ s^{-1} is obtained (Field & Försterling, 1986).

Overall, bromite is quickly and quantitatively transformed into bromate during typical ozonation conditions in water treatment. In real waters, ozone reactions with the DOM give rise to •OH and DOM also serves as •OH scavenger (Chapter 3). As •OH contributes significantly to bromate formation, its generation during ozonation of Br⁻-containing waters is a highly non-linear process for which predictions are very difficult. Some aspects are discussed below and in Chapter 14.

It is noteworthy that the reaction stops at bromate and does not continue up to perbromate. The reason for this is the high endothermicity of the reaction [spin allowed: $\Delta G^0(BrO_3^- + O_3 \rightarrow BrO_4^- + {}^1O_2) = +144$ kJ mol^{-1}; spin forbidden: $\Delta G^0(BrO_3^- + O_3 \rightarrow BrO_4^- + {}^3O_2) = +47$ kJ mol^{-1}, Naumov & von Sonntag (2011), unpublished results].

11.11 IODIDE

Ozone reacts very rapidly with I⁻ (Table 11.1). The reaction proceeds by O-transfer with an adduct as an intermediate [reactions (40) and (41)] (Muñoz et al., 2001).

$$I^- + O_3 \longrightarrow IOOO^- \longrightarrow IO^- + O_2 \qquad (40)/(41)$$

Both steps are exergonic (Table 11.2), but, in contrast to Br⁻, the first step is not reversible. Yet, the second step is only slightly exergonic and the adduct must have a certain lifetime. Evidence for this comes from the low yield of ${}^1O_2$ (14%) (Muñoz et al., 2001). As with Br⁻, this low ${}^1O_2$ yield is due to the heavy atom effect (see discussion in Section 11.1).

In slightly acidic solution and with some excess of I$^-$, the final product is I$_3^-$. Several equilibria have to be taken into account [equilibria (42)–(44), for equilibrium constants see Table 11.3].

$$IO^- + H^+ \rightleftarrows HOI \quad (42)$$
$$HOI + I^- + H^+ \rightleftarrows I_2 + H_2O \quad (43)$$
$$I_2 + I^- \rightleftarrows I_3^- \quad (44)$$

The situation is even more complex, as in both neutral and basic solutions major amounts of iodate (IO$_3^-$) are present in equilibrium. In the concentration range of typical drinking waters, this is, however, negligible.

HOI/IO$^-$ can be formed by ozonation but also in chlorination and chloramination processes. It reacts with DOM to form iodo-organic compounds (Krasner *et al.*, 2006; Bichsel & von Gunten, 1999b; Bichsel & von Gunten, 2000a). In drinking water or bottled water, some of the low molecular weight iodo-organic compounds can cause taste and odour problems [e.g. iodoform (Bichsel & von Gunten, 2000a; Bruchet, Duguet, 2004)] or are even of toxicological concern. Some of the iodo-organic compounds are substantially more toxic than the chloro- or bromo-analogues (Richardson *et al.*, 2008). The chance of forming iodo-organic compounds increases in the order O$_3$ < HOCl ≪ NH$_2$Cl, because the rate of the subsequent HOI/IO$^-$ reactions decreases in this order. This means that during chloramination, HOI/IO$^-$ has the longest lifetime and can react with DOM moieties (Bichsel & von Gunten, 1999b).

During ozonation, OI$^-$/HOI is quickly (Table 11.1) further oxidised to iodate in two steps, similar to the case of hypobromite, with a first step to iodite [reactions (45) and (46)] and a second step to iodate [reaction (47)].

$$O_3 + IO^- \rightarrow IO_2^- + O_2 \quad (45)$$
$$O_3 + HOI \rightarrow IO_2^- + H^+ + O_2 \quad (46)$$
$$O_3 + IO_2^- \rightarrow IO_3^- + O_2 \quad (47)$$

In contrast to HOCl and HOBr, HOI is quickly oxidised by ozone. Due to the high pK_a of HOI [pK_a = 10.4, (Bichsel & von Gunten, 2000b)], the reaction of ozone with HOI is the main pathway for its further oxidation towards iodate in the pH range relevant for water treatment. This oxidation has been tentatively written as an O-transfer, although, in analogy to the reactions of ClO$_2^-$ [reaction (26)] and BrO$_2^-$ [reaction (37)], an electron transfer is also conceivable. In contrast to bromate, iodate is not problematic from a human toxicological point of view. Iodate is even added to table salt as a food supplement for iodine in iodine-deficient areas such as Switzerland (Bürgi *et al.*, 2001).

The rate constant for the oxidation of iodite with ozone is not known, however, it is faster than the preceding reactions and therefore not rate limiting. Since the reactions of I$^-$ and its follow-up products with ozone are so fast, •OH radical reactions can be neglected (in contrast to Cl$^-$/Br$^-$ plus ozone systems). For the same reason, pre-ozonation provides an excellent opportunity to mitigate iodine-derived problems in water supply systems (for an example see Section 11.18).

11.12 NITRITE

Ozone reacts rapidly with nitrite (Table 11.1). The main product is nitrate. The O-transfer requires an adduct as an intermediate, and the released oxygen must be in its singlet state [reactions (48) and (49)]. Indeed, the ^{1}O$_2$ yield has been found at 96% (Muñoz *et al.*, 2001).

$$NO_2^- + O_3 \rightarrow {}^-OOONO_2 \rightarrow NO_3^- + {}^1O_2 \quad (48)/(49)$$

The reaction seems to be complex. The structure of the first adduct ($\Delta G° = -5$ kJ mol^{-1}) is far away from that of the final products, nitrate and $^1O_2$. It has to rearrange into a second adduct ($\Delta G° = -105$ kJ mol^{-1}) the structure of which is now very close to the final products to which it decays ($\Delta G° = -78$ kJ mol^{-1}) (Naumov *et al.*, 2010).

Ozone can also add to one of the oxygens [reaction (50), $\Delta G° = +58$ kJ mol^{-1})] leading to peroxynitrite [reactions (51), $\Delta G° = -65$ kJ mol^{-1} and (52), $\Delta G° = -128$ kJ mol^{-1}], a product that is formed with a 2.6% yield (Naumov *et al.*, 2010).

$$NO_2^- + O_3 \rightarrow ONOOOO^- \tag{50}$$

$$ONOOOO^- \rightarrow ONOO^- + {}^1O_2 \tag{51}$$

$$ONOOOO^- \rightarrow {}^\bullet NO + {}^3O_2 + O_2^{\bullet -} \tag{52}$$

Reaction (52) is followed by a recombination of the two radicals [reaction (53), $\Delta G° = -28$ kJ mol^{-1}].

$$^\bullet NO + O_2^{\bullet -} \rightarrow ONOO^- \tag{53}$$

An alternative route [reaction (54), $\Delta G° = +90$ kJ mol^{-1}] that would be analogous to one of the decay routes of the adduct formed in the peroxone process [reaction (18)] is markedly endergonic and thus excluded.

$$ONOOOO^- \rightarrow NO^- + 2{}^3O_2 \tag{54}$$

Of the two processes that lead to peroxynitrite, reaction (53) is not only favoured energetically but also produces $^3O_2$, in agreement with the observation that the $^1O_2$ yield is slightly below unity. For the reaction to peroxynitrite to be able to compete with the reactions to nitrate, there must be a substantial barrier to the first intermediate. The fact that the reaction of ozone with nitrite is four orders below diffusion controlled (Table 11.1) supports this conclusion.

In addition to these reactions, $^\bullet OH$ is also formed, which is likely to occur by electron transfer [reaction (55)]. The precursor of $^\bullet OH$ is $O_3^{\bullet -}$ (see section 11.3).

$$NO_2^- + O_3 \rightarrow NO_2 + O_3^{\bullet -} \tag{55}$$

The $^\bullet OH$ yield is near 4% (Naumov *et al.*, 2010).

11.13 AZIDE

The interest in the reaction of ozone with azide in aqueous solution dates back to 1929 (Gleu & Roell, 1929). The formation of a yellow compound that decomposed during the attempt to isolate it was reported. Upon heating such yellow solutions, gas evolved consisting of N_2, N_2O and O_2 in approximately equal concentrations. Arsenite and other reducing compounds were oxidised in these mildly alkaline solutions. A reaction with formaldehyde was also noticed. An excess of alkali prevented the formation of the yellow colour and also reduced the release of N_2 and N_2O. Much later, the formation of N_2O was confirmed and the oxidising species identified as peroxynitrite (Uppu *et al.*, 1996). N_2 was, however, not found among the reaction products. Singlet oxygen is formed in 17% yield (after correction for quenching) (Muñoz *et al.*, 2001). At this stage, it seems to be premature to come up with a well-founded mechanism, as reaction conditions seem to strongly influence the chemistry. Here, the complex chemistry of peroxynitrite (Mark *et al.*, 1996; Goldstein *et al.*, 2005) could be one of the reasons. The mechanistic suggestions depicted in reactions (56)–(61) are written to trigger new experiments *en route* to a better understanding of this interesting system.

$$N\equiv N=N^- \xrightarrow[(56)]{O_3} N\equiv N=N-O-O-O^- \xrightarrow{(57)} N\equiv N + \left[NO^\bullet + O_2^{\bullet-} \right]_{cage}$$

$$(58) \downarrow -{}^1O_2 \qquad\qquad (59)\downarrow$$

$$N\equiv N + NO^- \qquad\qquad O=N-O-O^-$$

$$\xrightarrow[(60)]{H^+} NOH \xrightarrow[(61)]{2\times} N_2O + H_2O$$

11.14 HYDROGEN SULFIDE

Bisulfide, HS⁻, reacts five orders of magnitude faster with ozone than H_2S [$pK_a(H_2S) = 7.0$] (Table 11.1). Thus, except at very low pH, the reaction of ozone is governed by HS⁻. At 16%, the yield of singlet oxygen (${}^1O_2$) is low (Muñoz *et al.*, 2001). This is in contrast to many organic sulfur compounds, where yields of 100% are common (with the exception of 1,4-dithiothreitol) (Table 9.1). This points to a complex chemistry. Formation of ${}^1O_2$ demands that an adduct is formed in the first step [reactions (62) and (63)].

$$HS^- + O_3 \rightarrow [HSOOO^-]_{TS} \rightarrow HSO^- + {}^1O_2 \qquad (62)/(63)$$

Sulfuric acid is formed as the only detectable final product, and 2.3 mol of ozone are required for the formation of 1 mol of sulfuric acid (Overbeck, 1995), and this surprisingly low value has been confirmed as 2.4 (Mark *et al.*, 2011). In the case where ozone would only react by O-transfer, a stoichiometry of 4 mol would be required. Thus, further oxidation equivalents must be provided in these reactions to account for the observed stoichiometry. There is the possibility that free-radical reactions are induced and the additional oxidation equivalents are provided by the dissolved O_2. Yet, there is also another possibility.

The overall reaction (62)/(63) is exergonic with $\Delta G^0 = -136$ kJ mol⁻¹ (Mark *et al.*, 2011). The product of reaction (62), the ozone adduct, is not a stable intermediate. It rather must be considered as a transition state (TS). In one of the calculations its structure and standard Gibbs free energy of formation has been obtained as a local minimum ($\Delta G^0 = -111$ kJ mol⁻¹). Avoiding this local minimum, the reaction proceeds in the calculations by internal cyclisation and O-transfer giving rise to the anion of sulfinic peracid, HS(O)OO⁻ [reaction (64)].

$$[HSOOO^-]_{TS} \rightarrow HS(O)OO^- \qquad (64)$$

This reaction is much more exergonic ($\Delta G^0 = -244$ kJ mol⁻¹) than the formation of ${}^1O_2$. Reaction (64) is facilitated by the positive charge at sulfur (Figure 11.2).

The identification of sulfinic peracid as a likely intermediate in the reaction of ozone with HS⁻ can now explain, why the stoichiometry of oxidation of HS⁻/H_2S by ozone is only ~2.4 rather than 4.0 if a series of simple O-transfer reactions were to occur [for a detailed discussion see Mark *et al.* (2011)].

An ozone treatment has been proposed for the remediation of H_2S/HS⁻-containing waters (Overbeck, 1995; Solisio *et al.*, 1999). Ozonation is quite frequently applied in the mineral water industry for the mitigation of sulfide-induced odour problems (another example is shown below for I⁻-induced odour problems). Because the ozone plus HS⁻ reaction is so fast, HS⁻ is rapidly consumed even considering ozone reactions with the water matrix, that is, only low ozone doses are required and formation of by-products such as bromate is not a problem. This, together with the fact that sulfuric acid is the only product, makes ozonation feasible both from an economic and human health point of view.

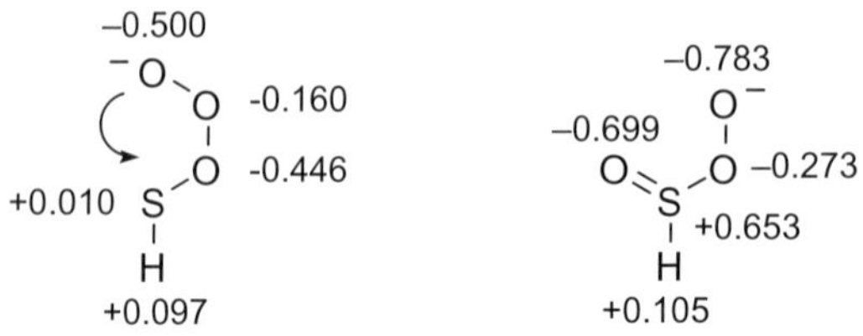

Figure 11.2 Charge distributions in the transition state of the reaction of ozone with HS⁻ as taken from a local minimum and its cyclisation product. The arrow indicates the direction of the O-transfer.

11.15 HYDROGEN SULFITE

In aqueous solution, SO_2 is in equilibrium with H_2SO_3 [$pK_a(H_2SO_3) = 1.9$], HSO_3^- [$pK_a(HSO_3^-) = 7.0$] and SO_3^{2-}. Compared to the reactions of sulfite and bisulfite (Table 11.1), the reaction of SO_2 must be very slow (Erickson et al., 1977), and thus HSO_3^- and SO_3^{2-} are the targets for the ozone reaction. Details of the reaction mechanisms are not known.

11.16 BROMATE FORMATION AND MITIGATION IN WATER TREATMENT

Bromate is the most relevant oxidation/disinfection by-product formed upon ozonation because of its toxicological relevance (see above). The low drinking water standard may not be fully justified according to some preliminary experiments on the fate of bromate in gastric juices (Keith et al., 2006a, b; Cotruvo et al., 2010). This, however, does not alter the situation in which this low value must not be exceeded in drinking water treatment using ozone.

Bromate is formed in Br⁻-containing waters through a complicated mechanism including both ozone and •OH reactions (Haag & Hoigné, 1983a; von Gunten & Hoigné, 1994; von Gunten et al., 1996; von Gunten & Oliveras, 1998; Westerhoff et al., 1998a, b; Song et al., 1996) (Chapters 11 and 14). HOBr/BrO⁻ is a crucial intermediate, because its further transformation is relatively slow. Therefore, its lifetime is sufficiently long for it to undergo various side reactions, which are summarised in Figure 11.3.

In presence of ammonia, it can form monobromamine which is only slowly oxidised by ozone (Chapter 8) to nitrate and bromide (see below). Therefore, the transient monobromamine concentration is significant and allows reduction of the rate of bromate formation. In fact, ammonia addition is one of the proposed methods for bromate minimisation (Song et al., 1997; Hofmann & Andrews, 2001; Pinkernell & von Gunten, 2001).

Another important sink for the HOBr intermediate is its reaction with H_2O_2. It may occur in H_2O_2-based AOPs [reaction (65), $k = 7.6 \times 10^8 \, M^{-1} \, s^{-1}$] (von Gunten & Oliveras, 1997).

$$HOBr + HO_2^- \rightarrow Br^- + {}^1O_2 + H_2O \tag{65}$$

Due to this reaction, bromate formation is lower in AOPs such as O_3/H_2O_2 compared to the conventional ozonation process. However, even in the presence of H_2O_2, bromate cannot be entirely suppressed during ozonation (von Gunten & Oliveras, 1998). In contrast, it is completely suppressed in the AOP UV/H_2O_2. From γ-radiolytic experiments, it has been concluded that HOBr is a key intermediate towards bromate formation even if •OH is the only oxidant (von Gunten & Oliveras, 1998) (see also Chapter 14).

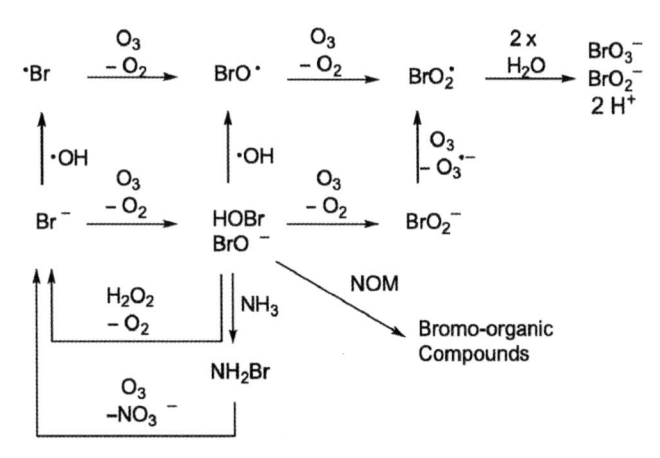

Figure 11.3 Simplified bromate formation mechanism during ozonation of bromide containing waters. The possible reactions of HOBr with ammonia, hydrogen peroxide and NOM are also shown. Adapted from von Gunten & Hoigné, 1996.

Finally, DOM also has an important influence on bromate formation. Its main role is related to its effect on ozone chemistry and •OH formation during ozonation (Chapter 3), but it is also important because the reaction of HOBr with NOM moieties leads to bromo-organic compounds.

Bromate removal, once formed, has been investigated with several processes such as activated carbon filtration, Fe(II) addition, UV treatment, photocatalytic and biological processes (Hijnen et al., 1995; Siddiqui et al., 1994, 1996; Mills et al., 1996; Asami et al., 1999; Bao et al., 1999; Gordon et al., 2002; Peldszus et al., 2004; van Ginkel et al., 2005; Kirisits et al., 2000, 2001). Activated carbon which is often in place after ozonation is only efficient when fresh. Initially, a chemical bromate reduction occurs. Thereafter, bromate is only reduced biologically under anaerobic conditions. However, such conditions are not easily met in the oxygen-rich environment after ozonation. UV-based processes for bromate removal are typically very energy intensive and chemical reduction is highly inefficient. Therefore up until now, no bromate reduction processes have been implemented in full-scale water treatment systems. The focus of mitigation is on the minimisation of its formation during ozonation. Based on the mechanistic understanding of bromate formation (Figure 11.3, Chapters 11, 14), various minimisation methods are available today and are summarised in Table 11.5 (Krasner et al., 1993; Kruithof et al., 1993; Galey et al., 2000; Song et al., 1997; Pinkernell & von Gunten, 2001; Hofmann & Andrews, 2001; Buffle et al., 2004; Neemann et al., 2004).

The chlorine–ammonia process is particularly interesting and is the result of the empirical observation that bromate formation can be substantially reduced in systems in which a pre-chlorination is followed by ammonia addition and by ozonation (Neemann et al., 2004). The mechanism of this process has been explained by the following reaction mechanism (Buffle et al., 2004) [reaction (66), $k = 1.55 \times 10^3 \, M^{-1} \, s^{-1}$

(Kumar & Margerum, 1987); reaction (67), $k = 8 \times 10^7 \, \text{M}^{-1} \, \text{s}^{-1}$ (Haag et al., 1984); reaction (68), $k = 40 \, \text{M}^{-1} \, \text{s}^{-1}$ (Haag et al., 1984); reaction (69), $k \gg 40 \, \text{M}^{-1} \, \text{s}^{-1}$ (Haag et al., 1984)].

$$\text{HOCl} + \text{Br}^- \longrightarrow \text{HOBr} + \text{Cl}^- \tag{66}$$

$$\text{HOBr} + \text{NH}_3 \longrightarrow \text{NH}_2\text{Br} + \text{H}_2\text{O} \tag{67}$$

$$\text{O}_3 + \text{NH}_2\text{Br} \longrightarrow Y \tag{68}$$

$$Y + 2\,\text{O}_3 \longrightarrow 2\,\text{H}^+ + \text{NO}_3^- + \text{Br}^- + 3\,\text{O}_2 \tag{69}$$

Table 11.5 Bromate mitigation strategies

Method	Reduction of BrO_3^- formation compared to conventional ozonation	Problems
pH depression	~Factor of 2	High alkalinity waters
Ammonia addition	~Factor of 2	Residual ammonia has to be removed in biological filtration, ammonia addition to drinking waters is forbidden in many European countries
Chlorine–ammonia process	~Factor of 10	Residual ammonia, ammonia addition to drinking waters is forbidden in many European countries, formation of halo-organic compounds in pre-chlorination
Peroxone process	Depends largely on point of H_2O_2 addition	Disinfection efficiency might be significantly reduced; mostly suited for bromate minimisation in oxidation processes when disinfection is not intended

The advantage of this process is the efficient masking of bromide as monobromamine (NH_2Br) prior to ozonation. NH_2Br only reacts slowly with ozone (Chapter 8) and releases bromide (Figure 11.3). Therefore, especially during the initial phase of ozonation for which the •OH radical impact is high, bromide is not available for the oxidation towards bromate. Due to the acid–base speciation of HOCl, the rate of reaction (66) is pH-dependent (only HOCl reacts with bromide, $pK_a(\text{HOCl}) = 7.5$). Furthermore, the reaction of HOCl with DOM may lead to chloro-organic compounds. Therefore, the free chlorine contact time has to be optimised for maximum bromide oxidation and acceptable formation of chloro-/bromo-organics. Figure 11.4 shows an example of the application of the chlorine–ammonia process in Lake Zurich and Lake Greifensee water and the trade-off between bromate minimisation and trihalomethane (THM) formation.

In these experiments, both waters were buffered at pH 8, spiked with 590 µg/L Br^-, pre-chlorinated for 5 min, followed by the addition of 300 µg/L NH_3 and analysed 60 min after ozone addition (1.5 mg/L ozone, Lake Zurich water; 3 mg/L ozone, Lake Greifensee water). Exact ozone exposures were not determined for each data point but preliminary experiments showed that the above doses correspond to similar ozone exposures of (10 ± 1.5) mg min L^{-1} in both waters.

In ammonia-containing raw waters, chlorine is consumed quickly by ammonia [$k(\text{HOCl} + \text{NH}_3) \approx 3 \times 10^6 \, \text{M}^{-1} \, \text{s}^{-1}$ (Deborde & von Gunten, 2008)], and the NH_2Cl formed in this reaction only oxidizes Br^- slowly to HOBr [$k(\text{NH}_2\text{Cl} + \text{Br}^-) = 0.28 \, \text{M}^{-1} \, \text{s}^{-1}$ at pH 7 (Trofe et al., 1980)]. Nevertheless, bromate formation can be reduced by the chlorine–ammonia process. An additional mechanism has been suggested, namely the reaction of chlorine with organic amines leading to the corresponding chloramines. They seem to have a lower •OH yield in their reaction with ozone during post-ozonation (Buffle & von Gunten, 2006). For a given ozone exposure, the corresponding •OH exposure will thus be lower than in conventional ozonation, and hence less bromate will be formed.

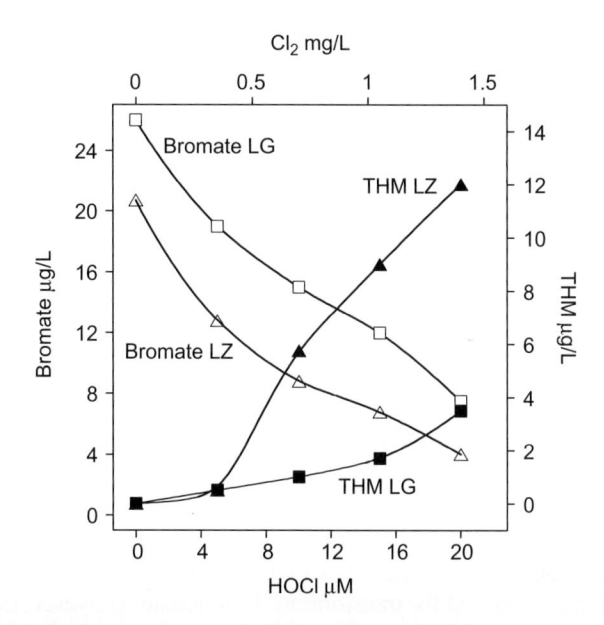

Figure 11.4 Bromate minimisation and sum of trihalomethane (THM) formation for the chlorine–ammonia process for various pre-chlorination doses in Lake Zurich (LZ) and Lake Greifensee (LG) water. Reprinted (adapted) with permission from Buffle *et al.*, 2004. Copyright (2004) American Chemical Society.

Bromate minimisation in the peroxone process has been optimised by the HiPOx® process (Bowman, 2005) in which ozone is added in small portions dosed at various points along the hydraulic pathway through a reactor to reduce residual ozone concentrations to very low levels. In excess of H_2O_2, HOBr, which now becomes a decisive intermediate, is reduced back to Br^-. However, this process is only applicable for oxidation and not for disinfection, because ozone concentrations are kept at a minimum.

11.17 BROMIDE-CATALYSED REACTIONS

The first reported case of a Br^--catalysis during ozonation is an improved ammonia oxidation which is shown in Figure 11.5a (Haag *et al.*, 1984). At circum-neutral pH, acceleration of ammonia oxidation is caused by the difference of the apparent second-order rate constant for the reaction of ozone with NH_3/NH_4^+ ($k_{app} \approx 0.025$ M^{-1} s^{-1}) and NH_2Br ($k = 40$ M^{-1} s^{-1}) (Haag *et al.*, 1984). Since the formation of HOBr from the oxidation of Br^- is reasonably fast ($k = 160$ $M^{-1} s^{-1}$, see above) and the reaction of HOBr with NH_3/NH_4^+ is very fast ($k = 8 \times 10^7 M^{-1}$ s^{-1}, see above), the oxidation of NH_2Br by ozone is the rate-limiting step in ammonia oxidation. Compared to the direct oxidation of NH_3/NH_4^+ by ozone (Chapter 8), the Br^--catalysed process is several orders of magnitude faster. Another Br^--catalysed reaction is the formation of *N*-nitrosodimethylamine (NDMA) during ozonation of dimethylsulfamide (DMS)-containing waters (Figure 11.5b). This is discussed in more detail in Chapter 8.

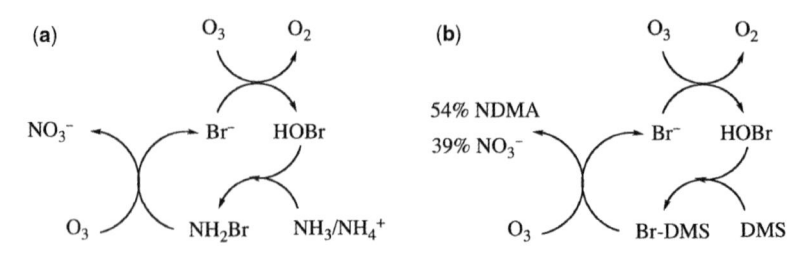

Figure 11.5 Bromide-catalysed oxidation of (a) ammonia and (b) dimetylsulfamide (DMS) during ozonation. Adapted from von Gunten et al., 2010 with permission.

11.18 MITIGATION OF IODIDE-RELATED PROBLEMS

Typical I$^-$ concentrations in water resources are fairly low (≤ 10 µg/L) meaning that iodide is not a problem generally. However in special geological formations or in coastal areas, I$^-$ concentrations can be elevated and depending on the oxidative treatment, problems with iodo-organic compounds can arise from the reaction of HOI/OI$^-$ with DOM. Because the reaction of I$^-$ to iodate with ozone is a fast process with a very short lifetime of HOI (see above), pre-ozonation provides an excellent opportunity to mitigate I$^-$-derived problems in the distribution system. The following example will illustrate this in more detail.

In a Swiss mineral water ([I$^-$] = 30 µg/L), occasionally iodine-derived taste and odour problems were observed which led to massive consumer complaints. In addition to I$^-$, the same water also contains relatively high Br$^-$ concentrations (≥ 100 µg/L). To mitigate this problem, introduction of an ozonation step into the process line was proposed for transforming I$^-$ to iodate. However, the ozone dose had to be adjusted for a complete transformation of I$^-$ to iodate without any significant bromate formation. Iodate is a stable sink of iodine which hinders the formation of iodo-organic compounds. Based on the reaction kinetics shown above and in Chapter 14, bromate formation during ozonation is several orders of magnitude slower than iodate formation. This shows the high selectivity of ozone for I$^-$ oxidation. Therefore, based on laboratory experiments, a plug-flow reactor with a residence time of 30 s and an ozone residual concentration of ≤ 0.15 mg/L at the effluent was selected. The residual ozone was quenched in an activated carbon filter.

Table 11.6 shows the results of the full-scale treatment train for iodide, iodate, bromide and bromate concentrations, from raw water to bottled water.

Table 11.6 Ozonation of a Swiss mineral water for mitigation of iodide-derived taste and odour problems. Bromide and iodide concentrations in the raw water and resulting iodate levels (µg/L) in the bottled water (Bichsel & von Gunten, 1999a; Salhi & von Gunten, 1999)

Anion	Raw water	Ozonated water	Bottled water
Bromide	101	102	101
Bromate	< d.l.	< d.l.	< d.l.
Iodide	32	n.d.	n.d.
Iodate	< d.l.	46	46

d.l.: detection limit (bromate, iodate: 0.1 µg/L), n.d.: not detected.

It is evident that I^- is completely oxidised to iodate, while the Br^- concentration remains constant. In the bottled water samples, bromate levels were always < 2 μg/L and iodate was stable for several months. This process is a promising approach for solving I^--derived drinking water problems. It could be applied as a pre-treatment step in drinking waters containing elevated I^- levels to avoid formation of iodo-organic compounds in post-disinfection processes such as chlorination and chloramination.

Chapter 12
Reactions with metal ions

12.1 REACTIVITY OF METAL IONS

The first detailed report on the reactivity of ozone towards metals and metal ions was by *Schönbein* (Schönbein, 1854), and this early report is still worthwhile reading. Reported rate constants are compiled in Table 12.1.

Table 12.1 Compilation of ozone rate constants with metal ions. Published rate constants are rounded to significant figures

Compound	pK_a	pH	$k/M^{-1}s^{-1}$	Reference
Ag^+			0.035	Noyes et al., 1937
			0.04	Reisz et al., 2012b
$As(OH)_3$	9.2		5.5×10^5	Dodd et al., 2006b
$AsO(OH)_2^-$	12.2		1.5×10^8	Dodd et al., 2006b
Co^{2+}			0.62	Hill, 1949
			0.66 at 0°C	Yeatts & Taube, 1949
Cu^{2+}			no reaction	Yeatts & Taube, 1949
Fe^{2+}		0–2	8.2×10^5	Lögager et al., 1992
		0–3	8.5×10^5	Jacobsen et al., 1997
$Fe(CN)_6^{4-}$			1.2×10^6	Flyunt et al., 2003a
Mn^{2+}			1.5×10^3	Jacobsen et al., 1998a
			1.8×10^3	Jacobsen et al., 1998a
			3.9×10^3	Tyupalo & Yakobi, 1980
		7	1.3×10^3	Reisz et al., 2008
		7	1.0×10^3	Reisz et al., 2008
Mn(II)oxalate			760	Reisz et al., 2008
Mn(II)polyphosphate			1.4×10^4	Reisz et al., 2008

(Continued)

Table 12.1 Compilation of ozone rate constants with metal ions. Published rate constants are rounded to significant figures (*Continued*)

Compound	pK$_a$	pH	k/M^{-1}s^{-1}	Reference
Pb^{2+}			no reaction	Reisz & von Sonntag, unpublished
Pb(OH)$^+$			slow	Reisz & von Sonntag, unpublished
Pb(OH)$_2$			~1 × 10^3	Reisz & von Sonntag, unpublished
SeO(OH)$_2$	2.8		negligible	von Gunten & Jaeggi, unpublished
SeO$_2$(OH)$^-$	8.5		~0.5	von Gunten & Jaeggi, unpublished
SeO$_3^{2-}$			1.7 × 10^4	von Gunten & Jaeggi, unpublished
Sn^{2+}			2 × 10^9	Utter et al., 1992
Sn(butyl)$_3$Cl			4–6	Sein et al., 2009
Sn(butyl)$_3$OSn(butyl)$_3$			7	Sein et al., 2009
Sn(butyl)$_2$Cl$_2$			0.5–0.7	Sein et al., 2009

The kinetics of the reactions of ozone with metal ions in their low oxidation states have been largely studied with As(III), Fe(II) and Mn(II). These reactions are governed by O-transfer reactions and ozone adducts are likely intermediates. With As(III), which can be present as As(OH)$_3$ and As(OH)$_2^-$ in the near neutral pH range, the monoanion reacts faster by nearly three orders of magnitude, and this reflects the general behaviour of electrophilic ozone. For Fe^{2+}, the rate constant has only been measured in acid solutions, and based on the above one may speculate that, in analogy to the oxidation with oxygen (King, 1998), the rate may go up in less acidic and neutral solutions where contributions of Fe(OH)$^+$ and Fe(OH)$_2$ dominate. Complexation has a marked effect on the rate of reaction, as shown for Mn(II). The iron complex Fe(CN)$_6^{4-}$ does not exchange its ligands readily, and there is no free site for ozone to add on. Thus, Fe(II) has to undergo a one-electron transfer reaction.

12.2 ARSENIC

In neutral solution As(III) is largely undissociated [equilibria (1) and (2), pK$_a$ (As(OH)$_3$) = 9.2; pK$_a$(AsO(OH)$_2^-$) = 12.2].

$$As(OH)_3 \rightleftharpoons AsO(OH)_2^- + H^+ \quad (1)$$
$$AsO(OH)_2^- \rightleftharpoons AsO_2(OH)^{2-} + H^+ \quad (2)$$

As(III) is readily oxidised by ozone, whereas As(OH)$_3$ reacts much more slowly [reaction (3)] than the monoanion, AsO(OH)$_2^-$ [reaction (4), cf. Table 12.1].

$$As(OH)_3 + O_3 \rightarrow AsO(OH)_3 + O_2 \quad (3)$$
$$AsO(OH)_2^- + O_3 \rightarrow AsO_2(OH)_2 + O_2 \quad (4)$$

Arsenic acid [As^{V}O(OH)$_3$] is a stronger acid than arsenous acid [pK$_{a1}$ = 2.2, pK$_{a2}$ = 6.9 and pK$_{a3}$ = 11.5]. These pK$_a$ values are very similar to those of the related phosphoric acid. The oxidation of As(III) and the concomitant formation of deprotonated arsenic acid species are desired in water treatment because As(V) adsorbs much better on iron(III)(hydr)oxide-containing filters than As(III) (Ruhland et al., 2003a, b; Roberts et al., 2004; Voegelin et al., 2010). Therefore, the combination of ozonation with iron(III)(hydr)oxide filtration is a powerful tool for arsenic mitigation in drinking and mineral waters.

12.3 COBALT

Cobalt has two oxidation states, Co(II) and Co(III). The lower oxidation state reacts with ozone (Table 12.1). It has been suggested that in this slow reaction $^{\bullet}$OH radicals are generated (Hill, 1949). As the reduction potential $E_0(Co^{3+}/Co^{2+}) = 1.8$ V is considerably higher than $E_0(O_3/O_3^{\bullet-}) = 1.03$ V (Chapter 2), this suggestion has to be put on a better footing. Techniques are now available for doing this (Chapter 14).

12.4 COPPER

A reaction of Cu^{2+} with ozone is not observed (Yeatts & Taube, 1949), although there is a higher oxidation state, Cu(III) (Asmus et al., 1978; Ulanski & von Sonntag, 2000).

12.5 IRON

Originally, it was thought that Fe^{2+} reacts with ozone by electron transfer (Tyupalo & Dneprovskii, 1981), but this view has since been corrected in favour of an O-transfer [reaction (5)] (Lögager et al., 1992; Jacobsen et al., 1997)].

$$Fe^{2+} + O_3 \longrightarrow FeO^{2+} + O_2 \tag{5}$$

The Fe(IV) species that is formed in reaction (5) is a strong oxidant and reacts rapidly with Fe^{2+} [reaction (6), $k = 7.2 \times 10^4$ M^{-1} s^{-1} (Jacobsen et al., 1997), 1.4×10^5 M^{-1} s^{-1} (Lögager et al., 1992)] and Mn^{2+} [reaction (7), $k = 1 \times 10^4$ M^{-1} s^{-1} (Jacobsen et al., 1998b)].

$$FeO^{2+} + Fe^{2+} + 2H^+ \longrightarrow 2Fe^{3+} + H_2O \tag{6}$$

$$FeO^{2+} + Mn^{2+} + 2H^+ \longrightarrow Fe^{3+} + Mn^{3+} + H_2O \tag{7}$$

The latter reaction is of relevance for the further oxidation of manganese to MnO_4^- in the ozonation of drinking waters and wastewaters that contain both Fe^{2+} and Mn^{2+} (see below).

Reliable rate constants are only available for acid solutions (Table 12.1). Due to a higher electron density at iron, $Fe(OH)^+$ is expected to react more rapidly than Fe^{2+} and hence dominates the Fe(II) oxidation kinetics in such waters. In the case of oxidation of iron(II) by O_2, the rate of its oxidation at circum-neutral pH is largely controlled by the $Fe(OH)_2$ species [$k(O_2 + Fe^{2+}) = 5.5 \times 10^{-5}$ M^{-1} s^{-1}, $k[O_2 + Fe(OH)^+] = 4.2 \times 10^2$ M^{-1} s^{-1}, $k[O_2 + Fe(OH)_2] = 5.2 \times 10^7$ M^{-1} s^{-1} (King, 1998)]. In carbonate-containing waters, carbonate and hydroxo-carbonate complexes also play an important role (King, 1998).

In $Fe(CN)_6^{4-}$, iron is fully complexed by tightly-bound cyano ligands, and ozone can no longer form an adduct. Thus, ozone has to react by one-electron transfer [reaction (8)] (Flyunt et al., 2003a).

$$Fe(CN)_6^{4-} + O_3 \longrightarrow Fe(CN)_6^{3-} + O_3^{\bullet-} \tag{8}$$

The $^{\bullet}$OH yield is only about 70% (note that $O_3^{\bullet-}$ gives rise to $^{\bullet}$OH, Chapter 11), and it has been tentatively suggested that in competition with electron transfer, ozone may also react with the ligands.

12.6 LEAD

The browning of paper strips soaked with basic lead acetate was used by *Schönbein* for detecting ozone (Schönbein, 1854). Yet, only very recently [Reisz & von Sonntag (2011), unpublished] this reaction has found some further attention. Depending on pH, Pb(II) is present as Pb^{2+}, $Pb(OH)^+$, $Pb(OH)_2$ or $Pb(OH)_3^-$ (Carell & Olin, 1960). In acid solution, where Pb^{2+} dominates, barely any reaction is observed overnight [Reisz, Naumov & von Sonntag, 2011, unpublished results]. Even at pH 9, where $Pb(OH)^+$ and $Pb(OH)_2$ dominate, $k_{obs} \approx 330\ M^{-1}\ s^{-1}$ (Table 12.1) is still low. Based on the data on As(III), one has to conclude that the reaction is mainly given by the less positively charged species. Under such conditions $\alpha[Pb(OH)_2]$ is near 30%, and we conclude that $Pb(OH)^+$, like Pb^{2+}, is of little reactivity and that $k[Pb(OH)_2 + O_3]$ must be near $10^3\ M^{-1}\ s^{-1}$.

In this context, some quantum-chemical calculations have been carried out. According to these, addition of ozone to Pb^{2+} is endergonic ($\Delta G^0 = +22\ kJ\ mol^{-1}$) and the Pb–O bond in this adduct is extremely long (2.61 Å). For $Pb(OH)^+$, no stable ozone adduct could be calculated, and in these preliminary calculations, triplet (ground-state) O_2 was released in an endergonic reaction ($\Delta G^0 = +13\ kJ\ mol^{-1}$). Only when one proceeds to $Pb(OH)_2$, does the overall reaction also become exergonic for the spin-allowed formation of singlet oxygen [reaction (9), $\Delta G^0 = -11\ kJ\ mol^{-1}$].

$$Pb(OH)_2 + O_3 \longrightarrow OPb(OH)_2 + {}^1O_2 \tag{9}$$

These calculations reflect the experimental observations well.

12.7 MANGANESE

Ozone reacts markedly slower (Table 12.1) with Mn^{2+} than with Fe^{2+} [reaction (10)]. Complexation of Mn^{2+} by polyphosphate or oxalate markedly influences the rate of reaction (Table 12.1).

$$Mn^{2+} + O_3 \longrightarrow MnO^{2+} + O_2 \tag{10}$$

The MnO^{2+} species reacts rapidly with water [reaction (11), $k > 5\ s^{-1}$ (Reisz et al., 2008)] yielding a colloidal MnO_2 solution. This prevents a ready further oxidation by ozone to MnO_4^-.

$$MnO^{2+} + H_2O \longrightarrow MnO_2 + 2\,H^+ \tag{11}$$

As a consequence, in distilled water and in neutral solutions, only small amounts of MnO_4^- are formed. In acid solution or when Mn^{2+} is complexed, however, the reaction proceeds to MnO_4^- when ozone is in large excess. Otherwise and with an excess of Mn^{2+}, Mn(III) is the final product [reaction (12)].

$$MnO^{2+} + Mn^{2+} + 2\,H^+ \longrightarrow 2\,Mn^{3+} + H_2O \tag{12}$$

Oxidation to MnO_4^- is most interesting, since as much as four times more ozone is required than the stoichiometry suggests. The reason for this is that ozone may also act as a reductant with some of the high oxidation states of manganese. To visualise this, the highly exergonic ($\Delta G^0 = -200\ kJ\ mol^{-1}$) reaction (13) may be drawn. There may be other intermediates that could react similarly.

$$MnO^{2+} + O_3 \longrightarrow Mn^{2+} + 2\,O_2 \tag{13}$$

A reaction where ozone acts as a reductant has also been observed in reactions with hypobromite ($BrO^- + O_3 \rightarrow Br^- + 2\,O_2$ in competition with $BrO^- + O_3 \rightarrow BrO_2^- + O_2$) and with hypochlorite (Chapter 11). Another example seems to be the reaction of ozone with Ag(II) (see below).

In drinking water and in wastewater, the formation of substantial amounts of MnO_4^- may be observed (quite noticeable when such waters turn pink upon ozonation). Under such conditions, it cannot be the acidic environment or large amounts of complexing agents which cause the substantial formation of MnO_4^- (bicarbonate buffer induces MnO_4^- formation only at elevated concentrations), and thus there must be an additional route. Waters that contain Mn(II) typically also contain Fe(II). Fe(II) reacts rapidly with ozone and the fraction of Fe(IV) that is not scavenged by Fe(II) then oxidises Mn(II) [reaction (7)]. The Mn(III) thus formed can be more readily oxidised to MnO_4^- than the very short-lived MnO^{2+} intermediate (Reisz et al., 2008).

Colloidal MnO_2 has very interesting properties. It is a much stronger oxidant than ground technical MnO_2 (birnesite). It reacts most rapidly with H_2O_2 under violent O_2 formation, a reaction not given by birnesite (von Sonntag, unpublished results). The most complex reaction involves, in the first step, the activation of the MnO_2 solution by forming Mn(III) centres ($k = 700 \, M^{-1} \, s^{-1}$) (Baral et al., 1985). This activated solution catalyses the further decomposition of H_2O_2 without being much consumed. The MnO_2 solution also bleaches indigotrisulfonic acid ($k > 10^7 \, M^{-1} \, s^{-1}$) while permanganate reacts much more slowly ($k = 1.3 \times 10^3 \, M^{-1} \, s^{-1}$) (Reisz et al., 2008). For further studies on the formation and reactivity of MnO_2 colloids see Baral et al. (1986) and Lume-Pereira et al. (1985).

12.8 SELENIUM

In aquatic environmental systems, selenium is present as selenite ($Se^{IV}O_3^{2-}$) or selenate ($Se^{VI}O_4^{2-}$). Selenite reacts stoichiometrically with ozone by O-transfer to selenate [reaction (14)].

$$SeO_3^{2-} + O_3 \longrightarrow SeO_4^{2-} + O_2 \tag{14}$$

The reaction is highly pH-dependent because only SeO_3^{2-} reacts rapidly with ozone, whereas $SeO_2(OH)^-$ reacts rather slowly and $SeO(OH)_2$ practically not at all (Table 12.1).

The oxidation of selenite to selenate by ozone is of interest for selenite-containing mineral waters. Selenite can be slowly (within several weeks) transformed by micro-organisms to selenium-organic compounds which cause taste and odour problems. Once selenium is present in its higher oxidised form, selenium-organic compounds are no longer formed.

12.9 SILVER

In water, silver can exist in four oxidation states: Ag(0) (colloidal silver) Ag(I), Ag(II) and Ag(III). Over the whole pH range of interest Ag(I) is present as Ag^+. Ag(II) dominated only below pH 5.35 by Ag^{2+} and at pH > 8.35 $Ag(OH)_2$ dominates (Asmus et al., 1978). Ag^{2+} is a strong oxidant ($E^0 = +2.0 \, V$) (Blumberger et al., 2004), but the other Ag(II) species must be less oxidising. When Ag(II) is generated by pulse radiolysis at pH 7, the species decays by second-order kinetics, and it has been assumed that Ag(II) disproportionates into Ag^+ and Ag(III) (Pukies et al., 1968). As further details are not known, it is also conceivable that Ag(II) forms a dimer that would be equivalent to an Ag(I)/Ag(III) binuclear complex.

The reaction of ozone with Ag^+ is slow (Table 12.1) and complex (Reisz et al., 2012b). Generally, the positive charges (low electron density at the site of reaction) impede the formation of an adduct by the electrophilic ozone. Concomitant proton release from the hydration shell [four water molecules in the case of Ag^+ (Blumberger et al., 2004)] may facilitate the process [reaction (15)].

$$Ag(H_2O)_4^+ + O_3 \rightleftarrows {}^-OOOAg(H_2O)_3(OH)^+ + H^+ \tag{15}$$

For releasing singlet O₂ in a spin-allowed reaction, the energetics are often very unfavourable. Silver, as a heavy atom, can mediate the release of (triplet) ground state O_2 [e.g. reaction (16)].

$$^-OOOAg(H_2O)_3(OH)^+ \longrightarrow OAg(H_2O)_3OH + {}^3O_2 \tag{16}$$

As the slow reaction of Ag^+ with ozone proceeds, the kinetics continue at the slow rate of the initial reaction (Figure 12.1, squares) or suddenly accelerate as in the majority of experiments (Figure 12.1, triangles). Concomitant with the sudden acceleration, colloidal Ag(0) is formed, which can be monitored at 500 nm [cf. (Stamplecoskie & Scaiano, 2010)] (Reisz et al., 2012b). The decomposition of ozone on Ag(0) nanoparticles has been reported (Morozov et al., 2011).

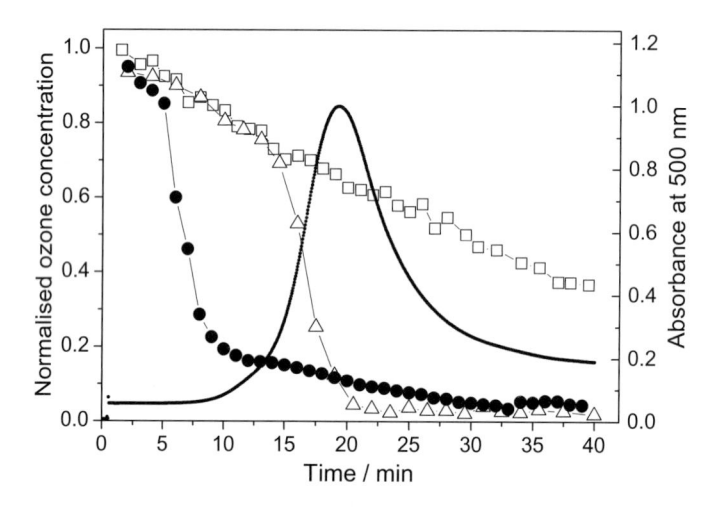

Figure 12.1 Decay of ozone (close to 100 µM in the various experiments), followed by titrating remaining ozone with 3-butenol (open squares and triangles). The solid line indicates Ag(0) formation measured at 500 nm and relates to the ozone decay shown in triangles. Data in closed circles have been obtained in the presence of 1 mM acetate. According to Reisz et al., 2012b.

The precursor of Ag(0) has been considered to be Ag(II) [in equilibrium with Ag^+ and Ag(III)], for example, the exergonic reaction (17) (Reisz et al., 2012b).

$$AgOH^+ + O_3 \longrightarrow Ag(0) + 2\,O_2 + H^+ \tag{17}$$

In this reaction, ozone acts as a reducing agent. Such a reaction is not without precedent (cf. Section 12.7 and Chapter 11).

Ag(0) formation is accelerated by the addition of acetate. Here, Ag(II) formation is favoured by the intramolecular electron transfer reaction (18).

$$(HO)_2AgOC(O)CH_3 + H_2O \longrightarrow CO_2 + Ag(OH)_2 + {}^\bullet CH_3 \tag{18}$$

Oxalate also accelerates the ozone plus Ag$^+$ reactions, but without giving rise to Ag(0). Ag(III) oxalate decomposes by intramolecular electron transfer to Ag$^+$ without Ag(II) as an intermediate [reaction (19)].

$$Ag[OC(O)C(O)O]^+ \longrightarrow Ag^+ + 2\,CO_2 \qquad (19)$$

A well-documented intramolecular electron transfer process of a transition metal ion oxalate in its high oxidation state is Mn(III) (Taube, 1947, 1948a, b).

12.10 TIN

The reaction of ozone with Sn^{2+} is close to diffusion controlled, much faster than any other transition metal ion (Table 12.1). In an organic solvent, this reaction gives rise to $^1O_2$ in high yields (Emsenhuber et al., 2003). This is a strong indication of an adduct being formed in the first step as depicted in reaction (20).

$$Sn^{2+} + O_3 \rightleftarrows \;^-OOOSn^{3+} \longrightarrow OSn^{2+} + {}^1O_2 \qquad (20)/(21)$$

One may assume that OSn^{2+} formed in reaction (21) subsequently hydrolyses giving rise to SnO$_2$. Experiments confirming this point have not been carried out.

Tributyltin reacts somewhat faster with ozone (Table 12.1) than 2-propanol (Table 10.1). Somehow, tin must be involved in its reactivity as has been proposed for tetrabutyl tin in carbon tetrachloride solutions (Alexandrov et al., 1971). In tributyltin, the electron density at tin is lower. This is not a favourable situation for a reaction of an electrophilic agent such as ozone. Indeed, dibutyltin whose electron density at tin is even lower reacts an order of magnitude more slowly [$k = 0.5 - 0.7$ M^{-1} s^{-1} (Sein et al., 2009)].

Tributyl tin chloride gives rise to dibutyl tin chloride as the main product. Further products are formaldehyde, acetaldehyde, propanal and butanal (Sein et al., 2009).

12.11 METAL IONS AS MICROPOLLUTANTS

Worldwide, about 50 million people suffer from high levels of arsenic in their drinking water, in some severe cases exceeding the limiting permissible value of 10 µg/l by a factor of 100. Countries most affected are Bangladesh, India (West Bengal), Vietnam, China, Taiwan, Argentina, Chile, Mexico and the West of the USA (Smedley & Kiniburgh, 2002). Some suggestions as to the potential reasons for the toxicity of arsenic have been made (Hughes, 2002). Arsenic remediation typically involves an oxidation of As(III) to As(V) and adsorption of the latter on iron hydroxides (e.g. in flocculation with FeCl$_3$) (Katsoyiannis et al., 2008; Karcher et al., 1999) or on iron(III)(hydr)oxide-based adsorbers (Driehaus, 2002). In the oxidation of As(III) to As(V), ozone may be of some advantage as the reaction is very fast (Table 12.1). Nevertheless, this solution is only applicable in industrialised countries. In developing countries pre-chlorination is a feasible alternative, since As(III) is quickly oxidised by chlorine [pH 7: k_{app}(As(III) + chlorine) = 2.6 × 10^5 M^{-1} s^{-1} (Dodd et al., 2006b)].

Iron(II) and manganese(II) are common water contaminants in reduced source waters. Iron(II) is mainly an aesthetic problem, whereas manganese has also been discussed in relation to its neurotoxicity especially for children (WHO, 2004; Bouchard et al., 2011). Iron(II) and manganese(II) removal is carried out by oxidation followed by precipitation and separation of the Fe(III) and Mn(III,IV) (hydr)oxides. Oxidation is mostly performed with oxygen in abiotic or microbiologically mediated processes (Mouchet, 1992). These processes are often sufficiently fast for iron(II) (Stumm & Lee, 1961). At low pH or in the presence of organic matter forming strong complexes with iron(II), however, a stronger oxidant such as ozone may be needed (Theis & Singer, 1974). At circum-neutral pH, abiotic Mn(II) oxidation with oxygen is very slow (Morgan, 2005) and therefore, other oxidants such as ozone must be applied. In this case, permanganate may be formed (see above) which has to be removed by a post-filtration step containing activated carbon or another reducing material (Singer, 1990).

Organotin compounds such as tributyltin have been widely used as an anti-fouling agent in ship paints and as biocides in wood preservatives and in textiles. The use of these substances has not been allowed since the late 1990s in most European and North American countries (Chau et al., 1997), but they are still observed in effluents of WWTPs (European Commission, 2000). Their biodegradation is very slow and thus they persist in the environment [for their detection see Takeuchi et al., (2000)]. They have endocrinic properties and their effect on aquatic life is dramatic (Duft et al., 2003a) (Chapter 4). As their rate of reaction with ozone is low (Table 12.1), a direct reaction with ozone is ineffective in their elimination from contaminated waters and only the •OH route can reduce their levels (Chapter 14).

Chapter 13

Reactions with free radicals

13.1 REACTIVITY OF RADICALS

Ozone reacts fast with most free radicals, in some cases close to diffusion-controlled. The rate constants reported thus far are compiled in Table 13.1.

Table 13.1 Compilation of ozone rate constants with free radicals. Published rate constants are rounded to significant figures

Radical	$k/M^{-1} s^{-1}$	Reference
Bromine atom ($^{\bullet}$Br)	1.5×10^{8} [a]	von Gunten & Oliveras, 1998
Carboxymethyl radical ion (1-) ($^{\bullet}CH_2C(O)O^-$)	2×10^{9}	Sehested et al., 1987
Chlorine dioxide ($ClO_2^{\bullet}$)	1.05×10^{3}	Kläning et al., 1985
	1.1×10^{3}	Hoigné et al., 1985
Dichloro radical anion ($Cl_2^{\bullet -}$)	9×10^{7}	Bielski, 1993
(Dioxido)trioxidosulfate(dot-) ($^-O_3SOO^{\bullet}$)	1.6×10^{5}	Lind et al., 2003
Hydrated electron (e_{aq}^-)	3.6×10^{10}	Bahnemann & Hart, 1982
	3.6×10^{10}	Sehested et al., 1983
Hydrogen atom ($^{\bullet}$H)	2×10^{10}	Sehested et al., 1983
Hydroxyl ($^{\bullet}$OH)	3×10^{9}	Bahnemann & Hart, 1982
	1.1×10^{8}	Sehested et al., 1984
2-Hydroxy-2-methylpropylperoxyl ($^{\bullet}OOCH_2C(CH_3)_2OH$)	1.8×10^{4}	Lind et al., 2003
2-Peroxylacetic acid ($HOC(O)CH_2OO^{\bullet}$)	$\geq 2.7 \times 10^{4}$	Lind et al., 2003
2-Peroxylacetate $^-OC(O)CH_2OO^{\bullet}$	7.8×10^{3}	Lind et al., 2003
2-Peroxylchloroacetate ($^-OC(O)CHClOO^{\bullet}$)	1.6×10^{4}	Lind et al., 2003
1-Peroxylpropan-2-one ($CH_3C(O)CH_2OO^{\bullet}$)	$\geq 7.3 \times 10^{4}$	Lind et al., 2003
Superoxide ($O_2^{\bullet -}$)	1.5×10^{9}	Sehested et al., 1983
	1.6×10^{9}	Bühler et al., 1984
TEMPO	10^{7}	Muñoz & von Sonntag, 2000a

[a] Estimated by kinetic modelling.

For a considerable number of radicals, standard Gibbs free energies of their reactions with ozone have been calculated (Naumov & von Sonntag, 2011a). In several cases, radicals react with ozone by giving rise to adducts. Only the adduct of the hydrated electron to ozone, $O_3^{\bullet-}$, is sufficiently long-lived to become detectable on the microsecond time scale, that is, by pulse radiolysis (Elliot & McCracken, 1989). For more short-lived adducts, evidence for their formation can be obtained by quantum-mechanical calculations as potential minima on the reaction path. In the majority of cases, however, no such potential minimum is obtained. "Adducts" are then, at best, bona-fide transition states *en route* to the products. The outcome of these calculations is compiled in Tables 13.2–13.5. In its reaction with free radicals, ozone has to compete with O_2, a well-known free-radical scavenger. Yet, as will be discussed below, O_2 reacts only with very limited radical types rapidly and irreversibly. Thus, ozone reacts with many free radicals effectively, despite the fact that O_2 is typically present in large excess over ozone.

13.2 OZONE REACTIONS WITH REDUCING RADICALS

The reduction potentials of e_{aq}^-, $H^\bullet$, $CO_2^{\bullet-}$ and $O_2^{\bullet-}$ at pH 7 are -2.78 V, -2.4 V, -1.8 V and -0.33 V (-0.18 V, correcting for O_2 solubility, cf. Chapter 2), respectively, while ozone has a reduction potential of $+1.1$ V (Wardman, 1989). Thus, there is a substantial driving force for an electron transfer. With the hydrated electron, ozone reacts at a diffusion-controlled rate by addition [reaction (1)].

$$O_3 + e_{aq}^- \longrightarrow O_3^{\bullet-} \tag{1}$$

In water, $O_3^{\bullet-}$ is relatively long-lived (Elliot & McCracken, 1989), and dissociative electron transfer ($O_3 + e_{aq}^- \longrightarrow O^{\bullet-} + O_2$) can be excluded as an alternative reaction path. The subsequent decay of $O_3^{\bullet-}$, an abundant precursor of $^\bullet OH$ in ozone reactions, is discussed in Chapter 11.

$H^\bullet$ also reacts by addition [reaction (2)], as indicated by quantum-chemical calculations (Table 13.2). An electron transfer process ($O_3 + H^\bullet \longrightarrow O_3^{\bullet-} + H^+$) is less likely.

$$O_3 + H^\bullet \longrightarrow HO_3^\bullet \tag{2}$$

Table 13.2 Reactions of reducing free radicals with ozone in aqueous solution. Calculated standard Gibbs free energies (ΔG^0, kJ mol^{-1}) of the overall reaction and of radical–ozone adducts (if formed) and their decay (Naumov & von Sonntag, 2011a)

Reactants	Products	Overall reaction	Adduct formation	Adduct decay
e_{aq}^-, O_3	$O_3^{\bullet-}$	(-374; pH > 11)*		see (Merényi *et al.*, 2010a)
$^\bullet H$, O_3	$^\bullet OH$, O_2	-411	-373	-38 [-47 (Naumov & von Sonntag, 2011b)]
$CO_2^{\bullet-}$, O_3	CO_2, $O_3^{\bullet-}$	-315	**	
$CO_2^{\bullet-}$, O_3	$CO_3^{\bullet-}$, O_2	-370	**	
$O_2^{\bullet-}$, O_3	$O_3^{\bullet-}$, O_2	-176	**	

*Calculated from reported reduction potentials (Wardman, 1989); **an adduct cannot be calculated as a potential minimum on the reaction path.

$HO_3^\bullet$ is a strong acid [$pK_a(HO_3^\bullet) = -2$], and rapid deprotonation [reaction (3), $k \approx 10^{12}$ s^{-1}] and decay into $^\bullet OH + O_2$ [reaction (4)], for which there is also a high driving force (Table 13.2), eventually lead to O_2 and $^\bullet OH$ (see Chapter 10).

$$HO_3^\bullet \rightleftarrows O_3^{\bullet -} + H^+ \quad (3)$$
$$HO_3^\bullet \rightarrow {}^\bullet OH + O_2 \quad (4)$$

Based on the reduction potentials of $CO_2^{\bullet -}$ and ozone (see above), there is a considerable driving force for an electron transfer [reaction (5)], but an O-transfer [reaction (6)] is not only energetically (Table 13.2) but also kinetically favoured.

$$CO_2^{\bullet -} + O_3 \rightarrow CO_2 + O_3^{\bullet -} \quad (5)$$
$$CO_2^{\bullet -} + O_3 \rightarrow CO_3^{\bullet -} + O_2 \quad (6)$$

Evidence for this is obtained by the fact that no chain reaction of any importance is observed in the γ-radiolysis of formate solutions in the presence of ozone (Lind et al., 2003).

The reaction of $O_2^{\bullet -}$ with ozone is the only one of the reducing radicals that takes place by electron transfer [reaction (7)].

$$O_3 + O_2^{\bullet -} \rightarrow O_3^{\bullet -} + O_2 \quad (7)$$

Its conjugate acid, the $HO_2^\bullet$ radical, often behaves like a peroxyl radical (for ozone reactions with peroxyl radicals see below), and as such is not a typical reducing radical. It is, however, in equilibrium with its conjugate base, $O_2^{\bullet -}$, [equilibrium (8); $pK_a(HO_2^\bullet) = 4.8$ (Bielski et al., 1985)], and even in slightly acid solutions the ozone reaction is dominated by $O_2^{\bullet -}$.

$$HO_2^\bullet \rightleftarrows O_2^{\bullet -} + H^+ \quad (8)$$

13.3 OZONE REACTIONS WITH CARBON-CENTERED RADICALS

Oxygen, which is always present when ozone is applied as an oxidant is typically in excess of ozone. Simple carbon-centred radicals react very rapidly with O_2 ($k = 2 \times 10^9 \, M^{-1} \, s^{-1}$) (von Sonntag & Schuchmann, 1997) as does ozone (Sehested et al., 1992). Due to this competition, ozone stands a chance of reacting with this type of radical significantly only as long as it is present at concentrations comparable to that of O_2. O_2 reacts with alkyl radicals by addition, and for simple carbon-centred radicals the ensuing peroxyl radicals are stable [cf. reaction (9), for exceptions see below, for their reactions see Chapter 14].

$$^\bullet CH_3 + O_2 \rightarrow {}^\bullet OOCH_3 \quad (9)$$

The corresponding reaction with ozone also gives rise to an adduct [cf. reaction (10)], but this adduct is highly unstable (Table 13.3) and releases O_2, thereby giving rise to an oxyl radical [cf. reaction (11)].

$$^\bullet CH_3 + O_3 \rightarrow CH_3OOO^\bullet \rightarrow CH_3CO^\bullet + O_2 \quad (10/11)$$

For larger alkyl radicals, adducts could not be calculated, and the reaction runs through to the oxyl radical (Table 13.3).

Oxyl radicals are very short-lived undergoing 1,2-H-shift and β-fragmentation reactions (Chapter 14). Thus, their reactions with ozone can be neglected.

Table 13.3 Reactions of carbon-centred free radicals and phenoxyl radicals with ozone in aqueous solution. Compilation of products and calculated standard Gibbs free energies (ΔG^0, kJ mol^{-1}) of the overall reaction and of radical–ozone adducts (if formed) and their decay (Naumov & von Sonntag, 2011a)

Reactants	Products	Overall reaction	Adduct formation	Adduct decay
•CH$_3$, O$_3$	CH$_3$O•, O$_2$	−340	−259	−81
•CH$_2$C(O)O$^-$, O$_3$	•CCH$_2$C(O)O$^-$, O$_2$	−344	*	
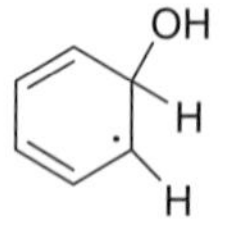 [hydroxycyclohexadienyl radical], O$_3$	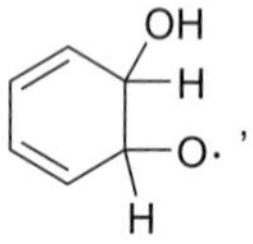 [hydroxycyclohexadienyl-OO•], O$_2$	−259	*	
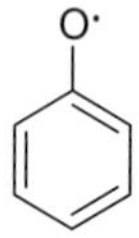 [phenoxyl radical], O$_3$	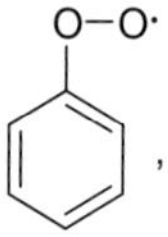 [phenyl-O-O•], O$_2$	−85	*	
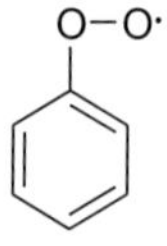 [phenyl-O-O•]	 [phenyl•], O$_2$	+61		
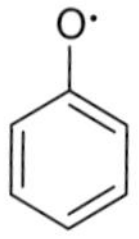 [phenoxyl radical], O$_3$	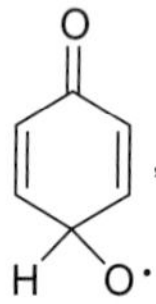 [cyclohexadienone with H, O•], O$_2$	−187	*	
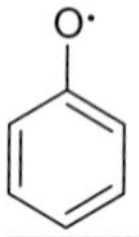 [phenoxyl radical], O$_3$	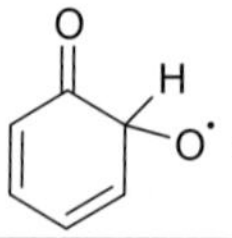 [cyclohexadienone with H, O•], O$_2$	−191	*	

*An adduct cannot be calculated as a potential minimum on the reaction path.

Hydroxycylohexadienyl radicals must play an important role, notably in wastewater ozonation, where ozone is converted to a considerable extent into •OH (Chapter 3). It reacts predominantly with the aromatic moieties of the organic matter giving rise to hydroxycylohexadienyl radicals. In contrast to O$_2$, which reacts reversibly with these radicals (Chapter 14), ozone must react irreversibly [reaction (12)] without a detectable adduct as an intermediate (Table 13.3). In this respect, the hydroxycyclohexadienyl radicals differ from simple alkyl radicals, for which an, albeit very labile, adduct has been calculated [cf. reactions (10)/(11) and Table 13.3].

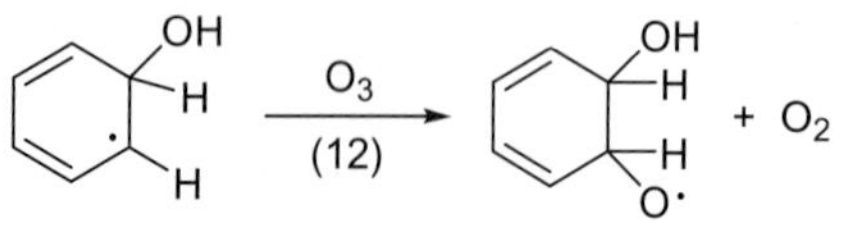

[hydroxycyclohexadienyl radical] $\xrightarrow{O_3}$ [hydroxycyclohexadienyl-O•] + O$_2$ (12)

Phenoxyl radicals are intermediates in the ozonolysis of phenols (Chapter 7), and phenolic compounds are generated in wastewater (Chapter 3) in the course of ozone treatment. They barely react with O$_2$

(see below). Thus their reactions with ozone deserve some consideration. They are commonly written with the free spin at oxygen, but the overwhelming spin density is at carbon at the *ortho* and *para* positions of the benzene ring [(Naumov & von Sonntag, 2011a, SI)]. Recombination of two phenoxyl radicals at oxygen is strongly endergonic and does not take place, but recombination does occur by forming C–O and C–C linkages (Benn *et al.*, 1979; Ye & Schuler, 1989; Jin *et al.*, 1993). Thus, phenoxyl radicals largely behave as carbon-centred radicals. Although ozone addition at oxygen and concomitant O_2 loss is exergonic [reaction (13)], an addition at carbon [reaction (14)] is even more so (Table 13.3).

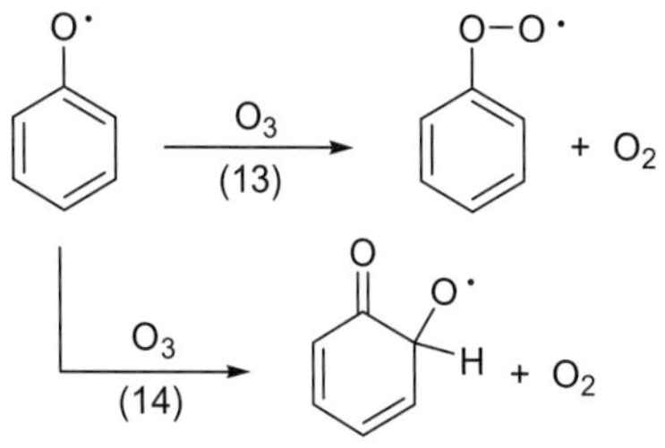

Since ozone is strongly electrophilic (Chapter 7), an addition at carbon seems to be more likely than an addition at oxygen. Hence reaction (14) may be preferred over reaction (13). O_2 loss from the phenylperoxyl radical is markedly endergonic (Table 13.3), and such peroxyl radicals are indeed well documented (Mertens & von Sonntag, 1994a; Naumov & von Sonntag, 2005). No adducts have been detected as intermediates in the calculations of reactions (13) and (14).

13.4 OZONE REACTIONS WITH OXYGEN-CENTERED RADICALS

Here, we deal with radicals of strongly varying reactivity towards DOM, the highly reactive •OH (for its reactions see Chapter 14), and the practically non-reactive peroxyl and nitroxyl radicals. In their reactivity, alkoxyl radicals (RO•) resemble •OH. Yet in water, they undergo very rapid unimolecular reactions (Chapter 14) that compete effectively with a reaction with ozone and thus they are not considered here. Calculated standard Gibbs free energies of their ozone reactions are compiled in Table 13.4.

The reaction of •OH with ozone has been formulated as an O-transfer [reaction (15)]. This reaction is fast (Table 13.1) and strongly exergonic (Table 13.4). An adduct could not be calculated.

$$\bullet OH + O_3 \longrightarrow HO_2^\bullet + O_2 \tag{15}$$

In the absence of •OH scavengers, this reaction turns into a chain reaction in which ozone is decomposed into O_2 (Chapter 3).

A number of peroxyl radicals have been generated radiolytically, and their reactions with ozone have been studied (Lind *et al.*, 2003). From an evaluation of the data, rate constants have been calculated and are compiled in Table 13.1. They centre near $10^4 \, M^{-1} \, s^{-1}$, except for $^-O_3SOO^\bullet$ which reacts an order of magnitude faster (Table 13.1). For the methylperoxyl radical, a short-lived ozone adduct has been calculated [reaction (16)] that releases O_2 [reaction (17)] (Table 13.4). With other peroxyl radicals, such adducts are only bona fide transition states.

$$CH_3OO^\bullet + O_3 \rightleftarrows CH_3OOOOO^\bullet \longrightarrow CH_3O^\bullet + 2\,O_2 \tag{16/17}$$

Table 13.4 Reactions of oxygen-centred free radicals with ozone in aqueous solution. Compilation of products and calculated standard Gibbs free energies (ΔG^0, kJ mol^{-1}) of the overall reaction and of radical–ozone adducts (if formed) and their decay (Naumov & von Sonntag, 2011a)

Reactants	Products	Overall reaction	Adduct formation	Adduct decay
•OH, O_3	$HO_2^\bullet$, O_2	−229	*	
$CH_3O^\bullet$, O_3	$CH_3OO^\bullet$, O_2	−190	*	
$CH_3OO^\bullet$, O_3	$CH_3O^\bullet$, 2 O_2	−238	−19	−219
$HOC(O)CH_2OO^\bullet$, O_3	$HOC(O)CH_2O^\bullet$, 2 O_2	−249	*	
$^-OC(O)CH_2OO^\bullet$, O_3	$^-OC(O)CH_2O^\bullet$, 2 O_2	−272	*	
$^-OC(O)CHClOO^\bullet$, O_3	$^-OC(O)CHClO^\bullet$, 2 O_2	−362	*	
$CH_3C(O)CH_2OO^\bullet$, O_3	$CH_3C(O)CH_2O^\bullet$, 2 O_2	−255	*	
$HOC(CH_3)_2CH_2OO^\bullet$, O_3	$HOC(CH_3)_2CH_2O^\bullet$, 2 O_2	−398	*	
$^-O_3SOO^\bullet$, O_3	$^-O_3SO^\bullet$, 2 O_2	−253	*	
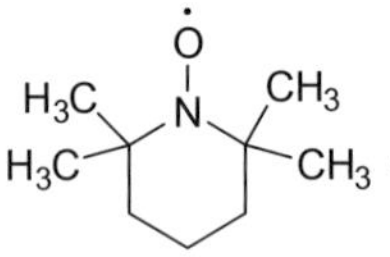 TEMPO, O_3	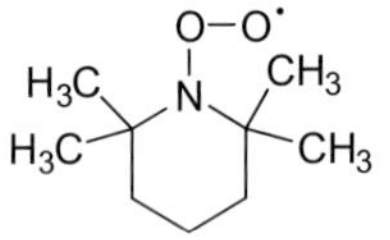 TEMPO-peroxyl adduct, O_2	−46	*	
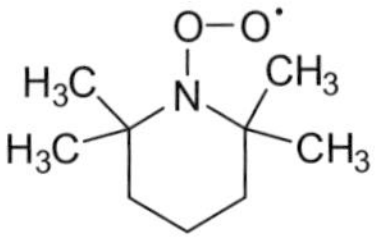 TEMPO-peroxyl,	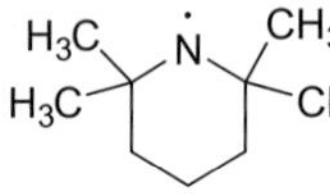 TEMPO, O_2	−51		
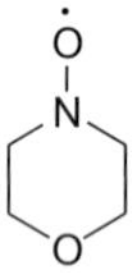 morpholine-N-oxyl, O_3	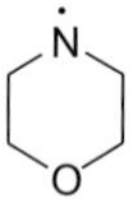 morpholine-N•, 2 O_2	−96	*	

*An adduct cannot be calculated as a potential minimum on the reaction path.

The stable nitroxyl radical TEMPO, often used as ESR standard, reacts rapidly with ozone (Table 13.1). In the first step, one molecule of O_2 is released [reaction (18)] without a detectable adduct as an intermediate. The nitrogen-centred peroxyl radical thus formed is unstable and loses O_2 [reaction (19)]. For standard Gibbs free energies see Table 13.4.

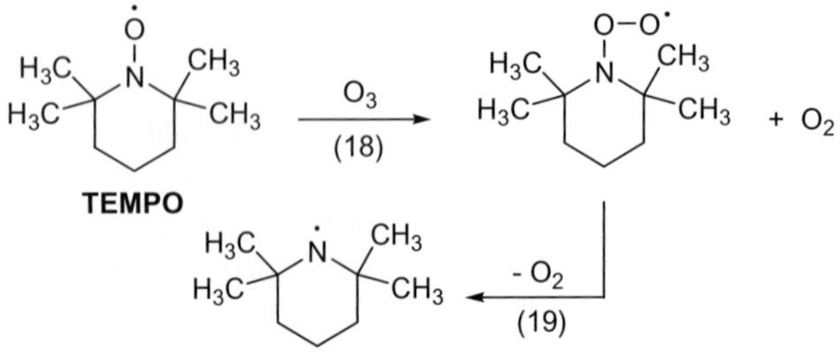

TEMPO + O_3 →(18) nitrogen peroxyl + O_2 →(19) [−O_2] nitrogen-centred radical

The negative standard Gibbs free energy of reaction (19) is in agreement with the general observation, that nitrogen-centred radicals do not react with O_2 (von Sonntag & Schuchmann, 1997). In its reaction with ozone, the morpholine-derived nitroxyl radical shows the same overall standard Gibbs free energy for the release of two molecules of O_2 (Table 13.4). In contrast to the TEMPO system, an intermediate peroxyl radical cannot be calculated here.

13.5 OZONE REACTIONS WITH NITROGEN- AND SULFUR-CENTRED RADICALS

Nitrogen-centred radicals are intermediates in the reaction of ozone with primary and secondary amines, as all amines give rise to $^{\bullet}OH$ (Chapter 8). The precursor of the latter is $O_3^{\bullet-}$, and an amine radical cation must be the corresponding product. Except for tertiary amine radical cations, amine radical cations are in equilibrium with the aminyl radicals [cf. equilibrium (20)].

$$R_2N(H)^{\bullet+} \rightleftarrows R_2N^{\bullet} + H^+ \qquad (20)$$

Rate constants for the reaction of aminyl radicals with ozone are as yet not known, but the exergonicity of this reaction is remarkable (cf. Table 13.5), and this reaction (21) may be quite fast.

$$R_2N^{\bullet} + O_3 \rightarrow [R_2NOOO^{\bullet}] \rightarrow R_2NO^{\bullet} + O_2 \qquad (21)/(22)$$

Table 13.5 Reactions of nitrogen- and sulfur-centred free radicals with ozone in aqueous solution. Compilation of products and calculated standard Gibbs free energies (ΔG^0, kJ mol^{-1}) of the overall reaction (Naumov & von Sonntag, 2011a)

Reactants	Products	Overall reaction
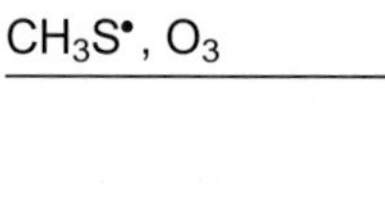 TMP aminyl radical, O_3	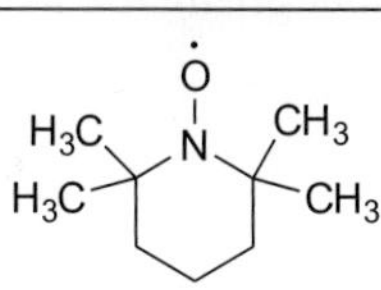 TEMPO, O_2	−331
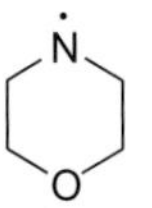 morpholine aminyl radical, O_3	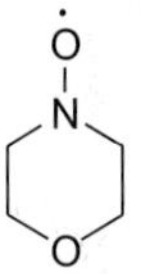 morpholine nitroxyl, O_2	−320
$CH_3S^{\bullet}$, O_3	$CH_3SO^{\bullet}$, O_2	−356

Considering the fast reactions (18)/(19), reactions (21)/(22) would give rise to an ozone consuming chain reaction without forming products other than O_2. It has been suggested that such a cycle plays a role in the ozone chemistry of diclofenac to explain the surprisingly low diclofenac consumption (Sein et al., 2008). Further data are required to substantiate this suggestion. In contrast to the $^{\bullet}OH/HO_2^{\bullet}$ chain reaction, this ozone destroying chain reaction stands a chance of being much shorter and always of importance, as some aminyl radicals can rearrange into the corresponding carbon-centred radicals, for example, reaction (23) ($\Delta G^0 = -31$ kJ mol^{-1}).

$$CH_3-CH_2-^{\bullet}NH \rightleftarrows CH_3-^{\bullet}CH-NH_2 \qquad (23)$$

Such reactions are analogous to the well-documented 1,2-H shift reactions of alkoxyl radicals [cf. reaction (24), $\Delta G^0 = -36$ kJ mol^{-1}].

$$CH_3-CH_2-O^\bullet \rightleftarrows CH_3-{}^\bullet CH-OH \qquad (24)$$

The 1,2-H shift reactions of alkoxyl radicals are very fast (in the order of 10^6 s^{-1}), and as the driving force for reaction (23) is lower by 5 kJ mol^{-1}, its rate may be lower. The rate of reaction (23) will determine whether or not the above chain reaction will become effective.

In the thiol system, in contrast, the equilibrium heteroatom-centred vs. carbon-centred radical lies on the side of the thiyl radical [reaction (25), $\Delta G^0 = +26$ kJ mol^{-1}].

$$CH_3-CH_2-S^\bullet \rightleftarrows CH_3-{}^\bullet CH-SH \qquad (25)$$

Moreover, the reaction of a thiyl radical with O_2 is also slightly endergonic [reaction (26), $\Delta G^0 = +15$ kJ mol^{-1}].

$$CH_3S^\bullet + O_2 \rightleftarrows CH_3SOO^\bullet \qquad (26)$$

This causes the thiyl radical/O_2 reaction to become reversible (Zhang et al., 1994; Naumov & von Sonntag, 2005).

Thiols react very rapidly with ozone without giving rise to thiyl radicals (Chapter 9). Therefore, reaction (27) is of little consequence in practice.

$$CH_3S^\bullet + O_3 \rightarrow CH_3SO^\bullet + O_2 \qquad (27)$$

Reaction (27) is highly exergonic (Table 13.5) and thus irreversible. This is in contrast to the reaction of thiyl radicals with O_2 (see above) and, again, is a striking example for how differently ozone and O_2 may behave in their radical scavenging properties.

13.6 OZONE REACTIONS WITH HALOGEN-CENTRED RADICALS

The reactions of ozone with halide ions are discussed in Chapter 11. They do not directly give rise to free radicals. Ozone-induced free-radical formation only sets in when the oxidation proceeds to chlorite and bromite (Chapter 11). Yet, $^\bullet$OH radicals, which are abundantly formed in ozone reactions with DOM (Chapter 3), give rise in their reactions with Br$^-$ to Br$^\bullet$ (Chapter 14), and the $^\bullet$OH route in ozone reactions can contribute to bromate formation in drinking water ozonation (Chapter 11). Although the reaction of $^\bullet$OH with Cl$^-$ to Cl$^\bullet$ becomes effective only in acid solution (Chapter 14), Cl$^\bullet$ reactions are included here for discussing similarities and differences when compared to Br$^\bullet$. In the presence of halide ions, the halogen atoms are in equilibrium with the corresponding three-electron bonded complexes such as Cl$_2^{\bullet-}$ and Br$_2^{\bullet-}$ (Chapter 14). Rate constants of halogen-derived radicals are compiled in Table 13.1 and calculated standard Gibbs free energies in Table 13.6.

Table 13.6 Reactions of halogen-containing free radicals with ozone in aqueous solution. Compilation of products and calculated standard Gibbs free energies (ΔG^0, kJ mol^{-1}) of the overall reaction and of radical–ozone adducts (if formed) and their decay (Naumov & von Sonntag, 2011a)

Reactants	Products	Overall reaction	Adduct formation	Adduct decay
•Cl, O$_3$	ClO•, O$_2$	−204	*	
Cl$_2$•−, O$_3$	ClO•, O$_2$, Cl$^-$	−133	*	
ClO•, O$_3$	OClO•, O$_2$	−119	*	
ClO•, O$_3$	ClOO•, O$_2$	−229	+5, see text	
OClO•, O$_3$	ClO$_3$•, O$_2$	+9	*	
OClO•, O$_3$	ClO•, 2 O$_2$	−311	*	
2 ClO$_3$•	O$_3$Cl–ClO$_3$	+109		
2 ClO$_3$•	ClO$_4^-$, ClO$_2^+$	−40	*	
OClO• + ClO$_3$•	O$_3$Cl–ClO$_2$	+93		
OClO• + ClO$_3$•	O$_2$Cl–OClO$_2$	+29		
OClO• + ClO$_3$•	ClO• + O$_2$ + OClO•	−222	*	
Br•, O$_3$	BrO•, O$_2$	−149	*	
BrO•, O$_3$	BrOO•, O$_2$	−268	*	
BrOO•	Br• + O$_2$	−11	*	
BrO•, O$_3$	OBrO•, O$_2$	−47	*	
2 OBrO•	OBrOOBrO	+11		
2 OBrO•	O$_2$BrOBrO	+130		
OBrOOBrO	2 •Br, 2 O$_2$	−282		
OBrO•, O$_3$	BrO• + 2 O$_2$	−380		
OBrO•, O$_3$	O$_2$BrO•, O$_2$	+81	+102	−1
O$_2$BrO•, O$_3$	BrO$_4$•, O$_2$	+93	+114	−21
BrO$_3$•	BrO•, O$_2$	−506		
BrO$_4$•	OBrO•, O$_2$	−413		

*An adduct cannot be calculated as a potential minimum on the reaction path.

With ozone, •Cl and Cl$_2$•− undergo O-transfer without adducts as potential minima on the reaction paths [reactions (28) and (29)].

$$\text{Cl}^\bullet + \text{O}_3 \rightarrow \text{ClO}^\bullet + \text{O}_2 \tag{28}$$

$$\text{Cl}_2^{\bullet-} + \text{O}_3 \rightarrow \text{ClO}^\bullet + \text{O}_2 + \text{Cl}^- \tag{29}$$

A rate constant is only available for Cl$_2$•− (Table 13.1). As expected, the reaction becomes less exergonic, when Cl• is complexed to Cl$^-$ (Table 13.6). The Cl$_2$•− reaction resembles the reaction of •OH with ozone. In the both cases, rate constants are well below the diffusion-controlled limit (Table 13.1) despite the high exergonicity of these reactions. As an explanation, one may tentatively suggest that the pronounced electrophilicity of ozone may play a role in preventing, by electrostatic repulsion, a favourable transition state in reactions of nitrogen-, oxygen- and halogen-centred radicals and thus reducing the rate of reaction (cf. also TEMPO and peroxyl radicals, Table 13.1).

For the subsequent oxidation of the ClO• radical by ozone, two exergonic reactions may be written. Reaction (30) gives rise to (stable) chlorine dioxide, OClO•, but a reaction to its isomer, the unstable chlorineperoxyl radical, ClOO•, is even more exergonic [reaction (31)/(32), Table 13.6). Without additional information, a decision as to which of them is kinetically favoured cannot be made.

$$ClO^• + O_3 \rightarrow OClO^• + O_2 \tag{30}$$

$$ClO^• + O_3 \rightarrow ClOO^• + O_2 \rightleftarrows Cl^• + 2O_2 \tag{31/32}$$

The ClOO• species has a calculated Cl–OO• bond length of 2.331 Å and ClO–O• bond length of 1.888 Å. The latter is practically identical to the O=O bond length, 1.186 Å. This indicates that the ClOO• species detected in these calculations must be a kind of van der Waals complex of Cl• with O_2, and the value of $\Delta G^0 = +5$ kJ mol^{-1} cannot be related to the equilibrium constant of reaction (32). This is in contrast to the only seemingly analogous equilibrium HOOO• $\rightleftarrows$ •OH + O_2 ($\Delta G^0 = -47$ kJ mol^{-1}), where HO–OO• (1.520 Å) and HOO–O• (1.243 Å) bond lengths are quite different from those of the final products (Naumov & von Sonntag, 2011b).

In the reaction of chlorine dioxide with ozone, there may be two competing reactions [reactions (33) and (34)].

$$OClO^• + O_3 \rightarrow ClO_3^• + O_2 \tag{33}$$

$$OClO^• + O_3 \rightarrow OCl^• + 2O_2 \tag{34}$$

The observed reaction is only moderately fast (Table 13.6), and this is compatible with reaction (33) being slightly endergonic. Reaction (34) is strongly exergonic (Table 13.6) but complex. Thus, reaction (33) may compete successfully in the case where there is substantial activation energy for reaction (34). Formation of $^1O_2$ in uncommonly high yield has been reported for this reaction in CCl$_4$ solutions (Chertova et al., 2000).

The ClO$_3^•$ radical cannot be further oxidised by ozone and has to decay bimolecularly. Among the two potential combination reactions [reaction (35), Cl–Cl 2.5 Å] and (36), the high endergonicity of reaction (35) (Table 13.6) prevents this reaction from taking place, and the bimolecular decay of ClO$_3^•$ radical must occur via an asymmetrical recombination, reaction (36). Upon trying to calculate the standard Gibbs free energies of this reaction, the program proceeded automatically toward perchlorate ion and ClO$_2^+$ [reaction (37), Table 13.6). ClO$_2^+$ reacts with water yielding chlorate [reaction (38)].

$$2 ClO_3^• \rightarrow O_3Cl-ClO_3 \tag{35}$$

$$2 ClO_3^• \rightarrow [O_3ClOClO_2] \rightarrow ClO_4^- + ClO_2^+ \tag{36/37}$$

$$ClO_2^+ + H_2O \rightarrow ClO_3^- + 2H^+ \tag{38}$$

Perchlorate formation has been intriguing the scientific community for a long time (Rao et al., 2010). ClO$_3^•$ has been envisaged as precursor, but details have remained obscure. It is important to note that perchlorate is not an important product in the reaction of ClO$^-$/ClO$_2^-$ with ozone (Haag & Hoigné, 1983b), and perchlorate formation must be a minor side reaction. This raises the question, whether bimolecular decay of ClO$_3^•$ may compete with a termination with OClO• radicals in the case where there is a sufficiently high steady-state concentration of OClO• radicals present in equilibrium (note the relatively slow reaction of OClO• with ozone, Table 13.1). This reaction could eventually give rise to two molecules of chlorate ion in the case where the dimers were sufficiently long-lived to react with water. Yet, reaction (39) is markedly endergonic (Table 13.6) with a long Cl—Cl bond (1.980 Å), and this reaction is most likely reversible. Reaction (40) is less endergonic but also potentially reversible.

$$ClO_3^* + OClO^• \rightleftarrows O_3Cl-ClO_2 \tag{39}$$

$$ClO_3^* + OClO^• \rightleftarrows O_2ClO-ClO_2 \tag{40}$$

Moreover, a highly exergonic termination via the oxygens, proceeding directly, with a peroxidic arrangement only as a likely transition state, to OCl$^•$ and OClO$^•$ plus O_2, could also take place [reactions (41) and (42)]. To which extent these reactions contribute to chlorate formation has still to be explored.

$$ClO_3^* + OClO^• \rightarrow [O_2ClO-OClO]_{TS} \rightarrow ClO_2^* + O_2 + ClO^• \tag{41}/(42)$$

The reaction of the Br$^•$ atom with ozone is strongly exergonic [reaction (43), Table 13.6], and for the ensuing BrO$^•$ one can envisage the likewise exergonic reactions (44)–(48).

$$Br^• + O_3 \rightarrow BrO^• + O_2 \tag{43}$$

$$BrO^• + O_3 \rightarrow BrOO^• + O_2 \tag{44}$$

$$BrOO^• \rightleftarrows Br^• + O_2 \tag{45}$$

$$BrO^• + O_3 \rightarrow OBrO^• + O_2 \tag{46}$$

Preference will be determined by kinetics, that is, whether ozone addition at oxygen or at bromine is preferred.

In the ozonation of Br$^-$-containing waters, there is a second route to OBrO$^•$. There is now strong evidence that the current concept of bromate formation (von Gunten, 2003b) has to be revised. The reaction of bromite with ozone does not give rise to bromate by O-transfer but is an electron transfer reaction that yields OBrO$^•$ and $O_3^{•-}$ (Chapter 11). Formation and decay of OBrO$^•$ has been studied by pulse radiolysis (Buxton & Dainton, 1968) concluding that the bimolecular decay leading to a Br_2O_4 intermediate is (i) reversible, (ii) decays by reacting with water and (iii) with OH$^-$. There are two conceivable dimers [reactions (47) and (48)].

$$2\,OBrO^• \rightleftarrows OBrOOBrO \tag{47}$$

$$2\,OBrO^• \rightleftarrows O_2BrOBrO \tag{48}$$

While the symmetrical dimerisation is only mildly endergonic (Table 13.6) and could account for the reported reversibility, the asymmetrical one is strongly endergonic and has to be disregarded.

There is, however, also the possibility that the symmetrical dimer decays according to reaction (49).

$$OBrOOBrO \rightarrow 2\,Br^• + 2\,O_2 \tag{49}$$

This is the route taken upon prolonged optimisation. Yet in water, where an OH$^-$-induced component of the decay of OBrO$^•$ radicals is observed, this reaction seems not to be kinetically favoured despite the high exergonicity (Table 13.6).

The bimolecular decay of OBrO$^•$ to Br_2O_4 is reversible, but Br_2O_4 reacts rapidly with water and OH$^-$ and thus sufficiently high OBrO$^•$ concentrations should not build up for reactions of OBrO$^•$ with ozone to become relevant. Yet in the case where it would, the energetically favoured (Table 13.6) route is reaction (50), while reaction sequence (51)/(52) may not compete (Table 13.6).

$$OBrO^• + O_3 \rightarrow BrO^• + 2\,O_2 \tag{50}$$

$$OBrO^• + O_3 \rightarrow O_2BrOOO^• \rightarrow O_2BrO^• + O_2 \tag{51}/(52)$$

Oxidation of $O_2BrO^\bullet$ to $O_3BrO^\bullet$ [reactions (53)/(54)] is endergonic (Table 13.6), and if $O_2BrO^\bullet$ and $O_3BrO^\bullet$ were formed they would undergo the strongly exergonic $^3O_2$-releasing reactions (55) and (56) (Table 13.6).

$$O_2BrO^\bullet + O_3 \rightarrow O_3BrOOO^\bullet \rightarrow O_3BrO^\bullet + O_2 \qquad (53)/(54)$$
$$O_2BrO^\bullet \rightarrow BrO^\bullet + O_2 \qquad (55)$$
$$O_3BrO^\bullet \rightarrow OBrO^\bullet + O_2 \qquad (56)$$

Thus it seems that $Br^\bullet$ is not oxidised by ozone beyond $OBrO^\bullet$.

Chapter 14
Reactions of hydroxyl and peroxyl radicals

14.1 INTRODUCTORY REMARKS

Hydroxyl radicals are abundant by-products of ozone reactions (cf. Chapters 3, 7, 8, and 11) and are generated in high yields in the reaction of ozone with the DOM of the water matrix of drinking water and wastewater (Chapter 3). They assist in eliminating ozone-refractory micropollutants, as their reactivity is generally high, often close to the diffusion-controlled limit [for a compilation of rate constants see Buxton *et al.* (1988)]. While in pollution abatement, their action is beneficial, they may also contribute to the formation of undesired disinfection by-products such as bromate (Chapters 3, 11 and below). First, the general reactions of •OH are discussed [for a more detailed discussion see von Sonntag (2006)]. In technical applications, ozone is always introduced together with a large excess of O_2. The radicals that are formed upon •OH attack react with ozone (Chapter 13), but many of them also react with O_2, thereby forming peroxyl radicals. With O_2 in excess, the formation and reactions of peroxyl radicals are of major importance in ozone reactions in aqueous solution. There is a considerable wealth of data on those reactions, and peroxyl radical chemistry in aqueous solution is now fairly well understood. Some aspects are discussed below [for more detailed reviews see von Sonntag & Schuchmann (1991, 1997) and von Sonntag (2006)]. At the end of this chapter, the role of •OH in disinfection by-product formation and micropollutant abatement is addressed.

14.2 HYDROXYL RADICAL REACTIONS

There are three types of reactions, addition reactions, H-abstraction reactions and electron transfer reactions. The most common and fastest reactions are the addition reactions. Many of them are close to diffusion-controlled. When this reaction pathway is not available, •OH may also react by H-abstraction. Electron transfer reactions are rare, and the high reduction potential of •OH [2.3 V at pH 7 (Wardman, 1989)] is not a measure of its high reactivity. As with many other reactions, •OH reactions are often governed by kinetics rather than by thermodynamics (see below), and this has to be recalled when thermodynamic data are used for discussing probabilities of reaction pathways.

14.2.1 Addition reactions

The •OH radical adds readily to C–C, C–N and S–O double bonds (with four-valent sulfur) but not to C–O double bonds [e.g. reactions (1) (Söylemez & von Sonntag, 1980) and (2) (Veltwisch *et al.*, 1980)].

$$\cdot OH + CH_2=CH_2 \rightarrow {}^\bullet CH_2-CH_2OH \qquad (1)$$
$$\cdot OH + (CH_3)_2S=O \rightarrow (CH_3)_2S(OH)O^\bullet \qquad (2)$$

Despite this, there is a high regioselectivity in the reaction of the electrophilic (Anbar et al., 1966) •OH with C—C double bonds carrying different electron-donating or electron-withdrawing substituents. The first intermediate must be a π-complex as inferred from the kinetics of •OH with benzene at very high temperatures (Ashton et al., 1995). The π-complex is in equilibrium with benzene and •OH [equilibrium (3)] and subsequently collapses into a σ-complex, the •OH adduct proper [reaction (4)]. In the π-complex, the •OH radical is guided to its final position, and substituents at the C—C double bond may have a strong directing effect.

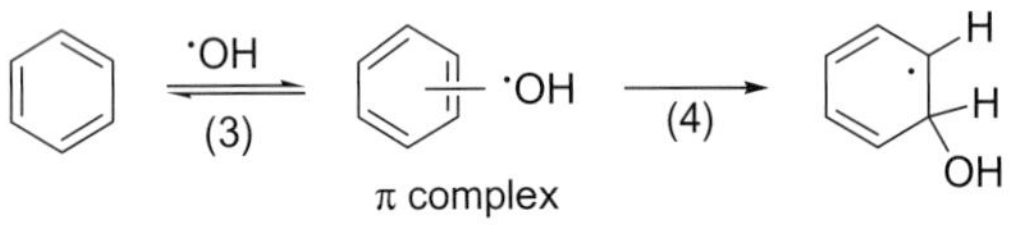

The site of addition is dominated by kinetics rather than by thermodynamics, that is, the preferred •OH addition may not lead to the thermodynamically favoured radical. A case in point is the •OH chemistry of the nucleobases (von Sonntag, 2006; Naumov & von Sonntag, 2008a). Another striking example is the reaction of •OH with toluene. Here, formation of a π-complex and subsequent collapse into the σ-complex causes a preferred addition to the ring (92%) [reaction (5)], and only 8% undergo H-abstraction from the methyl group [reaction (6), formation of the thermodynamically favoured benzyl radical] (Christensen et al., 1973).

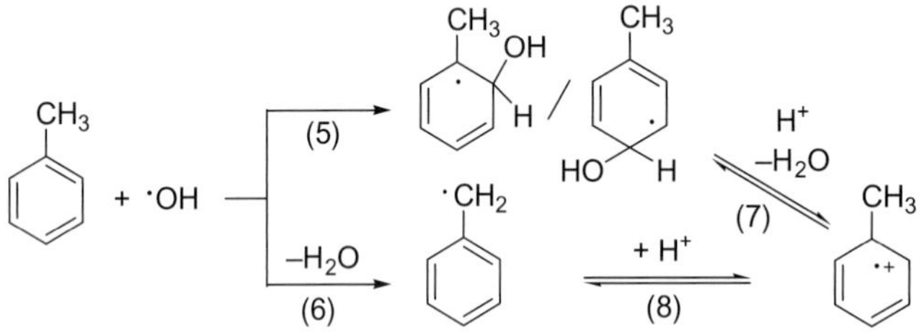

Upon proton catalysis, the thermodynamic equilibrium can be established by converting the •OH adduct into the benzyl radical via a radical cation [reactions (7) and (8)].

Addition reactions also can take place to the lone electron pairs of electron-rich heteroatoms such as in amines, sulfides, disulfides, halide and pseudo-halide ions. Under such conditions, three-electron intermediates are formed [e.g. reactions (9) and (10), for a review see Asmus (1979)].

$$\cdot OH + (CH_3)_2S \rightarrow (CH_3)_2S \therefore OH \qquad (9)$$
$$\cdot OH + Cl^- \rightleftarrows HO \therefore Cl^- (HOCl^{\bullet-}) \qquad (10)$$

Experimental evidence for these three-electron-bonded intermediates have been obtained by pulse radiolysis, but in some cases, such as with the amines (Das & von Sonntag, 1986), they are too short-lived for detection and may only be inferred by analogy.

Many oxidation reactions of transition metal ions are in fact also addition reactions [e.g. reaction (11)] (Asmus et al., 1978).

$$\cdot OH + Tl^+ \rightarrow TlOH^+ \qquad (11)$$

The product to be expected for an electron transfer is often only observed in acid solution such as reactions (12) and (13).

$$HOCl^{\bullet-} + H^+ \longrightarrow Cl^{\bullet} + H_2O \tag{12}$$

$$TlOH^+ + H^+ \longrightarrow Tl^{2+} + H_2O \tag{13}$$

Reactions (10) [$K_{10} = 1.4$ M (Jayson et al., 1973)] and (12) [$k_{12} = 5 \times 10^{10}$ M^{-1} s^{-1}, cf. (Buxton et al., 2000)] are of importance for Cl$^-$-containing waters, and these data may be used for assessing to what extent $^{\bullet}$OH and Cl$^{\bullet}$ play a role at a given pH [note that $^{\bullet}$OH is drawn out of equilibrium (10) by $^{\bullet}$OH-reactive substrates other than Cl$^-$ present in such waters, see below; for a discussion of the reactivity of Cl$^{\bullet}$ in aqueous solution see (Buxton et al., 1998; Buxton et al., 1999; Buxton et al., 2000)].

Reversibility of $^{\bullet}$OH addition is also observed with other systems such as Cu^{2+} in acid solution [equilibrium (14)] (Ulanski & von Sonntag, 2000).

$$Cu^{2+} + {}^{\bullet}OH \rightleftarrows Cu(OH)^{2+} \tag{14}$$

14.2.2 H-abstraction reactions

The H-abstraction reactions are typically somewhat slower than addition reactions. Here, the R—H bond dissociation energy (BDE) plays a major role (for a compilation of BDEs see Table 14.1).

Table 14.1 Compilation of some R—H BDEs (unit: kJ mol^{-1}; 1 kcal mol^{-1} = 4.18 kJ mol^{-1})

R—H bond	BDE/kJ mol^{-1}	Reference
C—H (aromatic)	465	Berkowitz et al., 1994
C—H (vinylic)	465	Berkowitz et al., 1994
C—H (primary)	423	Berkowitz et al., 1994
C—H (primary in MeOH)	402	Berkowitz et al., 1994
C—H (secondary)	412	Berkowitz et al., 1994
C—H (secondary in EtOH)	396	Reid et al., 2003
C—H (tertiary)	403	Berkowitz et al., 1994
C—H (tertiary in 2-PrOH)	393	Reid et al., 2003
C—H (allylic)	367	Berkowitz et al., 1994
C—H (benzylic)	367	Berkowitz et al., 1994
C—H (pentadienylic)	343	Schöneich et al., 1990
C—H (in peptides)	330–370	Reid et al., 2003
O—H (alcoholic)	436	Berkowitz et al., 1994
O—H (water)	499	Berkowitz et al., 1994
O—H (phenolic)	360	Parsons, 2000
O—H (hydroperoxidic)	366	Golden et al., 1990
S—H (thiolic)	366	Armstrong, 1999
S—H (thiophenolic)	330	Armstrong, 1999

With the weak RS—H bond, the rate of reaction is diffusion controlled (in fact, H-abstraction may be preceded by addition to sulfur), but with C–H bonds there is some grading, tertiary > secondary > primary. With 2-propanol, for example, the tertiary hydrogen competes favourably [reaction (15), 86%] over the abstraction of one of the six primary ones [reaction (16), 14%] (Asmus et al., 1973).

$$^{\bullet}OH + (CH_3)_2C(OH)H \rightarrow H_2O + {}^{\bullet}C(CH_3)_2OH \qquad (15)$$

$$^{\bullet}OH + (CH_3)_2COH \rightarrow H_2O + {}^{\bullet}CH_2C(CH_3)OH \qquad (16)$$

An H-abstraction from the OH group [reaction (17)] can usually be neglected (here 3%) due to its high BDE, as a rapid 1,2-H shift converts the oxyl radical into the thermodynamically favoured carbon-centred radical [reaction (18); for competing reactions of the oxyl radical see section on peroxyl radicals and (von Sonntag, 2006)].

$$^{\bullet}OH + HC(CH_3)_2OH \rightarrow H_2O + HC(CH_3)_2O^{\bullet} \qquad (17)$$

$$HC(CH_3)_2O^{\bullet} \rightarrow {}^{\bullet}C(CH_3)_2OH \qquad (18)$$

Addition and H-abstraction reactions may be used for the study of the same intermediate. A case in point is the $^{\bullet}C(O)NH_2$ radical, which is formed from formamide by H-abstraction and from cyanide by addition and subsequent rearrangement of the $^{\bullet}OH$ adduct radical (Muñoz et al., 2000). Similarly, there are two ways to generate α-hydroxyalkyl radicals. One is by H-abstraction [e.g. reaction (15)]. The other is by reacting the corresponding carbonyl compound with the solvated electron and protonation of the thus-formed radical anion. In the absence of O_2, carbon-centred radicals decay by dimerisation and disproportionation. The latter occurs typically by H-transfer, but some more complex reactions have also been reported for $CO_2^{\bullet-}$ and $CHO^{\bullet-}$ (Flyunt et al., 2001b; Wang et al., 1996).

14.2.3 Electron transfer reactions

At pH 7, the one-electron reduction potential of the $^{\bullet}OH$ radical is 2.3 V (Wardman, 1989). Thus in principle, it could undergo one-electron oxidation reactions quite readily. Yet, this reaction is often kinetically disfavoured, and addition reactions occur instead. A case in point is reaction (11). An interesting example is the reaction of phenol and its halogen-substituted derivatives. The reduction potential of phenol at pH 7 is 0.86 V (Merényi et al., 1988). Yet, phenoxyl radicals [deprotonation products of the phenol radical cation, $pK_a = -2$ (Dixon & Murphy, 1976)] are formed in very minor primary yield (Mvula et al., 2001). This even holds for a reaction of $^{\bullet}OH$ with phenolates. Addition is the dominating reaction, unless the preferred sites of addition (the *ortho* and *para* positions) are blocked by a bulky substituent (Table 14.2) (Fang et al., 2000).

Table 14.2 Addition vs. electron transfer in the reaction of $^{\bullet}OH$ with the anions of halogenated phenols according to Fang et al. (2000)

Phenolate	Addition/%	Electron transfer/%
Pentafluorophenolate	73	27
Pentachlorophenolate	47	53
Pentabromophenolate	27	73
2,4,6-Triiodophenolate	3	97

14.3 DETERMINATION OF •OH RATE CONSTANTS

Most of the available •OH rate constants [for a compilation see Buxton et al. (1988)] were determined by pulse radiolysis. Here, an N$_2$O-saturated water containing the compound in question at low concentrations is subjected to a short electron pulse of a few nanoseconds to 1 μs duration. Under such conditions, the energy of the electron beam (energy 1–10 MeV) is absorbed by the solvent. This causes ionisations [reaction (19)] and excitations [reaction (20)].

$$H_2O + \text{ionising radiation} \longrightarrow H_2O^{\bullet+} + e^- \tag{19}$$

$$H_2O + \text{ionising radiation} \longrightarrow H_2O^* \tag{20}$$

The water radical cation is a very strong acid and rapidly (within pulse duration) deprotonates, giving rise to an •OH radical [reaction (21)]. The excited water molecule decomposes into •H and •OH [reaction (22)]. The electron formed in reaction (19) becomes solvated by water [reaction (23)], and the solvated electron [for rate constants see Buxton et al. (1988)] is converted into further •OH in the presence of N$_2$O [reaction (24)].

$$H_2O^{\bullet+} \longrightarrow {}^{\bullet}OH + H^+ \tag{21}$$

$$H_2O^* \longrightarrow {}^{\bullet}OH + H^{\bullet} \tag{22}$$

$$e^- + nH_2O \longrightarrow e_{aq}^- \tag{23}$$

$$e_{aq}^- + N_2O + H_2O \longrightarrow {}^{\bullet}OH + N_2 + OH^- \tag{24}$$

Of the two radicals, •OH (90%) dominates over H• (10%). Moreover, •OH is considerably more reactive than H• [for rate constants see Buxton et al. (1988)], and the kinetics of the build-up of transients caused by the reaction of •OH with the substrate (N$_2$O does not react with •OH) can usually be separated from the small contribution of the H• atom. The •OH radical only absorbs far into the UV, and one has to rely in such measurements on the formation of products (often short-lived transients). In such a case, k_{obs} of product build-up is plotted as a function of the substrate concentration. From the slope of this plot, the •OH rate constant is calculated ($k = k_{obs}/[\text{substrate}]$). When the product of •OH with the substrate has no marked absorption in the accessible wavelength region, one can measure the •OH rate constant by competition with a substrate that gives rise to a strong signal and for which the rate constant is well established. [Fe(CN)$_6$]$^{4-}$ ($k = 1.05 \times 10^{10}$ M^{-1} s^{-1}, product: [Fe(CN)$_6$]$^{3-}$) or thymine ($k = 6.5 \times 10^9$ M^{-1} s^{-1}, product: •OH adduct radicals) may serve this purpose Buxton et al. (1988).

In competition kinetics, the product of the competitor (P) at constant competitor (C) concentration and constant •OH concentration is measured as a function of varying substrate (S) concentrations. As the •OH rate constant of the competitor (k_C) is known, that of the substrate (k_S) can be determined based on equations (25)–(27).

$$S + {}^{\bullet}OH \longrightarrow X\,(\text{not detected}) \tag{25}$$

$$C + {}^{\bullet}OH \longrightarrow P\,(\text{detected}) \tag{26}$$

$$[P] = [{}^{\bullet}OH] \times k_C \times [C]/(k_C \times [C] + k_S \times [S]) \tag{27}$$

At a given •OH concentration (kept constant throughout these experiments) and in the absence of the substrate, the product concentration is $[P]_0$. One can then write equation (28).

$$[P] = [P]_0 \times k_C \times [C]/(k_C \times [C] + k_S \times [S]) \tag{28}$$

This equation is then transformed into equation (29).

$$[P]_0/[P] = 1 + (k_S \times [S])^{-1} \tag{29}$$

By plotting $[P]_0/[P]$ -1 vs. $[S]^{-1}$ one obtains a straight line whose slope is $1/k_S$.

An alternative competition kinetics approach is based on the measurement of the decrease in two compounds, a reference compound (R) with a known second order rate constant and a target compound (X) with an unknown second order rate constant [e.g. Huber et al. (2003)]. The •OH rate constant for the target compound X can be determined by equations (30)–(34)

$$R + {}^\bullet OH \longrightarrow P1 \text{ (not detected)} \tag{30}$$

$$X + {}^\bullet OH \longrightarrow P2 \text{ (not detected)} \tag{31}$$

In an AOP, the time-dependent decrease of R and X can be formulated by equations (32) and (33).

$$\ln \frac{[R]_t}{[R]_0} = k_R \times [{}^\bullet OH] \times t \tag{32}$$

$$\ln \frac{[X]_t}{[X]_0} = k_X \times [{}^\bullet OH] \times t \tag{33}$$

Rearranging equations (32) and (33) results in equation (34).

$$\ln \frac{[X]_t}{[X]_0} = \frac{k_X}{k_R} \times \ln \frac{[R]_t}{[R]_0} \tag{34}$$

A plot of the logarithm of the relative decrease of X versus the logarithm of the relative decrease of R results in a straight line with a slope that corresponds to the ratio of k_X and k_R. Because k_R is known, k_X can be calculated.

For the determination of an, as yet, unknown •OH rate constant by these approaches, any AOP may be used in principle, but certain precautions have to be taken. For example when generating •OH by the Fe^{2+}/H_2O_2, O_3/H_2O_2 or UV/H_2O_2 processes, one has to take into account that H_2O_2 reacts also, albeit slowly, with •OH ($k = 2.7 \times 10^7$ M^{-1} s^{-1}) and this additional •OH scavenging capacity would have to be taken into account when setting up a similar equation system as the one shown above. Obviously, the potential error in the value thus obtained is higher. One also has to ensure that with UV none of the components that we are interested in, S, C, P, R and X are photolysed under such conditions. One also has to ensure that H_2O_2 is exposed to a constant photon flux, that is, no UV absorbing compounds build up or that this inner-filter effect is constant for all experiments. With the peroxone process (O_3/H_2O_2), there is the additional problem that O_3 must not react with S, C, P, R or X at an appreciable rate. In the case of ultrasound as AOP, competition kinetics for the determination of •OH rate constants fail in the first case, as the accumulation of substrates, here C and S, at the surface of cavitating bubbles strongly depends on their lipophilic/hydrophilic properties (Henglein & Kormann, 1985; Mark et al., 1998; von Sonntag et al., 1999).

14.4 DETECTION OF •OH IN OZONE REACTIONS

In the reaction of ozone with drinking water and wastewater, the DOM present gives rise to the formation of •OH radicals (Chapter 3). Their presence has been detected by adding small amounts of *para*-chlorobenzoic acid (pCBA) and following its decay as a function of the •OH (O_3) concentration (Elovitz & von Gunten,

1999; Elovitz et al., 2000a). Based on this approach, the R_{ct} concept has been developed allowing a prediction of micropollutant degradation during ozonation processes (Chapter 3). pCBA reacts rapidly with •OH ($k = 5.0 \times 10^9$ M^{-1} s^{-1}) (Buxton et al., 1988) but very slowly with ozone ($k = 0.15$ M^{-1} s^{-1}) (Neta et al., 1988). Ozone reacts rapidly with the DOM present in drinking water and in wastewater (for its decay kinetics see Chapter 3), and thus pCBA is protected against destruction by ozone. The DOM also reacts with •OH (for rate constants see Chapter 3) in competition with a reaction with pCBA. However, here the competition is not as biased toward a reaction with the DOM due to the much higher rate constant of pCBA for its reaction with •OH.

In other AOPs, that is, AOPs such as UV/H$_2$O$_2$ or the Fenton process, which are not based on a highly reactive agent such as ozone, other •OH dosimeters are commonly in use, such as the terephthalic acid (Fang et al., 1996) and the salicylic acid (Ingelmann-Sundberg et al., 1991; Coudray et al., 1995; Bailey et al., 1997) dosimeters. They both have in common that upon •OH reaction they form hydroxylated products, 2-hydroxyterephthalate and 2,3-dihydroxybenzoic/2,5-dihydroxybenzoic acids, respectively. 2-Hydroxyterephthalate fluoresces, and thus its yield can be determined without a prior separation by HPLC. The 2,3-dihydroxybenzoic and 2,5-dihydroxybenzoic acids must be separated, and their yields should be practically equal (von Sonntag, 2006). Otherwise, other processes leading to these products are potentially (also) involved. In using these, the fact that the •OH radicals do not lead directly to these products, but that •OH adducts are formed in the first step and that these have to be further oxidised to the desired products, is often neglected. This additional step allows other reactions to compete. This is shown for the terephthalate dosimeter that has been investigated in some detail [reactions (35)–(43)] (Fang et al., 1996).

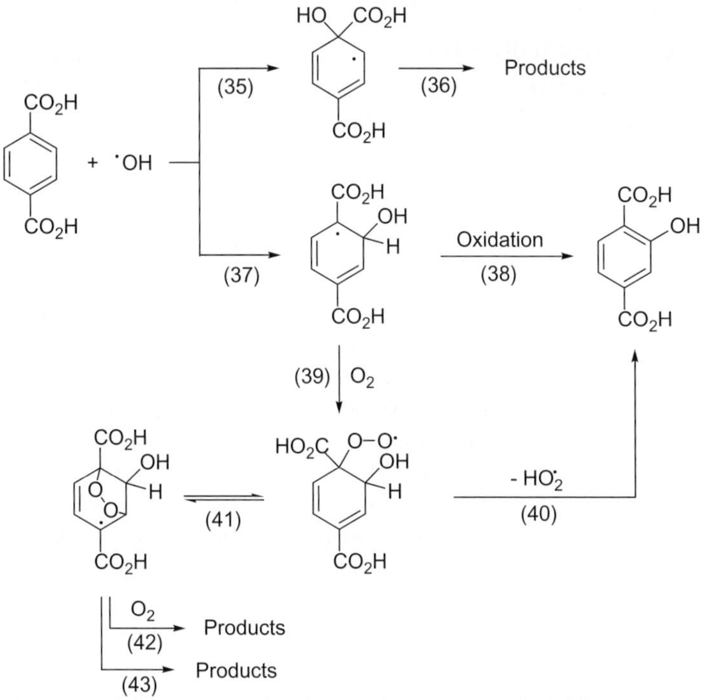

The yield of 2-hydroxyterephthalate is not 100% with respect to •OH for several reasons. First, there is a competition between an *ipso* addition and an *ortho* addition [reactions (35) and (37)], and only the latter can lead to the desired product. Second, the hydroxycylohexadienyl radical has to be oxidised [reaction (38)]. $Fe(CN)_6^{3-}$ is often a sufficiently strong oxidant to achieve full oxidation (Buxton et al., 1986), but in the present system a stronger oxidant such as $IrCl_6^{2-}$ is required. In its presence, the 2-hydroxytherephthalate yield is 84%. With O_2 as oxidant, the 2-hydroxytherephthalate yield is much lower (35%), and this is now due to a complex peroxyl radical chemistry [cf. reactions (40)–(43)], which has been elucidated with the parent system, benzene (Pan et al., 1993).

14.5 DETERMINATION OF •OH YIELDS IN OZONE REACTIONS

While the above methods can detect •OH radicals, the quantification of •OH yields is only possible in cases where the •OH scavenging rate of the system is known. This is often not the case. A case in point may be an unusual wastewater such as an effluent from an industrial production line. Thus, for an estimation of the •OH yield a detection system is required that allows us to overrun the scavenging rate of the system. This can be done with tertiary butanol (tBuOH). This is miscible with water, has a high •OH rate constant [$k = 6 \times 10^8$ $M^{-1} s^{-1}$ (Buxton et al., 1988)] and barely reacts with ozone [$k = 1 \times 10^{-3} M^{-1} s^{-1}$, Chapter 10]. The major reactions that take place upon •OH attack on tBuOH in the presence of O_2, which is always in excess over O_3 in ozonation processes, are depicted by reactions (44)–(54) (Schuchmann & von Sonntag, 1979).

$$\bullet OH + (CH_3)_3COH \rightarrow H_2O + \bullet CH_2C(CH_3)_2OH \tag{44}$$

$$HOC(CH_3)_2CH_2^\bullet + O_2 \rightarrow HOC(CH_3)_2CH_2OO^\bullet \tag{45}$$

$$2\,HOC(CH_3)_2CH_2OO^\bullet \rightarrow (HOC(CH_3)_2CH_2OO)_2 \tag{46}$$

$$(HOC(CH_3)_2CH_2OO)_2 \rightarrow HOC(CH_3)_2C(O)H + O_2 + HOC(CH_3)_2CH_2OH \tag{47}$$

$$(HOC(CH_3)_2CH_2OO)_2 \rightarrow 2\,HOC(CH_3)_2C(O)H + H_2O_2 \tag{48}$$

$$(HOC(CH_3)_2CH_2OO)_2 \rightarrow 2\,HOC(CH_3)_2CH_2O^\bullet + O_2 \tag{49}$$

$$HOC(CH_3)_2CH_2O^\bullet \rightarrow \bullet C(CH_3)_2OH + H_2C=O \tag{50}$$

$$\bullet C(CH_3)_2OH + O_2 \rightarrow \bullet OOC(CH_3)_2OH \tag{51}$$

$$\bullet OOC(CH_3)_2OH \rightarrow (CH_3)_2C=O + HO_2^\bullet \tag{52}$$

$$\bullet OOC(CH_3)_2OH + OH^- \rightarrow (CH_3)_2C=O + O_2^{\bullet -} + H_2O \tag{53}$$

$$HOC(CH_3)_2CH_2O^\bullet \rightarrow HOC(CH_3)_2\bullet CH(OH) \tag{54}$$

The tBuOH-derived radical formed in reaction (44) reacts with O_2 giving rise to the corresponding peroxyl radical [reaction (45)]. They decay bimolecularly [reaction (46)]. The tetroxide thus formed may decay by various pathways, the Russell-type reaction that gives O_2, one molecule of 2-hydroxy-2-methyl-propanal and 2-hydroxy-2-methylpropanol [reaction (47)]. The Bennett-type reaction (48) yields two molecules 2-hydroxy-2-methyl-propanal and H_2O_2. The pathway that has oxyl radicals [reactions (49)–(54)] as intermediates essentially yields acetone and formaldehyde. The 1,2-Hshift reaction (54) seems to be minor compared to the β-fragmentation reaction (50) [for reviews on the chemistry of peroxyl radicals in aqueous solution see von Sonntag & Schuchmann (1997) and von Sonntag (2006)]. The aldehydes 2-hydroxy-2-methyl-propanal and formaldehyde are readily quantified by HPLC as their 2,4-dinitrophenylhydrazones, while acetone quantification using this assay is fraught with considerable errors (Flyunt et al., 2003a). Formaldehyde can also be sensitively determined spectrophotometrically using the Hantzsch reaction (Nash, 1953). As a rule of thumb, the •OH yield is twice the formaldehyde yield.

There is an additional •OH assay in the ozone reaction. Dimethyl sulfoxide (DMSO) reacts only slowly with ozone ($k = 1.8$ $M^{-1} s^{-1}$, Chapter 9), while its reaction with •OH is fast [$k = 7 \times 10^9$ $M^{-1} s^{-1}$

(Buxton et al., 1988)]. It gives rise to methanesulfinic acid and a methyl radical in 92% yield [reaction (55), cf. also reaction (2)] (Veltwisch & Asmus, 1982).

$$\bullet OH + (CH_3)_2S{=}O \rightarrow CH_3S(O)OH + \bullet CH_3 \tag{55}$$

The methyl radicals are converted by O_2 into the corresponding peroxyl radicals which partly oxidise methanesulfinic acid to methanesulfonic acid (Flyunt et al., 2001a) in competition with their bimolecular decay (Schuchmann & von Sonntag, 1984a). Thus, another major product is formaldehyde (Yurkova et al., 1999). The sum of methanesulfinic and methanesulfonic acids can be determined by ion chromatography (Yurkova et al., 1999). As some of the methanesulfinic acid is always oxidised to methanesulfonic acid, it does not suffice to determine only the methanesulfinic acid yield by either derivatisation to a coloured compound (Smith et al., 1990; Babbs & Steiner, 1990) or chromatography with electrochemical detection (Jahnke, 1999), as suggested. For the application of the DMSO assay in quantifying $\bullet OH$ radical yields in UV/H_2O_2 and pulsed electrical discharge systems see Lee et al. (2004) and Sahni & Locke (2006). The DMSO assay in ozone reactions has the disadvantage of the much higher ozone rate constant of DMSO and its product methanesulfinic acid (Chapter 9) as compared to tBuOH, but it has the advantage that one does not have to worry about the cross-termination reactions of other radicals with the tBuOH derived peroxyl radicals, which may modify the formaldehyde yield. The two assays agree well, as has been shown in the ozonolysis of phenols (Mvula & von Sonntag, 2003).

14.6 FORMATION OF PEROXYL RADICALS

Simple carbon-centred radicals react with O_2 at close to diffusion-controlled rates [reaction (56), $k = 2 \times 10^9 \, M^{-1} \, s^{-1}$]. This reaction is practically irreversible.

$$R^\bullet + O_2 \rightarrow ROO^\bullet \tag{56}$$

Bisallylic radicals, such as the $\bullet OH$ adducts to aromatic compounds react reversibly with O_2 (Pan & von Sonntag, 1990; Pan et al., 1993; Fang et al., 1995; Naumov & von Sonntag, 2005). This would give them some advantage to react with ozone in competition (Chapter 13). Nitrogen- and oxygen-centred radicals do not react with O_2. This also holds for heteroatom-centred radicals such as the phenoxyl (Jin et al., 1993) and guanyl (von Sonntag, 1994) radicals, although there is considerable free spin also at carbon atoms. To what extent they would react with ozone is as yet not known (for the reactions of radicals with ozone see Chapter 13).

14.7 REDOX PROPERTIES OF PEROXYL RADICALS AND REACTION WITH OZONE

Simple alkylperoxyl radicals are weak oxidants [reduction potential at pH 7, $E^7 = 0.77$ V) (Merényi et al., 1994)]. Electron-withdrawing substituents in the α-position to the peroxyl radical function enhance the reduction potential substantially [$E^7(Cl_3COO^\bullet) = 1.6$ V] and hence the rate of oxidation (e.g. of ascorbate ion). The acetylperoxyl radical (Schuchmann & von Sonntag, 1988) is the most strongly oxidising peroxyl radical known thus far. In contrast, $O_2^{\bullet-}$ is a reducing species [$E^7 = -0.33$ V (Wardman, 1989); the reduction potential relates, by definition, to O_2-saturated solutions. For comparison with other values that are based on 1 M, a value of -0.18 V should be taken (Wardman, 1991), see also Chapter 2]. It reacts rapidly with ozone [reaction (57)], and thus its interaction with other peroxyl radicals can be neglected under the conditions considered here.

$$O_2^{\bullet-} + O_3 \rightarrow O_2 + O_3^{\bullet-} \tag{57}$$

Reaction (57) gives rise to •OH ($O_3^{•−} + H_2O \rightarrow$ •OH + OH$^−$; Chapter 13). Yet, the enhanced •OH production also enhances peroxyl radical formation. In this context, it may be mentioned that other peroxyl radicals also react with ozone, albeit not in a redox reaction. A case in point is the reaction of the tertiary butanol derived peroxyl radical (Chapter 13).

14.8 UNIMOLECULAR DECAY OF PEROXYL RADICALS

Adequately substituted peroxyl radicals may eliminate $HO_2^•/O_2^{•−}$. Typical examples are α-hydroxyalkylperoxyl radicals and their corresponding anions [reactions (58)–(60)].

$$•OOCR_2OH \rightleftarrows •OOCR_2O^− + H^+ \quad (58)$$

$$•OOCR_2OH \rightarrow R_2C=O + HO_2^• \quad (59)$$

$$•OOCR_2O^− \rightarrow R_2C=O + O_2^{•−} \quad (60)$$

The rate of the $HO_2^•$ elimination reaction (59) may be slow and varies markedly (many orders of magnitude) with the substituents R (Bothe et al., 1978; von Sonntag & Schuchmann, 1997), but the $O_2^{•−}$ elimination reaction (60) is so fast, that this process is typically determined by the rate of the deprotonation reaction (58) ($k = 2 \times 10^{10}$ M$^{−1}$ s$^{−1}$ when OH$^−$-induced, two orders of magnitude slower when phosphate buffer induced) (Bothe et al., 1983). With α-aminoalkylperoxyl radicals the $O_2^{•−}$ elimination can already proceed with the help of the lone electron pair, while with α-amidoperoxyl radicals, the deprotonated peroxyl radical is sufficiently long-lived to establish equilibrium (61) and measure the rate of $O_2^{•−}$ elimination (62) by pulse radiolysis (Mieden et al., 1993).

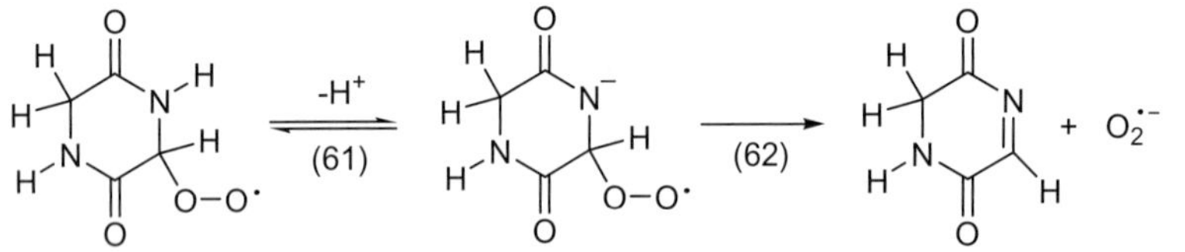

The slow $HO_2^•$ elimination encountered with many of the peroxyl radicals mentioned above implies that at high radical concentrations which one may encounter, for example, with electron beam irradiation, these peroxyl radicals may decay by second order kinetics rather than eliminating $HO_2^•/O_2^{•−}$. Thus the product distribution may be both dose rate (here: prevailing ozone concentration) and pH dependent. A case in point is the peroxyl radical chemistry of uracil (Schuchmann & von Sonntag, 1983).

When electron-donating substituents are present in the α-position to the peroxyl radical function, $O_2^{•−}$ release by an S_N1 mechanism is also possible, e.g. the rapid release of $O_2^{•−}$ from the 1,1-dimethoxyethylperoxyl radical (Schuchmann et al., 1990).

The H-abstractive power of peroxyl radical is only low, and intramolecular H-abstractions may be of relevance only at long peroxyl radical lifetimes. For example, the rate of reaction (63) is only near 1 s$^{−1}$ (Schuchmann & von Sonntag, 1982).

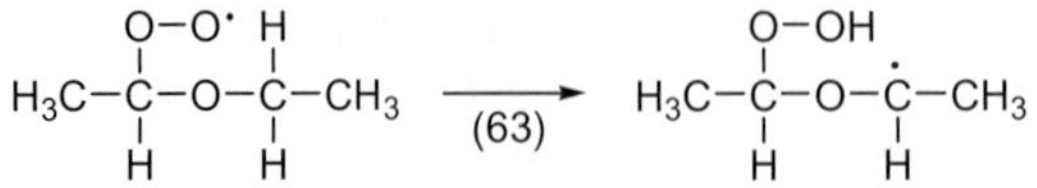

In ozone reactions, such slow peroxyl reactions are likely to be of minor importance; as long as ozone is still present they may interact with ozone (Chapter 13).

The intermolecular analogue of reaction (63) is the chain autoxidation reaction that can generally be neglected under AOP conditions (low substrate concentrations).

Peroxyl radicals may also add intramolecularly to C–C double bonds, when a six-membered transition state sets favourable conditions for this reaction to proceed [reaction (64)] (Porter et al., 1980, 1981).

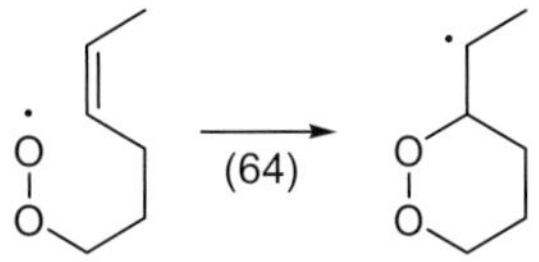

As this reaction is relatively slow, an intermolecular analogue is of no importance for typical AOP applications for the same reason as autoxidations do not take place to any significant extent.

14.9 BIMOLECULAR DECAY OF PEROXYL RADICALS

In their bimolecular decay, peroxyl radicals first form short-lived tetroxides [cf. reaction (46), well established in organic solvents (Bartlett & Guaraldi, 1967; Adamic et al., 1969; Bennett et al., 1970; Howard & Bennett, 1972; Howard, 1978; Furimsky et al., 1980)], which may either revert to the peroxyl radicals or undergo a series of reactions, depending on their structure. Tertiary peroxyl radicals, which decay generally more slowly than primary and secondary peroxyl radicals, show only one decay route, the formation of oxyl radicals and O_2. With primary and secondary peroxyl radicals, the route to oxyl radicals also plays a role [reaction (65)], but the hydrogens can also take part in the decay of the tetroxides [reaction (66); Russell mechanism (Russell, 1957)] and [reaction (67); Bennett mechanism (Bennett & Summers, 1974; Bothe & Schulte-Frohlinde, 1978)].

$$R_2HCOOOOCR_2H \longrightarrow 2R_2C(O^\bullet)H + O_2 \tag{65}$$
$$R_2HCOOOOCR_2H \longrightarrow R_2C{=}O + O_2 + R_2C(OH)H \tag{66}$$
$$R_2HCOOOOCR_2H \longrightarrow 2\,R_2C{=}O + H_2O_2 \tag{67}$$

The latter two reactions may be written as concerted reactions with a six-membered transition state (Russell mechanism) and an arrangement consisting of two five-membered rings (Bennett mechanism).

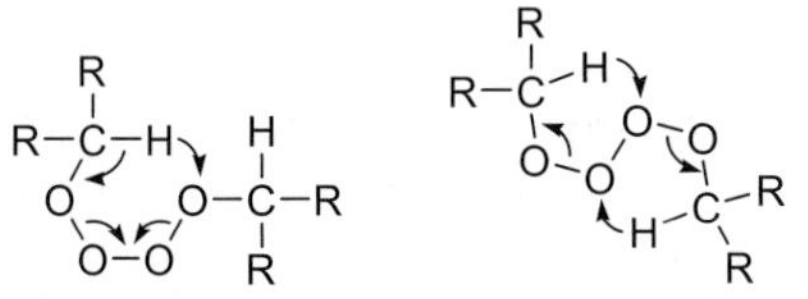

Russell mechanism Transition state   **Bennett mechanism Transition state**

$HO_2^\bullet/O_2^{\bullet-}$ radicals terminate forming H_2O_2 plus O_2 [reactions (68), $k = 8.6 \times 10^5\,M^{-1}\,s^{-1}$, and (69), $1.02 \times 10^8\,M^{-1}\,s^{-1}$ (Bielski et al., 1985)]. A bimolecular decay of two $O_2^{\bullet-}$ radicals does not take place to any significant extent.

$$2\,HO_2^\bullet \longrightarrow H_2O_2 + O_2 \tag{68}$$
$$HO_2^\bullet + O_2^{\bullet-} \longrightarrow HO_2^- + O_2 \tag{69}$$

Reaction (69) competes with the (reversible) formation of HO_4^- [reaction (70)], and this explains why the observed reaction (69) is relatively slow (Merényi et al., 2010b).

$$HO_2^\bullet + O_2^{\bullet-} \rightleftharpoons HO_4^- \qquad (70)$$

$HO_2^\bullet/O_2^{\bullet-}$ also may interact with other peroxyl radicals. As yet, there is little information on such reactions. Note that during ozonation, $O_2^{\bullet-}$ reacts rapidly with ozone (Chapter 13), and much of the above may not be relevant under such conditions.

14.10 REACTIONS OF OXYL RADICALS

The oxyl radicals formed in the bimolecular decay of the peroxyl radicals [reaction (65)] and in the reaction of peroxyl radicals with ozone (Chapter 13) are highly reactive intermediates. Tertiary oxyl radicals undergo a rapid (in the order of 10^6 s^{-1}) β-fragmentation [reaction (71)] (Rüchardt, 1987; Gilbert et al., 1981; Erben-Russ, 1987).

$$R_3CO^\bullet \longrightarrow R_2C{=}O + R^\bullet \qquad (71)$$

In the primary and secondary oxyl radicals, a 1,2-H shift often competes successfully [reaction (72); for mechanistic aspects see Konya et al. (2000)].

$$R_2C(H)O^\bullet \longrightarrow {}^\bullet CR_2(OH) \qquad (72)$$

The driving forces (standard Gibbs free energies) for a number of such reactions are compiled in Table 14.3.

Table 14.3 Calculated standard Gibbs free energies ($\Delta G°$/kJ mol^{-1}) for the decay of some oxyl radicals by β-fragmentation and by 1,2-H shift in water (DFT, Jaguar programme, Naumov & von Sonntag, 2011, unpublished)

Reaction	$\Delta G°$/ kJ mol^{-1}
$(CH_3)_3CO^\bullet \rightarrow {}^\bullet CH_3 + (CH_3)_2C{=}O$	−70 [−57 (gas phase)]
$(tBu, Et, Me)CCO^\bullet \rightarrow {}^\bullet CH_3 + (tBu, Et)C{=}O$	−94
$(tBu, Et, Me)CCO^\bullet \rightarrow {}^\bullet CH_2CH_3 + (tBu, Me)C{=}O$	−110
$(tBu, Et, Me)CCO^\bullet \rightarrow {}^\bullet C(CH_3)_3 + (Me, Et)C{=}O$	−154
$CF_3CF_2O^\bullet \rightarrow {}^\bullet CF_3 + F_2C{=}O$	−97
$CF_3CCl_2O^\bullet \rightarrow {}^\bullet CF_3 + Cl_2C{=}O$	−88
$CCl_3CCl_2O^\bullet \rightarrow {}^\bullet CCl_3 + Cl_2C{=}O$	−151
$CCl_3CCl_2O^\bullet \rightarrow {}^\bullet Cl + CCl_3C(O)Cl$	−86
$^-OC(O)Cl_2CO^\bullet \rightarrow CO_2^{\bullet-} + Cl_2C{=}O$	*
$CF_3CCl_2O^\bullet \rightarrow {}^\bullet Cl + CF_3C(O)Cl$	−71
$CF_3CF_2O^\bullet \rightarrow {}^\bullet F + CF_3C(O)F$	+55
$CH_3O^\bullet \rightarrow {}^\bullet CH_2OH$	−35
$CH_3CH_2O^\bullet \rightarrow {}^\bullet CH(CH_3)OH$	−36
$CH(CH_3)_2O^\bullet \rightarrow {}^\bullet C(CH_3)_2OH$	−48

*This oxyl radical decomposes without barrier upon structure optimisation.

The α-hydroxyalkyl radicals formed in reaction (72) subsequently give rise to α-hydroxyalkylperoxyl radicals (Berdnikov *et al.*, 1972; Gilbert *et al.*, 1976, 1977; Schuchmann & von Sonntag, 1981), the source of $HO_2^{\bullet}/O_2^{\bullet-}$ in these systems (see above).

The β-fragmentation of acyloxyl radicals is even several orders of magnitude faster than that of tertiary alkoxyl radicals (Hilborn & Pincock, 1991).

14.11 INVOLVEMENT OF •OH RADICALS IN CHLORATE AND BROMATE FORMATION

14.11.1 Chlorate formation

Because chloride is oxidised very slowly by ozone (Chapter 11), chlorate is only formed in significant concentrations when ozonation follows a pre-treatment with chlorine or chlorine dioxide (von Gunten, 2003b; Haag & Hoigné, 1983b). The chlorine–ozone reaction system is discussed in detail in Chapter 11; for ozone reactions with halogen-centred radicals see Chapter 13.

Because the reactions in the hypochlorite-ozone system are relatively slow, •OH reactions may play an important role under typical treatment conditions with relatively low chlorine concentrations entering an ozonation reactor after pre-chlorination.

The reaction of chloride with •OH is given by a pH-dependent equilibrium [reactions (73) and (74); $k_{73} = 4.3 \times 10^9 \, M^{-1} \, s^{-1}$ (Kläning & Wolff, 1985), $k_{-73} = 6 \times 10^9 \, s^{-1}$ (Kläning & Wolff, 1985), $k_{74} = 2.1 \times 10^{10} \, M^{-1} \, s^{-1}$ (Kläning & Wolff, 1985), $k_{-74} = 2.5 \times 10^5 \, s^{-1}$ (McElroy, 1990)].

$$Cl^- + {}^{\bullet}OH \rightleftarrows HOCl^{\bullet-} \tag{73}$$

$$HOCl^{\bullet-} + H^+ \rightleftarrows Cl^{\bullet} + H_2O \tag{74}$$

Based on this reaction sequence with the two equilibria, the apparent second order rate constant for the reaction of chloride with •OH to Cl• can be calculated as a function of pH. The apparent second order rate constant is expressed as in equation (75).

$$k_{app} = \frac{k_{73}}{k_{-73}} \left(k_{74}[H^+] + k_{-74} \right) \tag{75}$$

This is illustrated in Figure 14.1. It is evident that the formation of Cl• is only relevant at pH < 3. Therefore, for typical water treatment conditions during ozonation of drinking waters and wastewaters, the formation of chlorinated organic by-products is minimal.

In waters with higher salinity, the $HOCl^{\bullet-}$ species may further react with chloride and bromide [reactions (76), $k = 1.0 \times 10^4 \, M^{-1} \, s^{-1}$ (Grigorev *et al.*, 1987) and (77), $k = 1.0 \times 10^9 \, M^{-1} \, s^{-1}$ (estimated, (Matthew & Anastasio, 2006)].

$$HOCl^{\bullet-} + Cl^- \rightarrow Cl_2^{\bullet-} + OH^- \tag{76}$$

$$HOCl^{\bullet-} + Br^- \rightarrow BrCl^{\bullet-} + OH^- \tag{77}$$

The conversion of •OH to these more selective radicals ($Cl_2^{\bullet-}$ and $BrCl^{\bullet-}$) may lead to a better abatement of electron-rich micropollutants such as phenols [$k(Cl_2^{\bullet-} + phenol) = 2.5 \times 10^8 \, M^{-1} \, s^{-1}$ (Alfassi *et al.*, 1990)] in saline waters (Grebel *et al.*, 2010). For a critical compilation of rate for chlorine and bromine species in saline waters see Grebel *et al.* (2010) and references therein.

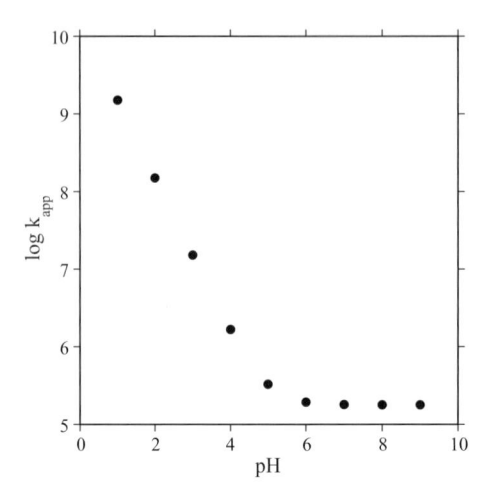

Figure 14.1 pH dependence of the second order rate constant for the reaction $Cl^- + {}^\bullet OH \rightarrow {}^\bullet Cl + OH^-$.

In waters containing hypochlorite, ${}^\bullet OH$ contributes significantly to its oxidation [reactions (78), $k = 9 \times 10^9$ M^{-1} s^{-1} (Buxton & Subhani, 1972) and (79), $k_{79} < k_{78}$].

$$OCl^- + {}^\bullet OH \rightarrow ClO^\bullet + OH^- \tag{78}$$
$$HOCl + {}^\bullet OH \rightarrow ClO^\bullet + H_2O \tag{79}$$

The reaction product $ClO^\bullet$ undergoes a self-reaction with pre-equilibrium (Cl_2O_2) and a decay to OCl^- and ClO_2^- [reactions (80), $2k = 1.5 \times 10^{10}$ M^{-1} s^{-1} (Buxton & Subhani, 1972) and (81)].

$$2\,ClO^\bullet \rightleftarrows Cl_2O_2 \tag{80}$$
$$Cl_2O_2 + H_2O \rightarrow ClO^- + ClO_2^- + 2H^+ \tag{81}$$

The overall $ClO^\bullet$ self-reaction can be formulated as follows [reaction (82), $k = 2.5 \times 10^9$ M^{-1} s^{-1} (Kläning & Wolff, 1985)].

$$2\,ClO^\bullet + H_2O \rightarrow ClO^- + ClO_2^- + 2H^+ \tag{82}$$

Chlorite can be further oxidised by ${}^\bullet OH$ to chlorine dioxide and finally chlorate which is the end product during ozonation and AOPs of chlorine-containing waters [reactions (83), $k = 4.2 \times 10^9$ M^{-1} s^{-1} (Kläning et al., 1985) and (84), $k = 4.0 \times 10^9$ M^{-1} s^{-1} (Kläning et al., 1985)].

$$ClO_2^- + {}^\bullet OH \rightarrow ClO_2^\bullet + OH^- \tag{83}$$
$$ClO_2^\bullet + {}^\bullet OH \rightarrow ClO_3^- + H^+ \tag{84}$$

14.11.2 Bromate formation

In the following section, the most important bromine reactions occurring in ${}^\bullet OH$-based AOPs at near neutral pH are discussed. For a complete compilation of all reactions occurring in the ${}^\bullet OH$-bromine system see Matthew & Anastasio (2006) and references therein.

The ozone reactions occurring during ozonation of Br^--containing waters are discussed in Chapter 11 and the reactions between ozone and bromine-containing radicals in Chapter 13. The oxidation of Br^-

by •OH to •Br proceeds through equilibrium (85) [reactions (85), $k_{85} = 1.06 \times 10^{10}$ M^{-1} s^{-1}, $k_{-85} = 3.3 \times 10^7$ s^{-1} (Zehavi & Rabani, 1972)] followed by a decay and reformation of BrOH$^{•-}$ [equilibrium (86), $k_{86} = 4.2 \times 10^6$ s^{-1} (Zehavi & Rabani, 1972), $k_{-86} = 1.3 \times 10^{10}$ M^{-1} s^{-1} (Lind et al., 1991)].

$$Br^- + {}^{•}OH \rightleftarrows BrOH^{•-} \tag{85}$$

$$BrOH^{•-} \rightleftarrows Br^{•} + OH^- \tag{86}$$

BrOH$^{•-}$ and Br$^{•}$ react further with Br$^-$ [reaction (87), $k = 1.9 \times 10^8$ M^{-1} s^{-1} (Zehavi & Rabani, 1972), and equilibrium (88), $k_{88} \approx 10^{10}$ M^{-1} s^{-1} (Zehavi & Rabani, 1972), $k_{-88} = 6.6 \times 10^3$ s^{-1} modelled by (Matthew & Anastasio, 2006)].

$$BrOH^{•-} + Br^- \rightarrow Br_2^{•-} + OH^- \tag{87}$$

$$Br^{•} + Br^- \rightleftarrows Br_2^{•-} \tag{88}$$

In natural waters, Br$^{•}$ can also react with carbonate/bicarbonate [reactions (89), $k = 2.0 \times 10^6$ M^{-1} s^{-1}, and (90), $k = 1.0 \times 10^6$ M^{-1} s^{-1} (estimated values) (Matthew & Anastasio, 2006)]

$$Br^{•} + CO_3^{2-} \rightarrow Br^- + CO_3^{•-} \tag{89}$$

$$Br^{•} + HCO_3^- \rightarrow Br^- + H^+ + CO_3^{•-} \tag{90}$$

The Br$_2^{•-}$ species undergoes further reactions, leading to Br$_3^-$ and HOBr, which are crucial in the context of bromate formation during ozonation (Chapter 11) [reactions (91), $k = 3.4 \times 10^9$ M^{-1} s^{-1} (Ershov et al., 2002), (92), $k > 1.0 \times 10^9$ M^{-1} s^{-1} (Wagner & Strehlow, 1987) and (93), $k = 1.1 \times 10^5$ M^{-1} s^{-1} (Huie et al., 1991)]

$$2\,Br_2^{•-} \rightarrow Br_3^- + Br^- \tag{91}$$

$$Br_2^{•-} + {}^{•}OH \rightarrow HOBr + OH^- \tag{92}$$

$$Br_2^{•-} + CO_3^{2-} \rightarrow CO_3^{•-} + 2\,Br^- \tag{93}$$

Br$_3^-$ is in equilibrium with Br$_2$ and Br$^-$ [equilibrium (94), $k_{94} = 5.5 \times 10^7$ s^{-1}, $k_{-94} = 9.6 \times 10^8$ M^{-1} s^{-1} (Ershov, 2004)]. Br$_2$ is in equilibrium with HOBr and Br$^-$ [equilibrium (95), $K = 3.5 \times 10^{-9}$ M^2 (Beckwith et al., 1996)].

$$Br_3^- \rightleftarrows Br_2 + Br^- \tag{94}$$

$$Br_2 + H_2O \rightleftarrows HOBr + H^+ + Br^- \tag{95}$$

As shown in Chapter 11, HOBr ($pK_a = 8.8$) is a crucial intermediate during ozonation of Br$^-$-containing waters, and most mitigation strategies are based on HOBr minimisation. HOBr/OBr$^-$ can be further oxidised by •OH [reactions (96), $k = 2 \times 10^9$ M^{-1} s^{-1}, and (97), $k = 4.5 \times 10^9$ M^{-1} s^{-1} (Buxton & Dainton, 1968)].

$$HOBr + {}^{•}OH \rightarrow BrO^{•} + H_2O \tag{96}$$

$$OBr^- + {}^{•}OH \rightarrow BrO^{•} + OH^- \tag{97}$$

The BrO$^{•}$ species may also be formed by carbonate radicals [reaction (98), $k = 4.3 \times 10^7$ M^{-1} s^{-1} (Buxton & Dainton, 1968)].

$$CO_3^{•-} + OBr^- \rightarrow BrO^{•} + CO_3^{2-} \tag{98}$$

During ozonation, the BrO• species can also be formed from a reaction of Br• with ozone (Chapter 13). It undergoes disproportionation [reaction (99), $k = 4.5 \times 10^9$ M^{-1} s^{-1} (Buxton & Dainton, 1968)].

$$2\,BrO^\bullet + H_2O \longrightarrow OBr^- + BrO_2^- + 2H^+ \tag{99}$$

Subsequently, several oxidation steps lead to bromate which is undesired in drinking waters (Chapter 11) [reaction (100), $k = 2 \times 10^9$ M^{-1} s^{-1}, reaction (101), $k = 1.1 \times 10^8$ M^{-1} s^{-1} (Buxton & Dainton, 1968), and reaction (102), $k = 2 \times 10^9$ M^{-1} s^{-1} (Field et al., 1972)].

$$BrO_2^- + {}^\bullet OH \longrightarrow BrO_2^\bullet + OH^- \tag{100}$$

$$BrO_2^- + CO_3^{\bullet -} \longrightarrow BrO_2^\bullet + CO_3^{2-} \tag{101}$$

$$BrO_2^\bullet + {}^\bullet OH \longrightarrow BrO_3^- + H^+ \tag{102}$$

Therefore, bromate can be formed by •OH radical reactions only. This was demonstrated by γ-radiolysis experiments in which bromate increased as a function of irradiation time in N$_2$O-saturated solutions (Figure 14.2). Figure 14.2 also shows that bromate is formed only when the H$_2$O$_2$ concentration decreases to zero. This shows that HOBr is an intermediate during the •OH-induced bromate formation since this species is reduced by H$_2$O$_2$ (Chapter 11). Therefore, no bromate is formed in the AOP UV/H$_2$O$_2$, because a H$_2$O$_2$ is always in excess as long as •OH radicals are formed.

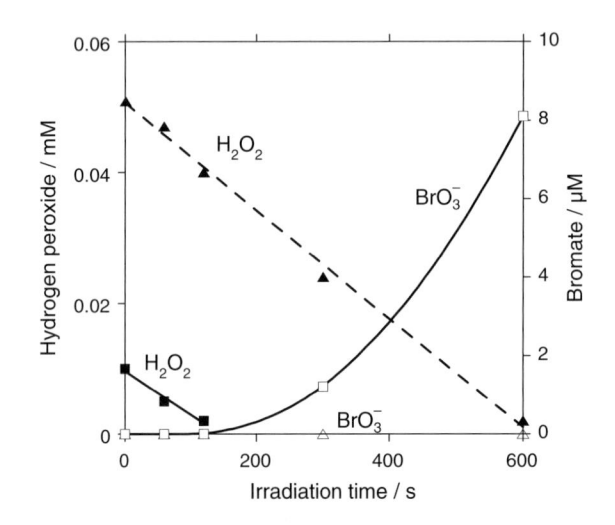

Figure 14.2 γ-Radiolysis of Br$^-$- and H$_2$O$_2$-containing solutions. The evolution of bromate (open symbols) and H$_2$O$_2$ (closed symbols) is shown as a function of the irradiation time (dose rate 0.58 Gy s^{-1}) at pH 7. [Br$^-$]$_0$ = 50 μM, [H$_2$O$_2$]$_0$: 50 μM (closed triangles), 10 μM (closed squares), 2 mM phosphate, N$_2$O-saturated solutions. Reprinted with permission from von Gunten & Oliveras (1998). Copyright (1998) American Chemical Society.

From the above reaction sequence, it can be seen that carbonate alkalinity can have a direct influence on bromate formation kinetics and pathways through various reactions with carbonate radicals. Finally, DOM also has an important influence on bromate formation due to its role related to its effect on the ozone chemistry and •OH formation during ozonation [Chapter 3, (Pinkernell & von Gunten, 2001)].

14.12 DEGRADATION OF OZONE-REFRACTORY MICROPOLLUTANTS BY •OH/PEROXYL RADICALS

The products of the degradation of ozone-refractory micropollutants have only been studied in detail in some cases. Examples are methyl-t-butyl ether (MTBE) and atrazine. For others, only a brief account can be given on likely reactions based on the large body of other studies on •OH/peroxyl radical reactions in aqueous solution. For the compounds discussed below, •OH rate constants are compiled in Table 14.4. For a discussion of the efficiency of the elimination of ozone-refractory micropollutants see also Chapter 3.

Table 14.4 Compilation of •OH rate constants for some selected ozone-refractory micropollutants

Compound	$k/M^{-1} s^{-1}$	Reference
Amidotrizoic acid	9.6×10^8	Jeong et al., 2010
	5.4×10^8	Real et al., 2009
Atrazine	3×10^9	Acero et al., 2000
1,4-Dioxane	2.5×10^9–3.1×10^9	Buxton et al., 1988
Geosmin	7.8×10^9	Peter & von Gunten, 2007
Ibuprofen	7.4×10^9	Huber et al., 2003
Iomeprol	Similar to Iopromide	Huber et al., 2005
Iopromide	3.3×10^9	Huber et al., 2003
Iopamidol	Similar to Iopromide	Huber et al., 2005
2-Methylisoborneol	5.1×10^9	Peter & von Gunten, 2007
Methyl tertiary butyl ether (MTBE)	1.6×10^9–1.9×10^9	Cooper et al., 2009
N-Nitrosodimethylamine (NDMA)	4.5×10^8	Lee et al., 2007
Perfluorooctanoic acid (PFOA)	No significant reaction	Plumlee et al., 2009
	$\leq 10^5$	Vecitis et al., 2009
Perfluorooctanesulfonic acid (PFOS)	No significant reaction	Plumlee et al., 2009
Tetrachloroethene (PER)	2.3×10^9	Köster & Asmus, 1971
Trichloroethene (TRI)	4.0×10^9	Köster & Asmus, 1971
Tri-n-butyl phosphate (TnBP)	$(2.8 \pm 0.8) \times 10^9$	Pocostales et al., 2010
	6.4×10^9	Watts & Linden, 2009
Tris-2-chloroisopropyl phosphate (TCPP)	$(7 \pm 2) \times 10^8$	Pocostales et al., 2010
	1.98×10^8	Watts & Linden, 2009

14.12.1 Saturated aliphatic compounds

Hydroxyl radicals react with saturated aliphatic compounds by H-abstraction. There is a considerable preference for hydrogens activated by a neighbouring heteroatom [cf. reactions (15) and (16)]. Such a preference also holds for •OH attack at methoxy on the gasoline additive MTBE [56% (Acero et al., 2001), 60% (Stefan et al., 2000)]. The rate constant for the reaction of MTBE with •OH radicals is given in Table 14.4. Details of the ensuing chemistries have been reviewed in detail (Cooper et al., 2009) and remediation technologies have been discussed (Baus et al., 2005). For a discussion of MTBE removal in multibarrier treatment see Chapter 5.

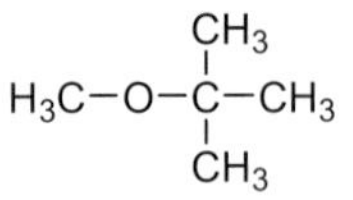

Methyl-*t*-butyl ether (MBTE)

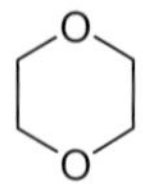

1,4-Dioxane

The •OH-induced degradation of 1,4-dioxane in the presence of O_2 has been studied in detail (Stefan & Bolton, 1998) and discussed on the basis of mechanisms shown above. The radicals formed upon •OH attack are reducing radicals and, therefore, also react rapidly with oxidants other than O_2 (Nese et al., 1995).

There are many other ozone-refractory micropollutants, notably the taste and odour compounds 2-methylisoborneol and geosmin (for a compilation of taste and odour compounds see Chapter 5). Their odour threshold is as low as a few ng/L.

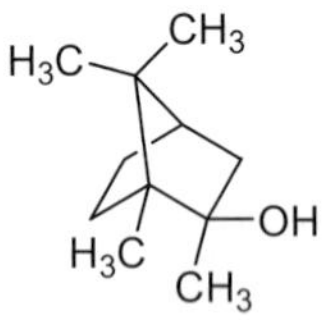

2-Methylisoborneol 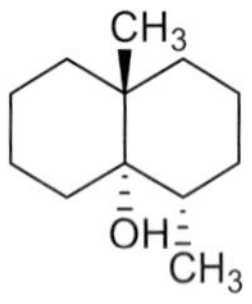 **Geosmin**

2-Methylisoborneol and geosmin react faster with •OH than MTBE (Table 14.4). There are many different hydrogens available for •OH attack, and hence there must be a multitude of •OH-induced degradation products. They will all be more polar and less volatile than their parents which leads to a loss of their taste and odour properties (Peter & von Gunten, 2007).

There is another important group of ozone-refractory organic micropollutants, namely the organic phosphates. They serve as plasticisers and additives in the clothing industry such as tri-*n*-butyl phosphate (TnBP) or as flame retardants such as tris-2-chloroisopropyl phosphate (TCPP) [for their occurrence in waters of various origin see Andresen & Bester (2006), Andresen et al. (2004) and Meyer & Bester (2004); for their degradation in wastewater see Pocostales et al. (2010), Chapter 3.

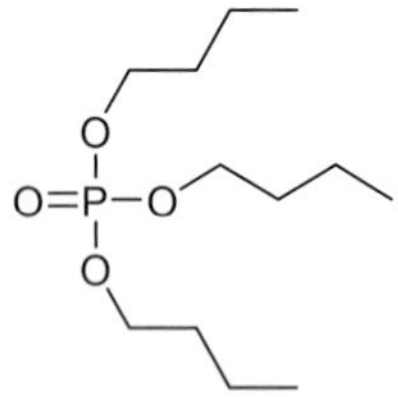

Tri-n-butyl phosphate (TnBP)

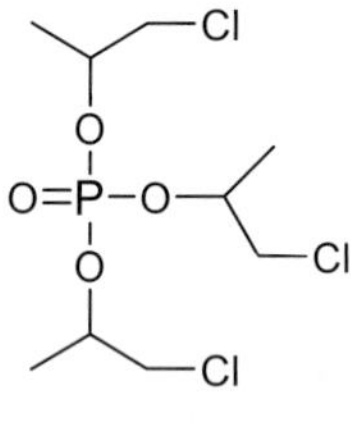

Tris-2-chloroisopropyl phosphate (TCPP)

Their •OH-induced degradation products have not been determined, but there are detailed studies on trimethyl and triisopropyl phosphates (Schuchmann & von Sonntag, 1984b; Schuchmann et al., 1984, 1995). H-abstraction of the α-hydrogen and subsequent peroxyl radical reactions lead to labile compounds that readily hydrolyse [e.g. reactions (103) and (104)].

$$(CH_3O)_2P(O)CH_2OH + H_2O \rightarrow (CH_3O)P(O)OH + CH_2O \tag{103}$$

$$(CH_3O)_2P(O)C(O)H + H_2O \rightarrow (CH_3O)P(O)OH + HC(O)OH \tag{104}$$

The rates of these reactions have been determined by pulse radiolysis.

N-Nitrosodimethylamine (NDMA) reacts only slowly with ozone and with •OH (Table 14.4). Therefore, the transformation of NDMA in ozone-based AOPs is very slow. One of the main products from NDMA oxidation by •OH is methylamine (Lee et al., 2007b).

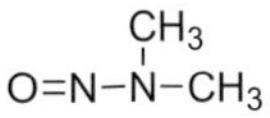

N-Nitrosodimethylamine
NDMA

NDMA can be efficiently degraded by direct UV photolysis to mainly dimethylamine, nitrate and nitrite (Stefan & Bolton, 2002). For similar UV fluences, the combination of UV with H_2O_2 does not contribute significantly to the degradation rate (Sharpless & Linden, 2003). The dimethylamine formed from NDMA is unproblematic for further NDMA formation in, for example, postchloramination, because the concentration will be very low (NDMA is typically present in the nanograms per litre range) and the NDMA yield from dimethylamine in NDMA formation potential tests with monochloramine is of the order of 3% (Lee et al., 2007a). NDMA can also be partially or fully removed by biological filtration (Schmidt & Brauch, 2008).

14.12.2 Aromatic compounds

Abundant aromatic ozone-refractory micropollutants ($k_{ozone} < 10$ $M^{-1} s^{-1}$) are the x-ray contrast media such as amidotrizoic acid, iomeprol, iopamidol and iopromide, the pharmaceutical ibuprofen, the musk fragrance tonalide (AHTN) (Chapter 7) as well as the triazine herbicides such as atrazine (see below).

Second order rate constants for the reaction of x-ray contrast media with •OH have been determined, with amidotrizoic acid having the lowest value ($k = 0.96 \times 10^9$ $M^{-1} s^{-1}$ (Jeong et al., 2010) and the others having values in a small range between $2-3.4 \times 10^9$ $M^{-1}s^{-1}$ (Huber et al., 2003; Jeong et al., 2010). The x-ray contrast media compounds are characterised by three iodine atoms at the aromatic ring and three further substituents for solubilisation. Iodine is especially bulky and the addition of •OH to the aromatic ring will preferentially occur at the α-positions to the iodines (Fang et al., 2000). Moreover, there will be a considerable contribution of H-abstraction from the side chains. This must lead to a complex peroxyl radical chemistry that is difficult to predict.

The •OH-induced peroxyl radical chemistry of ibuprofen is easier to predict.

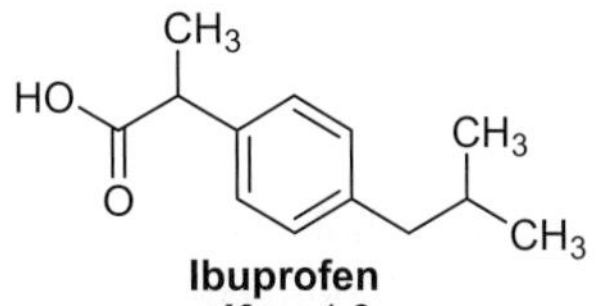

Ibuprofen
pK_a = 4.9

The rate constants for the reaction of ibuprofen with ozone and •OH are 9.6 $M^{-1} s^{-1}$ and 7.4×10^9 $M^{-1} s^{-1}$, respectively (Huber et al., 2003). These values warrant that ibuprofen is mainly degraded by •OH during ozonation. This will largely resemble the degradation of its parent benzene, that is, one would expect mainly hydroxylation reactions as depicted above for terephthalate [reactions (35)–(43)]. The resulting phenols will react very rapidly with ozone and thus will be further degraded (Chapter 7).

The •OH-induced reactions of atrazine have been studied in quite some detail (Tauber & von Sonntag, 2000; Acero et al., 2000) [for reviews on the degradation of herbicides by ozone and AOPs see (Ikehata & El-Din, 2005a, b)].

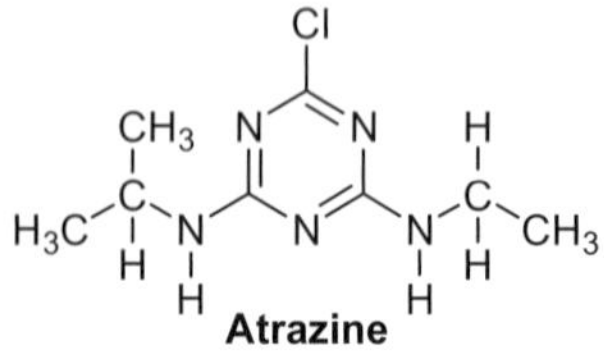

Atrazine

The reaction of atrazine with ozone is slow ($k = 6$ M^{-1} s^{-1}), and it will mainly react with •OH during ozonation ($k = 3 \times 10^9$ M^{-1} s^{-1}) (Acero et al., 2000). The aromatic ring is so electron-deficient, that the reactivity of the triazine ring is low and hence reacts rather slowly with •OH (Tauber & von Sonntag, 2000). This allows H-abstraction reactions at the side chains to become of importance (ca. 60% of •OH attack) (Tauber & von Sonntag, 2000). The products that are formed upon •OH attack at the ring have not yet been elucidated, but the reactions that follow an attack at the side chains are known in quite some detail. The major products are deethyl- and deisopropyl-atrazine (and the corresponding keto compounds, acetaldehyde and acetone) in a 2.6:1 ratio. An example for the reactions that follow an •OH attack at the α-position of the ethyl group and subsequent O_2 addition is given by reactions (105)–(108).

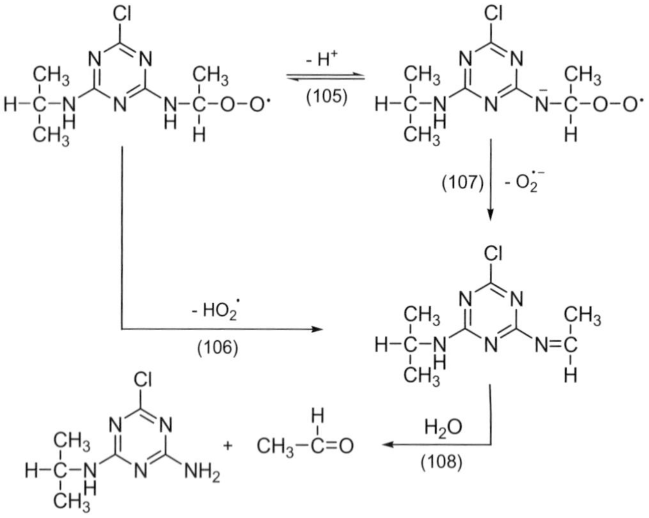

Kinetic data for all the relevant reactions have been determined by pulse radiolysis (Tauber & von Sonntag, 2000) and by competition kinetics in O_3/H_2O_2 systems (Acero et al., 2000). The hydrolysis of the Schiff base is OH^--catalysed and is so slow [$k(OH^-) = 5.2$ M^{-1} s^{-1}] that it can be followed conveniently after γ-irradiation. For a study on atrazine degradation using electrolytically generated •OH see Balci et al. (2009).

14.12.3 Chlorinated olefins

Trichloroethene (TRI) and tetrachloroethene (PER) barely react with ozone (Chapter 6), but react readily with •OH (Table 14.4).

Trichloroethene
TRI

Tetrachloroethene
PER

Attack of •OH on the symmetrical molecule PER gives rise to a single radical [reaction (109)].

$$\bullet OH + Cl_2C=CCl_2 \longrightarrow HOCCl_2Cl_2C^\bullet \qquad (109)$$

The observation of trichloroacetic acid TCA (in the presence of O_2) as an important product has led to the suggestion that a rapid rearrangement [HOCCl$_2$C(Cl$_2$) $\longrightarrow$ Cl$_3$CC(OH,H)•] takes place (Glaze, 1986; Mao et al., 1993) in competition with the fast HCl elimination of the geminal chlorohydrin function [reaction (110), $k > 7 \times 10^5$ s^{-1} (Köster & Asmus, 1971; Mertens & von Sonntag, 1994b)].

$$HOCCl_2Cl_2C^\bullet \longrightarrow HCl + ClC(O)Cl_2C^\bullet \qquad (110)$$

A product study in the absence of O_2 showed that TCA is not a product under such conditions and that the rapid release of HCl must only be due to reaction (110) (Mertens & von Sonntag, 1994b).

In fact, the resulting radicals recombine [reaction (111); $k = 6.9 \times 10^8$ M^{-1} s^{-1}].

$$2 \, ClC(O)Cl_2C^\bullet \longrightarrow Cl(O)CCCl_2CCl_2C(O)Cl \qquad (111)$$

Upon hydrolysis of one of the acyl chloride functions [reaction (112), $k = 5$ s^{-1}], an acid is formed that cyclises rapidly upon forming tetrachlorosuccinic anhydride (reaction not shown).

$$Cl(O)CCCl_2CCl_2C(O)Cl + H_2O \longrightarrow Cl(O)CCCl_2CCl_2C(O)O^- + 2H^+ + Cl^- \qquad (112)$$

This anhydride hydrolyses more slowly and could no longer be determined by pulse radiolysis with conductometric detection [for a comparison, hydrolysis of succinic anhydride takes place at $k = 1.8 \times 10^{-3}$ s^{-1} (Eberson, 1964)]. In γ-radiolysis, the ClC(O)Cl$_2$C• radicals have a lifetime of 0.1 s, and under such conditions tetrachlorosuccinic anhydride was not detected. From this it follows that either tetrachlorosuccinic anhydride hydrolyses much faster than succinic anhydride or the ClC(O)Cl$_2$C• radicals have hydrolysed prior to recombination.

In a much slower reaction tetrachlorosuccinic acid decarboxylates [reaction (113)], $k = 6.2 \times 10^{-5}$ s^{-1} at 55°C, $E_a = 115$ kJ mol^{-1}].

$$^-O(O)CCCl_2CCl_2C(O)O^- \longrightarrow CO_2 + HCl + Cl_2C=CCl(O)O^- \qquad (113)$$

In the presence of O_2, the ClC(O)Cl$_2$C• radicals are converted into the corresponding peroxyl radicals [reaction (114)].

$$ClC(O)Cl_2C^\bullet + O_2 \longrightarrow ClC(O)Cl_2COO^\bullet \qquad (114)$$

Data from a preliminary γ-radiolytic product study (Mertens, 1994) are given in Table 14.5.

Table 14.5 Products and their G values (unit: 10^{-7} mol J^{-1}) in the γ-radiolysis of tetrachloroethene (1 mM) in N$_2$O/O$_2$-(4:1)-saturated aqueous solutions. The conversion into mol product per mol •OH is based on G(•OH) = 5.6 × 10^{-7} mol J^{-1}, the contribution of •H is neglected

Product	G value/10^{-7} mol J^{-1}	Yield per mol •OH
Chloride	33.1	5.9
Carbon dioxide	10.4	1.85
Trichloroacetic acid (TCA)	2.1	0.38
Dichloroacetic acid (DCA)	< 0.1	0.02

From these data, it is seen that a short chain reaction must prevail (the Cl$^-$ yield is above four times the •OH yield). Moreover, the formation of TCA shows that chlorine atoms are generated in the course of the reaction.

The acyl-substituted, quasi-primary peroxyl radicals may undergo hydrolysis [reaction (115)].

$$ClC(O)Cl_2COO^\bullet + H_2O \rightarrow {}^-OC(O)Cl_2COO^\bullet + 2H^+ + Cl^- \tag{115}$$

Depending on the rate of •OH production, bimolecular termination of the peroxyl radicals will be by the primary (quasi-primary) peroxyl radicals at high initiation rate such as by electron beam irradiation, a mixture of primary (quasi-primary) plus secondary peroxyl radicals at moderate •OH generation rate, and by the secondary peroxyl radicals at low •OH generation rate. This may strongly influence the product distribution. At high initiation rate one may write reaction (116).

$$2\,ClC(O)Cl_2COO^\bullet \rightarrow 2\,ClC(O)Cl_2CO^\bullet + O_2 \tag{116}$$

The ensuing oxyl radical may undergo two β-elimination reactions [reactions (117) and (118)/(119), $\Delta G^0(119) = +24$ kJ mol^{-1} (Naumov & von Sonntag, 2011, unpublished), note that the analogous (reverse) reaction of •OH with CO is close to diffusion controlled (Buxton et al., 1988)].

$$ClC(O)Cl_2CO^\bullet \rightarrow ClC(O)C(O)Cl + {}^\bullet Cl \tag{117}$$
$$ClC(O)Cl_2CO^\bullet \rightarrow {}^\bullet C(O)Cl + Cl_2CO \tag{118}$$
$$^\bullet C(O)Cl \rightleftarrows CO + {}^\bullet Cl \tag{119}$$

One of the products of reaction (117) is oxalyl chloride. It hydrolyses rapidly [$k > 350$ s^{-1} (Dowideit, 1996; Prager et al., 2001)]. Once the first acyl chloride function has been hydrolysed, the remaining one will decompose into CO$_2$, CO plus HCl [reactions (120) and (121)] (Staudinger, 1908).

$$ClC(O)C(O)Cl + H_2O \rightarrow 2H^+ + {}^-OC(O)C(O)Cl + Cl^- \tag{120}$$
$$^-OC(O)C(O)Cl \rightarrow CO_2 + CO + Cl^- \tag{121}$$

In most AOP applications, the lifetime of free radicals will be near 100 ms rather than a few ms as in electron beam treatment. This implies that the acyl chloride-derived peroxyl radicals stand a good chance of hydrolysing prior to recombination. As a consequence, one has to take the involvement of the hydrolysed peroxyl radicals into account [reactions (122) and (123)].

$$^-OC(O)Cl_2COO^\bullet + ClC(O)Cl_2COO^\bullet$$
$$\rightarrow\ ^-OC(O)Cl_2COOOOCCl_2C(O)Cl \tag{122}$$
$$\rightarrow CO_2 + 2\,Cl_2CO + CO + Cl^- \tag{123}$$

In contrast to reactions (116)–(119), this termination reaction does not liberate $^\bullet Cl$ atoms. Similarly, oxylradical formation in the bimolecular decay would give rise to $CO_2^{\bullet -}$ [reaction (124), cf. Table 14.3] and subsequently to $O_2^{\bullet -}$ [reaction (125), $k = 2 \times 10^9\,M^{-1}\,s^{-1}$].

$$^-OC(O)Cl_2CO^\bullet \rightarrow CO_2^{\bullet -} + C(O)Cl_2 \tag{124}$$
$$CO_2^{\bullet -} + O_2 \rightarrow CO_2 + O_2^{\bullet -} \tag{125}$$

Such reactions break the (short) $^\bullet Cl$-promoted chain reaction. The formation of $^\bullet Cl$ is responsible for TCA formation. Here, $^\bullet Cl$ will have to add to PER [reaction (126)].

$$^\bullet Cl + Cl_2C{=}CCl_2 \rightarrow Cl_3CCl_2C^\bullet \tag{126}$$

Subsequent reactions, abbreviated in reaction (127) give rise to trichloroacetyl chloride which hydrolyses very rapidly [reaction (128), $k > 350\,s^{-1}$ (Dowideit, 1996)]

$$Cl_3CCl_2C^\bullet + O_2 \rightarrow Cl_3CCl_2COO^\bullet \rightarrow Cl_3CCl_2CO^\bullet$$
$$\rightarrow Cl_3C(O)Cl + {}^\bullet Cl \tag{127}$$
$$Cl_3C(O)Cl + H_2O \rightarrow Cl_3C(O)OH + HCl \tag{128}$$

The pentachloroethyl radical is also an intermediate in the photolysis of PER (Mertens & von Sonntag, 1995), as a chlorine atom is released in the photolytic step [reaction (129), for the formation and optical properties of the ensuing vinylperoxyl radicals see (Mertens & von Sonntag, 1994a; Naumov & von Sonntag, 2005)].

$$Cl_2C{=}CCl_2 + h\nu \rightarrow {}^\bullet CCl{=}CCl_2 + {}^\bullet Cl \tag{129}$$

As the molar absorption coefficient of PER at 254 nm ($\varepsilon = 225\,M^{-1}\,cm^{-1}$) is much higher than that of H_2O_2 ($\varepsilon = 20\,M^{-1}\,cm^{-1}$) and the quantum yield is also high [Φ(product forming processes) $= 0.31$], PER photolysis (Mertens & von Sonntag, 1995) must always contribute to product formation when H_2O_2/UV is used as AOP. In the presence of DOM, DOM will act as $^\bullet Cl$ scavenger, as will bicarbonate and carbonate. Thus formation of TCA, a highly undesired by-product, may no longer be significant at high concentrations of these scavengers [for model studies see Mertens & von Sonntag (1995)]. It is interesting to note that the ratios of rate constants for $^\bullet OH$ and $^\bullet Cl$ with carbonate and bicarbonate differ substantially. While for $^\bullet OH$ this ratio is 46, it is only 2.3 for $^\bullet Cl$ (Mertens, 1994).

The chemistries of the peroxyl radicals derived from TRI will be very similar to those discussed above for PER. In Table 14.6, the products and their yields are compiled [for data on 1,1-dichloroethene, *cis*-1,2-dichloroethene, and dichloromethane see Mertens (1994)].

Table 14.6 Products and their G values (unit: 10^{-7} mol J^{-1}) in the γ-radiolysis of trichloroethene (1 mM) in N$_2$O/O$_2$-(4:1)-saturated aqueous solutions. The conversion into mol product per mol •OH is based on G(•OH) = 5.6×10^{-7} mol J^{-1}, the contribution of •H is neglected. Source: Mertens (1994)

Product	G value/10^{-7} mol J^{-1}	Yield per mol •OH
Chloride	22.6	4.0
Carbon dioxide	8.9	1.6
Formic acid	5.5	1.0
Glyoxylic acid	0.4	0.07
Dichloroacetic acid (DCA)	1.7	0.3
H$_2$O$_2$ (final)	1.6	0.3

14.12.4 Perfluorinated compounds

The environmentally persistent perfluorinated surfactants perfluorooctanoic acid (PFOA) and perfluorooctanesulfonic acid (PFOS) do not react with ozone [for their fate in drinking water treatment see Tagagi *et al.* (2011)].

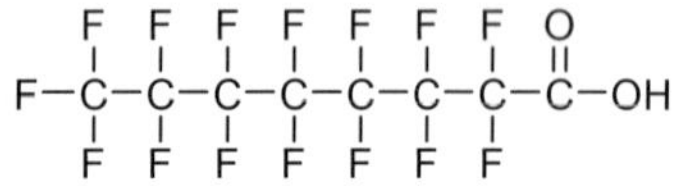

F—C—C—C—C—C—C—C—C(=O)—OH   **PFOA**

F—C—C—C—C—C—C—C—C—S(=O)$_2$—OH   **PFOS**

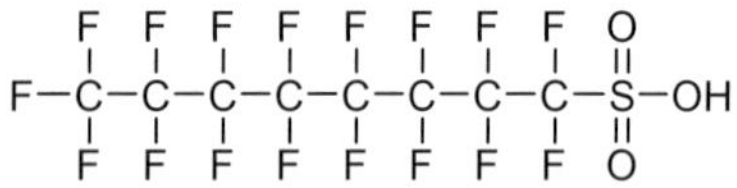

Even their oxidative degradation by •OH is most inefficient (Hori *et al.*, 2004; Schröder & Meesters, 2005), as ET from the anion, the only feasible reaction, is not a favoured •OH reaction anyway (see above) and is further hampered by the strong electron-withdrawing effects of fluorine atoms. SO$_4^{•-}$ is a better electron acceptor but even with this agent the rate of reaction is low and other reactions compete quite effectively (Hori *et al.*, 2005). The present state of technology in removing these surfactants has recently been reviewed (Vecitis *et al.*, 2009; Lutze *et al.*, 2011b).

References

Acero J. L. and von Gunten U. (2000). Influence of carbonate on the ozone/hydrogen peroxide based advanced oxidation process for drinking water treatment. *Ozone: Sci Eng*, **22**, 305–328.

Acero J. L. and von Gunten U. (2001). Characterization of oxidation processes: ozonation and the AOP O_3/H_2O_2. *J Am Water Works Ass*, **93**(10), 90–100.

Acero J. L., Stemmler K. and von Gunten U. (2000). Degradation kinetics of atrazine and its degradation products with ozone and OH radicals: a predictive tool for drinking water treatment. *Environ Sci Technol*, **34**, 591–597.

Acero J. L., Haderlein S. B., Schmidt T. C. and von Gunten U. (2001). MTBE oxidation by conventional ozonation and the combination ozone/hydrogen peroxide: efficiency of the processes and bromate formation. *Environ Sci Technol*, **35**, 1016–1024.

Acero J. L., Piriou P. and von Gunten U. (2005). Kinetics and mechanisms of formation of bromophenols during drinking water chlorination: assessment of taste and odor development. *Wat Res*, **39**, 2979–2993.

Adamic K., Howard J. A. and Ingold K. U. (1969). Absolute rate constants for hydrocarbon autoxidation. XVI. Reactions of peroxy radicals at low temperatures. *Can J Chem*, **47**, 3803–3808.

Aerni H.-R., Kobler B., Rutishauser B. V., Wettstein F. E., Fischer R., Giger W., Hungerbühler A., Marazuela M. D., Peter A., Schönenberger R., Vögeli A. C., Suter M. J. F. and Eggen R. I. L. (2004). Combined biological and chemical assessment of estrogenic activities in wastewater treatment plant effluents. *Anal Bioanal Chem*, **378**, 688–696.

Ahel M., Giger W. and Koch M. (1994). Behaviour of alkylphenol polyethoxylate surfactants in the aquatic environment -I. Occurrence and transformation in sewage treatment. *Wat Res*, **28**, 1131–1142.

Akhlaq M. S., Schuchmann H.-P. and von Sonntag C. (1990). Degradation of the polysaccharide alginic acid: a comparison of the effects of UV light and ozone. *Environ Sci Technol*, **24**, 379–383.

Alexandrov Yu A., Tarunin B. I. and Shushunov V. A. (1971). Kinetics of reactions between ozone and tetraethyl compounds of silicon, tin and lead. *Kinet Catal (Engl Transl)*, **12**, 802–805.

Alfassi Z. B., Huie R. E., Neta P. and Shoute L. C. T. (1990). Temperature dependence of the rate constants for reaction of inorganic radicals with organic reductants. *J Phys Chem*, **94**, 8800–8805.

Allen A. O., Hochanadel C. J., Ghormley J. A. and Davis T. W. (1952). Decomposition of water and aqueous solutions under mixed fast neutron and gamma radiation. *J Phys Chem*, **56**, 575–586.

Allpike B. P., Heitz A., Joll C. A., Kagi R. I., Abbt-Braun G., Frimmel F. H., Brinkmann T., Her N. and Amy G. (2005). Size exclusion chromatography to characterize DOC removal in drinking water treatment. *Environ Sci Technol*, **39**, 2334–2342.

Al-Sheikhly M. I., Hissung A., Schuchmann H.-P., Schuchmann M. N., von Sonntag C., Garner A. and Scholes G. (1984). Radiolysis of dihydrouracil and dihydrothymine in aqueous solutions containing oxygen; first- and second-order reactions of the organic peroxyl radicals; the role of isopyrimidines as intermediates. *J Chem Soc, Perkin Trans 2*, 601–608.

Alum A., Yoon Y., Westerhoff P. and Abbaszadegan M. (2004). Oxidation of bisphenol A, 17β-estradiol, and 17α-ethinyl estradiol and byproduct estrogenicity. *Environ Toxicol*, **19**, 257–264.

Amichai O. and Treinin A. (1969). Chemical reactivity of O(^{3}P) atoms in aqueous solution. *Chem Phys Lett*, **3**, 611–613.

Anbar M., Meyerstein D. and Neta P. (1966). The reactivity of aromatic compounds toward hydroxyl radicals. *J Phys Chem*, **70**, 2660–2662.

Andersen H., Siegrist H., Halling-Sörensen B. and Ternes T. (2003). Fate of estrogens in a municipal sewage treatment plant. *Environ Sci Technol*, **37**, 4021–4026.

Andreozzi R. and Marotta R. (1999). Ozonation of *p*-chlorophenol in aqueous solution. *J Hazard Mater*, **B69**, 303–317.

Andreozzi R., Insola A., Caprio V. and D'Amore M. G. (1991). Ozonation of pyridine in aqueous solution. Mechanistic and kinetic aspects. *Wat Res*, **25**, 655–659.

Andreozzi R., Insola A., Caprio V. and D'Amore M. G. (1992). Quinoline ozonation in aqueous solution. *Wat Res*, **26**, 639–643.

Andreozzi R., Caprio V., Insola A. and Tufano V. (1996). Measuring ozonation rate constants in gas-liquid reactors under the kinetic-diffusional transition regime. *Chem Eng Comn*, **143**, 195–204.

Andreozzi R., Caprio V., Marotta R. and Vogna D. (2003). Paracetamol oxidation from aqueous solution by means of ozonation and H$_2$O$_2$/UV system. *Wat Res*, **37**, 993–1004.

Andreozzi R., Canterino M., Marotta R. and Paxeus N. (2005). Antibiotic removal from wastewaters: the ozonation of amoxicillin. *J Hazard Mater*, **122**, 243–250.

Andresen J. and Bester K. (2006). Elimination of organophosphate ester flame retardants and plasticizers in drinking water purification. *Wat Res*, **40**, 621–629.

Andresen J. A., Grundmann A. and Bester K. (2004). Organophosphorus flame retardants and plasticisers in surface waters. *Sci Total Environ*, **332**, 155–166.

Andrzejewski P., Kasprzyk-Hordern B. and Nawrocki J. (2008). N-nitrosodimethylamine (NDMA) formation during ozonation of dimethylamine-containing waters. *Wat Res*, **42**, 863–870.

Anglada J. M., Crehuet R. and Bofill J. M. (1999). The ozonolysis of ethylene: a theoretical study of the gas-phase reaction mechanism. *Chem Eur J*, **5**, 1809–1822.

Armstrong D. A. (1999). Thermochemistry of sulfur radicals. In: *S-Centered Radicals*, Z. B. Alfassi (ed.), Wiley, New York, pp. 27–61.

Asami M., Aizawa T., Morioka T., Nishijima W., Tabata A. and Magara Y. (1999). Bromate removal during transition from new granular activated carbon (GAC) to biological activated carbon (BAC). *Wat Res*, **33**, 2797–2804.

Asano T., Burton F. L., Leverenz H. L., Tsuchihashi R. and Tchobanoglous G. (2007). *Water Reuse: Issues, Technologies, and Applications*. McGraw-Hill, New York.

Ashton L., Buxton G. V. and Stuart C. R. (1995). Temperature dependence of the rate of reaction of OH with some aromatic compounds in aqueous solution. *J Chem Soc, Faraday Trans*, **91**, 1631–1633.

Asmus K.-D. (1979). Stabilization of oxidized sulfur centers in organic sulfides. Radical cations and odd-electron sulfur–sulfur bonds. *Acc Chem Res*, **12**, 436–442.

Asmus K.-D., Möckel H. and Henglein A. (1973). Pulse radiolytic study of the site of OH$^\cdot$ radical attack on aliphatic alcohols in aqueous solution. *J Phys Chem*, **77**, 1218–1221.

Asmus K.-D., Bonifacic M., Toffel P., O'Neill P., Schulte-Frohlinde D. and Steenken S. (1978). On the hydrolysis of Ag^{II}, Tl^{II}, Sn^{III} and Cu^{III}. *J Chem Soc, Faraday Trans 1*, **74**, 1820–1826.

Assalin M. R., de Moraes S. G., Queiroz S. N. C., Ferracini V. L. and Duran N. (2010). Studies on degradation of glyphosate by several oxidative chemical processes: ozonation, photolysis and heterogeneous photocatalysis. *J Environ Health B*, **45**, 89–94.

Babbs C. F. and Steiner M. G. (1990). Detection and quantitation of hydroxyl radical using dimethyl sulfoxide as a molecular probe. *Meth Enzymol*, **186**, 137–147.

Babuna F. G., Camur S., Alaton I. A., Okay O. and Iskender G. (2009). The application of ozonation for the detoxification and biodegradability improvement of a textile auxiliary: naphthalene sulfonic acid. *Desalination*, **249**, 682–686.

References

Bader H. and Hoigné J. (1981). Determination of ozone in water by the indigo method. *Wat Res*, **15**, 449–456.

Bader H. and Hoigné J. (1982). Determination of ozone in water by the indigo method: a submitted standard method. *Ozone: Sci Eng*, **4**, 169–176.

Bader H., Sturzenegger V. and Hoigné J. (1988). Photometric method for the determination of low concentrations of hydrogen peroxide by the peroxidase catalysed oxidation of *N,N*-diethyl-*p*-phenylenediamine (DPD). *Wat Res*, **22**, 1109–1115.

Bagno A., Boso R. L., Ferrari N. and Scorrano G. (1995). Steric effects on the solvation of protonated di-*tert*-butyl ketone an phenyl *tert*-butyl ketone. *J Chem Soc, Chem Commun*, 2053–2054.

Bahnemann D. and Hart E. J. (1982). Rate constants of the reaction of the hydrated electron and hydroxyl radical with ozone in aqueous solution. *J Phys Chem*, **86**, 252–255.

Bahr C., Schumacher J., Ernst M., Luck F., Heinzmann B. and Jekel M. (2007). SUVA as control parameter for the effective ozonation of organic pollutants in secondary effluent. *Wat Sci Tech*, **55**, 267–274.

Bailey P. S. (1972). Organic groupings reactive toward ozone mechanisms in aqueous media. In: *Ozone in Water and Wastewater Treatment*, F. L. Evans (ed.), Ann Arbor Science, Ann Arbor, Michigan, pp. 29–59.

Bailey P. S. (1978). *Ozonation in Organic Chemistry*. Vol. I. Olefinic Compounds. Academic Press, New York.

Bailey P. S. (1982). *Ozonation in Organic Chemistry*. Vol. 2. Nonolefinic Compounds. Academic Press, New York.

Bailey S. M., Fauconnet A.-L. and Reinke L. A. (1997). Comparison of salicylate and *d*-phenylalanine for detection of hydroxyl radicals in chemical and biological reactions. *Redox Report*, **3**, 17–22.

Balci B., Oturan N., Cherrier R. and Oturan M. A. (2009). Degradation of atrazine in aqueous medium by electrolytically generated hydroxyl radicals. A kinetic and mechanistic study. *Wat Res*, **43**, 1924–1934.

Banker R., Carmeli S., Werman M., Teltsch B., Porat R. and Sukenik A. (2001). Uracil moiety is required for toxicity of the cyanobacterial hepatotoxin cylindrospermopsin. *J Toxicol Environm Health A*, **62**, 281–288.

Bao M. L., Griffini O., Santianni D., Barbieri K., Burrini D. and Pantani F. (1999). Removal of bromate ion from water using granular activated carbon. *Wat Res*, **33**, 2959–2970.

Baral S., Lume-Pereira C., Janata E. and Henglein A. (1985). Chemistry of colloidal manganese dioxide. 2. Reaction with O_2^- and H_2O_2. (Pulse radiolysis and stop flow studies.). *J Phys Chem*, **89**, 5779–5783.

Baral S., Lume-Pereira C., Janata E. and Henglein A. (1986). Chemistry of colloidal manganese oxides. 3. Formation in the reaction of hydroxyl radicals with Mn^{2+} ions. *J Phys Chem*, **90**, 6025–6028.

Barrett J., Mansell A. L. and Ratcliffe R. J. M. (1968). Cage effects in the photolysis of hydrogen peroxide in alcohol-water mixtures. *Chem Commun*, 48–49.

Barron E., Deborde M., Rabouan M., Mazellier P. and Legube B. (2006). Kinetic and mechanistic investigations of progesterone reaction with ozone. *Wat Res*, **40**, 2181–2189.

Bartlett P. D. and Guaraldi G. (1967). Di-*t*-butyl trioxide and di-*t*-butyl tetroxide. *J Am Chem Soc*, **89**, 4799–4801.

Bauer D., D'Ottone L. and Hynes A. J. (2000). O^1D quantum yields from O_3 photolysis in the near UV region between 305 and 375 nm. *Phys Chem Chem Phys*, **2**, 1421–1424.

Baus C., Hung H. W., Sacher F., Fleig M. and Brauch H.-J. (2005). MTBE in drinking water production – occurrence and efficiency of treatment technologies. *Acta Hydrochim Hydrobiol*, **33**, 118–132.

Beckwith R. C., Wang T. X. and Margerum D. W. (1996). Equilibrium and kinetics of bromine hydrolysis. *Inorg Chem*, **35**, 995–1000.

Bell R. P. (1966). The reversible hydration of carbonyl compounds. In: *Advances in Physical and Organic Chemistry*, V. Gold (ed.), Academic Press, London, pp. 1–28.

Beltrán F. J. (2004). *Ozone Reaction Kinetics for Water and Wastewater Systems*. Lewis Publishers, Boca Raton.

Beltrán F. J., González M., Rivas J. and Marin M. (1994). Oxidation of mecoprop in water with ozone and ozone combined with hydrogen peroxide. *Ind Eng Chem Res*, **33**, 125–136.

Beltrán F. J., Encinar J. M. and Alonso M. A. (1998). Nitroaromatic hydrocarbon ozonation in water: 1: single ozonation. *Ind Eng Chem Res*, **37**, 25–31.

Beltrán F. J., Rodríguez E. M. and Romero M. T. (2006). Kinetics of the ozonation of muconic acid in water. *J Hazard Mater B*, **138**, 534–538.

Beltrán F. J., Pocostales P., Alvarez P. and Aguinaco A. (2009). Ozone-activated carbon mineralization of 17-α-ethinylestradiol aqueous solution. *Ozone: Sci Eng*, **31**, 422–427.

Benbelkacem H., Mathé S. and Debellefontaine H. (2004). Taking mass transfer limitation into account during ozonation of pollutants reacting fairly quickly. *Wat Sci Tech*, **49**, 25–30.

Benitez F. J., Real F. J., Acero J. L. and Garcia C. (2007). Kinetics of the transformation of phenyl-urea herbicides during ozonation of natural waters: rate constants and model predictions. *Wat Res*, **41**, 4073–4084.

Benn R., Dreeskamp H., Schuchmann H.-P. and von Sonntag C. (1979). Photolysis of aromatic ethers. *Z Naturforsch*, **34b**, 1002–1009.

Benner J. and Ternes T. A. (2009a). Ozonation of metoprolol: elucidation of oxidation pathways and major oxidation products. *Environ Sci Technol*, **43**, 5472–5480.

Benner J. and Ternes T. A. (2009b). Ozonation of propranolol: formation of oxidation products. *Environ Sci Technol*, **43**, 5086–5093.

Benner J., Salhi E., Ternes T. and von Gunten U. (2008). Ozonation of reverse osmosis concentrate: kinetics and efficiency of beta blocker oxidation. *Wat Res*, **42**, 3003–3012.

Bennett D. A., Yao H. and Richardson D. E. (2001). Mechanism of sulfide oxidation by peroxomonocarbonate. *Inorg Chem*, **40**, 2996–3001.

Bennett J. E. and Summers R. (1974). Product studies of the mutual termination reactions of sec-alkylperoxy radicals: evidence for non-cyclic termination. *Can J Chem*, **52**, 1377–1379.

Bennett J. E., Brown D. M. and Mile B. (1970). Studies by electron spin resonance of the reactions of alkylperoxy radicals. Part 2. Equilibrium between alkylperoxy radicals and tetroxide molecules. *Trans Faraday Soc*, **66**, 397–405.

Benotti M. J., Trenholm R. A., Vanderford B. J., Holady J. C., Stanford B. D. and Snyder S. A. (2009). Pharmaceuticals and endocrine disrupting compounds in U.S. drinking water. *Environ Sci Technol*, **43**, 597–603.

Berdnikov V. M., Bazhin N. M., Fedorov V. K. and Polyakov O. V. (1972). Isomerization of the ethoxyl radical to the α-hydroxyethyl radical in aqueous solution. *Kinet Catal (Engl Transl)*, **13**, 986–987.

Berkowitz J., Ellison G. B. and Gutman D. (1994). Three methods to measure RH bond energies. *J Phys Chem*, **98**, 2744–2765.

Bichsel Y. and von Gunten U. (1999a). Determination of iodide and iodate by ion chromatography with postcolumn reaction and UV/Visible detection. *Anal Chem*, **71**, 34–38.

Bichsel Y. and von Gunten U. (1999b). Oxidation of iodide and hypoiodous acid in the disinfection of natural waters. *Environ Sci Technol*, **33**, 4040–4045.

Bichsel Y. and von Gunten U. (2000a). Formation of iodo-trihalomethanes during disinfection and oxidation of iodide-containing waters. *Environ Sci Technol*, **34**, 2784–2791.

Bichsel Y. and von Gunten U. (2000b). Hypoiodous acid: kinetics of the buffer-catalysed disproportionation. *Wat Res*, **34**, 3197–3203.

Biedenkapp D., Hartshorn L. G. and Bair E. J. (1970). The O(^{1}D)+H$_2$O reaction. *Chem Phys Lett*, **5**, 379–380.

Bielski B. H. J. (1970). Kinetics of deuterium sequioxide in heavy water. *J Phys Chem*, **74**, 3213–3216.

Bielski B. H. J. (1993). A pulse radiolysis study of the reaction of ozone with Cl$_2^{\cdot-}$ in aqueous solutions. *Radiat Phys Chem*, **41**, 527–530.

Bielski B. H. J. and Schwarz H. A. (1968). The absorption spectra and kinetics of hydrogen sesquioxide and the perhydroxyl radical. *J Phys Chem*, **72**, 3836–3841.

Bielski B. H. J., Cabelli D. E., Arudi R. L. and Ross A. B. (1985). Reactivity of HO$_2$/O$_2^-$ radicals in aqueous solution. *J Phys Chem Ref Data*, **14**, 1041–1100.

Blank B., Henne A., Laroff G. P. and Fischer H. (1975). Enol intermediates in photoreduction and type I cleavage reactions of aliphatic aldehydes and ketones. *Pure Appl Chem*, **41**, 475–494.

Blumberger J., Bernasconi L., Tavernelli I., Vuilleumier R. and Sprik M. (2004). Electronic structure and solvation of copper and silver ions: a theoretical picture of a model aqueous redox reaction. *J Am Chem Soc*, **126**, 3928–3938.

Böhme A. (1999). Ozone technology of German industrial enterprises. *Ozone: Sci Eng*, **21**, 163–176.

Bonefeld-Jörgensen E. C., Long M., Hofmeister M. V. and Vinggaard A. M. (2007). Endocrine-disrupting potential of bisphenol A., bisphenol A dimethacrylate, 4-*n*-nonylphenol, and 4-*n*-octylphenol *in vitro*: new data and a brief review. *Environ Health Perspect*, **115**(Suppl. 1), 69–76.

Bosshard F., Riedel K., Schneider T., Geiser C., Bucheli M. and Egli T. (2010). Protein oxidation and aggregation in UVA-irradiated *Escherichia coli* cells as signs of accelerated cellular senescence. *Environ Microbiol*, **12**, 2931–2945.

Bothe E. and Schulte-Frohlinde D. (1978). The bimolecular decay of the α-hydroxymethylperoxyl radicals in aqueous solution. *Z Naturforsch B*, **33**, 786–788.

Bothe E. and Schulte-Frohlinde D. (1980). Reaction of dihydroxymethyl radical with molecular oxygen in aqueous solutions. *Z Naturforsch B*, **35b**, 1035–1039.

Bothe E., Behrens G. and Schulte-Frohlinde D. (1977). Mechanism of the first order decay of 2-hydroxypropyl-2-peroxyl radicals and of O_2^- formation in aqueous solution. *Z Naturforsch B*, **32**, 886–889.

Bothe E., Schulte-Frohlinde D. and von Sonntag C. (1978). Radiation chemistry of carbohydrates. Part 16. Kinetics of HO_2 elimination from peroxyl radicals derived from glucose and polyhydric alcohols. *J Chem Soc, Perkin Trans 2*, 416–420.

Bothe E., Schuchmann M. N., Schulte-Frohlinde D. and von Sonntag C. (1983). Hydroxyl radical-induced oxidation of ethanol in oxygenated aqueous solutions. A pulse radiolysis and product study. *Z Naturforsch B*, **38**, 212–219.

Botzenhart K., Tarcson G. M. and Ostruschka M. (1993). Inactivation of bacteria and coliphages by ozone and chlorine dioxide in a continuous flow reactor. *Water Sci Tech*, **27**, 363–370.

Bouchard M. F., Sauve S., Barbeau B., Legrand M., Brodeur M. E., Bouffard T., Limoges E., Bellinger D. C. and Mergler D. (2011). Intellectual impairment in school-age children exposed to manganese from drinking water. *Environ Health Perspect*, **119**, 138–143.

Bowman R. H. (2005). Hipox advanced oxidation of TBA and MTBE in groundwater. In: *Contaminated Soils, Sediments and Water: Science in the Real World*, E. J. Calabrese, P. T. Kostecki and J. Dragun (eds), Vol. 9, Springer, New York, pp. 299–313.

Boyd A. W., Willis C. and Cyr R. (1970). New determination of stoichiometry of the iodometric method for ozone analysis at pH 7. *Anal Chem*, **42**, 670–672.

Braslavsky S. E. and Rubin M. B. (2011). The history of ozone VIII. Photochemical formation of ozone. *Photochem Photobiol Sci*, **10**, 1515–1520.

Brezonik P. N. and Fulkerson-Brekken J. (1988). Nitrate-induced photolysis in natural waters: controls on concentrations of hydroxyl radical photo-intermediates by natural scavenging agents. *Environ Sci Technol*, **32**, 3004–3010.

Broadwater W. T., Hoehn R. C. and King P. H. (1973). Sensitivity of three selected bacterial species to ozone. *Appl Microbiol*, **26**, 391–393.

Broséus R., Vincent S., Aboufadl K., Daneshvar A., Sauvé S., Barbeau B. and Prévost M. (2009). Ozone oxidation of pharmaceuticals, endocrine disruptors and pesticides during drinking water treatment. *Wat Res*, **43**, 4707–4717.

Bruchet A. and Duguet J. P. (2004). Role of oxidants and disinfectants on the removal, masking and generation of tastes and odours. *Wat Sci Tech*, **49**, 297–306.

Brusa M. A., Churio M. S., Grela M. A., Bertolotti S. G. and Previtali C. M. (2000). Reaction volume and reaction enthalpy upon aqueous peroxodisulfate dissociation: $S_2O_8^{2-} \rightarrow 2\ SO_4^{\bullet-}$. *Phys Chem Chem Phys*, **2**, 2383–2387.

Bucher G. and Scaiano J. C. (1994). Laser flash photolysis of pyridine N-oxide: kinetic studies of atomic oxygen [O(3P)] in solution. *J Phys Chem*, **98**, 12471–12473.

Buffle M.-O. and von Gunten U. (2006). Phenols and amine induced HO$^\bullet$ generation during the initial phase of natural water ozonation. *Environ Sci Technol*, **40**, 3057–3063.

Buffle M.-O., Galli S. and von Gunten U. (2004). Enhanced bromate control during ozonation: the chlorine-ammonia process. *Environ Sci Technol*, **38**, 5187–5195.

Buffle M.-O., Schumacher J., Meylan S., Jekel M. and von Gunten U. (2006a). Ozonation and advanced oxidation of wastewater: effect of O_3 dose, pH, DOM and $^\bullet$OH-scavengers on ozone decomposition and HO$^\bullet$ generation. *Ozone: Sci Eng*, **28**, 247–259.

Buffle M.-O., Schumacher J., Salhi E., Jekel M. and von Gunten U. (2006b). Measurement of the initial phase of ozone decomposition in water and wastewater by means of a continuous quench-flow system: application to disinfection and pharmaceutical oxidation. *Wat Res*, **40**, 1884–1894.

Bühler R. E., Staehelin J. and Hoigné J. (1984). Ozone decomposition in water studied by pulse radiolysis. 1. HO_2/O_2^- and HO_3/O_3^- as intermediates. *J Phys Chem*, **88**, 2560–2564.

Bünning G. and Hempel D. C. (1999). Anlagenkonzeption und Inaktivierungskinetik zur Wasserdesinfektion mittels Ozon (Setup of equipment and inactivation kinetics for disinfection in aqueous solutions with ozone). *GWF Wasser-Abwasser*, **140**, 173–181.

Burger J. D. and Liebhafsky H. A. (1973). Thermodynamic data for aqueous iodine solutions at various temperatures – an exercise in analytical chemistry. *Anal Chem*, **45**, 600–602.

Bürgi H., Schaffner T. and Seiler J. P. (2001). The toxicology of iodate: a review of the literature. *Thyroid*, **11**, 449–455.

Busetti F., Linge K. L. and Heitz A. (2009). Analysis of pharmaceuticals in indirect potable reuse systems using solid-phase extraction and liquid chromatography-tandem mass spectrometry. *J Chromatogr A*, **1216**, 5807–5818.

Butkovic V., Klasinc L., Orhanović M., Turk J. and Güsten H. (1983). Reaction rates of polynuclear aromatic hydrocarbons with ozone in water. *Environ Sci Technol*, **17**, 546–548.

Buxton G. V. and Dainton F. (1968). The radiolysis of aqueous solutions of oxybromine compounds; the spectra and reactions of BrO and BrO_2. *Proc R Soc A*, **304**, 427–439.

Buxton G. V. and Subhani M. S. (1972). Radiation chemistry and photochemistry of oxychlorine ions. Part 1. Radiolysis of aqueous solutions of hypochlorite and chlorite ions. *J Chem Soc, Faraday Trans 1*, **68**, 947–957.

Buxton G. V., Langan J. R. and Lindsay Smith J. R. (1986). Aromatic hydroxylation. 8. A radiation chemical study of the oxidation of hydroxycyclohexadienyl radicals. *J Phys Chem*, **90**, 6309–6313.

Buxton G. V., Greenstock C. L., Helman W. P. and Ross A. B. (1988). Critical review of rate constants for reactions of hydrated electrons, hydrogen atoms and hydroxyl radicals (OH/O^-) in aqueous solution. *J Phys Chem Ref Data*, **17**, 513–886.

Buxton G. V., Bydder M. and Salmon G. A. (1998). Reactivity of chlorine atoms in aqueous solution. Part 1: the equilibrium $Cl^{\bullet}+Cl^-=Cl_2^{\bullet-}$. *J Chem Soc, Faraday Trans*, **94**, 653–657.

Buxton G. V., Bydder M. and Salmon G. A. (1999). The reactivity of chlorine atoms in aqueous solution Part II. The equilibrium $SO_4^{\bullet-}+Cl^-=Cl^{\bullet}+SO_4^{2-}$. *Phys Chem Chem Phys*, **1**, 269–273.

Buxton G. V., Bydder M., Salmon G. A. and Williams J. E. (2000). The reactivity of chlorine atoms in aqueous solution Part III. The reactions of $Cl^{\bullet}$ with solutes. *Phys Chem Chem Phys*, **2**, 237–245.

Cain W. S., Schmidt R. and Wolkoff P. (2007). Olfactory detection of ozone and D-limonene: reactants in indoor spaces. *Indoor Air*, **17**, 337–347.

Callender T. and Davis L. C. (2002). Nitrification inhibition using benzotriazoles. *J Hazard Substance Res*, **4**, 2-1–2-16.

Cancilla D. A., Baird C., Geis S. W. and Cossi S. R. (2003). Studies of the environmental fate and effect of aircraft deicing fluids: detection of 5-methyl-1H-benzotriazole in the fathead minnow (*Pimephales promelas*). *Environ Toxicol Chem*, **22**, 134–140.

Canonica S., Kohn T., Mac M., Real F. J., Wirz J. and von Gunten U. (2005). Photosensitizer method to determine rate constants for the reaction of carbonate radical with organic compounds. *Environ Sci Technol*, **39**, 9182–9188.

Cao N., Yang M., Zhang Y., Hu J., Ike M., Hirotsuji J., Matsui H., Inoue D. and Sei K. (2009). Evaluation of wastewater reclamation technologies based on in vitro and in vivo bioassays. *Sci Total Environ*, **407**, 1588–1597.

Caprio V. and Insola A. (1985). Aniline and anilinium ion ozonation in aqueous solution. *Ozone: Sci Eng*, **7**, 169–179.

Carell B. and Olin A. (1960). Studies on the hydrolysis of metal ions. 31. The complex formation between Pb^{2+} and OH^- in Na^+ (OH^-, ClO_4^-) medium. *Acta Chem Scand*, **14**, 1999–2008.

Carmichael W. W., Briggs D. D. and Peterson M. A. (1979). Pharmacology of Anatoxin-a, produced by the freshwater cyanophyte *Anabena Flos-Aquae* NRC-44-1. *Toxicology*, **17**, 229–236.

Cataldo F. (2008). Ozone decomposition of patulin – a mycotoxin and food contaminant. *Ozone: Sci Eng*, **30**, 197–201.

Cerkovnik J. and Plesnicar B. (1993). Characterization and reactivity of hydrogen trioxide (HOOOH): a reactive intermediate formed in the low-temperature ozonation of 3-ethylanthrahydroquinone. *J Am Chem Soc*, **115**, 12169–12170.

Chaiket T., Singer P. C., Miles A., Moran M. and Palotta C. (2002). Effectiveness of coagulation, ozonation, and biofiltration in controlling DBPs. *J Am Water Works Ass*, **94**(12), 81–95.

Chang C.-N., Ma J.-S. and Zing F.-F. (2002). Reducing the formation of disinfection by-products by pre-ozonation. *Chemosphere*, **46**, 21–30.

Chau Y. K., Maguire R. C., Brown M., Yang F. and Batchelor S. P. (1997). Occurrence of organotin compounds in the Canadian aquatic environment five years after the regulation of antifouling uses of tributyltin. *Wat Qual Res J Canada*, **32**, 453–521.

Chelkowska K., Grasso D., Fábián I. and Gordon G. (1992). Numerical simulations of aqueous ozone decomposition. *Ozone: Sci Eng*, **14**, 33–49.

Chen W. R., Wu C., Elovitz M. S., Linden K. G. and Suffet I. H. (2008). Reactions of thiocarbamate, triazine and urea herbicides, RDX and benzenes on EPA Contaminant Candidate List with ozone and with hydroxyl radicals. *Wat Res*, **42**, 137–144.

Chertova Y. S., Avzyanova E. V., Timerganzin K. K., Khalizov A. F., Sherreshovets V. V. and Imashev U. B. (2000). The formation of singlet molecular oxygen in the interaction of chlorine dioxide with ozone. *Russ J Phys Chem*, **74** (Suppl. 3), S473–S475.

Chia Y. (1958). Chemistry of +1 iodine in alkaline solution. Dissertation, University of California, Berkeley.

Chin A. and Bérubé P. R. (2005). Removal of disinfection by-product precursors with ozone-UV advanced oxidation process. *Wat Res*, **39**, 2136–2144.

Cho M., Chung H. and Yoon J. (2002). Effect of pH and importance of ozone initiated radical reactions in inactivating *Bacillus subtilis* spores. *Ozone: Sci Eng*, **24**, 145–150.

Choi J. and Valentine R. S. (2002). Formation of *N*-nitrosodimethylamine (NDMA) from reaction of monochloramine: a new disinfection by-product. *Wat Res*, **36**, 817–824.

Christensen H. C., Sehested K. and Hart E. J. (1973). Formation of benzyl radicals by pulse radiolysis of toluene in aqueous solutions. *J Phys Chem*, **77**, 983–987.

Chuang T. J., Hoffman G. W. and Eisenthal K. B. (1974). Picosecond studies of the cage effect and collision induced predissociation of iodine in liquids. *Chem Phys Lett*, **25**, 201–205.

Collins A. R., Dobson V. L., Dusinska M., Kennedy G. and Stetina R. (1997). The comet assay: what can it really tell us? *Mutation Res*, **375**, 183–193.

Cooper W. J., Cramer C. J., Martin N. H., Mezyk S. P., O'Shea K. E. and von Sonntag C. (2009). Free radical mechanism for the treatment of methyl *tert*-butyl ether (MTBE) *via* advanced oxidation/reductive processes in aqueous solution. *Chem Rev*, **109**, 1302–1345.

Cotruvo J. A., Keith J. D., Bull R. J., Pacey G. E. and Gordon G. (2010). Bromate reduction in simulated gastric juice. *J Am Water Works Ass*, **102**(11), 77–86.

Coudray C., Talla M., Martin S., Fatome M. and Favier A. (1995). HPLC-electrochemical determination of salicylate hydroxylation products as an in vivo marker of oxidative stress. *Anal Biochem*, **227**, 101–111.

Criegee R. (1975). Mechanismus der Ozonolyse (Mechanisms of the ozonolysis). *Angew Chem*, **87**, 765–771.

Czapski G. and Bielski B. H. J. (1963). The formation and decay of H_2O_3 and HO_2 in electron-irradiated aqueous solutions. *J Phys Chem*, **67**, 2180–2184.

Czapski G., Lymar S. V. and Schwarz H. A. (1999). Acidity of the carbonate radical. *J Phys Chem A*, **103**, 3447–3450.

D'Alessandro N., Bianchi G., Fang X., Jin F., Schuchmann H.-P. and von Sonntag C. (2000). Reaction of superoxide with phenoxyl-type radicals. *J Chem Soc, Perkin Trans 2*, 1862–1867.

Dahi E. (1976). Physicochemical aspects of disinfection of water by means of ultrasound and ozone. *Wat Res*, **10**, 677–684.

Dalmagro J., Yunes R. A. and Simionation E. L. (1994). Mechanism of reaction of azobenzene formation from aniline and nitrosobenzene in basic conditions. General base catalysis by hydroxide ion. *J Phys Org Chem*, **7**, 399–402.

Dantas R. F., Canterino M., Marotta R., Sans C., Esplugas S. and Andreozzi R. (2007). Bezafibrate removal by means of ozonation: primary intermediates, kinetics and toxicity assessment. *Wat Res*, **41**, 2525–2532.

Darbre P. D. and Harvey P. W. (2008). Paraben esters: review of recent studies of endocrine toxicity, absorption, esterase and human exposure, and discussion of potential human health risks. *J Appl Toxicol*, **28**, 561–578.

Das S. and von Sonntag C. (1986). Oxidation of trimethylamine by OH radicals in aqueous solution, as studied by pulse radiolysis, ESR and product analysis. The reactions of the alkylamine radical cation, the aminoalkyl radical and the protonated aminoalkyl radical. *Z Naturforsch B*, **41**, 505–513.

Das S., Schuchmann M. N., Schuchmann H.-P. and von Sonntag C. (1987). The production of the superoxide radical anion by the OH radical-induced oxidation of trimethylamine in oxygenated aqueous solution. The kinetics of the hydrolysis of (hydroxymethyl)dimethylamine. *Chem Ber*, **120**, 319–323.

Davi M. L. and Gnudi F. (1999). Phenolic compounds in surface water. *Wat Res*, **33**, 3213–3219.

Deborde M. and von Gunten U. (2008). Reactions of chlorine with inorganic and organic compounds during water treatment – kinetics and mechanisms: a critical review. *Wat Res*, **42**, 13–51.

Deborde M., Rabouan S., Duguet J.-P. and Legube B. (2005). Kinetics of aqueous ozone-induced oxidation of some endocrine disruptors. *Environ Sci Technol*, **39**, 6086–6092.

Deborde M., Rabouan M., Mazellier P., Duguet J.-P. and Legube B. (2008). Oxidation of bisphenol A by ozone in aqueous solution. *Wat Res*, **42**, 4299–4308.

de Laat J., Maouala-Makata P. and Dore M. (1996). Constantes cinetiques de reaction de l'ozone moleculaire et des radicaux hydroxyles sur quelques phenyl-urees et acetamides (Rate constants of ozone and hydroxyl radicals with several phenyl-ureas and acetamides). *Environ Technol*, **17**, 707–716.

Del Vecchio R. and Blough N. V. (2004). On the origin of the optical properties of humic substances. *Environ Sci Technol*, **38**, 3885–3891.

Dillon D., Combes R., McConville M. and Zeiger E. (1992). Ozone is mutagenic in *Salmonella*. *Environ Mol Mutagen*, **19**, 331–337.

Dixon W. T. and Murphy D. (1976). Determination of the acidity constants of some phenol radical cations by means of electron spin resonance. *J Chem Soc, Faraday Trans 2*, **72**, 1221–1230.

Dodd M. C. (2008). Characterization of ozone-based oxidative treatment as a means of eliminating target-specific biological activities of municipal wastewater-borne antibacterial compounds. Dissertation No 17934, ETH, Zürich.

Dodd M. C., Buffle M.-O. and von Gunten U. (2006a). Oxidation of antibacterial molecules by aqueous ozone: moiety-specific kinetics and application to ozone-based wastewater treatment. *Environ Sci Technol*, **40**, 1969–1977.

Dodd M. C., Vu N. D., Ammann A., Le V. C., Kissner R., Pham H. V., Cao T. H., Berg M. and von Gunten U. (2006b). Kinetics and mechanistic aspects of As(III) oxidation by aqueous chlorine, chloramines, and ozone: relevance for drinking water treatment. *Environ Sci Technol*, **40**, 3285–3292.

Dodd M. C., Zuleeg S., von Gunten U. and Pronk W. (2008). Ozonation of source-separated urine from resource recovery and waste mineralization: process modelling, reaction chemistry, and operational considerations. *Environ Sci Technol*, **42**, 9329–9337.

Dodd M. C., Kohler H.-P. and von Gunten U. (2009). Oxidation of antibacterial compounds by ozone and hydroxyl radical: elimination of biological activity during aqueous ozonation processes. *Environ Sci Technol*, **43**, 2498–2504.

Dodd M. C., Rentsch D., Singer H. P., Kohler H.-P. E. and von Gunten U. (2010). Transformation of β-lactam antibacterial agents during aqueous ozonation: reaction pathways and quantitative bioassay of biologically-active oxidation products. *Environ Sci Technol*, **44**, 5940–5948.

Do-Quang Z., Roustan M. and Duguet J.-P. (2000). Mathematical modeling of theoretical *Cryptosporidium* inactivation in full-scale ozonation reactors. *Ozone: Sci Eng*, **22**, 99–111.

Dore M., Langlais B. and Legube B. (1980). Mechanism of the reaction of ozone with soluble aromatic pollutants. *Ozone: Sci Eng*, **2**, 39–54.

Dowideit P. (1996). Der Abbau chlorierter Olefine und verwandter Verbindungen mit Ozon, Ozon/UV und Ozon/Wasserstoffperoxid in wässeriger Lösung (Degradation of chlorinated olefins and related compounds with ozone, ozone/UV and ozone/hydrogen peroxide in aqueous solution). Dissertation Ruhr-Universität, Bochum.

Dowideit P. and von Sonntag C. (1998). The reaction of ozone with ethene and its methyl- and chlorine-substituted derivatives in aqueous solution. *Environ Sci Technol*, **32**, 1112–1119.

Dowideit P., Mertens R. and von Sonntag C. (1996). The non-hydrolytic decay of formyl chloride into CO and HCl in aqueous solution. *J Am Chem Soc*, **118**, 11288–11292.

Driedger A., Staub E., Pinkernell U., Marinas B., Köster W. and von Gunten U. (2001). Inactivation of *Bacillus subtilis* spores and formation of bromate during ozonation. *Wat Res*, **35**, 2950–2960.

Driehaus W. (2002). Arsenic removal – experience with the GEH process in Germany. *Wat Sci Tech*, **2**, 275–280.

Dubeau H. and Chung Y. S. (1982). Genetic effects of ozone. Induction of point mutation and genetic recombination in *Saccharomyces cerevisiae*. *Mutation Res*, **102**, 249–259.

Duft M., Schulte-Oehlmann U., Tillmann M., Markert B. and Oehlmann J. (2003a). Toxicity of triphenyltin and tributyltin to the freshwater mudsnail *Potomopyrgus Antipodarum* in a new sediment biotest. *Environ Toxicol Chem*, **22**, 145–152.

Duft M., Schulte-Oehlmann U., Wetje L., Tillmann M. and Oehlmann J. (2003b). Stimulated embryo production as a parameter of estrogenic exposure via sediments in the freshwater mudsnail *Potamopyrgus antipodarum*. *Aquatic Toxicol*, **64**, 437–449.

Duft M., Schmitt C., Bachmann J., Brandelik C., Schulte-Oehlmann U. and Oehlmann J. (2007). Prosobranch snails as test organisms for the assessment of endocrine active chemicals – an overview and a guideline proposal for a reproduction test with the freshwater mudsnail *Potomopyrgus antipodarum*. *Ecotoxicology*, **16**, 169–182.

Eaton A. D., Clesceri L. S., Rice E. W., Greenberg A. E. and Franson M. A. H. (2005). *Standard Methods for the Examination of Water and Wastewater*. American Public Health Organization, Washington.

Eberson L. (1964). Studies on cyclic anhydrides I. Rate constants and activation energies for the solvolysis of alkylsubstituted cyclic anhydrides in aqueous solution by the pH-stat method. *Acta Chem Scand*, **18**, 534–542.

Eigen M. (1963). Protonenübertragung, Säure-Base-Katalyse und enzymatische Hydrolyse. Teil I: Elementarvorgänge (Proton transfer, acid-base catalysis and enzymatic hydrolysis. Part I: Elementary processes). *Angew Chem*, **75**, 489–508.

Eigen M. and Kustin K. (1962). The kinetics of halogen hydrolysis. *J Am Chem Soc*, **84**, 1355–1361.

Eliasson B., Hirth M. and Kogelschatz U. (1987). Ozone synthesis from oxygen in dielectric barrier discharges. *J Phys D: Appl Phys*, **20**, 1421–1437.

Elliot A. J. and McCracken D. R. (1989). Effect of temperature on $O^{\bullet-}$ reactions and equilibria: a pulse radiolysis study. *Radiat Phys Chem*, **33**, 69–74.

Ellis K. V. (1991). Water disinfection: a review with some consideration of the requirements of the third world. *Crit Rev Env Contr*, **20**, 341–407.

Elovitz M. S. and von Gunten U. (1999). Hydroxyl radical/ozone ratios during ozonation processes. I. The R_{ct} concept. *Ozone: Sci Eng*, **21**, 239–260.

Elovitz M. S., von Gunten U. and Kaiser H.-P. (2000a). Hydroxyl radical/ozone ratios during ozonation processes. II. The effect of temperature, pH, alkalinity, and DOM properties. *Ozone: Sci Eng*, **22**, 123–150.

Elovitz M. S., von Gunten U. and Kaiser H.-P. (2000b). The influence of dissolved organic matter character on ozone decomposition rates and R_{ct}. In: *Natural Organic Matter and Disinfection By-Products. Characterization and Control in Drinking Water*, S. E. Barrett, S. W. Krasner and G. L. Amy (eds), American Chemical Society, Washington, DC, pp. 248–269.

Emsenhuber M., Pöchlauer P., Aubry J. M., Nardello V. and Falk H. (2003). Evidence for the generation of singlet oxygen ($^1O_2$, $^1\Delta_g$) from ozone promoted by inorgnic salts. *Monatsh Chem*, **134**, 387–391.

Engdahl A. and Nelander B. (2002). The vibrational spectrum of H_2O_3. *Science*, **295**, 482–483.

Erben-Russ M., Michel C., Bors W. and Saran M. (1987). Absolute rate constants of alkoxyl radical reactions in aqueous solution. *J Phys Chem*, **91**, 2362–2365.

Erickson R. E., Yates L. M., Clark R. L. and McEwen D. (1977). The reaction of sulfur dioxide with ozone in water and its possible atmospheric significance. *Atmos Environ*, **11**, 813–817.

Ershov B. G. (2004). Kinetics, mechanism and intermediates of some radiation-induced reactions in aqueous solutions. *Russ Chem Rev*, **73**, 101–113.

Ershov B. G., Kelm M., Gordeev A. V. and Janata E. (2002). A pulse radiolysis study of the oxidation of Br^- by $Cl_2^{\bullet-}$ in aqueous solution: formation and properties of $ClBr^{\bullet-}$. *Phys Chem Chem Phys*, **4**, 1872–1875.

Escher B. I. and Fenner K. (2011). Recent advances in environmental risk assessment of transformation products. *Environ Sci Technol*, **45**, 3835–3847.

Escher B. I., Bramaz N. and Ort C. (2009). Monitoring the treatment efficiency of a full scale ozonation on a sewage treatment plant with a mode-of-action based test battery. *J Environ Monit*, **11**, 1836–1846.

Escher B. I., Baumgartner R., Koller M., Treyer K., Lienert J. and McArdell C. S. (2011). Environmental toxicology and risk assessment of pharmaceuticals from hospital wastewater. *Wat Res*, **45**, 75–92.

European Commission (2000). Towards the establishment of a priority list of substances for further evaluation of their role in endocrinic disruption, Final Report, European Commission, Delft.

Evans F. L. (1972). *Ozone in Water and Wastewater Treatment*. Ann Arbor Sciences, Ann Arbor.
Facile N., Barbeau B., Prévost M. and Koudjonou B. (2000). Evaluating bacterial aerobic spores as a surrogate for *Giardia* and *Cryptosporidium* inactivation by ozone. *Wat Res*, **34**, 3238–3246.
Fahlenkamp H., Nöthe T., Ries T. and Peulen C. (2004). Untersuchungen zum Eintrag und zur Eliminierung von gefährlichen Stoffen in kommunalen Kläranlagen (Investigations on the input and elimination of hazardous compounds in municipal wastewater treatment plants). Rastatt, Greiserdruck.
Fang X., Pan X., Rahmann A., Schuchmann H.-P. and von Sonntag C. (1995). Reversibility in the reaction of cyclohexadienyl radicals with oxygen in aqueous solution. *Chem Eur J*, **1**, 423–429.
Fang X., Mark G. and von Sonntag C. (1996). OH-Radical formation by ultrasound in aqueous solutions – Part I: the chemistry underlying the terephthalate dosimeter. *Ultrasonics Sonochem*, **3**, 57–63.
Fang X., Schuchmann H.-P. and von Sonntag C. (2000). The reaction of the OH radical with pentafluoro-, pentachloro-, pentabromo- and 2,4,6-triiodophenol in water: electron transfer *vs.* addition to the ring. *J Chem Soc, Perkin Trans 2*, 1391–1398.
Fenton H. J. H. (1894). LXXIII.–Oxidation of tartaric acid in presence of iron. *J Chem Soc Transact (London)*, **65**, 899–910.
Fenton H. J. H. and Jackson H. (1899). The oxidation of polyhydric alcohols in the presence of iron. *J Chem Soc Transact (London)*, **75**, 1–11.
Fernández L. A., Hernández C., Bataller M., Véliz E., López A., Ledea O. and Padrón S. (2010). Cyclophosphamide degradation by advanced oxidation processes. *Wat Environ J*, **24**, 174–180.
Fick J., Lindberg R. H., Tysklind M., Haemig P. D., Waldenström J., Wallensten A. and Olsen B. (2007). Antiviral oseltamivir is not removed of degraded in normal sewage water treatment: implication for development of resistance by influenza A virus. *PLoS ONE*, **2**, e986.
Field R. J. and Försterling H.-D. (1986). On the oxybromine chemistry rate constants with cerium ions on the Field-Körös-Noyes mechanism of the Belusov-Zhabotinksii reaction: the equilibrium $HBrO_2 + BrO_3^- + H^+ = 2 BrO_2^{\bullet} + H_2O$. *J Phys Chem*, **90**, 5400–5407.
Field R. J., Körös E. and Noyes R. M. (1972). Oscillations in chemical systems. II. Thorough analysis of temporal oscillation in the bromate-cerium-malonic acid system. *J Am Chem Soc*, **94**, 8649–8664.
Finch G. R. and Fairbairn N. (1991). Comparative inactivation of poliovirus type 3 and MS2 coliphage in demand-free phosphate buffer by using ozone. *Appl Environ Microbiol*, **57**, 3121–3126.
Finch G. R., Smith D. W. and Stiles M. E. (1988). Dose response of *Escherichia coli* in ozone demand-free phosphate buffer. *Wat Res*, **22**, 1563–1570.
Finch G. R., Black E. K., Labatiuk C. W., Gyürék L. and Belosevic M. (1993). Comparison of *Giardia lamblia* and *Giardia muris* cysts inactivation by ozone. *Appl Environ Microbiol*, **59**, 3674–3680.
Fleming I. (2002). *Frontier Orbitals and Organic Chemical Reactions*. Wiley, Chichester.
Flyunt R., Makogon O., Schuchmann M. N., Asmus K.-D. and von Sonntag C. (2001a). The OH-radical-induced oxidation of methanesulfinic acid. The reactions of the methylsulfonyl radical in the absence and presence of dioxygen. *J Chem Soc, Perkin Trans 2*, 787–792.
Flyunt R., Schuchmann M. N. and von Sonntag C. (2001b). A common carbanion intermediate in the recombination and proton-catalysed disproportionation of the carboxyl radical anion, $CO_2^{\bullet-}$, in aqueous solution. *Chem Eur J*, **7**, 796–799.
Flyunt R., Theruvathu J. A., Leitzke A. and von Sonntag C. (2002). The reaction of thymine and thymidine with ozone. *J Chem Soc, Perkin Trans 2*, 1572–1582.
Flyunt R., Leitzke A., Mark G., Mvula E., Reisz E., Schick R. and von Sonntag C. (2003a). Determination of $^{\bullet}OH$ and $O_2^{\bullet-}$, and hydroperoxide yields in ozone reactions in aqueous solutions. *J Phys Chem B*, **107**, 7242–7253.
Flyunt R., Leitzke A. and von Sonntag C. (2003b). Characterisation and quantitative determination of (hydro)peroxides formed in the radiolysis of dioxygen-containing systems and upon ozonolysis. *Radiat Phys Chem*, **67**, 469–473.
Forni L., Bahnemann D. and Hart E. J. (1982). Mechanism of the hydroxide ion initiated decomposition of ozone in aqueous solution. *J Phys Chem*, **86**, 255–259.

Fuchs F. (1985a). Gelchromatographische Trennung von organischen Wasserinhaltsstoffen. Teil 1: Durchführung der Messungen (Separation of organic substances in water by gel chromatography. Part I: Execution of the measurements). *Vom Wasser*, **64**, 129–144.

Fuchs F. (1985b). Gelchromatographische Trennung von organischen Wasserinhaltsstoffen. Teil II: Ergebnisse der Trennungen bei Oberflächenwässern und Abwässern (Separation of organic substances by gel chromatography. Part II: results of the separations with surface waters and sewage waters). *Vom Wasser*, **65**, 93–105.

Fuchs F. (1985c). Gelchromatographische Trennung von organischen Wasserinhaltsstoffen. Teil III: Untersuchungen zu den Wechselwirkungen zwischen Gelmatrix, Probesubstanz und Elutionsmittel (Separation of organic substances in water by gel chromatography. Part III: investigations with regard to the interactions between gelmatrix, solute and elution liquid). *Vom Wasser*, **65**, 129–136.

Fujii T., Yashiro M. and Tokiwa H. (1997). Proton and Li$^+$ cation interaction with H_2O_3 and H_2O/O_2: *Ab initio* molecular orbital study. *J Am Chem Soc*, **119**, 12280–12284.

Furimsky E., Howard J. A. and Selwyn J. (1980). Absolute rate constants for hydrocarbon autoxidation. 28. A low temperature kinetic electron spin resonance study of the self-reactions of isopropylperoxy and related secondary alkylperoxy radicals in solution. *Can J Chem*, **58**, 677–680.

Galey C., Gatel D., Amy G. and Cavard J. (2000). Comparative assessment of bromate control options. *Ozone: Sci Eng*, **22**, 267–278.

Gallard H. and von Gunten U. (2002). Chlorination of natural organic matter: kinetics of chlorination and THM formation. *Wat Res*, **36**, 65–74.

Gallard H., von Gunten U. and Kaiser H.-P. (2003). Prediction of the disinfection and oxidation efficiency of full-scale ozone reactors. *J Wat Supply Res Technol – Aqua*, **52**, 277–290.

Games L. M. and Staubach J. A. (1980). Reaction of nitrilotriacetate with ozone in model and natural waters. *Environ Sci Technol*, **14**, 571–576.

Garcia-Ac A., Broséus R., Vincent S., Barbeau B., Prévost M. and Sauvé S. (2010). Oxidation kinetics of cyclophosphamide and methotrexate by ozone in drinking water. *Chemosphere*, **79**, 1056–1063.

Garland J. A., Elzerman A. W. and Penkett S. A. (1980). The mechanism for dry deposition of ozone to seawater surfaces. *J Geophys Res*, **85**, 7488–7492.

Geering F. (1999). Ozone application. The state-of-the-art in Switzerland. *Ozone: Sci Eng*, **21**, 187–200.

Gerecke A. C. and Sedlak D. L. (2003). Precursors of *N*-nitrosodimethylamine in natural waters. *Environ Sci Technol*, **37**, 1331–1336.

Gerrity D. and Snyder S. (2011). Review of ozone for water reuse applications: toxicity, regulations, and trace organic contaminant oxidation. *Ozone: Sci Eng*, **33**, 253–266.

Ghosh G. C., Nakada N., Yamashita N. and Tanaka H. (2009). Oseltamivir carboxylate, the active metabolite of oseltamivir phosphate (tamiflu), detected in sewage discharge and river water in Japan. *Environ Health Perspect*, **118**, 103–107.

Giamalva D. H., Church D. F. and Pryor W. A. (1986). Kinetics of ozonation. 5. Reactions of ozone with carbon-hydrogen bonds. *J Am Chem Soc*, **108**, 7678–7681.

Giger W. and Sigg L. (1997). Hommage to Werner Stumm, Kurt Grob und Jürg Hoigné. *Chimia*, **51**, 859–860.

Gilbert B. C., Holmes R. G. G., Laue H. A. H. and Norman R. O. C. (1976). Electron spin resonance studies. Part L. Reactions of alkoxyl radicals generated from alkylhydroperoxides and titanium (III) ion in aqueous solution. *J Chem Soc, Perkin Trans 2*, 1047–1052.

Gilbert B. C., Holmes R. G. G. and Norman R. O. C. (1977). Electron spin resonance studies. Part LII. Reactions of secondary alkoxyl radicals. *J Chem Res (S)*, 1.

Gilbert B. C., Marshall D. R., Norman R. O. C., Pineda N. and Williams P. S. (1981). Electron spin resonance studies. Part 61. The generation and reactions of the *t*-butoxyl radical in aqueous solution. *J Chem Soc, Perkin Trans 2*, 1392–1400.

Gilbert E. (1977). Über die Wirkung von Ozon auf Maleinsäure, Fumarsäure und deren Oxidationsprodukte in wäßriger Lösung (On the effect of ozone on maleic acid, fumaric acid and their oxidation products in aqueous solution). *Z Naturforsch B*, **32**, 1308–1313.

Gilbert E. (1980). Reaction of ozone with *trans-trans* muconic acid in aqueous solution. *Wat Res*, **14**, 1637–1643.

Gilbert E. (1981). Photometrische Bestimmung niedriger Ozonkonzentrationen in Wasser mit Hilfe von Diäthyl-p-phenylendiamin (DPD) (Photometric determination of low ozone concentrations in water by diethyl-*p*-phenylenediamine (DPD)). *GWF Wasser-Abwasser*, **122**, 410–416.

Gilbert E. and Hoffmann-Glewe S. (1983). Ozonbestimmung mit Diäthylphenylendiamin (DPD) in wäßriger Lösung in Gegenwart von Manganoxiden (Determination of ozone with diethylphenylenediamine (DPD) in aqueous solution in presence of manganese oxides). *GWF Wasser-Abwasser*, **124**, 469–473.

Gilbert E. and Hoigné J. (1983). Messung von Ozon in Wasserwerken; Vergleich der DPD- und Indigo-Methode (Ozone measurement in water works; comparison of the DPD- and the indigo-method). *GWF Wasser-Abwasser*, **124**, 527–531.

Gilbert E. and Zinecker H. (1980). Ozonation of aromatic amines in water. *Ozone: Sci Eng*, **2**, 65–74.

Glaze W. H. (1986). Reaction products of ozone: a review. *Environ Health Perspect*, **69**, 151–157.

Glaze W. H., Peyton G. R., Lin S., Huang R. Y. and Burleson J. L. (1982). Destruction of pollutants in water with ozone in combination with ultraviolet radiation. 2. Natural trihalomethane precursors. *Environ Sci Technol*, **16**, 454–458.

Gleu K. and Roell E. (1929). Die Einwirkung von Ozon auf Alkaliazid (The effect of ozone on alkali azide). *Z Anorg Allg Chem*, **179**, 233–266.

Glindemann D., Dietrich A., Staerk H.-J. and Kuschk P. (2006). The two odors of iron when touched or pickled: (skin) carbonyl compounds and organophosphines. *Angew Chem, Int Ed*, **45**, 7006–7009.

Golden D. M., Bierbaum V. M. and Howard C. J. (1990). Comments on "Reevaluation of the bond dissociation energies DH_{DBE} for H-OH, H-OOH, H-O$^-$, H-O$^\bullet$, H-OO$^-$, and H-OO$^\bullet$". *J Phys Chem*, **94**, 5413–5415.

Goldman R. C., Fesik S. W. and Doran C. C. (1990). Role of protonated and neutral forms of macrolides in binding to ribosomes from gram-positive and gram-negative bacteria. *Antimicrob Agents Chemother*, **34**, 426–431.

Goldstein S., Lind J. and Merényi G. (2005). Chemistry of peroxynitrites as compared to peroxynitrates. *Chem Rev*, **105**, 2457–2470.

Gordon A. J. and Ford R. A. (1972). *The Chemist's Companion: A Handbook of Practical Data, Techniques and References*. Wiley, New York.

Gordon G., Gauw R. D., Emmert G. L., Walters B. D. and Bubnis B. (2002). Chemical reduction methods for bromate ion removal. *J Am Water Works Ass*, **94**(2), 91–98.

Gottschalk C., Libra J. A. and Saupe A. (2010). *Ozonation of Water and Wastewater. A Practical Guide to Understanding Ozone and its Application*. Wiley-VCH, Weinheim.

Grebel J. E., Pignatello J. J. and Mitch W. A. (2010). Effect of halide ions and carbonates on organic contaminant degradation by hydroxyl radical-based Advanced Oxidation Processes in saline waters. *Environ Sci Technol*, **44**, 6822–6828.

Grigorev A. E., Makarov I. E. and Pikaev A. K. (1987). Formation of $Cl_2^{\bullet-}$ in the bulk solution during radiolysis of concentrated aqueous solutions of chloride. *High Energy Chem*, **21**, 99–102.

Grubbé E. H. (1933). Priority in the therapeutic use of X-rays. *Radiology*, **21**, 156–162.

Gujer W. and von Gunten U. (2003). A stochastic model of an ozonation reactor. *Wat Res*, **37**, 1667–1677.

Günther K. (2002). Östrogen-aktive Nonylphenole in Lebensmitteln (Oestrogenic nonylphenols in food). *GIT Labor-Fachzeitschrift*, **9**, 960–962.

Gurol M. D. and Bremen W. M. (1985). Kinetics and mechanism of ozonation of free cyanide species in water. *Environ Sci Technol*, **19**, 804–809.

Gurol M. D. and Nekouinaini S. (1984). Kinetic behavior of ozone in aqueous solutions of substituted phenols. *Ind Eng Chem Fundam*, **23**, 54–60.

Gurol M. D. and Vatistas R. (1987). Oxidation of phenolic compounds by ozone and ozone + u.v. radiation: a comparative study. *Wat Res*, **21**, 895–900.

Gyürék L. L. and Finch G. R. (1998). Modeling water treatment chemical disinfection kinetics. *J Environ Eng*, **124**, 783–793.

Haag W. R. and Hoigné J. (1983a). Ozonation of bromide-containing waters: kinetics of formation of hypobromous acid and bromate. *Environ Sci Technol*, **17**, 261–267.

Haag W. R. and Hoigné J. (1983b). Ozonation of water containing chlorine or chloramines. Reaction products and kinetics. *Wat Res*, **17**, 1397–1402.

Haag W. R. and Yao C. C. D. (1992). Rate constants for reaction of hydroxyl radicals with several drinking water contaminants. *Environ Sci Technol*, **26**, 1005–1013.

Haag W. R. and Yao C. C. D. (1993). Ozonation of U.S. drinking water sources: OH concentrations and oxidation competition values. In: *Proceedings of the Eleventh Ozone World Congress* 2 (S-17), 119–126, San Francisco, 1993.

Haag W. R., Hoigné J. and Bader H. (1982). Ozonung bromidhaltiger Trinkwässer: Kinetik der Bildung sekundärer Bromverbindungen (Ozonation of bromide-containing drinking waters: Kinetics of the formation of secondary bromo-compounds). *Vom Wasser*, **59**, 238–251.

Haag W. R., Hoigné J. and Bader H. (1984). Improved ammonia oxidation by ozone in the presence of bromide ion during water treatment. *Wat Res*, **18**, 1125–1128.

Hahn J., Lachmann G. and Pienaar J. J. (2000). Kinetics and simulation of the decomposition of ozone in acidic aqueous solutions. *S-Afr Tydskr Chem*, **53**, 132–138.

Hall R. M. and Sobsey M. D. (1993). Inactivation of hepatitis A virus and MS2 by ozone and ozone – hydrogen peroxide in buffered water. *Water Sci Tech*, **27**, 371–378.

Hamelin C. and Chung Y. S. (1989). Repair of ozone-induced DNA lesions in *Escherichia coli* B cells. *Mutation Res*, **214**, 253–255.

Hammes F., Salhi E., Köster O., Kaiser H.-P., Egli T. and von Gunten U. (2006). Mechanistic and kinetic evaluation of organic disinfection by-product and assimilable organic carbon (AOC) formation during the ozonation of drinking water. *Wat Res*, **40**, 2275–2286.

Hammes F., Meylan S., Salhi E., Köster O., Egli T. and von Gunten U. (2007). Formation of assimilable organic carbon (AOC) and specific natural organic matter (NOM) fractions during ozonation of phytoplankton. *Wat Res*, **41**, 1447–1454.

Hammes F., Berney M., Wang Y. Y., Vital M., Köster O. and Egli T. (2008). Flow-cytometric total bacterial cell counts as a descriptive microbiological parameter for drinking water treatment processes. *Wat Res*, **42**, 269–277.

Hammes F., Berney M. and Egli M. (2011). Cultivation-independent assessment of bacterial viability. In: *High Resolution Microbial Single Cell Analytics*, S. Müller and T. Bley (eds), Springer, Berlin, pp. 123–150.

Hancock G. and Tyley P. L. (2001). The near-uv photolysis of ozone: quantum yields of O(^{1}D) between 305 and 329 nm at temperatures from 227–298 K, and the room temperature quantum yield of O(^{3}P$_2$) between 308 and 310 nm, measured by resonance enhanced multiphoton ionisation. *Phys Chem Chem Phys*, **3**, 4984–4990.

Hart D. S., Davis L. C., Erickson L. E. and Callender T. M. (2004). Sorption and partitioning parameters of benzotriazole compounds. *Microchem J*, **77**, 9–17.

Hart E. J., Sehested K. and Holcman J. (1983). Molar absorptivities of ultraviolet and visible bands of ozone in aqueous solutions. *Anal Chem*, **55**, 46–49.

Hayes T., Haston K., Tsui M., Hoang A., Haeffele C. and Vonk A. (2002). Feminization of male frogs in the wild. *Nature*, **419**, 895–896.

Heilker E. (1979). The Mülheim process for treating Ruhr river water. *J Am Water Works Ass*, **71**(11), 623–627.

Hemmi M., Fusoaka Y., Tomioka H. and Kurihara M. (2010). High performance RO membranes for desalination and wastewater reclamation and their operation results. *Wat Sci Tech*, **62**, 2134–2140.

Henglein A. and Kormann C. (1985). Scavenging of OH radicals produced in the sonolysis of water. *Int J Radiat Biol*, **48**, 251–258.

Henschel K.-P., Wenzel A., Diedrich M. and Fliedner A. (1997). Environmental hazard assessment of pharmaceuticals. *Regul Toxicol Pharmacol*, **25**, 220–225.

Her N., Amy G., Foss O., Cho J., Yoon Y. and Kosenka P. (2002). Optimization of method for detecting and characterizing NOM by HPLC-size exclusion chromatography with UV and on-line DOC detection. *Environ Sci Technol*, **36**, 1069–1076.

Herron J. T. and Huie R. E. (1969). Rates of reaction of atomic oxygen. II. Some C_2 to C_8 alkanes. *J Phys Chem*, **73**, 3327–3337.

Hijnen W. A. M., Voogt R., Veenendaal H., van der Jagt H. and van der Kooij D. (1995). Bromate reduction by denitrifying bacteria. *Appl Environ Microbiol*, **61**, 239–244.

Hilborn J. W. and Pincock J. A. (1991). Rates of decarboxylation of acyloxy radicals formed in the photocleavage of substituted 1-naphthylmethyl alkanoates. *J Am Chem Soc*, **113**, 2683–2686.

Hill G. R. (1949). The kinetics of the oxidation of cobaltous ion by ozone. *J Am Chem Soc*, **71**, 2434–2435.

Höbel B. and von Sonntag C. (1998). OH-radical induced degradation of ethylenediaminetetraacetic acid (EDTA) in aqueous solution: a pulse radiolysis study. *J Chem Soc, Perkin Trans 2*, 509–513.

Hoerger C. C., Wettstein F. E., Hungerbühler K. and Bucheli T. D. (2009). Occurrence and origin of estrogenic isoflavones in Swiss river waters. *Environ Sci Technol*, **43**, 6151–6157.

Hoff J. C. and Geldreich E. E. (1981). Comparison of the biocidal efficiency of alternative disinfectants. *J Am Water Works Ass*, **73**(1), 40–44.

Hoffmann M. R. (1986). On the kinetics and mechanism of oxidation of aquated sulfur dioxide by ozone. *Atmos Environ*, **20**, 1145–1154.

Hofmann R. and Andrews R. C. (2001). Ammoniacal bromamines: a review of their influence on bromate formation during ozonation. *Wat Res*, **35**, 599–604.

Hoigné J. (1994). Characterization of water quality criteria for ozonation processes part I: minimal set of analytical data. *Ozone: Sci Eng*, **16**, 113–120.

Hoigné J. (1998). Chemistry of aqueous ozone and transformation of pollutants by ozonation and advanced oxidation processes. In: *The Handbook of Environmental Chemistry. Vol. 5, Part C, Quality and Treatment of Drinking Water*, J. Hrubec (ed.), Springer Verlag, Heidelberg, pp. 83–141.

Hoigné J. and Bader H. (1975). Ozonation of water: role of hydroxyl radicals as oxidizing intermediates. *Science*, **190**, 782–783.

Hoigné J. and Bader H. (1978). Ozonation of water: kinetics of oxidation of ammonia by ozone and hydroxyl radicals. *Environ Sci Technol*, **12**, 79–84.

Hoigné J. and Bader H. (1979). Ozonation of water: "oxidation-competition values" of different types of water used in Switzerland. *Ozone: Sci Eng*, **1**, 357–372.

Hoigné J. and Bader H. (1983a). Rate constants of reactions of ozone with organic and inorganic compounds in water – I. Non-dissociating organic compounds. *Wat Res*, **17**, 173–183.

Hoigné J. and Bader H. (1983b). Rate constants of the reactions of ozone with organic and inorganic compounds in water. – II. Dissociating organic compounds. *Wat Res*, **17**, 185–194.

Hoigné J. and Bader H. (1988). The formation of trichloronitromethane (chloropicrin) and chloroform in a combined ozonation/chlorination treatment of drinking water. *Wat Res*, **22**, 313–319.

Hoigné J. and Bader H. (1994). Characterization of water quality criteria for ozonation processes. Part II: lifetime of added ozone. *Ozone: Sci Eng*, **16**, 121–134.

Hoigné J., Bader H., Haag W. R. and Staehelin J. (1985). Rate constants of reactions of ozone with organic and inorganic compounds in water III. Inorganic compounds and radicals. *Wat Res*, **19**, 993–1004.

Hollender J., Zimmermann S. G., Koepke S., Krauss M., McArdell C. S., Ort C., Singer H., von Gunten U. and Siegrist H. (2009). Elimination of organic micropollutants in a municipal wastewater treatment plant upgraded with a full-scale post-ozonation followed by sand filtration. *Environ Sci Technol*, **43**, 7862–7869.

Hori H., Hayakawa E., Einaga H., Kutsuna S., Koike K., Ibusuki T., Kiatagawa H. and Arakawa R. (2004). Decomposition of environmentally persistent perfluorooctanoic acid in water by photochemical approaches. *Environ Sci Technol*, **38**, 6118–6124.

Hori H., Yamamoto A., Hayakawa E., Taniyasu S., Yamashita N. and Kutsuna S. (2005). Efficient decomposition of environmentally persistent perfluorocarboxylic acids by use of persulfate as a photochemical oxidant. *Environ Sci Technol*, **39**, 2383–2388.

Howard J. A. (1978). Self-reactions of alkylperoxy radicals in solution (1). In: *Organic Free Radicals*, W. A. Pryor (ed.), ACS Symposium Ser. 69. American Chemical Society, Washington, pp. 413–432.

Howard J. A. and Bennett J. E. (1972). The self-reaction of *sec*-alkylperoxy radicals: a kinetic electron spin resonance study. *Can J Chem*, **50**, 2374–2377.

Hoyer O., Lüsse B. and Bernhardt H. (1985). Isolation and characterization of extracellular organic matter (EOM) from algae. *Z Wasser-Abwasser-Forsch*, **18**, 76–90.

Hu L., Martin H. M., Arce-Bulted O., Sugihara M. N., Keating K. A. and Strathmann T. J. (2009). Oxidation of carbamazepine by Mn(VII) and Fe(VI): reaction kinetics and mechanism. *Environ Sci Technol*, **43**, 509–515.

Hua G. and Reckhow D. A. (2007). Comparison of disinfection byproduct formation from chlorine and alternative disinfectants. *Wat Res*, **41**, 1667–1678.

Huber M. M., Canonica S., Park G.-Y. and von Gunten U. (2003). Oxidation of pharmaceuticals during ozonation and advanced oxidation processes. *Environ Sci Technol*, **37**, 1016–1024.

Huber M. M., Ternes T. A. and von Gunten U. (2004). Removal of estrogenic activity and formation of oxidation products during ozonation of 17α-ethinylestradiol. *Environ Sci Technol*, **38**, 5177–5186.

Huber M. M., Göbel A., Joss A., Hermann M., Löffler D., McArdell C. S., Ried A., Siegrist H., Ternes T. A. and von Gunten U. (2005). Oxidation of pharmaceuticals during ozonation of municipal waste water effluents: a pilot study. *Environ Sci Technol*, **39**, 4290–4299.

Huber S. A. and Frimmel F. H. (1991). Flow injection analysis of organic and inorganic carbon in the low-ppb range. *Anal Chem*, **63**, 2122–2130.

Huber S. A. and Frimmel F. H. (1992). A new method for the characterization of organic carbon in aquatic systems. *Int J Environ Anal Chem*, **49**, 49–57.

Huber S. A. and Frimmel F. H. (1996). Gelchromatographie mit Kohlenstoffdetektion (LC-OCD): Ein rasches und aussagekräftiges Verfahren zur Charakterisierung hydrophiler organischer Wasserinhaltsstoffe (Gel chromatography with carbon detection (LC-OCD): A rapid and meaningful procedure for the characterisation of hydrophilic organic water components). *Vom Wasser*, **86**, 277–290.

Huber S. A., Gremm T. and Frimmel F. H. (1990). Chromatographische Trennung natürlicher organischer Wasserinhaltsstoffe mit UV-, Fluoreszenz- und DOC/TOC-Detektion ohne Probenvoranreicherung: Auswahl geeigneter Trennsysteme (Chromatographic separation of natural organic substances in water by UV-, fluorescence- and DOC/TOC-detection with sample concentration: Choice of suitable separation systems). *Vom Wasser*, **75**, 331–342.

Huggett D. B., Khan I. A., Foran C. M. and Schlenk D. (2003). Determination of beta-adrenergic receptor blocking pharmaceuticals in United States wastewater effluent. *Environ Pollut*, **121**, 199–205.

Hughes M. F. (2002). Arsenic toxicity and potential mechanisms of action. *Toxicol Lett*, **133**, 1–16.

Huie R., Clifton C. L. and Neta P. (1991). Electron transfer reaction rates and equilibria of the carbonate and sulfate radical anions. *Radiat Phys Chem*, **38**, 477–481.

Hunt N. K. and Marinas B. J. (1997). Kinetics of *Escherichia coli* inactivation with ozone. *Wat Res*, **31**, 1355–1362.

Ikehata K. and El-Din M. G. (2005a). Aqueous pesticide degradation by ozonation and ozone-based advanced oxidation processes: a review (Part I). *Ozone: Sci Eng*, **27**, 83–114.

Ikehata K. and El-Din M. G. (2005b). Aqueous pesticide degradation by ozonation and ozone-based advanced oxidation processes: a review (Part II). *Ozone: Sci Eng*, **27**, 173–202.

Ikemizu K., Orita M., Sagiike M., Morooka S. and Kato Y. (1987). Ozonation of organic refractory compounds in water in combination with UV radiation. *J Chem Eng Jpn*, **20**, 369–374.

Ingelman-Sundberg M., Kaur H., Terelius Y., Persson J.-O. and Halliwell B. (1991). Hydroxylation of salicylate by microsomal fractions and cytochrome *P*-450. Lack of production of 2,3-dihydroxybenzoate unless hydroxyl radical formation is permitted. *Biochem J*, **276**, 753–757.

Ishizaki K., Shinriki N. and Ueda T. (1984). Degradation of nucleic acids with ozone. V. Mechanism of action of ozone on deoxyribonucleoside 5′-monophosphates. *Chem Pharm Bull*, **32**, 3601–3606.

Jacangelo J. G., Trussell R. R. and Watson M. (1997). Role of membrane technology in drinking water treatment in the United States. *Desalination*, **113**, 119–127.

Jacobsen F., Holcman J. and Sehested K. (1997). Activation parameters of ferryl ion reactions in aqueous acid solutions. *Int J Chem Kinet*, **29**, 17–24.

Jacobsen F., Holcman J. and Sehested K. (1998a). Oxidation of manganese(II) by ozone and reduction of manganese(III) by hydrogen peroxide in acidic solution. *Int J Chem Kinet*, **30**, 207–214.

Jacobsen F., Holcman J. and Sehested K. (1998b). Reactions of the ferryl ion with some compounds found in cloud water. *Int J Chem Kinet*, **30**, 215–221.

Jahnke L. S. (1999). Measurement of hydroxyl radical-generated methane sulfinic acid by high-performance liquid chromatography and electrochemical detection. *Anal Biochem*, **269**, 273–277.

Janna H., Scrimshaw M. D., Williams R. J., Churchley J. and Sumpter J. P. (2011). From dishwasher to tap? Xenobiotic substances benzotriazole and tolyltriazole in the environment. *Environ Sci Technol*, **45**, 3858–3864.

Jans U. (1996). Radikalbildung aus Ozon in atmosphärischen Wassern – Einfluss von Licht, gelösten Stoffen und Russpartikeln (Formation of free radicals from ozone in air-containing waters – Influence of light, dissolved compound and soot particles). Dissertation No 11814, ETH, Zürich.

Jans U. and Hoigné J. (1998). Activated carbon and carbon black catalyzed transformation of aqueous ozone into OH radicals. *Ozone: Sci Eng*, **20**, 67–90.

Jarocki A., von Sonntag C. and Schmidt T. C. (2012). The •OH radical yield in the $H_2O_2+O_3$ (peroxone) reaction, in preparation.

Jayson G. G., Parsons B. J. and Swallow A. J. (1973). Some simple, highly reactive, inorganic chlorine derivatives in aqueous solution. *J Chem Soc, Faraday Trans*, **69**, 1597–1607.

Jeong J., Jung J., Cooper W. J. and Song W. (2010). Degradation mechanisms and kinetic studies for the treatment of X-ray contrast media compounds by advanced oxidation/reduction processes. *Wat Res*, **44**, 4391–4398.

Jiang T., Kennedy M. D., de Schepper V., Nam S.-N., Nopens I., Vanrolleghem P. A. and Amy G. (2010). Characterization of soluble microbial products and their fouling impacts in membrane bioreactors. *Environ Sci Technol*, **44**, 6642–6648.

Jin F., Leitich J. and von Sonntag C. (1993). The superoxide radical reacts with tyrosine-derived phenoxyl radicals by addition rather than by electron transfer. *J Chem Soc, Perkin Trans 2*, 1583–1588.

Jobling S., Casey D., Rodgers-Gray T., Oehlmann J., Schulte-Oehlmann U., Pawlowski S., Baunbeck T., Turner A. P. and Tyler C. R. (2004). Comparative responses of molluscs and fish to environmental estrogens and an estrogenic effluent. *Aquatic Toxicol*, **66**, 207–222.

Jones I. T. N. and Wayne R. P. (1970). The photolysis of ozone by ultraviolet ratiation IV. Effect of photolysis wavelength on primary step. *Proc Roy Soc London A*, **319**, 273–287.

Jonsson M., Lind J., Eriksen T. E. and Merényi G. (1994). Redox and acidity properties of 4-substituted aniline radical cations in water. *J Am Chem Soc*, **116**, 1423–1427.

Jonsson M., Lind J., Merényi G. and Eriksen T. E. (1995). N–H bond dissociation energies, reduction potentials and pK_as of multisubstituted anilines and aniline radical cations. *J Chem Soc, Perkin Trans 2*, 61–65.

Joss A., Siegrist H. and Ternes T. A. (2008). Are we about to upgrade wastewater treatment for removing organic micropollutants? *Wat Sci Tech.*, **57**, 251–255.

Jung J. Y., Oh B. S. and Kang J. W. (2008). Synergistic effect of sequential or combined use of ozone and UV radiation for the disinfection of *Bacillus subtilis* spores. *Wat Res*, **42**, 1613–1621.

Kaiga N., Takase O., Todo Y. and Yamanashi I. (1997). Corrosion resistance of ozone generator electrode. *Ozone: Sci Eng*, **19**, 169–178.

Kanofsky J. R. and Sima P. D. (1995). Reactive absorption of ozone by aqueous biomolecule solutions: implications for the role of sulfhydryl compounds as targets for ozone. *Arch Biochem Biophys*, **316**, 52–62.

Karcher S., Cáceres L., Jekel M. and Contreras R. (1999). Arsenic removal from water supplies in Northern Chile using ferric chloride coagulation. *J Chartered Inst Wat Environ Manage (JCIWEM)*, **13**, 164–169.

Kasprzyk-Hordern B., Dinsdale R. M. and Guwy A. J. (2009). The removal of pharmaceuticals, personal care products, endocrine disruptors and illicit drugs during wastewater treatment and its impact on the quality of receiving waters. *Wat Res*, **43**, 363–380.

Katsoyiannis I. A., Zikoudi A. and Hug S. J. (2008). Arsenic removal from groundwaters containing iron, ammonium, manganese and phosphate: a case study from a treatment unit in northern Greece. *Desalination*, **224**, 330–339.

Katsoyiannis I. A., Canonica S. and von Gunten U. (2011). Efficiency and the energy requirements for the transformation of organic micropollutants by ozone, O_3/H_2O_2 and UV/H_2O_2. *Wat Res*, **45**, 3811–3822.

Katzenelson E., Kletter B., Schechter H. and Shuval H. I. (1974). Inactivation of viruses and bacteria by ozone. In: *Chemistry of Water Supply, Treatment and Distribution*, A.-J. Rubin (ed.), Ann Arbor Science Publishers, Ann Arbor, pp. 409–421.

Katzenelson E., Koerner G., Biedermann N., Peleg M. and Shuval H. I. (1979). Measurement of the inactivation kinetics of poliovirus by ozone in a fast-flow mixer. *Appl Environ Microbiol*, **37**, 715–718.

Kavanagh F. (1947). Activities of twenty-two antibacterial substances against nine species of bacteria. *J Bacteriol*, **54**, 761–766.

Keith J. D., Pacey G. E., Cotruvo J. A. and Gordon G. (2006a). Experimental results from the reaction of bromate ion with synthetic and real gastric juices. *Toxicology*, **221**, 225–228.

Keith J. D., Pacey G. E., Cotruvo J. A. and Gordon G. (2006b). Preliminary data on the fate of bromate ion in simulated gastric juices. *Ozone: Sci Eng*, **28**, 165–170.

Khadhraoui M., Trabelsi H., Ksibi M., Bouguerra S. and Elleuch B. (2009). Discoloration and detoxicification of a Congo red dye solution by means of ozone treatment for a possible water reuse. *J Hazard Mater*, **161**, 974–981.

Kidd K. A., Blanchfield P. J., Mills K. H., Pallace V. P., Evans R. E., Lazorchak J. M. and Flick R. W. (2007). Collapse of a fish population after exposure to a synthetic estrogen. *Proc Nat Acad Sci USA*, **104**, 8897–8901.

Kilpatrick M. L., Herrick C. C. and Kilpatrick M. (1956). The decomposition of ozone in aqueous solution. *J Am Chem Soc*, **78**, 1784–1789.

Kim C. K., Gentile D. M. and Sproul O. J. (1980). Mechanism of ozone inactivation of bacteriophage f2. *Appl Environ Microbiol*, **39**, 210–218.

Kim J.-H., von Gunten U. and Marinas B. J. (2004). Simultaneous prediction of *Cryptosporidium parvum* oocyst inactivation and bromate formation during ozonation of synthetic waters. *Environ Sci Technol*, **38**, 2232–2241.

Kim J.-H., Elovitz M. S., von Gunten U., Shukairy H. M. and Marinas B. J. (2007a). Modeling *Cryptosporidium parvum* oocysts inactivation and bromate in a flow-through ozone contactor treating natural water. *Wat Res*, **41**, 467–475.

Kim S. D., Cho J., Kim I. S., Vanderford B. J. and Snyder S. A. (2007b). Occurrence and removal of pharmaceuticals and endocrine disruptors in South Korean surface, drinking, and waste water. *Wat Res*, **41**, 1013–1021.

King D. W. (1998). Role of carbonate speciation on the oxidation rate of Fe(II) in aquatic systems. *Environ Sci Technol*, **32**, 2997–3003.

Kirisits M. J., Snoeyink V. L. and Kruithof J. (2000). The reduction of bromate by granular activated carbon. *Wat Res*, **34**, 4250–4260.

Kirisits M. J., Snoeyink V. L., Inan H., Chee-Sanford J. C., Raskin L. and Brown J. C. (2001). Water quality factors affecting bromate reduction in biologically active carbon filters. *Wat Res*, **35**, 891–900.

Kirschner M. J. (1991). Ozone. In: *Ullmann's Encyclopedia of Industrial Chemistry*, B. Elvers, S. Hawkins and G. Schulz (eds), Verlag Chemie, Weinheim, pp. 349–357.

Kläning U. K. and Wolff T. (1985). Laser flash photolysis of HClO, ClO$^-$, HBrO, and BrO$^-$ in aqueous solution – reactions of Cl- and Br-atoms. *Ber Bunsenges Phys Chem*, **89**, 243–245.

Kläning U. K., Sehested K. and Wolff T. (1984). Ozone formation in laser flash photolysis of oxoacids and oxoanions of chlorine and bromine. *J Chem Soc, Faraday Trans 1*, **80**, 2969–2979.

Kläning U. K., Sehested K. and Holcman J. (1985). Standard Gibbs energy of formation of the hydroxyl radical in aqueous solution. Rate constants for the reaction $ClO_2^- + O_3 \rightleftarrows O_3^- + ClO_2$. *J Phys Chem*, **89**, 760–763.

Klopman G. (1968). Chemical reactivity and the concept of charge- and frontier-controlled reactions. *J Am Chem Soc*, **90**, 223–234.

Kogelschatz U. (2003). Dielectric-barrier discharges: their history, discharge physics, and industrial applications. *Plasma Chem Plasma Processing*, **23**, 1–46.

Kogelschatz U. and Baessler P. (1987). Determination of nitrous oxide and dinitrogen pentoxide concentrations in the output of air-fed ozone generators of high power density. *Ozone: Sci Eng*, **9**, 195–206.

Kogelschatz U., Eliasson B. and Hirth M. (1988). Ozone generation from oxygen and air: discharge physics and reaction mechanism. *Ozone: Sci Eng*, **10**, 367–377.

Koike K., Nifuku M., Izumi K., Nakamura S., Fujiwara S. and Horiguchi S. (2005). Explosion properties of highly concentrated ozone gas. *J Loss Prevent Process Industries*, **18**, 465–468.

Koller J. and Plesnicar B. (1996). Mechanism of the participation of water in the decomposition of hydrogen trioxide (HOOOH). A theoretical study. *J Am Chem Soc*, **118**, 2470–2472.

Koller J., Hodoscek M. and Plesnicar B. (1990). Chemistry of hydrotrioxides. A comparative ab initio study of the equilibrium structures of monomeric and dimeric hydrotrioxides (CH_3OOOH, $H_3SiOOOH$) and hydroperoxides (CH_3OOH, H_3SiOOH). Relative bond strengths in and the gas phase acidities of hydrotrioxides and hydroperoxides. *J Am Chem Soc*, **112**, 2124–2129.

Kolpin D. W., Furlong E. T., Meyer M. T., Thurman E. M., Zaugg S. D., Barber L. B. and Buxton H. T. (2002). Pharmaceuticals, hormones, and other organic wastewater contaminants in U.S. streams, 1999–2000: a national reconnaissance. *Environ Sci Technol*, **36**, 1202–1211.

Komanapalli I. R. and Lau B. H. S. (1996). Ozone-induced damage of *Escherichia coli K-12*. *Appl Microbiol Biotechnol*, **46**, 610–614.

Konsowa A. H., Ossman M. E., Chen Y. S. and Crittenden J. C. (2010). Decolorization of industrial wastewater by ozonation followed by adsorption on activated carbon. *J Hazard Mater*, **176**, 181–185.

Konya K. G., Paul T., Lin S., Lusztyk J. and Ingold K. U. (2000). Laser flash photolysis studies on the first superoxide thermal source. First direct measurements of the rates of solvent-assisted 1,2-hydrogen atom shifts and a proposed new mechanism for this unusual rearrangement. *J Am Chem Soc*, **122**, 1718–7527.

Koppenol W. H., Stanbury D. M. and Bounds P. L. (2010). Electrode potentials of partially reduced oxygen species, from dioxygen to water. *Free Radical Biol Med*, **49**, 317–322.

Kosaka K., Asami M., Konno Y., Oya M. and Kunikane S. (2009). Identification of antiyellowing agents as precursors of *N*-nitrosodimethylamine production on ozonation from sewage treatment plant effluent. *Environ Sci Technol*, **43**, 5236–5241.

Kosjek T., Andersen H. R., Kompare B., Ledin A. and Heath E. (2009). Fate of carbamazepine during water treatment. *Environ Sci Technol*, **43**, 6256–6261.

Köster R. and Asmus K.-D. (1971). Die Reaktionen chlorierter Äthylene mit hydratisierten Elektronen und OH-Radikalen in wässriger Lösung (The reaction of chlorinated ethylenes with hydrated electrons and OH radicals in aqueous solution). *Z Naturforsch*, **26b**, 1108–1116.

Kovac F. and Plesnicar B. (1979). The substituent effect on the thermal decomposition of acetal hydrotrioxides. Polar and radical decomposition paths. *J Am Chem Soc*, **101**, 2677–2681.

Krasner S. W. (2009). The formation and control of emerging disinfection by-products of health concern. *Phil Trans R Soc A*, **367**, 4077–4095.

Krasner S. W., Glaze W. H., Weinberg H. S., Daniel P. A. and Najm I. N. (1993). Formation and control of bromate during ozonation of waters containing bromide. *J Am Water Works Ass*, **85**(1), 73–81.

Krasner S. W., Weinberg H. S., Richardson S. D., Pastor S. J., Chinn R., Sclimenti M. J., Onstad G. D. and Thurston A. D. (2006). Occurrence of a new generation of disinfection byproducts. *Environ Sci Technol*, **40**, 7175–7185.

Kruithof J. C. and Masschelein W. J. (1999). State-of-the-art of the application of ozonation in BENELUX drinking water treatment. *Ozone: Sci Eng*, **21**, 139–152.

Kruithof J. C., Meijers R. T. and Schippers J. C. (1993). Formation, restriction of formation and removal of bromate. *Wat Supply*, **11**, 331–342.

Kruithof J. C., Kamp P. C. and Martijn B. J. (2007). UV/H_2O_2 treatment: a practical solution for organic contaminant control and primary disinfection. *Ozone: Sci Eng*, **29**, 273–280.

Kumar K. and Margerum D. W. (1987). Kinetics and mechanism of general-acid-assisted oxidation of bromide by hypochlorite and hypochlorous acid. *Inorg Chem*, **26**, 2706–2711.

Kümmerer K. (2009a). Antibiotics in the aquatic environment – a review – part I. *Chemosphere*, **75**, 417–434.

Kümmerer K. (2009b). Antibiotics in the aquatic environment – a review – part II. *Chemosphere*, **75**, 435–441.

Kümmerer K. (2010). Pharmaceuticals in the environment. *Annu Rev Environ Resour*, **35**, 57–75.

Lange F., Cornelissen S., Kubac D., Sein M. M., von Sonntag J., Hannich C. B., Golloch A., Heipieper H. J., Möder M. and von Sonntag C. (2006). Degradation of macrolide antibiotics by ozone: a mechanistic case study with clarithromycin. *Chemosphere*, **65**, 17–23.

Langlais B., Reckhow D. A. and Brink D. R. (1991). *Ozone in Water Treatment. Application and Engineering*. Lewis Publishers, Chelsea.

Larocque R. L. (1999). Ozone application in Canada. A state of the art review. *Ozone: Sci Eng*, **21**, 119–125.

Larsen T. A. and Gujer W. (1996). Separate management of anthropogenic nutrient solutions (human urine). *Wat Sci Tech*, **34**, 87–94.

Larson M. A. and Marinas B. J. (2003). Inactivation of *Bacillus subtilis* spores with ozone and monochloramine. *Wat Res*, **37**, 833–844.

Latch D. E., Packer J. L., Stender B. L., VanOverbeke J., Arnold W. A. and McNeill K. (2005). Aqueous photochemistry of triclosan: formation of 2,4-dichlorophenol, 2,8-dichlorodibenzo-*p*-dioxin, and oligomerization products. *Environ Toxicol Chem*, **24**, 517–525.

Lau T. K., Chu W. and Graham N. (2007). Reaction pathways and kinetics of butylated hydroxyanisole with UV, ozonation and UV/O$_3$ processes. *Wat Res*, **41**, 765–774.

Le Paloüe J. and Langlais B. (1999). State-of-the-art of ozonation in France. *Ozone: Sci Eng*, **21**, 153–162.

Lee C., Schmidt C., Yoon J. and von Gunten U. (2007a). Oxidation *N*-nitrosodimethylamine (NDMA) precursors with ozone and chlorine dioxide: kinetics and effect on NDMA formation potential. *Environ Sci Technol*, **41**, 2056–2063.

Lee C., Yoon J. and von Gunten U. (2007b). Oxidative degradation of *N*-nitrosodimethylamine by conventional ozonation and the advanced oxidation process ozone/hydrogen peroxide. *Wat Res*, **41**, 581–590.

Lee Y. and von Gunten U. (2010). Oxidative transformation of micropollutants during wastewater treatment: comparison of kinetic aspects of selective (chlorine, chlorine dioxide, ferrateVI, and ozone) and non-selective oxidants (hydroxyl radical). *Wat Res*, **44**, 555–566.

Lee Y. and von Gunten U. (2012). Quantitative structure-activity relationships (QSARs) for transformation of organic micropollutants during oxidative water treatment. *Wat Res*, in press.

Lee Y., Lee C. and Yoon J. (2004). Kinetics and mechanism of DMSO (dimethylsulfoxide) degradation by UV/H$_2$O$_2$ process. *Wat Res*, **38**, 2579–2588.

Lee Y., Escher B. I. and von Gunten U. (2008). Efficient removal of estrogenic activity during oxidative treatment of waters containing steroid estrogens. *Environ Sci Technol*, **42**, 6333–6339.

Legrini O., Oliveros E. and Braun A. M. (1993). Photochemical processes for water treatment. *Chem Rev*, **93**, 671–698.

Legube B., Guyon S., Sugimitsu H. and Doré M. (1983). Ozonation of some aromatic compounds in aqueous solution: styrene, benzaldehyde, naphthalene, diethylphthalate, ethyl and chloro benzenes. *Sci Eng*, **5**, 151–170.

Legube B., Sugimitsu H., Guyon S. and Doré M. (1986). Ozonation du naphthalene en milieu aqueux – II. Etudes cinetiques de la phase initiale de la reaction (Ozonation of naphthalene in aqueous solution – II. Kinetic studies of the initial phase of the reaction). *Wat Res*, **20**, 209–214.

Legube B., Guyon S. and Doré M. (1987). Ozonation of aqueous solution of nitrogen heterocyclic compounds: benzotriazoles, atrazine, and amitrole. *Ozone: Sci Eng*, **9**, 233–246.

Leitzke A. (2003). Mechanistische und kinetische Untersuchungen zur Ozonolyse von organischen Verbindungen in wässriger Lösung (Mechanistic and kinetic investigations on the ozonolysis of organic compounds in aqueous solution). Dissertation, University Duisburg-Essen, Duisburg.

Leitzke A. and von Sonntag C. (2009). Ozonolysis of unsaturated acids in aqueous solution: acrylic, methacrylic, maleic, fumaric and muconic acids. *Ozone: Sci Eng*, **31**, 301–308.

Leitzke A., Reisz E., Flyunt R. and von Sonntag C. (2001). The reaction of ozone with cinnamic acids – formation and decay of 2-hydroperoxy-2-hydroxy-acetic acid. *J Chem Soc, Perkin Trans 2*, 793–797.

Leitzke A., Flyunt R., Theruvathu J. A. and von Sonntag C. (2003). Ozonolysis of vinyl compounds, CH$_2$=CH-X, in aqueous solution – the chemistries of the ensuing formyl compounds and hydroperoxides. *Org Biomol Chem*, **1**, 1012–1019.

Lesko T. M., Colussi A. J. and Hoffmann M. R. (2004). Hydrogen isotope effects and mechanism of aqueous ozone and peroxone decompositions. *J Am Chem Soc*, **126**, 4432–4436.

Li J., Liu H., Zhao X., Qu J., Liu R. and Ru J. (2008). Effect of preozonation on the characteristic transformation of fulvic acid and its subsequent trichloromethane formation potential: presence or absence of bicarbonate. *Chemosphere*, **71**, 1639–1645.

Lienert J., Koller M., Konrad J., McArdell C. S. and Schuwirth N. (2011). Multiple-criteria decision analysis reveals high stakeholder preference to remove pharmaceuticals from hospital wastewater. *Environ Sci Technol*, **45**, 3848–3857.

Lin Y.-C. and Wu S. C. (2006). Effect of ozone exposure on inactivation of intra- and extracellular enterovirus 71. *Antiviral Res*, **70**, 147–153.

Lind J., Shen X., Eriksen T. E., Merényi G. and Eberson L. (1991). One-electron reduction of *N*-bromosuccinimide. Rapid expulsion of a bromine atom. *J Am Chem Soc*, **113**, 4629–4633.

Lind J., Merényi G., Johansson E. and Brinck T. (2003). The reaction of peroxyl radicals with ozone in water. *J Phys Chem A*, **107**, 676–681.

Linden K. G., Rosenfeldt E. J. and Kullman S. W. (2007). UV/H_2O_2 degradation of endocrine-disrupting chemicals in water evaluated via toxicity assays. *Wat Sci Tech*, **55**, 313–319.

Lisle J. T., Pyle P. H. and McFeters G. A. (1999). The use of multiple indices of physiological activity to access viability in chlorine disinfected *Escherichia coli* O157 : H7. *Lett Appl Microbiol*, **29**, 42–47.

Liu Q., Schurter L. M., Muller C. E., Aloisio S., Francisco J. S. and Margerum D. W. (2001). Kinetics and mechanisms of aqueous ozone reactions with bromide, sulfite, hydrogen sulfite, iodide, and nitrite ions. *Inorg Chem*, **40**, 4436–4442.

Loeb B. L., Thomsom C. M., Drago J., Takahara H. and Baig S. (2011). Worldwide ozone capacity for treatment of drinking water and wastewater: a review. *IOA IUVA World Congress & Exhibition*. VII-2.1-VII-2.20, Paris, May 23–27, 2011.

Lögager T., Holcman J., Sehested K. and Pedersen T. (1992). Oxidation of ferrous ions by ozone in acidic solution. *Inorg Chem*, **31**, 3523–3529.

López-López A., Pic J. S. and Debellefontaine H. (2007). Ozonation of azo dye in a semi-batch reactor: a determination of the molecular and radical contributions. *Chemosphere*, **66**, 2120–2126.

Lowndes R. (1999). State of the art for ozone – U.K. experience. *Ozone: Sci Eng*, **21**, 201–205.

Lume-Pereira C., Baral S., Henglein A. and Janata E. (1985). Chemistry of colloidal manganese dioxide. 1. Mechanism of reduction by an organic radical (A radiation chemical study). *J Phys Chem*, **89**, 5772–5778.

Lutze H., Naumov S., Peter A., Liu S., von Gunten U., von Sonntag C. and Schmidt T. C. (2011a). Ozonation of benzotriazoles: rate constants and mechanistic aspects, in preparation.

Lutze H., Panglisch S., Bergmann A. and Schmidt T. C. (2011b). Treatment options for the removal and degradation of poly- and perfluorinated chemicals. In: *The Handbook of Environmental Chemistry,* Vol. 17, T. P. Knepper and F. T. Lange (eds), Springer Verlag, Heidelberg, Berlin, pp. 103–125.

Macova M., Escher B. I., Reungoat J., Carswell S., Chue K. L., Keller J. and Mueller J. F. (2010). Monitoring the biological activity of micropollutants during advanced wastewater treatment with ozonation and activated carbon filtration. *Wat Res*, **44**, 477–492.

Mao Y., Schöneich C. and Asmus K.-D. (1993). Radical mediated degradation mechanisms of halogenated organic compounds as studied by photocatalysis at TiO_2 and by radiation chemistry. In: *TiO_2 Photocatalytic Purification and Treatment of Water and Air*, H. Al-Ekabi and D. F. Ollis (eds), Elsevier, Amsterdam, pp. 49–66.

Mark G., Schuchmann M. N., Schuchmann H.-P. and von Sonntag C. (1990). The photolysis of potassium peroxodisulphate in aqueous solution in the presence of *tert*-butanol: a simple actinometer for 254 nm radiation. *J Photochem Photobiol A: Chem*, **55**, 157–168.

Mark G., Korth H.-G., Schuchmann H.-P. and von Sonntag C. (1996). The photochemistry of aqueous nitrate revisited. *J Photochem Photobiol A: Chem*, **101**, 89–103.

Mark G., Tauber A., Laupert R., Schuchmann H.-P., Schulz D., Mues A. and von Sonntag C. (1998). OH-radical formation by ultrasound in aqueous solution – Part II. Terephthalate and Fricke dosimetry and the influence of various conditions on the sonolytic yield. *Ultrasonics Sonochem*, **5**, 41–52.

Mark G., Naumov S. and von Sonntag C. (2011). The reaction of ozone with bisulfide (HS^-) in aqueous solution – mechanistic aspects. *Ozone: Sci Eng*, **33**, 37–41.

Matsumoto N. and Watanabe K. (1999). Foot prints and future steps of ozone applications in Japan. *Ozone: Sci Eng*, **21**, 127–138.

Matthew B. M. and Anastasio C. (2006). A chemical probe technique for the determination of reactive halogen species in aqueous solution: Part 1 – bromide solutions. *Atmos Chem Phys*, **6**, 2423–2437.

McDowell D. C., Huber M. M., Wagner M., von Gunten U. and Ternes T. A. (2005). Ozonation of carbamazepine in drinking water: identification and kinetic study of major oxidation products. *Environ Sci Technol*, **39**, 8014–8022.

McElroy W. J. (1990). A laser photolysis study of the reaction of SO$_4^-$ with Cl$^-$ and the subsequent decay of Cl$_2^-$ in aqueous solution. *J Phys Chem*, **94**, 2435–2441.

McKay D. J. and Wright J. S. (1998). How long can you make an oxygen chain? *J Am Chem Soc*, **120**, 1003–1013.

McKenzie K. S., Sarr A. B., Mayura K., Bailey R. H., Miller D. R., Rogers T. D., Norred W. P., Voss K. A., Plattner R. D., Kubena L. F. and Phillips T. D. (1997). Oxidative degradation and detoxfication of mycotoxins using a novel source of ozone. *Food Chem Toxicol*, **35**, 807–820.

Merényi G., Lind J. and Shen X. (1988). Electron transfer from indoles, phenol, and sulfite (SO$_3^{2-}$) to chlorine dioxide (ClO$_2^•$). *J Phys Chem*, **92**, 134–137.

Merényi G., Lind J. and Engman L. (1994). One- and two-electron reduction potentials of peroxyl radicals and related species. *J Chem Soc, Perkin Trans 2*, 2551–2553.

Merényi G., Lind J., Naumov S. and von Sonntag C. (2010a). The reaction of ozone with hydrogen peroxide (peroxone process). A revision of current mechanistic concepts based on thermokinetic and quantum-mechanical considerations. *Environ Sci Technol*, **44**, 3505–3507.

Merényi G., Lind J., Naumov S. and von Sonntag C. (2010b). The reaction of ozone with the hydroxide ion. Mechanistic considerations based on thermokinetic and quantum-chemical calculations. The role of HO$_4^-$ in superoxide dismutation. *Chem Eur J*, **16**, 1372–1377.

Mertens R. (1994). Photochemie und Strahlenchemie von organischen Chlorverbindungen in wässriger Lösung (Photochemistry and radiation chemistry of organic chloro compounds in aqueous solution). Dissertation, Ruhr Universität, Bochum.

Mertens R. and von Sonntag C. (1994a). Determination of the kinetics of vinyl radical reactions by the characteristic visible absorption of vinyl peroxyl radicals. *Angew Chem, Int Ed Engl*, **33**, 1262–1264.

Mertens R. and von Sonntag C. (1994b). The reaction of the OH radical with tetrachloroethene and trichloroacetaldehyde (hydrate) in oxygen-free solution. *J Chem Soc, Perkin Trans 2*, 2181–2185.

Mertens R. and von Sonntag C. (1995). Photolysis (λ=254 nm) of tetrachloroethene in aqueous solution. *J Photochem Photobiol A*, **85**, 1–9.

Mertens R., von Sonntag C., Lind J. and Merényi G. (1994). A kinetic study of the hydrolysis of phosgene in aqueous solution by pulse radiolysis. *Angew Chem, Int Ed Engl*, **33**, 1259–1261.

Mestankova H., Escher B., Schirmer K., von Gunten U. and Canonica S. (2011). Evolution of algal toxicity during (photo)oxidative degradation of diuron. *Aquatic Toxicol*, **101**, 466–473.

Mestankova H., Schirmer K., Escher B. I., von Gunten U. and Canonica S. (2012). Removal of the antiviral agent oseltamivir and its biological activity by oxidation processes. *Environ Pollut*, **161**, 30–35.

Meunier L., Canonica S. and von Gunten U. (2006). Implications of sequential use of UV and ozone for drinking water quality. *Wat Res*, **40**, 1864–1876.

Meyer J. and Bester K. (2004). Organophosphate flame retardants and plasticizers in wastewater treatment plants. *J Environ Monit*, **6**, 599–605.

Meylan S., Hammes F., Traber J., Salhi E., von Gunten U. and Pronk W. (2007). Permeability of low molecular weight organics through nanofiltration membranes. *Wat Res*, **41**, 3968–3976.

Mieden O. J., Schuchmann M. N. and von Sonntag C. (1993). Peptide peroxyl radicals: base-induced O$_2^{•-}$ elimination versus bimolecular decay. A pulse radiolysis and product study. *J Phys Chem*, **97**, 3783–3790.

Mills A., Belghazi A. and Rodman D. (1996). Bromate removal from drinking water by semiconductor photocatalysis. *Wat Res*, **30**, 1973–1978.

Mitch W. A., Sharp J. O., Trussell R. R., Valentine R. L., Alvarez-Cohen L. and Sedlak D. L. (2003). *N*-Nitrosodimethylamine (NDMA) as a drinking water contaminant: a review. *Environ Engen Sci*, **20**, 389–404.

Mizuno T. and Tsuno H. (2010). Evaluation of solubility and the gas-liquid equilibrium coefficient of high concentration gaseous ozone to water. *Ozone: Sci Eng*, **32**, 3–15.

Morgan J. J. (2005). Kinetics of reaction between O$_2$ and Mn(II) species in aqueous solutions. *Geochim Cosmochim Acta*, **69**, 35–48.

Morooka S., Kusakabe K., Hayashi J.-I., Isomura K. and Ikemizu K. (1988). Decomposition and utilization of ozone in water treatment reactor with ultraviolet radiation. *Ind Chem Eng Res*, **27**, 2372–2377.

Morozov P. A., Abkhalimov E. V., Chalykh A. E., Pisarev S. A. and Ershov B. G. (2011). Interaction of silver nanoparticles with ozone in aqueous solution. *Colloid J*, **73**, 248–252.

Morris J. C. (1966). The acid ionization constant of HOCl from 5 to 35 °C. *J Phys Chem*, **70**, 3798–3805.

Mouchet P. (1992). From conventional to biological removal of iron and manganese in France. *J Am Water Works Ass*, **84**(4), 158–167.

Mouret J. F., Odin F., Polverelli M. and Cadet J. (1990). ^{32}P-postlabeling measurement of adenine-*N*-1-oxide in cellular DNA exposed to hydrogen peroxide. *Chem Res Toxicol*, **3**, 102–110.

Müller A., Weiss S. C., Beißwenger J., Leukhardt H. G., Schulz W., Seitz W., Ruck W. K. L. and Weber W. H. (2012). Identification of ozonation by-products of 4- and 5-methyl-1H-benzotriazol during the treatment of surface water to drinking water. *Wat Res*, **45**, 679–690.

Muñoz F. and von Sonntag C. (2000a). Determination of fast ozone reactions in aqueous solution by competition kinetics. *J Chem Soc, Perkin Trans 2*, 661–664.

Muñoz F. and von Sonntag C. (2000b). The reactions of ozone with tertiary amines including the complexing agents nitrilotriacetic acid (NTA) and ethylenediaminetetraacetic acid (EDTA) in aqueous solution. *J Chem Soc, Perkin Trans 2*, 2029–2033.

Muñoz F., Schuchmann M. N., Olbrich G. and von Sonntag C. (2000). Common intermediates in the OH-radical-induced oxidation of cyanide and formamide. *J Chem Soc, Perkin Trans 2*, 655–659.

Muñoz F., Mvula E., Braslavsky S. E. and von Sonntag C. (2001). Singlet dioxygen formation in ozone reactions in aqueous solution. *J Chem Soc, Perkin Trans 2*, 1109–1116.

Murray R. W., Lumma W. C. Jr. and Lin J. W. P. (1970). Singlet oxygen sources in ozone chemistry. Decomposition of oxygen-rich intermediates. *J Am Chem Soc*, **92**, 3205–3207.

Mvula E. (2002). The reactions of ozone with compounds relevant to drinking water processing: phenol and its derivatives. Dissertation, University of Namibia, Windhoek.

Mvula E. and von Sonntag C. (2003). Ozonolysis of phenols in aqueous solution. *Org Biomolec Chem*, **1**, 1749–1756.

Mvula E., Schuchmann M. N. and von Sonntag C. (2001). Reactions of phenol-OH-adduct radicals. Phenoxyl radical formation by water elimination *vs*. oxidation by dioxygen. *J Chem Soc, Perkin Trans 2*, 264–268.

Mvula E., Naumov S. and von Sonntag C. (2009). Ozonolysis of lignin models in aqueous solution: anisole, 1,2-dimethoxybenzene, 1,4-dimethoxybenzene and 1,3,5-trimethoxybenzene. *Environ Sci Technol*, **43**, 6275–6282.

Nahir T. M. and Dawson G. A. (1987). Oxidation of sulfur dioxide by ozone in highly dispersed water droplets. *J Atmos Chem*, **5**, 373–383.

Nanaboina V. and Korshin G. V. (2010). Evolution of absorption spectra of ozonated wastewater and its relationship with the degradation of trace-level organic species. *Environ Sci Technol*, **44**, 6130–6137.

Nangia P. S. and Benson S. W. (1980). Thermochemistry and kinetics of ozonation reactions. *J Am Chem Soc*, **102**, 3105–3115.

Nash T. (1953). The colorimetric estimation of formaldehyde by means of the Hantzsch reaction. *Biochem J*, **55**, 416–421.

Naumov S. and von Sonntag C. (2005). UV-visible absorption spectra of alkyl-, vinyl-, aryl- and thiylperoxyl radicals and some related radicals in aqueous solution: a quantum-chemical study. *J Phys Org Chem*, **18**, 586–594.

Naumov S. and von Sonntag C. (2008a). The energetics of rearrangement and water elimination reactions in the radiolysis of the DNA bases in aqueous solution (e_{aq}^- and OH attack). DFT calculations. *Radiat Res*, **169**, 355–363.

Naumov S. and von Sonntag C. (2008b). The reactions of bromide with ozone towards bromate and the hypobromite puzzle: a Density Functional Theory study. *Ozone: Sci Eng*, **30**, 339–343.

Naumov S. and von Sonntag C. (2010). Quantum chemical studies on the formation of ozone adducts to aromatic compounds in aqueous solution. *Ozone: Sci Eng*, **32**, 61–65.

Naumov S. and von Sonntag C. (2011a). Standard Gibbs free energies of reactions of ozone with free radicals in aqueous solution – Quantum-chemical calculations. *Environ Sci Technol*, **45**, 9195–9204.

Naumov S. and von Sonntag C. (2011b). The reaction of $^{\bullet}$OH with O_2, the decay of $O_3^{\bullet-}$ and the pK_a of $HO_3^{\bullet}$ – interrelated questions in aqueous free-radical chemistry. *J Phys Org Chem*, **24**, 600–602.

Naumov S., Mark G., Jarocki A. and von Sonntag C. (2010). The reactions of nitrite ion with ozone in aqueous solution – new experimental data and quantum-chemical considerations. *Ozone: Sci Eng*, **32**, 430–434.

NDMA (2009). NDMA and other nitrosamines – drinking water issues, California Department of Public Health http://www.cdph.ca.gov/certlic/drinkingwater/Pages/NotificationLevels.aspx.

Neemann J., Hulsey R., Rexing D. and Wert E. (2004). Controlling bromate formation during ozonation with chlorine and ammonia. *J Am Water Works Ass*, **96**(2), 26–29.

Nemes A., Fábián I. and van Eldik R. (2000). Kinetics and mechanism of the carbonate ion inhibited aqueous ozone decomposition. *J Phys Chem A*, **104**, 7995–8000.

Nese C., Schuchmann M. N., Steenken S. and von Sonntag C. (1995). Oxidation *vs.* fragmentation in radiosensitization. Reactions of α-alkoxyalkyl radicals with 4-nitrobenzonitrile and oxygen. A pulse radiolysis and product study. *J Chem Soc, Perkin Trans 2*, 1037–1044.

Neta P., Huie R. E., Mosseri S., Shastri L. V., Mittal J. P., Maruthamuthu P. and Steenken S. (1989). Rate constants for reduction of substituted methylperoxyl radicals by ascorbate ions and N,N,N',N'-tetramethyl-p-phenylenediamine. *J Phys Chem*, **93**, 4099–4104.

Neta P., Huie R. E. and Ross A. B. (1988). Rate constants for reactions of inorganic radicals in aqueous solution. *J Phys Chem Ref Data*, **17**, 1027–1284.

Neumann N. B., von Gunten U. and Gujer W. (2007). Uncertainty in prediction of disinfection performance. *Wat Res*, **41**, 2371–2378.

Nicoson J. S., Wang L., Becker R. H., Huff Hartz K. E., Muller C. E. and Margerum D. W. (2002). Kinetics and mechanism of the ozone/bromite and ozone/chlorite reactions. *Inorg Chem*, **41**, 2975–2980.

Niki E., Yamamoto Y., Saito T., Nagano K., Yokoi S. and Kamiya Y. (1983). Ozonization of organic compounds. VII. Carboxylic acids, alcohols and carbonyl compounds. *Bull Chem Soc Jpn*, **56**, 223–228.

Ning B., Graham N., Zhang Y. P., Nakonechny M. and El-Din M. G. (2007a). Degradation of endocrine disrupting chemicals by ozone/AOPs. *Ozone: Sci Eng*, **29**, 153–176.

Ning B., Graham N. J. D. and Zhang Y. (2007b). Degradation of octylphenol and nonylphenol by ozone – Part I: direct reaction. *Chemosphere*, **68**, 1163–1172.

Nissen F. (1890). Über die desinficirende Eigenschaft des Chlorkalks (About the disinfecting properties of chlorinated lime). *Z mediz Mikrobiol Imunol*, **8**, 62–77.

Nolte P. (1999). *Christian Friedrich Schönbein – Ein Leben für die Chemie (Christian Friedrich Schönbein – A Life for Chemistry)*. Stadt Metzingen und VHS-Arbeitskreis Stadtgeschichte, Metzinger Heimatblätter, Sonderreihe A.5, Metzingen.

Nöthe T. (2009). Zur Ozonierung von Spurenstoffen in Abwasser (On the ozonation of micropollutants in wastewater). Dissertation, Technische Universität, Dortmund.

Nöthe T., Hartmann D., von Sonntag J., von Sonntag C. and Fahlenkamp H. (2007). Elimination of the musk fragrances galaxolide and tonalide from wastewater by ozonation and concomitant stripping. *Wat Sci Tech*, **55**, 287–292.

Nöthe T., Fahlenkamp H. and von Sonntag C. (2009). Ozonation of wastewater: rate of ozone consumption and hydroxyl radical yield. *Environ Sci Technol*, **43**, 5990–5995.

Nöthe T., Launer M., von Sonntag C. and Fahlenkamp H. (2010). Computer-assisted determination of reaction parameters for the simulation of micropollutant abatement in wastewater ozonation. *Ozone: Sci Eng*, **32**, 424–429.

Nowotny N., Epp B., von Sonntag C. and Fahlenkamp H. (2007). Quantification and modeling of the elimination behavior of ecologically problematic wastewater micropollutants by adsorption on powdered and granulated activated carbon. *Environ Sci Technol*, **41**, 2050–2055.

Noyes A. A., Coryell C. D., Stitt F. and Kossiakoff A. (1937). Argentic salts in acid solution. IV. The kinetics of the reduction by water and the formation by ozone of argentic silver in nitric acid solution. *J Am Chem Soc*, **59**, 1316–1325.

Noyes R. M. (1954). A treatment of chemical kinetics with special applicability to diffusion controlled reactions. *J Chem Phys*, **22**, 1349–1359.

Nupen E. M., Roy D., Engelbrecht R. S., Wong P. K. Y. and Chian E. S. K. (1981). Inactivation of enteroviruses by ozone. *Water Sci Technol*, **13**, 1335–1336.

Oaks J. L., Gilbert M., Virani M. Z., Watson R. T., Meteyer C. U., Rideout B. A., Shivaprasad H. L., Ahmed S., Chauddry M. J. I., Arshad M., Mahmood S., Ali A. and Khan A. A. (2004). Diclofenac residues as the cause of vulture population decline in Pakistan. *Nature*, **427**, 630–633.

Oehlmann J., Schulte-Oehlmann U., Tillmann M. and Markert B. (2000). Effects of endocrine disruptors on prosobranch snails (Mollusca: Gastropoda) in the laboratory. Part I: bisphenol A and octylphenol as xeno-estrogens. *Ecotoxicology*, **9**, 383–397.

Oehlmann J., Schulte-Oehlmann U., Bachmann J., Oetken M., Lutz I., Kloas W. and Ternes T. A. (2006). Bisphenol A induces superfemininzation in the ramshorn snail *Marisa cornuarietis* (Gastropoda: Prosobranchia) at environmentally relevant concentrations. *Environ Health Perspect*, **114**, 127–133.

Oehlmann J., Di Benedetto P., Tillmann M., Duft M., Oetken M. and Schulte-Oehlmann U. (2007). Endocrine disruption in prosobranche molluscs: evidence and ecological relevance. *Ecotoxicology*, **16**, 29–43.

Oesper R. E. (1929a). Christian Friedrich Schönbein. Part I. Life and character. *J Chem Educ*, **6**, 432–440.

Oesper R. E. (1929b). Christian Friedrich Schönbein. Part II. Experimental labors. *J Chem Educ*, **6**, 677–685.

Oetken M., Nentwig G., Löffler D., Ternes T. and Oehlmann J. (2005). Effects of pharmaceuticals on aquatic invertebrates. Part I. The antiepileptic drug carbamazepine. *Arch Environ Contam Toxicol*, **49**, 353–361.

Okazaki S., Sugimitsu H., Niwa H., Kogoma M., Moriwaki T. and Inomata T. (1988). Ozone formation from the reaction of O_2-activated N_2 molecules and a new type of ozone generator with fine wire electrode. *Ozone: Sci Eng*, **10**, 137–151.

Olzmann M., Kraka E., Cremer D., Gutbrod R. and Andersson S. (1997). Energetics, kinetics, and product distributions of the reactions of ozone with ethene and 2,3-dimethyl-2-butene. *J Phys Chem A*, **101**, 9421–9429.

Oneby M. A., Bromley C. O., Borchardt J. H. and Harrison D. S. (2010). Ozone treatment of secondary effluent at U.S. municipal wastewater treatment plants. *Ozone: Sci Eng*, **32**, 43–55.

Onstad G. D., Strauch S., Meriluoto J., Codd G. A. and von Gunten U. (2007). Selective oxidation of key functional groups in cyanotoxins during drinking water ozonation. *Environ Sci Technol*, **41**, 4397–4404.

Ormad M. P., Miguel N., Lanao M., Mosteo R. and Ovelleiro J. L. (2010). Effect of application of ozone and ozone combined with hydrogen peroxide and titanium dioxide in the removal of pesticides from water. *Ozone: Sci Eng*, **32**, 25–32.

Ort C., Hollender J., Schaerer M. and Siegrist H. (2009). Model-based evaluation of reduction strategies for micropollutants from wastewater treatment plants in complex river networks. *Environ Sci Technol*, **43**, 3214–3220.

Ort C., Lawrence M. G., Reungoat J., Eaglesham G., Carter S. and Keller J. (2010). Determining the fraction of pharmaceutical residues in wastewater originating from a hospital. *Wat Res*, **44**, 605–615.

Ostling O. and Johanson K. J. (1984). Microelectrophoretic study of radiation-induced DNA damages in individual mammalian cells. *Biochem Biophys Res Commun*, **123**, 291–298.

Overbeck P. K. (1995). Ground water color and sulfide reduction with ozone. *Proc –Annu Conf, Am Water Works Assoc*, 241–249.

Oya M., Kosaka T., Asami M. and Kunikane S. (2008). Formation of *N*-nitrosodimethylamine (NDMA) by ozonation of dyes and related compounds. *Chemosphere*, **73**, 1724–1730.

Pan X.-M. and von Sonntag C. (1990). OH-Radical-induced oxidation of benzene in the presence of oxygen: $R^{\bullet} \rightleftarrows RO_2^{\bullet}$ equilibria in aqueous solution. A pulse radiolysis study. *Z Naturforsch B*, **45**, 1337–1340.

Pan X.-M., Schuchmann M. N. and von Sonntag C. (1993). Oxidation of benzene by the OH radical. A product and pulse radiolysis study in oxygenated aqueous solution. *J Chem Soc, Perkin Trans 2*, 289–297.

Paraskeva P. and Graham N. J. D. (2002). Ozonation of municipal wastewater effluents. *Wat Environ Res*, **74**, 569–581.

Parsons A. F. (2000). *Introduction to Free Radical Chemistry*. Blackwell Science, Oxford.

Peldszus S., Andrews S. A., Souza R., Smith F., Douglas I., Bolton J. and Huck P. M. (2004). Effect of medium-pressure UV irradiation on bromate concentrations in drinking water, a pilot-scale study. *Wat Res*, **38**, 211–217.

Perkin F. M. (1910). Mercury vapour lamps and action of ultra violet rays. *Trans Faraday Soc*, **6**, 199–204.

Petala M., Kokokiris L., Samaras P., Papadopoulos A. and Zouboulis A. (2009). Toxicological and ecotoxic impact of secondary and tertiary treated sewage effluents. *Wat Res*, **43**, 5063–5074.

Peter A. and von Gunten U. (2007). Oxidation kinetics of selected taste and odor compounds during ozonation of drinking water. *Environ Sci Technol*, **41**, 626–631.

Peter A., Köster O., Schildknecht A. and von Gunten U. (2009). Occurrence of dissolved and particle-bound taste and odor compounds in Swiss lake waters. *Wat Res*, **43**, 2191–2200.

Peyton G. R. and Glaze W. H. (1987). Mechanism of photolytic ozonation. In: *Photochemistry of Environmental Aquatic Systems*, R. G. Zika and W. J. Cooper (eds), ACS Symposium Series; American Chemical Society, Washington, DC, pp. 76–88.

Peyton G. R. and Glaze W. H. (1988). Destruction of pollutants in water with ozone in combination with ultraviolet radiation. 3. Photolysis of aqueous ozone. *Environ Sci Technol*, **22**, 761–767.

Peyton G. R., Huang F. Y., Burleson J. L. and Glaze W. H. (1982). Destruction of pollutants in water with ozone in combination with ultraviolet radiation. 1. General principles and oxidation of tetrachloroethylene. *Environ Sci Technol*, **16**, 448–453.

Pierpoint A. C., Hapeman C. J. and Torrents A. (2001). Linear free energy study of ring-substituted aniline ozonation for developing treatment of aniline-based pesticide wastes. *J Agr Food Chem*, **49**, 3827–3832.

Pinkernell U. and von Gunten U. (2001). Bromate minimization during ozonation: mechanistic considerations. *Environ Sci Technol*, **35**, 2525–2531.

Pinkernell U., Nowack B., Gallard H. and von Gunten U. (2000). Methods for the photometric determination of reactive bromine and chlorine species with ABTS. *Wat Res*, **34**, 4343–4350.

Piriou P., Soulet C., Acero J. L., Bruchet A., von Gunten U. and Suffet I. H. (2007). Understanding medicinal taste and odour formation in drinking waters. *Wat Sci Tech*, **55**, 85–94.

Plesnicar B. (1983). Organic polyoxides. In: *The Chemistry of Functional Groups. Peroxides*, S. Patai (ed.), Wiley, London, pp. 483–520.

Plesnicar B. (2005). Progress in the chemistry of dihydrogen trioxide (HOOOH). *Acta Chim Slov*, **52**, 1–12.

Plesnicar B., Kocjan D., Murovec S. and Azman A. (1976). An *ab initio* molecular orbital study of polyoxides. 2. Fluorine and alkyl polyoxides (F_2O_3, F_2O_4, $(CH_3)_2O_3$, $(CF_3)_2O_3$). *J Am Chem Soc*, **98**, 3143–3145.

Plesnicar B., Cerkovnik J., Koller J. and Kovac F. (1991). Chemistry of hydrotrioxides. Preparation, characterization, and reactivity of dimethylphenylsilyl hydrotrioxides. Hydrogen trioxide (HOOOH), a reactive intermediate in their thermal decomposition? *J Am Chem Soc*, **113**, 4946–4953.

Plesnicar B., Cerkovnik J., Tekavec T. and Koller J. (1998). On the mechanism of the ozonation of isopropyl alcohol: an experimental and density fuctional theoretical investigation. ^{17}O NMR spectra of hydrogen trioxide (HOOOH) and the hydrotrioxide of isopropyl alcohol. *J Am Chem Soc*, **120**, 8005–8006.

Plesnicar B., Cerkovnik J., Tekavec T. and Koller J. (2000). ^{17}O NMR spectroscopic characterisation and the mechanism of alkyl hydrotrioxides (ROOOH) and hydrogen trioxide (HOOOH) in the low-temperature ozonation of isopropyl alcohol and isopropyl methyl ether: water assisted decomposition. *Chem Eur J*, **6**, 809–819.

Plumlee M. H., McNeill K. and Reinhard M. (2009). Indirect photolysis of perfluorochemicals: hydroxyl radical-initiated oxidation of *N*-ethyl perfluorooctane sulfonamido acetate (*N*-EtFOSAA) and other perfluoroalkanesulfonamides. *Environ Sci Technol*, **43**, 3662–3668.

Pocostales P., Sein M. M., Knolle W., von Sonntag C. and Schmidt T. C. (2010). Degradation of ozone-refractory organic phosphates in wastewater by ozone and ozone/hydrogen peroxide (peroxone): the role of ozone consumption by dissolved organic matter. *Environ Sci Technol*, **44**, 8248–8253.

Ponec R., Yuzhakov G., Haas Y. and Samuni U. (1997). Theoretical analysis of the stereoselectivity in the ozonolysis of olefins. Evidence for a modified Criegee mechanism. *J Org Chem*, **62**, 2757–2762.

Porter N. A., Roe A. N. and McPhail A. T. (1980). Serial cyclization of peroxy free radicals: models for polyolefin oxidation. *J Am Chem Soc*, **102**, 7574–7576.

Porter N. A., Lehman L. S., Weber B. A. and Smith K. J. (1981). Unified mechanism for polyunsaturated fatty acid autoxidation. Competition of peroxy radical hydrogen atom abstraction, β-scission, and cyclization. *J Am Chem Soc*, **103**, 6447–6455.

Prager L., Dowideit P., Langguth H., Schuchmann H.-P. and von Sonntag C. (2001). Hydrolytic removal of the chlorinated products from the oxidative free-radical-induced degradation of chloroethylenes: acid chlorides and chlorinated acetic acids. *J Chem Soc, Perkin Trans 2*, 1641–1647.

Prasse C., Schlüsener M. P., Schulz R. and Ternes T. A. (2010). Antiviral drugs in wastewater and surface waters: a new pharmaceutical class of environmental relevance? *Environ Sci Technol*, **44**, 1728–1735.

Prasse C., Wagner M., Schulz R. and Ternes T. A. (2012). Oxidation of the antiviral drug acyclovir and its biodegradation product carboxy-acyclovir with ozone: Kinetics and identification of oxidation products. *Environ Sci Technol*, **46**, 2169–2178.

Pronk W. and Kaiser H.-P. (2008). Tomorrow's drinking water treatment. *Eawag News*, **65e**, 28–31.

Pronk W., Biebow M. and Boller M. (2006). Electrodialysis for recovering salts from a urine solution containing micropollutants. *Environ Sci Technol*, **40**, 2414–2420.

Pryor W. A., Giamalva D. and Church D. F. (1983). Kinetics of ozonation. 1. Electron-deficient alkenes. *J Am Chem Soc*, **105**, 6858–6861.

Pryor W. A., Giamalva D. H. and Church D. F. (1984). Kinetics of ozonation. 2. Amino acids and model compounds in water and comparison to rates in nonpolar solvents. *J Am Chem Soc*, **106**, 7094–7100.

Pukies J., Roebke W. and Henglein A. (1968). Pulsradiolytische Untersuchung einiger Elementarprozesse der Silberreduktion (Pulse radiolytic investigation of some elementary processes of the silver reduction). *Ber Bunsenges Phys Chem*, **72**, 842–847.

Qiang Z., Adams C. and Surampalli R. (2004). Determination of the ozonation rate constants for lincomycin and spectinomycin. *Ozone: Sci Eng*, **26**, 525–537.

Qin L., Tripathi G. N. R. and Schuler R. H. (1985). Radiation chemical studies of the oxidation of aniline in aqueous solution. *Z Naturforsch A*, **40**, 1026–1039.

Rabani J., Klug-Roth D. and Henglein A. (1974). Pulse radiolytic investigations of $OHCH_2O_2$ radicals. *J Phys Chem*, **78**, 2089–2093.

Rabinowitch E. (1937). Collision, co-ordination, diffusion and reaction velocity in condensed systems. *Trans Faraday Soc*, **33**, 1225–1233.

Rabinowitch E. and Wood W. C. (1936). The collision mechanism and the primary photochemical process in solutions. *Trans Faraday Soc*, **32**, 1381–1387.

Radjenovic J., Godehardt M., Petrovic M., Hein A., Farré M., Jekel M. and Brarceló D. (2009). Evidencing generation of persistent ozonation products of antibiotics roxithromycin and trimethoprim. *Environ Sci Technol*, **43**, 6808–6815.

Ragnar M., Eriksson T. and Reitberger T. (1999a). Radical formation in ozone reactions with lignin and carbohydrate model compounds. *Holzforschung*, **53**, 292–298.

Ragnar M., Eriksson T., Reitberger T. and Brandt P. (1999b). A new mechanism in the ozone reaction with lignin like structures. *Holzforschung*, **53**, 423–428.

Rakness K. L. (2005). Ozone in Drinking Water Treatment. American Water Works Association, Denver.

Rakness K., Gordon G., Langlais B., Masschelein W., Matsumoto N., Richard Y., Robson C. M. and Somiya I. (1996). Guideline for measurement of ozone concentration in the process gas form an ozone generator. *Ozone: Sci Eng*, **18**, 209–229.

Rakness K. L., Najm I., Elovitz M., Rexing D. and Via S. (2005). Cryptosporidium log-inactivation with ozone using effluent CT10, geometric mean CT10, extended integrated CT10 and extended CSTR calculations. *Ozone: Sci Eng*, **27**, 335–350.

Rakness K. L., Wert E. C., Elovitz M. and Mahoney S. (2010). Operator-friendly technique and quality control considerations for indigo colorometric measurement of ozone residual. *Ozone: Sci Eng*, **32**, 33–42.

Ramseier M. K. and von Gunten U. (2009). Mechanism of phenol ozonation – kinetics of formation of primary and secondary reaction products. *Ozone: Sci Eng*, **31**, 201–215.

Ramseier M. K., von Gunten U., Freihofer P. and Hammes F. (2011a). Kinetics of membrane damage to high (HNA) and low (LNA) nucleic acid bacterial clusters in drinking water by ozone, chlorine, chlorine dioxide, monochloramine, ferrate(VI), and permanganate. *Wat Res*, **45**, 1490–1500.

Ramseier M. K., Peter A., Traber J. and von Gunten U. (2011b). Formation of assimilable organic carbon during oxidation of natural waters with ozone, chlorine dioxide, chlorine, permanganate, and ferrate. *Wat Res*, **45**, 2002–2010.

Rao B., Anderson T. A., Redder A. and Jackson W. A. (2010). Perchlorate formation by ozone oxidation of aqueous chlorine/oxy-chlorine species: role of Cl_xO_y radicals. *Environ Sci Technol*, **44**, 2961–2967.

Real F. J., Benitez F. J., Acero J. L., Sagasti J. J. P. and Casas F. (2009). Kinetics of the chemical oxidation of the pharmaceuticals primidone, ketoprofen, and diatrizoate in ultrapure and natural waters. *Ind Eng Chem Res*, **48**, 3380–3388.

Reckhow D. A., Legube B. and Singer P. C. (1986). The ozonation of organic halide precursors: effect of bicarbonate. *Wat Res*, **20**, 987–998.

Reemtsma T. and These A. (2005). Comparative investigation of low-molecular-weight fulvic acids of different origin by SEC-Q-TOF-MS: new insights into structure and formation. *Environ Sci Technol*, **39**, 3507–3512.

Reemtsma T., These A., Springer A. and Linscheid M. (2006a). Fulvic acids as transition state of organic matter: indications from high resolution mass spectrometry. *Environ Sci Technol*, **40**, 5839–5845.

Reemtsma T., These A., Venkatachari P., Xia X., Hopke P. K., Springer A. and Linscheid M. (2006b). Identification of fulvic acids and sulfated and nitrated analogues in atmospheric aerosol by electrospray ionization Fourier transform ion cyclotron resonance mass spectrometry. *Anal Chem*, **78**, 8299–8304.

Reemtsma T., These A., Springer A. and Linscheid M. (2008). Differences in the molecular composition of fulvic acid size fractions detected by size-exclusion chromatography-on-line Fourier transform ion cyclotron resonance (FTICR-) mass spectrometry. *Wat Res*, **42**, 63–72.

Reid D. L., Shustov G. V., Armstrong D. A., Rauk A., Schuchmann M. N., Akhlaq M. S. and von Sonntag C. (2002). H-Atom abstraction from thiols by *C*-centered radicals. A theoretical and experimental study of reaction rates. *Phys Chem Chem Phys*, **4**, 2965–2974.

Reid D. L., Armstrong D. A., Rauk A., Nese C., Schuchmann M. N., Westhoff U. and von Sonntag C. (2003). H-atom abstraction by C-centered radicals from cyclic and acyclic dipeptides. A theoretical and experimental study of reaction rates. *Phys Chem Chem Phys*, **5**, 3278–3288.

Reisz E., Schmidt W., Schuchmann H.-P. and von Sonntag C. (2003). Photolysis of ozone in aqueous solution in the presence of tertiary butanol. *Environ Sci Technol*, **37**, 1941–1948.

Reisz E., Leitzke A., Jarocki A., Irmscher R. and von Sonntag C. (2008). Permanganate formation in the reactions of ozone with Mn(II): a mechanistic study. *J Wat Suppl Res Tech – AQUA*, **57**, 451–464.

Reisz E., Jarocki A., Naumov S., von Sonntag C. and Schmidt T. C. (2012a). Hydride transfer: a dominating reaction of ozone with tertiary butanol and formate ion in aqueous solution, in preparation.

Reisz E., Naumov S. and von Sonntag C. (2012b). The intriguing reaction of ozone with Ag(I) – formation of silver nanoparticles, in preparation.

Rennecker J. L., Marinas B. J., Owens J. G. and Rice E. W. (1999). Inactivation of *Cryptosporidium parvum* oocysts with ozone. *Wat Res*, **33**, 2481–2488.

Reungoat J., Macova M., Escher B. I., Carswell S., Mueller J. F. and Keller J. (2010). Removal of micropollutants and reduction of biological activity in a full scale reclamation plant using ozonation and activated carbon filtration. *Wat Res*, **44**, 625–637.

Rice R. G. (1999). Ozone in the United States of America – state-of-the-art. *Ozone: Sci Eng*, **21**, 99–118.

Richardson D. E., Yao H. R., Frank K. M. and Bennett D. A. (2000). Equilibria, kinetics, and mechanism in the bicarbonate activation of hydrogen peroxide: oxidation of sulfides by peroxomonocarbonate. *J Am Chem Soc*, **112**, 1729–1739.

Richardson S. D. and Ternes T. A. (2011). Water analysis: emerging contaminants and current issues. *Anal Chem*, **83**, 4614–4648.

Richardson S. D., Thurston A. D., Caughran T. V., Chen P. H., Collette T. W., Floyd T. L., Schenck K. M., Lykins B. W., Sun G.-R. and Majetich G. (1999a). Identification of new drinking water disinfection byproducts formed in the presence of bromide. *Environ Sci Technol*, **33**, 3378–3383.

Richardson S. D., Thurston A. D., Caughran T. V., Chen P. H., Collette T. W., Floyd T. L., Schenck K. M., Lykins B. W., Sun G.-R. and Majetich G. (1999b). Identification of new ozone disinfection byproducts in drinking water. *Environ Sci Technol*, **33**, 3368–3377.

Richardson S. D., Fasano F., Ellington J. J., Crumley F. G., Buettner K. M., Evans J. J., Blount B. C., Silva L. K., Waite T. J., Luther G. W., McKague A. B., Miltner R. J., Wagner E. D. and Plewa M. J. (2008). Occurrence and mammalian cell toxicity of iodinated disinfection byproducts in drinking water. *Environ Sci Technol*, **42**, 8330–8338.

Ried A., Mielcke J. and Wieland A. (2009). The potential use of ozone in municipal wastewater. *Ozone: Sci Eng*, **31**, 415–421.

Roberts L. C., Hug S. J., Ruettimann T., Billah M. M., Khan A. W. and Rahman M. T. (2004). Arsenic removal with iron (II) and iron(III) waters with high silicate and phosphate concentrations. *Environ Sci Technol*, **38**, 307–315.

Rodrigues G. S., Madkour S. A. and Weinstein L. H. (1996). Genotoxic activity of ozone in *tradescantia*. *Environ Exp Bot*, **36**, 45–50.

Rodríguez E., Onstad G. D., Kull T. P. J., Metcalf J. S., Acero J. L. and von Gunten U. (2007). Oxidative elimination of cyanotoxins: comparison of ozone, chlorine, chlorine dioxide and permanganate. *Wat Res*, **41**, 3381–3393.

Rodríguez E. M., Acero J. L., Spoof L. and Meriluoto J. (2008). Oxidation of MC-LR and -RR with chlorine and potassium permanganate: toxicity of the reaction products. *Wat Res*, **42**, 1744–1752.

Roustan M., Stambolieva D. P., Duguet J. P., Wable O. and Mallevialle J. (1991). Influence of hydrodynamics on Giardia inactivation by ozone. Study by kinetic and by "ct" approach. *Ozone: Sci Eng*, **13**, 451–462.

Roustan M., Line A., Duguet J. P., Mallevialle J. and Wable O. (1992). Practical design of a new ozone contactor – the Deep U-Tube. *Ozone: Sci Eng*, **14**, 427–438.

Roustan M., Beck C., Wable O., Duguet J. P. and Mallevialle J. (1993). Modeling hydraulics of ozone contactors. *Ozone: Sci Eng*, **15**, 213–226.

Roy D., Engelbrecht R. S., Wong P. K. Y. and Chian E. S. K. (1980). Inactivation of enteroviruses by ozone. *Progr Water Technol*, **12**, 819–836.

Roy D., Chian E. S. K. and Engelbrecht R. S. (1981a). Kinetics of enteroviral inactivation by ozone. *J Environ Eng Div Am Soc Chem*, **107**, 887–901.

Roy D., Wong P. K. Y., Engelbrecht R. S. and Chian E. S. K. (1981b). Mechanism of enteroviral inactivation by ozone. *Appl Environ Microbiol*, **41**, 718–723.

Roy D., Chian E. S. K. and Engelbrecht R. S. (1982a). Mathematical model for enterovirus inactivation by ozone. *Wat Res*, **16**, 667–673.

Roy D., Engelbrecht R. S. and Chian E. S. K. (1982b). Comparative inactivation of six enteroviruses by ozone. *J Am Water Works Ass*, **74**(12), 660–664.

Rubin M. B. (2001). The history of ozone. The Schönbein period, 1839–1868. *Bull Hist Chem*, **26**, 40–56.

Rubin M. B. (2002). The history of ozone. II. 1869–1899. *Bull Hist Chem*, **27**, 81–106.

Rubin M. B. (2003). History of ozone. Part III. C.D. Harries and the introduction of ozone into organic chemisty. *Helv Chim Acta*, **86**, 930–940.

Rubin M. B. (2004). The history of ozone. IV. The isolation of pure ozone and determination of its physical properties. *Bull Hist Chem*, **29**, 99–106.

Rubin M. B. (2007). The history of ozone. V. Formation of ozone from oxygen at high temperatures. *Bull Hist Chem*, **32**, 45–56.

Rubin M. B. (2008). The history of ozone. VI. Ozone of silca gel ("dry ozone"). *Bull Hist Chem*, **33**, 68–75.

Rubin M. B. (2009). The history of ozone. VII. The mythical spawn of ozone: antozone, oxozone, and ozohydrogen. *Bull Hist Chem*, **34**, 39–49.

Rüchardt C. (1987). Basic principles of reactivity in free radical chemistry. *Free Radical Res Commun*, **2**, 197–216.

Ruhland A., Karschunke K. and Jekel M. (2003a). Arseneliminierung aus Trinkwasser, Teil 1 (Elimination of arsenic from dinking water, Part 1). *bbr*, **03/03**, 53–61.

Ruhland A., Karschunke K. and Jekel M. (2003b). Arseneliminierung aus Trinkwasser, Teil 2 (Elimination of arsenic from dinking water, Part 2). *bbr*, **3/4**, 37–46.

Runnalls T. J., Margiotta-Casaluci L., Kugathas S. and Sumpter J. P. (2010). Pharmaceuticals in the aquatic environment: steroids and anti-steroids as high priorities for research. *Human Ecolog Risk Asess*, **16**, 1318–1338.

Russell G. A. (1957). Deuterium-isotope effects in the autoxidation of aralkyl hydrocarbons. Mechanism of the interaction of peroxy radicals. *J Am Chem Soc*, **79**, 3871–3877.

Sahni M. and Locke B. R. (2006). Quantification of hydroxyl radicals produced in aqueous phase pulsed electrical discharge reactors. *Ind Chem Eng Res*, **45**, 5819–5825.

Salem L. (1968). Intermolecular orbital theory of the interaction between conjugated systems. II Thermal and photochemical cycloadditions. *J Am Chem Soc*, **90**, 553–566.

Salhi E. and von Gunten U. (1999). Simultaneous determination of bromide, bromate and nitrite in low $\mu g\, l^{-1}$ levels by ion chromatography without sample pretreatment. *Wat Res*, **33**, 3239–3244.

Sánchez-Polo M., von Gunten U. and Rivera-Utrilla J. (2005). Efficiency of activated carbon to transform ozone into •OH radicals: influence of operational parameters. *Wat Res*, **39**, 3189–3198.

Sánchez-Polo M., Salhi E., Rivera-Utrilla J. and von Gunten U. (2006). Combination of ozone with activated carbon as an alternative to conventional advanced oxidation processes. *Ozone: Sci Eng*, **28**, 237–245.

Schlünzen F., Zarivach R., Harms J., Bashan A., Tocilj A., Albrecht R., Yonath A. and Franceschi F. (2001). Structural basis for the interaction of antibiotics with the peptidyl transferase centre in eubacteria. *Nature*, **413**, 814–821.

Schmidt C. K. and Brauch H.-J. (2008). *N,N*-Dimethylsulfamide as precursor for *N*-nitrosodimethylamine (NDMA) formation upon ozonation and its fate during drinking water treatment. *Environ Sci Technol*, **42**, 6340–6346.

Schönbein C. F. (1840). Recherches sur la nature de l'odeur qui se manifeste dans certaines actions chimiques (Research on the nature of the odor which appears in certain chemical reactions). *Compt Rend Acad Sci*, **10**, 706–710.

Schönbein C. F. (1844). *Ueber die Erzeugung des Ozons auf chemischem Wege (On the production of ozone by chemical methods)*. Schweighauser'sche Buchhandlung, Basel.

Schönbein C. F. (1854). Über verschiedene Zustände des Sauerstoffs (On various states of oxygen). *Ann Chem Pharm*, **89**, 257–300.

Schönbein C. F. (1859). Ueber die gegenseitige Katalyse einer Reihe von Oxyden, Superoxyden und Sauerstoffsäuren und die chemisch gegensätzlichen Zustände des in ihnen enthaltenen thätigen Sauerstoffes (On the mutual catalysis of a number of oxides, superoxides and oxygen acids and the opposing states of their active oxygen). *J Prakt Chem*, **77**, 129–149.

Schöneich C., Bonifacic M., Dillinger U. and Asmus K.-D. (1990). Hydrogen abstraction by thiyl radicals from activated C-H-bonds of alcohols, ethers and polyunsaturated fatty acids. In: *Sulfur-Centered Reactive Intermediates in Chemistry and Biology*, C. Chatgilialoglu and K.-D. Asmus (eds), Plenum, New York, pp. 367–376.

Schriver-Mazzuoli L. (2001). Ozone photochemistry in the condensed phase. *Phys Chem Earth C*, **26**, 495–503.

Schröder H. Fr and Meesters R. J. W. (2005). Stability of fluorinated surfactants in advanced oxidation processes – a follow up of degradation using flow injection-mass spectrometry, liquid chromatography-mass spectrometry and liquid chromatorgaphy-multiple stage mass spectrometry. *J Chromatogr A*, **1082**, 110–119.

Schuchmann M. N. and von Sonntag C. (1979). Hydroxyl radical-induced oxidation of 2-methyl-2-propanol in oxygenated aqueous solution. A product and pulse radiolysis study. *J Phys Chem*, **83**, 780–784.

Schuchmann H.-P. and von Sonntag C. (1981). Photolysis at 185 nm of dimethyl ether in aqueous solution: involvement of the hydroxymethyl radical. *J Photochem*, **16**, 289–295.

Schuchmann M. N. and von Sonntag C. (1982). Hydroxyl radical induced oxidation of diethyl ether in oxygenated aqueous solution. A product and pulse radiolysis study. *J Phys Chem*, **86**, 1995–2000.

Schuchmann M. N. and von Sonntag C. (1983). The radiolysis of uracil in oxygenated aqueous solutions. A study by product analysis and pulse radiolysis. *J Chem Soc, Perkin Trans 2*, 1525–1531.

Schuchmann H.-P. and von Sonntag C. (1984a). Methylperoxyl radicals: a study of the γ-radiolysis of methane in oxygenated aqueous solutions. *Z Naturforsch B*, **39**, 217–221.

Schuchmann M. N. and von Sonntag C. (1984b). Radiolysis of di- and trimethylphosphates in oxygenated aqueous solution: a model system for DNA strand breakage. *J Chem Soc, Perkin Trans 2*, 699–704.

Schuchmann M. N. and von Sonntag C. (1988). The rapid hydration of the acetyl radical. A pulse radiolysis study of acetaldehyde in aqueous solution. *J Am Chem Soc*, **110**, 5698–5701.

Schuchmann M. N. and von Sonntag C. (1989). Reactions of ozone with D-glucose in oxygenated aqueous solution – direct action and hydroxyl radical pathway. *J Water Supply Res Tech – Aqua*, **38**, 311–317.

Schuchmann M. N., Al-Sheikhly M., von Sonntag C., Garner A. and Scholes G. (1984a). The kinetics of the rearrangement of some isopyrimidines to pyrimidines studied by pulse radiolysis. *J Chem Soc, Perkin Trans 2*, 1777–1780.

Schuchmann M. N., Zegota H. and von Sonntag C. (1984b). Reactions of peroxyl radicals derived from alkyl phosphates in aqueous solution – model system for DNA strand breakage. In: *Oxygen Radicals in Chemistry and Biology*, W. Bors, M. Saran and D. Tait (eds), Walter de Gruyter, Berlin, pp. 629–635.

Schuchmann M. N., Zegota H. and von Sonntag C. (1985). Acetate peroxyl radicals, $^{\bullet}O_2CH_2CO_2^-$: a study on the γ-radiolysis and pulse radiolysis of acetate in oxygenated aqueous solutions. *Z Naturforsch B*, **40**, 215–221.

Schuchmann M. N., Schuchmann H.-P. and von Sonntag C. (1990). Hydroxyl radical induced oxidation of acetaldehyde dimethyl acetal in oxygenated aqueous solution. Rapid $O_2^{\bullet-}$ release from the $CH_3C(OCH_3)_2O_2^{\bullet}$ radical. *J Am Chem Soc*, **112**, 403–407.

Schuchmann M. N., Scholes M. L., Zegota H. and von Sonntag C. (1995). Reaction of hydroxyl radicals with alkyl phosphates and the oxidation of phosphatoalkyl radicals by nitro compounds. *Int J Radiat Biol*, **68**, 121–131.

Schulte-Oehlmann U., Tillmann M., Markert B., Oehlmann J., Watermann B. and Scherf S. (2000). Effects of endocrine disruptors on prosobranch snails (Mollusca: Gastropodia) in the laboratory. Part II. Triphenyltin as xeno-androgen. *Ecotoxicology*, **9**, 399–412.

Schultis T. and Metzger J. W. (2004). Determination of estrogenic activity by LYES-assay (yeast estrogenscreen-assay assisted by enzymatic digestion with lyticase). *Chemosphere*, **57**, 1649–1655.

Schulz M., Löffler D., Wagner M. and Ternes T. A. (2008). Transformation of the X-ray contrast medium iopromide in soil and biological wastewater treatment. *Environ Sci Technol*, **42**, 7207–7217.

Schumacher J. (2006). Ozonung zur weitergehenden Aufbereitung kommunaler Klärabläufe (Ozonation for enhanced treatment of municipal wastewater effluents). Dissertation, TU, Berlin.

Schumacher J., Pi Y. Z. and Jekel M. (2004a). Ozonation of persistant DOC in municipal WWTP effluent for groundwater recharge. *Wat Sci Tech*, **49**(4), 305–310.

Schumacher J., Stoffregen A., Pi Y. and Jekel M. (2004b). Der Einsatz des Rct-Konzepts zur Beschreibung der Oxidationsleistung von Ozon gegenüber von Kläranlagenabläufen (The application of the Rct-concept for the description of the extent of oxidation by ozone in wastewater effluents). *Vom Wasser*, **102**, 16–21.

Schwaiger J., Ferling H., Mallow U., Wintermayr H. and Negele R. D. (2004). Toxic effects of the non-steroidal anti-inflamatory drug diclofenac. Part I: Histopathological alterations and bioaccumulation in rainbow trout. *Aquatic Toxicol*, **68**, 141–150.

Schwarzenbach R. P., Escher B. I., Fenner K., Hofstetter T. B., Johnson C. A., von Gunten U. and Wehrli B. (2006). The challenge of micropollutants in aquatic systems. *Science*, **313**, 1072–1077.

Schwarzenbach R. P., Egli T., Hofstetter T. B., von Gunten U. and Wehrli B. (2010). Global water pollution and human health. *Annu Rev Environ Resour*, **35**, 109–136.

Scott D. B. M. and Lesher E. C. (1963). Effect of ozone on survival and permeability of *Escherichia coli*. *J Bacteriol*, **85**, 567–576.

Sedlak D. L. and von Gunten U. (2011). The chlorine dilemma. *Science*, **331**, 42–43.

Sehested K., Holcman J. and Hart E. J. (1983). Rate constants and products of the reactions of e^-, O_2^- and H with ozone in aqueous solutions. *J Phys Chem*, **87**, 1951–1954.

Sehested K., Holcman J., Bjergbakke E. and Hart E. J. (1984). A pulse radiolytic study of the reaction OH + O_3 in aqueous medium. *J Phys Chem*, **88**, 4144–4147.

Sehested K., Holcman J., Bjergbakke E. and Hart E. J. (1987). Ozone decomposition in aqueous acetate solutions. *J Phys Chem*, **91**, 2359–2361.

Sehested K., Corfitzen H., Holcman J. and Hart E. J. (1992). Decomposition of ozone in aqueous acetic acid solutions (pH 0–4). *J Phys Chem*, **96**, 1005–1009.

Sein M. M., Golloch A., Schmidt T. C. and von Sonntag C. (2007). No marked kinetic isotope effect in the peroxone ($H_2O_2/D_2O_2 + O_3$) reaction: mechanistic consequences. *Chem Phys Chem*, **8**, 2065–2067.

Sein M. M., Zedda M., Tuerk J., Schmidt T. C., Golloch A. and von Sonntag C. (2008). Oxidation of diclofenac with ozone in aqueous solution. *Environ Sci Technol*, **42**, 6656–6662.

Sein M. M., Schmidt T. C., Golloch A. and von Sonntag C. (2009). Oxidation of some typical wastewater contaminants (tributyltin, clarithromycin, metoprolol and diclofenac) by ozone. *Wat Sci Tech*, **58**, 1479–1485.

Seitz W., Jiang J. Q., Weber W. H., Lloyd B. J., Maier M. and Maier D. (2006). Removal of iodinated X-ray contrast media during drinking water treatment. *Environ Chem*, **3**, 35–39.

Shah A. D., Krasner S. W., Lee C. F. T., von Gunten U. and Mitch W. A. (2012). Trade-offs in disinfection byproduct formation associated with precursor preoxidation for control of *N*-nitrosodimethylamine formation. *Environ Sci Technol*, **46**, 4809–4818.

Sharpless C. M. and Linden K. G. (2003). Experimental and model comparisons of low- and medium-pressure Hg lamps for the direct and H_2O_2 assisted UV photodegradation of *N*-nitrosodimethylamine in simulated drinking water. *Environ Sci Technol*, **37**, 1933–1940.

References

Shinriki N., Ishizaki K., Ikehata A., Yoshizaki T., Nomura A., Miura K. and Mizuno Y. (1981). Degradation of nucleic acids with ozone. II. Degradation of yeast RNA, yeast phenylalanine tRNA and tobacco mosaic virus RNA. *Biochim Biophys Acta*, **655**, 323–328.

Shultz S., Baral H. S., Charman S., Cunnigham A. A., Das D., Ghalsasi G. R., Goudar M. S., Green R. E., Jones A., Nighot P., Pain D. J. and Prakash V. (2004). Diclofenac poisoning is widespread in diclining vulture populations accross the Indian subcontinent. *Proc R Soc Lond B (Suppl)*, **271**, S458–S460.

Siddiqui M., Amy G., Ozekin K., Zhai W. and Westerhoff P. (1994). Alternative strategies for removing bromate. *J Am Water Works Ass*, **86**(10), 81–96.

Siddiqui M., Zhai W., Amy G. and Mysore C. (1996). Bromate ion removal by activated carbon. *Wat Res*, **30**, 1651–1660.

Siddiqui M. S., Amy G. L. and Murphy B. D. (1997). Ozone enhanced removal of natural organic matter from drinking water sources. *Wat Res*, **31**, 3098–3106.

Singer A. C., Nunn M. A., Gould E. A. and Johnson A. C. (2007). Potential risks associated with the proposed widespread use of Tamiflu. *Environ Health Perspect*, **115**, 102–106.

Singer H., Müller S., Tixier C. and Pillonel L. (2002). Triclosan: occurrence and fate of a widely used biocide in the aquatic environment: field measurements in wastewater treatment plants, surface waters, and lake sediments. *Environ Sci Technol*, **36**, 4998–5004.

Singer P. C. (1990). Assessing ozonation research needs in water-treatment. *J Am Water Works Ass*, **82**(10), 78–88.

Smedley P. L. and Kiniburgh D. G. (2002). A review of the source, behaviour and distribution of arsenic in natural waters. *Appl Geochem*, **17**, 517–568.

Smeets P., van der Helm A. W. C., Dullemont Y. J., Rietfeld L. C., van Dijk J. C. and Medema G. J. (2006). Inactivation of *Escherichia coli* by ozone under bench-scale plug flow and full-scale hydraulic conditions. *Wat Res*, **40**, 3239–3248.

Smith G. D., Molina L. T. and Molina M. J. (2000). Temperature dependence of O(^{1}D) quantum yields from the photolysis of ozone between 295 and 338 nm. *J Phys Chem A*, **104**, 8916–8921.

Smith J. B., Cusumano J. C. and Babbs C. F. (1990). Quantitative effects of iron chelators on hydroxyl radical production by the superoxide-driven Fenton reaction. *Free Radical Res Commun*, **8**, 101–106.

Smith R. M. and Martell A. E. (1987). Critical stability constants, enthalpies and entropies for the formation of metal complexes of aminopolycarboxylic acids and carboxylic acids. *Sci Total Environ*, **64**, 125–147.

Snyder S. A., Westerhoff P., Yoon Y. and Sedlak D. L. (2003). Pharmaceuticals, personal care products, and endocrine disruptors in water: implications for the water industry. *Environ Engen Sci*, **20**, 449–469.

Snyder S. A., Wert E. C., Rexing D. J., Zegers R. E. and Drury D. D. (2006). Ozone oxidation of endocrine disruptors and pharmaceuticals in surface water and wastewater. *Ozone: Sci Eng*, **28**, 445–460.

Soares A., Guieysse B., Jefferson B., Cartmell E. and Lester J. N. (2008). Nonylphenol in the environment: a critical review on occurrence, fate, toxicity and treatment in wastewaters. *Env Int*, **34**, 1033–1049.

Solisio C., Del Borghi A. and De Faveri D. M. (1999). Odour emission control: a case of H_2S removal by oxidation with ozone. *Chem Biochem Eng Q*, **13**, 59–64.

Sommariva C. (2010). *Desalination and Advanced Water Treatment – Economics and Financing*. Balaban Publishers, Hopkinton, USA.

Song R., Donohoe C., Minear R., Westerhoff P., Ozekin K. and Amy G. (1996). Empirical modeling of bromate formation during ozonation of bromide-containing waters. *Wat Res*, **30**, 1161–1168.

Song R., Westerhoff P., Minear R. and Amy G. (1997). Bromate minimization during ozonation. *J Am Water Works Ass*, **89**(6), 69–78.

Sonntag H. (1890). Über die Bedeutung des Ozons als Desinficiens (On the importance of ozone as a disinfectant). *Z Mediz Mikrobiol Imunol*, **8**, 95–136.

Sontheimer H., Heilker E., Jekel M. R., Nolte H. and Vollmer F. H. (1978). The Mülheim process. *J Am Water Works Ass*, **70**(7), 393–396.

Soulard M., Bloc F. and Hatterer A. (1981). Diagrams of existence of chloramines and bromamines in aqueous solution. *J Chem Soc, Dalton Trans*, (12), 2300–2310.

Söylemez T. and von Sonntag C. (1980). Hydroxyl radical-induced oligomerization of ethylene in deoxygenated aqueous solution. *J Chem Soc, Perkin Trans 2*, 391–394.

Spengler P., Körner W. and Metzger J. W. (2001). Substances with estrogenic activity in effluents of sewage treatment plants in Southwestern Germany. 1. Chemical analysis. *Environ Toxicol Chem*, **20**, 2133–2141.

Staehelin J. and Hoigné J. (1982). Decomposition of ozone in water: rate of initiation by hydroxide ions and hydrogen peroxide. *Environ Sci Technol*, **16**, 676–681.

Staehelin J. and Hoigné J. (1985). Decomposition of ozone in water in the presence of organic solutes acting as promoters and inhibitors of radical chain reactions. *Environ Sci Technol*, **19**, 1206–1213.

Staehelin J., Bühler R. E. and Hoigné J. (1984). Ozone decomposition in water studied by pulse radiolysis. 2. OH and HO_4 as chain intermediates. *J Phys Chem*, **88**, 5999–6004.

Stalter D., Magdeburg A. and Oehlmann J. (2010a). Comparative toxicity assessment of ozone and activated carbon treated sewage effluents using an in vivo test battery. *Wat Res*, **44**, 2610–2620.

Stalter D., Magdeburg A., Weil M., Knacker T. and Oehlmann J. (2010b). Toxication or detoxication? In vivo toxicity assessment of ozonation as advanced wastewater treatment with the rainbow trout. *Wat Res*, **44**, 439–448.

Stalter D., Magdeburg A., Wagner M. and Oehlmann J. (2011). Ozonation and activated carbon treatment of sewage effluents: removal of endocrine activity and cytotoxicity. *Wat Res*, **45**, 1015–1024.

Stamplecoskie K. G. and Scaiano J. C. (2010). Light emitting diode irradiation can control the morphology and optical properties of silver nanoparticles. *J Am Chem Soc*, **132**, 1825–1827.

Stanley J. H. and Johnson J. D. (1979). Amperometric membrane electrode for measurement of ozone in water. *Anal Chem*, **51**, 2144–2147.

Staples C. A., Dorn P. B., Klecka G. M., O'Block S. T. and Harris L. R. (1998). A review of the environmental fate, effects, and exposures of bisphenol A. *Chemosphere*, **36**, 2149–2173.

Staudinger H. (1908). Oxalylchlorid. *Chem Ber*, **41**, 3558–3567.

Stefan M. I. and Bolton J. R. (1998). Mechanism of the degradation of 1,4-dioxane in dilute aqueous solution using the UV/hydrogen peroxide process. *Environ Sci Technol*, **32**, 1588–1595.

Stefan M. I. and Bolton J. R. (2002). UV direct photolysis of *N*-nitrosodimethylamine (NDMA): kinetic and product study. *Helv Chim Acta*, **85**, 1416–1426.

Stefan M. I., Mack J. and Bolton J. R. (2000). Degradation pathways during the treatmnt of methyl *tert*-butyl ether in the UV/H_2O_2 process. *Environ Sci Technol*, **34**, 650–658.

Stemmler K. and von Gunten U. (2000). OH radical-initiated oxidation of organic compounds in atmospheric water phases: Part 2. Reactions of peroxyl radicals with transition metals. *Atmos Environ*, **34**, 4253–4264.

Stemmler K., Glod G. and von Gunten U. (2001). Oxidation of metal-diethylenetriamine-pentaacetate (DTPA)-complexes during drinking water ozonation. *Wat Res*, **35**, 1877–1886.

Stumm W. (1956). Einige chemische Gesichtspunkte zur Wasserozonisierung (Some chemical aspects of the ozonation of water). *Schweizer Z Hydrol*, **18**, 201–207.

Stumm W. and Lee G. F. (1961). Oxygenation of ferrous iron. *Ind Eng Chem*, **53**, 143–146.

Suarez S., Dodd M. C., Omil F. and von Gunten U. (2007). Kinetics of triclosan oxidation by aqueous ozone and consequent loss of antibacterial activity: relevance to municipal wastewater ozonation. *Wat Res*, **41**, 2481–2490.

Sumpter J. P. and Johnson A. C. (2008). 10th anniversary perspective: reflections on endocrine disruption in the aquatic environment: from known knowns to unknown unknowns (and many things in between). *J Environ Monit*, **10**, 1476–1485.

Sun Q., Deng S. B., Huang J. and Yu G. (2008). Relationship between oxidation products and estrogenic activity during ozonation of 4-nonylphenol. *Ozone: Sci Eng*, **30**, 120–126.

Tabrizi M. T. F., Glasser D. and Hildebrandt D. (2011). Wastewater treatment of reactive dyestuffs by ozonation in a semi-batch reactor. *Chem Eng J*, **166**, 662–668.

Takagi S., Adachi F., Miyano K., Koizumi Y., Tanaka H., Watanabe I., Tanabe S. and Kannan K. (2011). Fate of perfluorooctanesulfonate and perfluorooctanoate in drinking water treatment processes. *Wat Res*, **45**, 3925–3932.

Takahashi N. (1990). Ozonation of several organic compounds having low molecular weight under ultraviolet irradiation. *Ozone: Sci Eng*, **12**, 1–18.

Takeuchi M., Mizuishi K. and Hobo T. (2000). Determination of organotin compounds in enviromental samples. *Anal Sci*, **16**, 349–359.

Taniguchi N., Takahashi K., Matsumi Y., Dylewski S. M., Geiser J. D. and Houston P. L. (1999). Determination of the heat of formation of O_3 using vacuum ultraviolet laser-induced fluorescence spectroscopy and two-dimensional product imaging techniques. *J Chem Phys*, **111**, 6350–6355.

Taniguchi N., Takahashi K. and Matsumi Y. (2000). Photodissociation of O_3 around 309 nm. *J Phys Chem A*, **104**, 8936–8944.

Taube H. (1947). Catalysis of the reaction of chlorine and oxalic acid. Complexes of trivalent manganese in solutions containing oxalic axid. *J Am Chem Soc*, **69**, 1418–1428.

Taube H. (1948a). Catalysis by manganic ion of the reaction of bromine and oxalic acid. Stability of manganic ion complexes. *J Am Chem Soc*, **70**, 3928–3935.

Taube H. (1948b). The interaction of manganic ion and oxalate. Rates, equilibria and mechanism. *J Am Chem Soc*, **70**, 1216–1220.

Taube H. (1957). Photochemical reactions of ozone in solution. *Trans Faraday Soc*, **53**, 656–665.

Tauber A. and von Sonntag C. (2000). Products and kinetics of the OH-radical-induced dealkylation of atrazine. *Acta Hydrochim Hydrobiol*, **28**, 15–23.

Tekle-Röttering A., Schmidt W., Schmidt T. C. and von Sonntag C. (2011). Kinetics and OH radical yield of the reaction of ozone with aniline and morpholine. *GDCH Conference Proceedings*.

Ternes T. A. (1998). Occurrence of drugs in German sewage treatment plants and rivers. *Wat Res*, **32**, 3245–3260.

Ternes T. A. and Joss A. (2006). *Human Pharmaceuticals, Hormones, and Fragrances: The Challenge of Micropollutants in Urban Water Management*. IWA, London.

Ternes T. A., Meisenheimer M., McDowell D., Sacher F., Brauch H.-J., Haist-Gulde B., Preuss G., Wilme U. and Zulei-Seibert N. (2002). Removal of pharmaceuticals during drinking water treatment. *Environ Sci Technol*, **36**, 3855–3863.

Ternes T. A., Stüber J., Hermann N., McDowell D., Ried A., Kampmann M. and Teiser B. (2003). Ozonation: a tool for removal of pharmaceuticals, contrast media and musk fragrancies from wastewater? *Wat Res*, **37**, 1976–1982.

Thacker J. (1975). Inactivation and mutation of yeast cells by hydrogen peroxide. *Mutation Res*, **33**, 147–156.

Thacker J. and Parker W. F. (1976). The induction of mutation in yeast by hydrogen peroxide. *Mutation Res*, **38**, 43–52.

Theis T. L. and Singer P. C. (1974). Complexation of iron(II) by organic-matter and its effect on iron(II) oxygenation. *Environ Sci Technol*, **8**, 569–573.

Theruvathu J. A., Flyunt R., Aravindakumar C. T. and von Sonntag C. (2001). Rate constants of ozone reactions with DNA, its constituents and related compounds. *J Chem Soc, Perkin Trans 2*, 269–274.

These A. and Reemtsma T. (2003). Limitations of electrospray ionization of fulvic and humic acids as visible from size exclusion chromatography with organic carbon mass spectrometric detection. *Anal Chem*, **75**, 6275–6281.

These A. and Reemtsma T. (2005). Structure-dependent reactivity of low molecular weight fulvic acid molecules during ozonation. *Environ Sci Technol*, **39**, 8382–8387.

These A., Winkler M., Thomas C. and Reemtsma T. (2004). Determination of molecular formulas and structural regularities of low molecular weight fulvic acids by size-exclusion chromatography with electrospray ionization quadrupole time-of-flight mass spectrometry. *Rapid Commun Mass Spectrom*, **18**, 1777–1786.

Thomas K. V., Balaam M., Hurst M., Nedyalkova Z. and Mekenyan O. (2004). Potency and characterization of estrogen-receptor agonists in United Kingdom estuarine sediments. *Environ Toxicol Chem*, **23**, 471–479.

Thurston-Enriquez J. A., Haas C. N., Jacangelo J. and Gerba C. P. (2005). Inactivatin of enteric adenovirus and feline calcivirus by ozone. *Wat Res*, **39**, 3650–3656.

Trofe T. W., Inman G. W. and Johnson J. D. (1980). Kinetics of monochloramine decomposition in the presence of bromide. *Environ Sci Technol*, **14**, 544–549.

Trukhacheva T. V., Gavrilov V. B., Malama G. A. and Astakhof V. A. (1992). Kinetic patterns of the death of microorganisms under the action of ozone. *Microbiology*, **61**, 467–471.

Turhan K. and Uzman S. (2007). The degradation products of aniline in the solutions with ozone and kinetic investigations. *Ann Chim*, **97**, 1129–1138.

Tyler C. R., Jobling S. and Sumpter J. P. (1998). Endocrine disruption in wildlife: a critical review of the evidence. *Crit Rev Toxicol*, **28**, 319–361.

Tyupalo N. F. and Dneprovskii Y. A. (1981). Studying the reactions of ozone with iron(II) ions in aqueous solution. *Russ J Inorg Chem*, **26**, 357–359.

Tyupalo N. F. and Yakobi Y. A. (1980). The reactions of ozone with manganese(II) and manganese(III) ions in sulphuric acid. *Russian J Inorg Chem*, **25**, 865–868.

Udert K. M., Larsen T. A., Biebow M. and Gujer W. (2003). Urea hydrolysis and precipitation dynamics in a urine-collecting system. *Wat Res*, **37**, 2571–2582.

Udovicic L., Mark F. and Bothe E. (1994). Yields of single-strand breaks in double-stranded calf-thymus DNA irradiated in aqueous solution in the presence of oxygen and scavengers. *Radiat Res*, **140**, 166–171.

Udovicic L., Mark F., Bothe E. and Schulte-Frohlinde D. (1991). Non-homogeneous kinetics in the competition of single-stranded calf-thymus DNA and low-molecular weight scavengers for OH radicals: a comparison of experimental data and theoretical models. *Int J Radiat Biol*, **59**, 677–697.

Ulanski P. and von Sonntag C. (2000). Stability constants and decay of aqua-copper(III) – a study by pulse radiolysis with conductometric and optical detection. *Eur J Inorg Chem*, 1211–1217.

Uppu R. M., Squadrito G. L., Cueto R. and Pryor W. A. (1996). Synthesis of peroxynitrite by azide – ozone reaction. *Meth Enzymol*, **269**, 311–321.

Urfer D., von Gunten U., Revelly P., Courbat R., Ramseier S., Jordan R., Kaiser H.-P., Walther J.-L., Gaille P. and Stettler R. (2001). Utilisation de l'ozone pour le traitement des eaux potables en Suisse (Use of ozone for treatment of drinking water in Switzerland). 3e partie: etude de cas specifiques. *Gas Wasser Abwasser*, **81**, 29–41.

Utter R. G., Burkholder J. B., Howard C. J. and Ravishankara A. R. (1992). Measurement of the mass accommodation coefficient of ozone on aqueous surfaces. *J Phys Chem*, **96**, 4973–4979.

van der Kooij D., Hijnen W. A. M. and Kruithof J. C. (1989). The effects of ozonation, biological filtration, and distribution on the concentration of easily assimilable organic carbon (AOC) in drinking water. *Ozone: Sci Eng*, **11**, 297–311.

van der Zee J., Dubbelman T. M. A. R. and van Steveninck J. (1987). The role of hydroxyl radicals in the degradation of DNA by ozone. *Free Radical Res Commun*, **2**, 279–284.

van Ginkel C. G., van Haperen A. M. and van der Togt B. (2005). Reduction of bromate to bromide coupled to acetate oxidation by anaerobic mixed microbial cultures. *Wat Res*, **39**, 59–64.

Vandenberg L. N., Maffini M. V., Sonnenschein C., Rubin B. S. and Soto A. M. (2009). Bisphenol-A and the great divide: a review of controversies in the field of endocrine disruption. *Endocrine Rev*, **30**, 75–95.

Vanderford B. J., Pearson R. A., Rexing D. J. and Snyder S. A. (2003). Analysis of endocrine disruptors, pharmaceuticals, and personal care products in water using liquid chromatography/tandem mass spectrometry. *Anal Chem*, **75**, 6265–6274.

Vecitis C. D., Park H., Cheng J., Mader B. T. and Hoffmann M. R. (2009). Treatment technologies for aqueous perfluorooctanesulfonate (PFOS) and perfluuorooctanoate (PFOA). *Front Environ Sci Engin China*, **3**, 129–151.

Vel Leitner N. K. and Roshani B. (2010). Kinetic of benzotriazole oxidation by ozone and hydroxyl radical. *Wat Res*, **44**, 2058–2066.

Veltwisch D. and Asmus K.-D. (1982). Methyl radical addition to nitroalkane aci-anions in aqueous solution: rate constants and optical absorption spectra. *J Chem Soc, Perkin Trans 2*, 1143–1145.

Veltwisch D., Janata E. and Asmus K.-D. (1980). Primary processes in the reactions of OH• radicals with sulphoxides. *J Chem Soc, Perkin Trans 2*, 146–153.

Vieno N. M., Härkki H., Tuhkanen T. and Kronberg L. (2007). Occurrence of pharmaceuticals in river water and their elimination in a pilot-scale drinking water treatment plant. *Environ Sci Technol*, **41**, 5077–5084.

Voegelin A., Kaegi R., Frommer J., Vantelon D. and Hug S. J. (2010). Effect of phosphate, silicate, and Ca on Fe (III)-precipitates formed in aerated Fe(II)- and As(III)-containing water studied by X-ray absorption spectroscopy. *Geochim Cosmochim Acta*, **74**, 164–186.

Vogna D., Marotta R., Napolitano A., Andreozzi R. and d'Ischia M. (2004). Advanced oxidation of the pharmaceutical drug diclofenac with UV/H_2O_2 and ozone. *Wat Res*, **38**, 414–422.

von Gunten U. (2003a). Ozonation of drinking water: Part I. Oxidation kinetics and product formation. *Wat Res*, **37**, 1443–1467.
von Gunten U. (2003b). Ozonation of drinking water: Part II. Disinfection and by-product formation. *Wat Res*, **37**, 1469–1487.
von Gunten U. (2008). Can the quality of drinking water be taken for granted? *Eawag News*, **65e**, 4–7.
von Gunten U. and Hoigné J. (1992). Factors controlling the formation of bromate during ozonation of bromide-containing waters. *J Water SRT - Aqua*, **41**, 299–304.
von Gunten U. and Hoigné J. (1994). Bromate formation during ozonation of bromide-containing waters: interaction of ozone and hydroxyl radical reactions. *Environ Sci Technol*, **28**, 1234–1242.
von Gunten U. and Hoigné J. (1996). Ozonation of bromide-containing waters: bromate formation through ozone and hydroxyl radicals. In: *Disinfection By-Products in Water Treatment. The Chemistry of Their Formation and Control*, R. C. Minear and G. L. Amy (eds), CRC Press, Boca Raton, pp. 187–206.
von Gunten U. and Oliveras Y. (1997). Kinetics of the reaction between hydrogen peroxide and hypobromous acid: implications on water treatment and natural systems. *Wat Res*, **31**, 900–906.
von Gunten U. and Oliveras Y. (1998). Advanced oxidation of bromide-containing waters: bromate formation mechanisms. *Environ Sci Technol*, **32**, 63–70.
von Gunten U. and Salhi E. (2003). Bromate in drinking water, a problem in Switzerland? *Ozone: Sci Eng*, **25**, 159–166.
von Gunten U., Hoigné J. and Bruchet A. (1995). Bromate formation during ozonation of bromide-containing waters. *Wat Supply*, **13**, 45–50.
von Gunten U., Bruchet A. and Costentin E. (1996). Bromate formation in advanced oxidation processes. *J Am Water Works Ass*, **88**(6), 53–65.
von Gunten U., Elovitz M. S. and Kaiser H.-P. (1999). Calibration of full-scale ozonation systems with conservative and reactive tracers. *J Water SRT - Aqua*, **48**, 250–256.
von Gunten U., Salhi E., Schmidt C. K. and Arnold W. A. (2010). Kinetics and mechanisms of N-nitrosodimethylamine formation upon ozonation of N,N,-dimethylsufamide-containing waters: bromide catalysis. *Environ Sci Technol*, **44**, 5762–5768.
von Sonntag C. (1969). Strahlenchemie von Alkoholen (Radiation chemistry of alcohols). *Top Curr Chem*, **13**, 333–365.
von Sonntag C. (1980). Free radical reactions of carbohydrates as studied by radiation techniques. *Adv Carbohydr Chem Biochem*, **37**, 7–77.
von Sonntag C. (1987). The Chemical Basis of Radiation Biology. Taylor and Francis, London.
von Sonntag C. (1988). Disinfection with UV-radiation. In: *Process Technologies for Water Treatment*, S. Stucki (ed.), Plenum Press, New York, pp. 159–179.
von Sonntag C. (1994). Topics in free-radical-mediated DNA damage: purines and damage amplification – superoxic reactions – bleomycin, the incomplete radiomimetic. *Int J Radiat Biol*, **66**, 485–490.
von Sonntag C. (2006). Free-Radical-Induced DNA Damage and Its Repair. A Chemical Perspective. Springer Verlag, Heidelberg, Berlin.
von Sonntag C. (2008). Advanced oxidation processes: mechanistic aspects. *Wat Sci Tech*, **58**, 1015–1021.
von Sonntag C. and Schuchmann H.-P. (1991). The elucidation of peroxyl radical reactions in aqueous solution with the help of radiation-chemical methods. *Angew Chem, Int Ed Engl*, **30**, 1229–1253.
von Sonntag C. and Schuchmann H.-P. (1997). Peroxyl radicals in aqueous solution. In: Peroxyl Radicals, Z. B. Alfassi (ed.), Wiley, Chichester, pp. 173–234.
von Sonntag C. and Schuchmann H.-P. (2001). Carbohydrates. In: *Radiation Chemistry: Present Status and Future Trends*, C. D. Jonah and B. S. M. Rao (eds), Elsevier, Amsterdam, pp. 481–511.
von Sonntag C., Mark G., Tauber A. and Schuchmann H.-P. (1999). OH radical formation and dosimetry in the sonolysis of aqueous solutions. *Adv Sonochem*, **5**, 109–145.
Voutsa D., Hartmann P., Schaffner C. and Giger W. (2006). Benzotriazoles, alkylphenols and bisphenol A in municipal wastewaters and in the Glatt River, Switzerland. *Environ Sci Pollut Res*, **13**, 333–341.
Wagner L. and Strehlow H. (1987). On the flash photolysis of bromide ion in aqueous solutions. *Ber Bunsenges Phys Chem*, **91**, 1317–1321.

Waldemer R. H. and Tratnyek P. G. (2006). Kinetics of contaminant degradation by permanganate. *Environ Sci Technol*, **40**, 1055–1061.

Wang W.-F., Schuchmann M. N., Bachler V., Schuchmann H.-P. and von Sonntag C. (1996). Termination of •CH$_2$OH/CH$_2$O•$^-$ radicals in aqueous solutions. *J Phys Chem*, **100**, 15843–15847.

Wang X., Huang X., Zuo C. and Hu H. (2004). Kinetics of quinoline degradation by O$_3$/UV in aqueous phase. *Chemosphere*, **55**, 733–741.

Wardman P. (1989). Reduction potentials of one-electron couples involving free radicals in aqueous solution. *J Phys Chem Ref Data*, **18**, 1637–1755.

Wardman P. (1991). The reduction potential of benzyl viologen: an important reference compound for oxidant/radical redox couples. *Free Radical Res Commun*, **14**, 57–67.

Watts M. J. and Linden K. G. (2009). Advanced oxidation kinetics of aqueous trialkyl phosphate flame retardants and plasticizers. *Environ Sci Technol*, **43**, 2937–2942.

Wayne R. P. (1987). The photochemistry of ozone. *Atmos Environ*, **21**, 1683–1694.

Weeks J. L. and Rabani J. (1966). The pulse radiolysis of deaerated aqueous carbonate solutions. I. Transient optical spectrum and mechanism. II. pK for OH radicals. *J Phys Chem*, **70**, 2100–2106.

Weiss S., Jakobs J. and Reemtsma T. (2006). Discharge of three benzotriazole corrosion inhibitors with municipal wastewater and improvements by membrane bioreactor treatment and ozonation. *Environ Sci Technol*, **40**, 7193–7199.

Wentworth P. Jr, Jones L. H., Wentworth A. D., Zhu X., Larsen N. A., Wilson I. A., Xu X., Goddard W. A. III, Janda K. D., Eschenmoser A. and Lerner R. A. (2001). Antibody catalysis of the oxidation of water. *Science*, **293**, 1806–1811.

Wentworth P. Jr, Wentworth A. D., Zhu X., Wilson I. A., Janda K. D., Eschenmoser A. and Lerner R. A. (2003). Evidence for the production of trioxygen species during antibody-catalyzed chemical modification of antigens. *Proc Nat Acad Sci USA*, **100**, 1490–1493.

Wert E. C., Rosario-Ortiz F. L., Drury D. D. and Snyder S. A. (2007). Formation of oxidation byproducts from ozonation of wastewater. *Wat Res*, **41**, 1481–1490.

Wert E. C., Rosario-Ortiz F. L. and Snyder S. A. (2009a). Effect of ozone on the oxidation of trace organic contaminants in wastewater. *Wat Res*, **43**, 1005–1014.

Wert E. C., Rosario-Ortiz F. L. and Snyder S. A. (2009b). Using ultraviolet absorbance and color to assess pharmaceutical oxidation during ozonation of wastewater. *Environ Sci Technol*, **43**, 4858–4863.

Westerhoff P., Song R., Amy G. and Minear R. A. (1997). Application of ozone decomposition models. *Ozone: Sci Eng*, **19**, 55–74.

Westerhoff P., Song R., Amy G. L. and Minear R. (1998a). NOM's role in bromine and bromate formation during ozonation. *J Am Water Works Ass*, **89**(2), 82–94.

Westerhoff P., Song R., Amy G. L. and Minear R. (1998b). Numerical kinetic models for bromide oxidation to bromine and bromate. *Wat Res*, **32**, 1687–1699.

Westerhoff P., Yoon Y., Snyder S. and Wert E. (2005). Fate of endocrine-disruptor, pharmaceutical, and personal care product chemicals during simulated drinking water treatment processes. *Environ Sci Technol*, **39**, 6649–6663.

Westerhoff P., Mezyk S. P., Cooper W. J. and Minakata D. (2007). Electron pulse radiolysis determination of hydroxyl radical rate constants with Suwanee River fulvic acid and other dissolved organic matter isolates. *Environ Sci Technol*, **41**, 4640–4646.

WHO (2004). Guidelines for Drinking Water Quality. World Health Organization, Geneva.

Wick A., Fink G., Joss A., Siegrist H. and Ternes T. A. (2009). Fate of beta blockers and pseucho-active drugs in conventional wastewater treatment. *Wat Res*, **43**, 1060–1074.

Wickramanayake G. B., Rubin A. J. and Sproul O. J. (1984a). Inactivation of *Giardia lamblia* cysts with ozone. *Appl Environ Microbiol*, **48**, 671–672.

Wickramanayake G. B., Rubin A. J. and Sproul O. J. (1984b). Inactivation of *Naegleria* and *Giardia* cysts in water by ozonation. *J Water Pollut Control Fed*, **56**, 983–988.

Wilkinson F., Helman W. P. and Ross A. B. (1995). Rate constants for the decay and reactions of the lowest electronically excited singlet state of molecular oxygen in solution. An expanded and revised compilation. *J Phys Chem Ref Data*, **24**, 663–1021.

Wine P. H. and Ravishankara A. R. (1982). O₃ photolysis at 248 nm and O(¹D₂) quenching by H₂O, CH₄, H₂, and N₂O: O(³P_J) yields. *Chem Phys*, **69**, 365–373.

Wirzinger G., Vogt C., Bachmann J., Hasenbank M., Liers C., Stark C., Ziebart S. and Oehlmann J. (2007). Imposex of the netted welk *Nassarius reticulatus* (Prosobrachia) in Brittany along a transect from a point source. *Cah Biol Mar*, **48**, 85–94.

Wols B. A., Hofman J. A. M. H., Uijttewaal W. S. J., Rietveld L. C. and van Dijk J. C. (2010). Evaluation of different disinfection calculation methods using CFD. *Environ Model Software*, **25**, 573–582.

Xiong F. and Graham N. J. D. (1992). Rate constants for herbicide degradation by ozone. *Ozone: Sci Eng*, **14**, 283–301.

Xu A. H., Li X. X., Xiong H. and Yin G. C. (2011). Efficient degradation of organic pollutants in aqueous solution with bicarbonate-activated hydrogen peroxide. *Chemosphere*, **82**, 1190–1195.

Xu P., Janex M. L., Savoye P., Cockx A. and Lazarova V. (2002). Wastewater disinfection by ozone: main parameters for process design. *Wat Res*, **36**, 1043–1055.

Xu X. and Goddard W. A. III (2002). Peroxone chemistry: formation of H₂O₃ and ring-(HO₂)(HO₃) from O₃/H₂O₂. *Proc Nat Acad Sci USA*, **99**, 15308–15312.

Yang L., Chen Z., Shen J., Xu Z., Liang H., Tian J., Ben Y., Zhai X., Shi W. and Li G. (2009). Reinvestigation of the nitrosamine-formation mechanism during ozonation. *Environ Sci Technol*, **43**, 5481–5487.

Yao C. C. D. and Haag W. R. (1991). Rate constants for direct reactions of ozone with several drinking water contaminants. *Wat Res*, **25**, 761–773.

Yao H. R. and Richardson D. E. (2000). Epoxidation of alkenes with bicarbonate-activated hydrogen peroxide. *J Am Chem Soc*, **122**, 3220–3221.

Ye M. Y. and Schuler R. H. (1989). Second order combination reactions of phenoxyl radicals. *J Phys Chem*, **93**, 1898–1902.

Yeatts L. R. B. and Taube H. (1949). The kinetics of the reaction of ozone and chloride ion in acid aqueous solution. *J Am Chem Soc*, **71**, 4100–4105.

Yunes R. A., Terenzani A. J. and do Amaral L. (1975). Kinetics and mechanism for azobenzene formation. *J Am Chem Soc*, **97**, 368–373.

Yurkova I. L., Schuchmann H.-P. and von Sonntag C. (1999). Production of OH radicals in the autoxidation of the Fe(II)-EDTA system. *J Chem Soc, Perkin Trans 2*, 2049–2052.

Zehavi D. and Rabani J. (1972). The oxidation of the bromide ion by hydroxyl radicals. *J Phys Chem*, **76**, 312–319.

Zehender F. (1952). Über die Ozonbestimmung bei der Trinkwasseruntersuchung (On the determination of ozone for drinking water examination). *Mitt Gebiete Lebensm Hyg*, **43**, 143–151.

Zehender F. and Stumm W. (1953). Die Ozonbestimmung bei der Trinkwasseruntersuchung (The determination of ozone for drinking water examination). *II Mitt Gebiete Lebensm Hyg*, **44**, 206–213.

Zhang X. J., Zhang N., Schuchmann H.-P. and von Sonntag C. (1994). Pulse radiolysis of 2-mercaptoethanol in oxygenated aqueous solution. Generation and reactions of the thiylperoxyl radical. *J Phys Chem*, **98**, 6541–6547.

Zhang X.-M. and Zhu Q. (1997). Olefinic ozonation electron transfer mechanism. *J Org Chem*, **62**, 5934–5938.

Zhou H. D. and Smith D. W. (1995). Evaluation of parameter estimation methods for ozone disinfection kinetics. *Wat Res*, **29**, 679–686.

Zimmermann S. G., Wittenwiler M., Hollender J., Krauss M., Ort C., Siegrist H. and von Gunten U. (2011). Kinetic assessment and modeling of an ozonation step for full-scale municipal wastewater treatment: micropollutant oxidation, by-product formation and disinfection. *Wat Res*, **45**, 605–617.

Zimmermann S. G., Schukat A., Benner J., von Gunten U. and Ternes T. A. (2012). Kinetic and mechanistic investigations of the oxidation of tramadol by ferrate and ozone. *Environ Sci Technol*, **46**, 876–884.

Index

Bold page numbers refer to ozone rate constants, italic ones to hydroxyl radical rate constants. When a page number is followed by "f" ("ff"), the next page (the following pages) should be inspected as well.

A

Abacavir, 104
AC, see activated carbon
Acebutolol, **131**
Acetaminophen, see paracetamol
Acetamidoacrylic acid, **81**
Acetate ion, *28*, **169**, *173*
    as inhibitor for ozone decay, 27f
    in the reaction of silver in with ozone, 210
Acetic acid, **169**, *176*
Acetic peracid, 88, *176*, 178
Acetone, **169**, *176*
    enol of, 177
    photolysis of, 177
N-Acetylglycine, **131**
N-α-Acetylhistidine, **131**
N-α-Acetyllysine, **131**
N-ε-Acetyllysine, **131**
N-(4)-Acetylsulfamethoxazole, **131**, 150
Aciclovir, **81**, 103f
Acrylamide, **81**
Acrylic acid, 20, **81**, 92ff, 103
Acrylonitrile, **81**, 91
Activated carbon (AC)
    adsorption of micropollutants on, 47, 64, 66, 68ff, 129, 157, 199
    biological filtration, 34, 64, 67, 72
    formation of hydroxyl radicals in reaction of ozone with, 42, 46f, 143
    in drinking water treatment, 66ff, 71, 212
    in wastewater treatment, 72, 76
Activation energy
    of ozone reactions, 19f, 165, 170, 174f, 179
Acyclovir, see aciclovir
ADDA group, see microcystin-LR
Adenine, 53f, **81**
Adenosine, 53, **81**, 102, 142
5′-Adenylic acid, **81**
Advanced Oxidation Process (AOP), 4, 28, 42ff, 71, 75, 77, 128, 153, 188f, 198f, 230f
    energy requirements, 73f
Aflatoxins, 127
AHTN, see tonalide
Alachlor, **131**, 155
Alanine, **131**
β-Alanine, **131**
Aldehydes
    constituents of AOC, 34f, 64, 72
    formation of, in olefin reactions, 85ff
    formation of, in tributyl tin degradation, 211
    reaction with hydrogen peroxide, 85, 88
Aldicarb, **161**, 167
Algae, 32, 34, 70, 106, 155
Alginic acid, 182

Alkalinity
    effect on ozone stability, 23, 25ff, 65
    effect on bromate formation, 240
    see also Carbonate/bicarbonate
Alkoxyl radicals, 217, 220, 237
Amide, 137
Amidotrizoic acid, **109**, 129, *241*, 243
Amikacin, **132**, 150
Amines, 16, 20, 137ff, 219
Amine radical cation, 141f, 219
Amino acids, 52, 137f
4-Aminophenylmethyl sulfone, **132**, 150
Aminoxide, 53, 139ff, 145f, 147f
Aminyl radical, 152, 219
    see also amine radical cation
    see also aniline radical cation
Amiodarone, 74, 148
Ammonia, **132**, 138f, 146f
    in bromate mitigation, 198ff
Amoxillin, 126f, 155, 167
AMPA, 153
Anatoxin-a, **81**, 105f, 155
Aniline, 114, 116, **132**, 137, 142, 143ff
Aniline radical cation, 145
Anisole, see methoxybenzene
Antimicrobial compounds, 60ff, 103, 124, 126, 147ff, 166
Antibiotic resistance, 60
Antiviral compounds, 62, 103f, 151
Anthrax, 2
AOC, see Assimilable Organic Carbon
AOP, see Advanced Oxidation Process
Arginine, **132**
Aromatic compounds, 86, 109ff, 233, 243f
    Hammett–plot of, 113
    ozonide formation from, 116
    radical cations derived from, 117
    singlet oxygen yield from, 118
Arrhenius parameters of ozone reactions, 19
Arrhenius plots
    of reaction of ozone with dimethyl sulfoxide, 165
    of reaction of ozone with formate ion, 174
    of reaction of ozone with 2-propanol, 179
    of reaction of ozone with 2-methyl-2-propanol (tertiary butanol), 175
Arsenic(III), 20, **205**, 206f, 211
Arylhydrocarbon receptor response test, 62
Asparagine, **132**
Aspartate ion, **132**
Aspergillus, 127
Assimilable Organic Carbon (AOC)
    constituents of, 33ff
    degradation of by biological sand filtration, 34f, 66, 72f
    formation of, in ozonation of DOM-containing waters, 33f, 68
Atenolol, 39, **132**, 148f
Atrazine, **41**, 47, 58, 74, **132**, 146, 153, *241*, 243f
Auramine, 159
Azide ion, **185**, 196f
Azithromycin, 61, **132**
Azobenzene, **132**, 144, 153f
Azoxybenzene, 154

B
Bacillus subtilis spores, **51**
BDE, see Bond dissociation energy
BDOC, see Biodegradable dissolved organic carbon
Bennett reaction, 232, 235
Benzaldehyde, 109
1-(2-Benzaldehyde)-4-hydro(1H,3H)-quinazoline-2-one, 102, **132**
Benzene, 86, **109**, 116f, 145
    Hammett plot for derivatives, 113ff
    reaction with hydroxyl radicals, 226
Benzenesulfonate, **109**
Benzo[a]pyrene, 129f, **130**
Benzoate ion, **109**
Benzophenone, 129
1,4-Benzoquinone, **81**, 121ff
Benzotriazole, **132**, 142, 153
Benzylamine, **132**
Berzelius, J., 2
Beta blocker, 21, 75, 148f
Bezafibrate, **39**, **109**, 128
BHT, see 2,6-di-*t*-butyl-4-methylphenol
Bicarbonate ion, *26*, **185**
    as inhibitor of ozone decay, 25ff
    hydroxyl radical reactivity pK of, 27
    pK_a value of, 27
Biochanin, 57
Biodegradable dissolved organic carbon (BDOC), see Assimilable organic carbon (AOC)
Biofilm growth, 75
Biological filtration, 34f, 67ff, 200
Bioluminescence inhibition test, 62f
Biopolymers, in size exclusion chromatography, 31f
Bisoprolol, 148f
Bisphenol A, 56f, 58ff, **109**, 125f
Boiling point of ozone, 8
Bond angle
    of O–O–O, in ozone, 8

Index

Bond dissociation energy (BDE), 11, 45, 172f, 227f
Bond length
   of O–O bond, in oxygen, 222
   of O–O bond, in ozone, 7
   of O–O bond, in singlet oxygen, 164
Borate ion, **185**
Bromate ion, **185**
   as disinfection by-product, 4, 36, 55, 156, 226
   formation of, 46, 52, 75, 192ff, 198ff, 223, 202f, 223, 238ff
   mitigation of, 198ff
Bromide ion, 35f, 52, 75, **185**, 190, 192ff, 198ff, 223, 238ff, *239*
   activation energy of ozone reaction, 20
   in NDMA formation, 36, 157ff, 243
   interference of, in ozone determination, 13f
   formation of singlet oxygen, 187
   reactions catalysed by, 157f, 201
   role in disinfection by-products formation, 35f
Bromine, 2, 59, 191, 193, 239
Bromine atom, **213**, 220f, 223, 239
Bromine dioxide, 194, 221, 223f
Bromite ion, **185**, 192ff, 223, *240*
Bromoform, 36, 156, **169**
Bromonitromethane, 35
Bromoorganic disinfection by-products, 35f
Bromophenol, 35, 66, 71
tBuOH, see 2-methyl-2-propanol
1-Butanol, **169**
tertiary Butanol (tBuOH), see 2-methyl-2-propanol
2-Butanone, **169**
Buten-3-ol, **81**, 89f, 95
   for determination of ozone rate constants, 16, 18
Butylamine, **132**
s-Butylamine, **132**
2-*t*-Butyl-4-methoxyphenol, 126
t-Butylphenol, 56f, 125
Butyrate ion, **169**
Butyric acid, **169**

C

CaDTPA^{3-}, **133**
Cage reaction, 141f, 174, 177f
Carbamazepine, 39, 58, **82**, 102
Carbofuran, **109**
Carbohydrates, 147, 180ff
   fraction containing, in DOM, 32f
Carbonate ion, *26*, **185**, 207, 239
   alkalinity, 23, 25, 65, 74, 239f
   see also Bicarbonate ion

Carbonate radical (CO$_3^{\cdot-}$), 27, 239f
Carbon dioxide radical anion (CO$_2^{\cdot-}$), 174, 247
   bimolecular decay of, 228
   reaction with ozone, 214f
   reduction potential of, 214
Carbon tetrachloride, **169**
Carboxone process, 42, 46
Carboxylic acids
   constituents of AOC, 33ff
Carboxymethyl radical anion, **213**
Catechol, **109**, 121ff
   ozonolysis products of, 124
Cephalexin, 62, **82**, 103, 166
Cephalexin (*R*)-sulfoxide, 103, 166
Cephalexin (*S*)-sulfoxide, 166
Chain reaction, 27, 88, 99, 145, 182, 215, 217, 219f, 246f
Charge transfer interaction, 37
Chemical oxygen demand (COD), 33
Chironomid toxicity test, 63f
Chloramination, 35f, 67, 75, 156, 195, 200, 203
Chlorate ion, **185**, 192, 222, 237f
Chlordane, **82**, 107
Chloride ion, **185**, 190f, 226, *237*, 237f
   activation energy of ozone reaction with, 20
Chlorination
   arsenic oxidation by, 211
   elimination of oestrogenicity by, 59
   formation of disinfection by-products upon, 35f, 156, 195
   history of, 2ff, 65f
   inactivation of bacteria by, 51f
   in drinking water treatment schemes, 67f
   taste and odour formation by, 35
Chlorine, 2f, 191, 201
Chlorine atom, 88, 220ff, 227, 237f, 246f
Chlorine dioxide, 52, 59, 67f, 102, 192, **213**, 222, 237f
Chlorite ion, **185**, 192, *238*
3-Chloroaniline, **132**
4-Chloroaniline, **132**
Chlorobenzene, **109**, 114
p-Chlorobenzoic acid (pCBA), 39, 42ff, 46f, 73f, 230f, **231**, *231* as ·OH probe in ozonation reactions, 28f, 43ff, 46f
5-Chlorobenzotriazole, **132**
Chloroethene, 90
Chloroform, **169**
Chlorophenol, 35, 66, 71
2-Chlorophenol, **109**
4-Chlorophenol, **109**
2-(4-Chlorophenoxy)-2-methylpropionic acid (clofibric acid), **110**, 128

Chloropicrin, see trichloronitromethane
Chlorotoluron, **133**
5-Chlorouracil, **82**, 101
Cholera, 2
Chromosome aberration, 50
Cinnamic acid, **82**, 98f
Ciprofloxacin, 61, **133**, 149
Clarithromycin, 61, **133**, 147
Clofibric acid, see 2-(4-chlorophenoxy)-
    2-methylpropionic acid
Cobalt(II), **205**, 207
COD, see Chemical oxygen demand
Colour removal, 65
Comet assay, 64
Competition kinetics, 15, 17f, 229f, 244
Completely stirred tank reactor (CSTR), 51
Copper(II), **205**, 207, 227
Creatine, **133**
Creatinine, **133**
2-Cresol, **110**
3-Cresol, **110**
4-Cresol, **110**
Criegee, R., 3, 84
Criegee mechanism, 84ff, 103, 106, 166
Cryptosporidium parvum oocysts, 4, **51**, 55
CSTR, see completely stirred tank reactor
CT-concept, 55
Cyanate ion, **185**
Cyanide ion, **186**, 228
Cyanotoxins, 77, 104f
Cybutryn, see irgarol
Cycloaddition, 85f, 113
β-Cyclocitral, 70, **82**, 106
Cyclohexanemethylamine, 133
Cyclohexylamine, **133**
Cyclopentanol, **169**
Cyclophosphamide, **133**, 155
Cylindrospermopsin, **82**, 105, 155
Cysteine anion, **161**
Cystine, **161**
Cytidine, 53, **82**, 102
Cytosine, 53, **82**, 102

D

2,4-D, see 2,4-dichlorophenoxyacetic acid
DABCO, see 1,4-diazabicyclo[2.2.2]octane
Daidzein, 57
DBPs, see Disinfection by-products
Decarboxylation, 33, 88, 98
Deethylatrazine, **133**

Deethyldeisopropylatrazine, **133**
Deisopropylatrazine, **133**
2′-Deoxyadenosine, 53, **82**, 102
2′-Deoxycytidine, 53, **82**, 102
5′-Deoxycytidylic acid, **82**
2′-Deoxyguanosine, **82**, 102
5′-Deoxyguanylic acid, **82**
2,4-Diamino-5-methylpyrimidine, **133**
1,4-Diazabicyclo[2.2.2]octane (DABCO), **133**, 138, 140f
Diazepam, **41**, *41*, **133**, 150f
1,2-Dibromoethene, 91
2,6-Di-*t*-butyl-4-methylphenol (BHT), 71, **110**, 130
Dibutyl tin, **206**, 211f
Dichloramine, **133**
1,3-Dichlorobenzene, **110**
1,4-Dichlorobenzene, **110**
1,1-Dichloroethene, **82**, 90, 247
cis-1,2-Dichloroethene, **82**, 90, 102, 247
trans-1,2-Dichloroethene, **82**, 90f
Dichloromaleic acid, **82**, 99
Dichloromethane, **169**, 247
2,3-Dichlorophenol, **110**
2,4-Dichlorophenol, **110**
2,4-Dichlorophenoxyacetic acid (2,4-D), **110**, 128
1,1-Dichloropropene, **82**, 90
Dichloro radical anion, **213**, 220f, 237
Diclofenac, 39, **133**, 142f, 145, 151f, 219
    ozone consumption by, 15ff
Diethylamine, **133**, 140
N,N-Diethylaniline, 140, 142
Diethylenetriaminepentaacetic acid (DTPA), **133**, 154
    CaDTPA^{3-}, **133**
    Fe(III)DTPA^{2-}, **133**
    Fe(III)(OH)DTPA^{3-}, **133**
    ZnDTPA^{3-}, **134**
Diethyl ether, **169**, 234
Diethyl malonate, **169**
N,N-Diethyl-*p*-phenylenediamine (DPD), 12f, 142
    hydroxyl radical formation by ozone, 13, 142
    in ozone determination, 12f
Diethyl-*o*-phthalate, **110**
Diethyl vinyl phosphonate, **82**, 91
Dihydrogen sulfide, **186**
Dihydrogen trioxide (H$_2$O$_3$)
    decay of, 183
    formation of, 176ff, 182ff
    ^{17}O-NMR of, 176
    pK_a value of, 183
    reactivity of, 184
3,4-Dihydroxycinnamic acid, **82**

Index

1,2-Dimethoxybenzene, 118ff
1,4-Dimethoxybenzene, **110**, 116, 118ff
　　singlet oxygen formation from, in reaction with ozone, 118
1,2-Dimethoxytoluene, **110**
N,N-Dimethylacetamide, **134**
Dimethylamine, **134**, 139f
　　as precursor in NDMA formation, 159
　　formation from NDMA photolysis, 243
4-Dimethylaminoantipyrine, **134**, 156
3-(Dimethylaminomethyl) indole, **134**
Dimethylaniline, **134**, 159
　　formation of NDMA from, in reaction with ozone, 159
　　singlet oxygen formation from, in reaction with ozone, 118
5,6-Dimethylbenzotriazole, **134**, 142
Dimethylchloramine, **134**
N,N-Dimethylcyclohexylamine, **134**
Dimethylethanolamine, **134**
Dimethylformamide, **134**
　　as precursor for NDMA formation, 159
3,5-Dimethylisoxazole, **134**
N,N-Dimethylphenylenediamine
　　as precursor in NDMA formation, 159
Dimethyl-*o*-phthalate, **110**
Dimethylsulfamide (DMS), **134**, 156ff, 201
Dimethyl sulfide
　　addition of ozone to, 163
　　reaction with hydroxyl radical, 226
Dimethyl sulfoxide (DMSO), 13, 123f, 146, **161**, 164f, *232*
　　activation energy of ozone reaction, 20
　　Arrhenius plot for ozone reaction with, 165
　　reaction with hydroxyl radical, 226, 232f
1,3-Dimethyluracil, **82**
　　singlet oxygen formation from, in reaction with ozone, 101
1,4-Dioxane, **169**, *241*, 242
(Dioxido)trioxidosulfate(dot-), **213**
Dioxin, 62
2,4-Dinitrotoluene, **110**
2,6-Dinitrotoluene, **110**
Dipole moment of ozone, 7
Disinfection
　　by chlorine, 2, 65
　　by ozone, 2, 49, 65ff
　　by solar radiation, 49
　　by UV radiation, 2, 49, 65
　　CT-concept in, 55
　　damage by ozone, 49f, 52ff
　　damage by ionising radiation in, 50, 64
　　kinetics of, 49ff
　　lag phase in, 51
　　of bacteria, 49ff, 72
　　of bacterial spores, 49ff
　　of drinking water, 55, 65, 66ff
　　of protozoa, 4, 49ff
　　of viruses, 49ff
　　of wastewater, 55, 65, 72
Disinfection by-products (DBPs)
　　formation of, 4, 35f, 67, 156, 195, 198ff, 202f
　　mitigation of, 35f
Dissolved air flotation, 76
Dissolved organic carbon (DOC), see also Dissolved organic matter (DOM)
Dissolved organic matter (DOM)
　　competition of, in micropollutant oxidation, 37ff, 73
　　cytoplasmic, 34
　　effect on bromate formation by, 199
　　hydrophobic fraction of, 31ff
　　in drinking water, 23, 25
　　in wastewater, 23
　　isolates, 25f, 37
　　formation of AOC from, 33f
　　mineralization of, 33
　　molecular weight distribution of, 31ff
　　ozone reactions with, 23f, 25f, 67
　　rate constant with hydroxyl radical, 26
　　SEC–OCD chromatogram of, 31f
　　UV/Vis absorption of, 36f
trans-1,2-Dithiane-4,5-diol, **161**ff
1,4-Dithiothreitol, **161**
Diuron, **134**, 155
　　toxicity elimination by ozone, 155
DMS, see dimethylsulfamide
DMSO, see dimethyl sulfoxide
DNA, **54**, *54*, **82**
　　damage of, 52ff, 105
　　repair of, 49f
DOC, see Dissolved organic carbon
DOM, see Dissolved organic matter
DPD, see *N,N*-Diethyl-*p*-phenylenediamine
Drinking water
　　treatment schemes, 66ff
DTPA, see diethylenetriaminepentaacetic acid
ΔG^0, see standard Gibbs free energy

E

E1, see Oestrone
E2, see Oestradiol

EDCs, see endocrine disrupting compounds
EDTA, see ethylenediaminetetracetic acid
EE2, see 17α-Ethinyloestradiol
Electron, hydrated, **213**, 214, 229
Electron transfer
   adduct formation vs., 117, 189
   in ozone chemistry of silver ion, 209f
   in reaction of amines with ozone, 141
   in reaction of bromite with ozone, 194
   in reaction of chlorite with ozone, 192
   in reaction of hydroxyl radical with
      halophenols, 228
   in reaction of nitrite with ozone, 196
   in reaction of olefins with ozone, 85
   to ozone by aromatic compounds, 117
Endocrine disrupting compounds (EDCs), 56ff
Endrin, **82**, 107
Energy requirement
   for desalination, 75
   for micropollutant abatement with ozone, 73f
   for micropollutant abatement with
      peroxone, 73f
   for micropollutant abatement with
      UV/peroxide, 73f
   for wastewater reclamation, 75
Enrofloxacin, 61, **134**, 149
EOM, see Extracellular organic matter
EPTC, see S-ethyldipropyl(thiocarbamate)
Equol, 57
Erythromycin, **147**
Escherichia coli (*E. coli*), **51**, 52, 72
Estrogen etc., see Oestrogen etc.
Ethanol, **169**, 172f
Ethene, **83**, 89f
17α-Ethinyloestradiol (EE2), 56, 58f, 60, 78f, *79*, **110**, 125
Ethylamine, 140
Ethylbenzene, **110**
S-Ethyl-N,N-dipropyl(thiocarbamate) (EPTC), **162**, 167
Ethylenediaminetetraacetic acid (EDTA), **134**, 138, 140, 154
   CaEDTA^{4-}, **134**, 140, 154
   Fe(III)EDTA^{3-}, **134**, 140, 154
Ethyl-N-piperazinecarboxylate, **134**
Explosion threshold of ozone, 8
Exposure (integral)
   of hydroxyl radicals in ozone reactions, 28ff
   of ozone, 29, 50ff, 55, 158
Extracellular organic matter (EOM), 32

F
Faraday, M., 2
Fe(III)DTPA^{2-}, **133**
FELST, see fish early life toxicity test
Fe(III)(OH)DTPA^{3-}, **133**
Feminisation, 56, 58
Fenton, H. J. H., 1
Fenton reaction, 1, 230f
Ferrate(VI)
   elimination of oestrogenicity by, 59
   reaction with carbamazepine, 102
Fish early life toxicity test (FELST), 64
Flocculation, 66f, 211
Flow cytometry, 52, 68f, 73
Flumequine, **83**
Fluoride ion, 189f
2-Fluoroaniline, **134**
3-Fluoroaniline, **134**
4-Fluoroaniline, **134**
Fluoroquinolones, 60f
Fluoxetine, 149
Formaldehyde, **170**, 211
   detection of, 16
Formate ion, **170**, 171ff
   activation energy of ozone reaction, 20
   Arrhenius plot for ozone reaction with, 174
Formic acid, **170**
Formic peracid, 88ff, 92
Formononetine, 57
Formyl bromide, 90
Formyl chloride, 90
Formyl phosphonate, 91
β-Fragmentation, 178f, 215, 232, 236f
Fuel cell, 1
Fumaric acid, **83**, 86, 95f
Fuscarum, 127

G
GAC, see granular activated carbon
Galaxolide (HHCB), **110**, 116, 128
Gel permeation chromatography (GPC), 31f
Gemfibrozil, **110**, 127f
Genistein, 57
Genotoxicity, 62ff, 105
Geosmin, 70, **170**, *241*, 242
Giardia lamblia cysts, 4, 51
Gibbs free energy; see standard Gibbs free
   energy (ΔG^0)
Glucose, **170**, 180f
Glutamate ion, **134**

Glutarate ion, **170**
Glutaric acid, **170**
Glutathione, 63, **162**
Glutathione peroxidase test, 63
Glutathione S-transferase test, 63
Glycine, **134**, 137, 140
Glyoxylate ion, **170**
Glyoxylic acid, **170**
Glyphosate, 153
Goethe, J.W., von, 4
GPC, see Gel permeation chromatography
Granular activated carbon (GAC), see also activated carbon, 71
Growth inhibition test, 61, 63
Guajac resin, 1
Guanine, 53ff, 103
Guanosine, **83**, 102f
Gun cotton, 1

H

HAA, see haloacetic acid
H-Abstraction
    by hydroxyl radical, 27, 227f, 241f
    by ozone, 171ff
    standard Gibbs free energies in, by ozone, 172f
Haem peroxidase test, 63
Halide ions, 190ff
    energetics of reactions with ozone, 190
Haloacetic acids (HAAs), 35f
Halonitromethanes, 35f
Hammett plot, 113, 115f
Hammond–Leffler postulate, 165
Hardenberg, F. P., Freiherr von, 4
Harries, C. F., 3, 84
HDMS, see 4,4′-hexamethylenebis (1,1′-dimethylsemicarbazide)
Heavy atom effect, 184, 187, 193, 210
Herbicides, 107, 128, 153, 155, 167, 244
2,2′,4,4′,5,5′-Hexachlorobiphenyl, **110**, 129
Hexachlorocyclopentadiene, **83**
Hexahydro-1,3,5-trinitro-1,3,5-triazine (RDX), **135**
4,4′-Hexamethylenebis(1,1′-dimethysemicarbazide) (HDMS), **156**
2-Hexenoic acid, **83**
cis-3-Hexen-1-ol, 70, **83**, 106
HHCB, see galaxolide
HiPOx process, 201
Histidine, **135**
History
    of chlorination, 1ff, 65f
    of disinfection with ozone, 65f
    of Fenton reaction, 1
    of ozone, 1ff, 65ff
    of UV disinfection, 1ff, 65
Hoigné, J., 3, 113
HOMO, highest occupied molecular orbital, 114f, 139, 163f
1,2-H shift, 215, 219f, 228, 236
Humics in SEC–OCD, 31f
Hydride transfer, 171ff, 175f
Hydrogen atom (·H), **213**, 214, 229
Hydrogen peroxide (H_2O_2), 1, 42, 45f, 73f, **186**, 199ff, 230, 240
    activation energy of ozone reaction, 20
    photolysis of, 46, 141, 174, 247
    ^{17}O-NMR of, 176
    pK_a value of, 42, 183, 186, 188
Hydrogen sulfide ion, 162, **186**, 197
    singlet oxygen formation in reaction of ozone with, 197
    rearrangement of ozone adduct of, 198
Hydroperoxide ion, **186**
    reaction with ozone, see also peroxone process, 188f
Hydroperoxyl radical ($HO_2^·$)
    bimolecular decay of, 188
    elimination of, 24, 27, 178f, 182, 234
    reduction potential of, 18
    pK_a value of, 24, 183, 187, 215
    see also Superoxide radical
Hydrophobic fraction of DOM, 31f
Hydroquinone, **110**, 121f
    ozonolysis products of, 123
Hydrotrioxides, 101, 117, 120, 176ff, 182f
Hydrotrioxyl radical ($HO_3^·$)
    decay of, 188, 215
    formation of, 214
    pK_a value of, 183f, 188f, 214
Hydroxide ion, **186**
    formation of hydroxyl radical in reaction of ozone with, 187f
α-Hydroxyalkylhydroperoxides
    formation of, 85ff
    detection of, 88f
Hydroxycyclohexadienyl radical
    intermediate in aromatic hydroxylation, 24, 226
    reaction of ozone with, 216
bis(2-Hydroxyethyl)disulfide, **162**, 163
Hydroxylamine, **135**, 141, 149
    formation of, 141, 147, 149

Hydroxylation of aromatic compounds, 24, 58, 117f, 120, 122
Hydroxyl radical (•OH), 213, 217, 225ff
   addition reactions of, 225ff
   detection and quantification of, 230ff
   determination of rate constants of, 229f
   electron transfer reactions by, 228f
   energy requirement for production of, 73f
   exposure of in ozone reactors, 28f, 201
   formation of, by ionising radiation, 229
   formation of, by reaction of ozone with activated carbon, 46f
   formation of, by reaction of ozone with aromatic compounds, 117f, 122
   formation of, by reaction of ozone with bromite, 194
   formation of, by reaction of ozone with chlorite, 192
   formation of, by reaction of ozone with formate ion, 174
   formation of, by reaction of ozone with hydroxide, 188
   formation of, by reaction of ozone with iron hexacyanoferrate(II), 207
   formation of, by reaction of ozone with nitrite ion, 196
   formation of, by reaction of ozone with nitrogen-containing compounds, 141ff, 219
   formation of, by reaction of ozone with 2-propanol, 180
   formation of, by reaction of ozone with superoxide radical, 27, 188
   formation of, by the peroxone process, 42ff, 188
   formation of, by UV photolysis of hydrogen peroxide, 174
   formation of, by UV photolysis of ozone, 45f
   formation of, by ozone in drinking water and wastewater, 24ff
   H-abstraction reactions by, 27, 227f, 241f, 244
   pK_a of, 183, 189
   rate constant for reaction with DOM, 26
   reaction with DNA, 54
   reaction with DOM, 27
   reactions with aromatic compounds, 118, 243ff
   reactions with chlorinated olefins, 245ff
   reactions with oestrogens, 58f
   reactions with perfluorinated compounds, 248
   reactions with saturated aliphatic compounds, 241ff
   reduction potential of, 18, 225, 228
   role of, in micropollutant elimination, 39ff, 70f, 184, 241
   role of, in ozone generators, 9
   role of, in bromate formation, 198ff, 238ff
   scavenging by bicarbonate, 26f
   scavenging by DOM, 26, 39f
   scavenging by 2-methyl-2-propanol (tBuOH), 27, *173*, 180
   steady-state concentration of, in water ozonation, 28
   yield of, 39f, 42f, 119f, 142f, 201, 232f
2-Hydroxypropene, 177f
Hypobromite ion, 13f, 157f, **186**, 191ff, 198ff, 239f
   activation energy of ozone reaction, 20
Hypobromous acid, see hypobromite ion
Hypochlorite ion, **186**, 191f, 200f, 238
Hypochlorous acid, see hypochlorite ion
Hypoiodite ion, **186**, 192, 194f, 202
Hypoiodous acid, see hypoiodite ion

I

Ibuprofen, 39, 78, *79*, **79**, 111, 128, *241*, 243
Imidazole, **135**
Iminodiacetic acid, **135**, 140
Indigotrisulfonic acid (Indigo), 14, **83**
   for ozone decay measurements, 15f
   for ozone determination, 1, 13f
   reaction with Mn(III), 14
   reaction with MnO_2 colloids, 14, 209
   reaction with permanganate, 14
Inhibitor concept, 27f
Iodate, **186**, 195, 202f
Iodide, 1, 11, 36, **186**, 190, 194f
   formation of iodate from, 195, 203
   formation of iodo-organic compounds, 195
   heavy atom effect in formation of singlet from oxygen, 187
   mitigation of problems induced by, 202f
3-Iodoaniline, **135**
4-Iodoaniline, **135**
Iodo-organic compounds, formation of, 195, 202f
Iomeprol, **111**, 129, *241*, 243
Ionising radiation, 49f, 69, 181, 229
β-Ionone, 70f, **71**, *71*, **83**, 106
Iopamidol, **111**, 129, *241,* 243
Iopromide, 39, **41**, *41*, **111**, 129, *241*, 243
IPMP, see 2-isopropyl-3-methoxypyrazine
Irgarol (Cybutryn), 167
Iron, 2, **205**, 207, 211
   determination of molar absorption coefficient of ozone, with, 11
   hydroxides of, in arsenic removal, 211
   smell of, 7
Iron hexacyanoferrate(II), **205**, 206f, 229
2-(4-Isobutylphenyl)propionic acid, see Ibuprofen
Isoleucine, **135**
Isopropylbenzene, **111**

Index

2-Isopropyl-3-methoxypyrazine (IPMP), 70f, **71**, *71*, **135**, 146
Isoproturon, **135**, 155
Isopyrimidine, 101

J
Japanese medaka embryo exposure test, 63

K
Ketones
  cause for toxicity in wastewater ozonation, 64
  constituents of AOC, 33ff
  formation of, in olefin reactions, 85ff
  reaction with hydrogen peroxide, 85, 88
Ketoprofen, 129, 129
Koch, R., 2

L
β-Lactam antibiotics, 61f, 103
  see also penicillin G and cephalexin
Lamivudine, 103
Lead(II), 2, **206**, 208
Lemna minor growth inhibition test, 63f
Leucine, **135**
Liebig, J. von, 1
Lincomycin, **135**, 150, 167
Linuron, **135**, 155
Lipid peroxidase test, 63
Liquid holding, 49
Lumbriculus variegatus test, 63f
LUMO, lowest occupied molecular orbital, 115, 163f
LYES, 58
Lysine, **135**

M
Macrolide antibiotics, 60f, 103, 147f
Maleic acid, **83**, 86, 95f
Malonate ion, **170**
Malonic acid, **170**
Manganese(II), **205**, 206f, 208f, 211f
  interference of, in ozone determination, 13
  in water treatment, 211f
  permanganate formation by ozone in reaction with, 208f, 211f
MCPA, see 2-methyl-4-chlorophenoxyacetic acid
MCPP, see 2-methyl-4-chlorophenoxyacetic acid
Mecoprop, see 2-methyl-4-chlorophenoxypropionic acid
Melting point of ozone, 8
Membrane damage, in disinfection of bacteria, 52f
Meprobamate, 154f

Metal ions
  reactions of, with ozone, 205ff
Methacrylic acid, **83**, 93ff
Methane, 171
Methanesulfinate, **162**, 164, 233
Methanol, **170**, 172f
  as promoter of ozone decay, 27f
  hydroxyl radical induced reactions of, 27
Methionine, **162**
Methotrexate, **135**, 152f
2-Methoxyaniline, **135**
Methoxybenzene, **111**, 114, 117f
Methoxychlor, **111**, 127f
4-Methoxycinnamic acid, **83**
3-Methoxy-4-hydroxy cinnamic acid, **83**
4-Methoxy-1-naphthalenesulfonic acid, **111**
N-Methylacetamide, **135**, 138
Methylamine, **135**
3-Methylaniline, **135**
Methylbenzoate, **111**
5-Methylbenzotriazole, **135**, 142
Methyl-*t*-butylether (MTBE), 70f, **71**, *71*, *241*, 242
Methylchloramine, **135**
2-Methyl-4-chlorophenoxyacetic acid (MCPA), **111**, 128
2-Methyl-4-chlorophenoxypropionic acid (MCPP, Mecprop), 39, **111**, 128
Methyldichloramine, **135**
Methylene blue, 159
4-Methylimidazole, **135**
2-Methylisoborneol (MIB), 70, **170**, *241*, 242
1-Methylnaphthalene, **111**
Methyl orange, 159
2-Methylphenol, **111**
3-Methylphenol, **111**
2-Methyl-2-propanol (tertiary butanol), 27f, 40, **170**, 172f, *173*
  Arrhenius plot for ozone reaction with, 175
  as hydroxyl radical scavenger, 27,
  as inhibitor for ozone decay, 27f
  for quantifying hydroxyl radical yields, 40, 232f
1-Methylpyrrolidine, **136**
4-Methylsulfonylaniline, **136**
6-Methyluracil, **83**, 101
Methyl violet B, 159
Metolachlor, **136**, 155
Metoprolol, 21f, 39, **75**, *75*, **136**, 148f
MIB, see 2-methylisoborneol
Microcystin-LR, **83**, 104f
Micro-organisms, inactivation by ozone, 49ff

Micropollutants
  adsorption of, on activated carbon, 66, 68ff, 71
  biological activity of, 38
  elimination of, 37ff, 43, 71
  metal ions as, 211f
  mineralization of, 33, 38, 55
  occurrence of, 70
  oxidation of, energy requirements for, 73f
  ozone-refractory, 40ff, 241ff
  toxicology of transformation products, 55f, 62ff
  transformation of, 38, 55f, 70
  with aromatic functions, reaction with ozone, 124ff
  with nitrogen-containing functions, reaction with ozone, 146ff
  with olefinic functions, reaction with ozone, 102ff
  with sulfur functions, reaction with ozone, 166ff
Mineralization of dissolved organic matter, 25, 33
Molinate, **162**, 167
Monobromamine, **136**, 198ff
Monochloramine, 52, **136**
  consumption of by ozone, 15
  role of in formation of iodo-organics, 195
Morpholine, **136**, 142, 145, 218f
MTBE, see methyl-*t*-butylether
cis,cis-Muconic acid, **83**, 96ff, 121ff
cis,trans-Muconic acid, **83**
trans,trans-Muconic acid, **83**, 96
Muconic products, 116, 120f, 123f, 128
Mülheim process, 66, 68
Musk fragrances (see also tonalide and galaxolide), 116, 128
Mutations, induced by ozone, 50, 52
Mycotoxin, 106, 127

N

Naphthalene, **111**, 127, 129, 149
Naphthalene sulfonic acid, 127
Naproxen, **111**, 127
NDMA, see *N*-nitrosodimethylamine
Neurotoxicity, 62, 76, 211
Nevirapine (NVP), 151
Nitrate ion, 146, 158, **186**, 187, 195f
Nitric acid, 158
  role in ozone generation, 9
Nitrilotriacetic acid (NTA), **136**, 140, 154
Nitrite ion, **186**, 195f, 243
3-Nitroaniline, **136**
4-Nitroaniline, **136**
Nitrobenzene, **111**, 114, 144
4-Nitrocinnamic acid, **83**
Nitroform anion, 119, 121, **136**

Nitrogen
  excited states of, 9
  role in ozone generation, 9
4-Nitrophenol, **111**
Nitrosobenzene, 144
N-Nitrosodimethylamine (NDMA), 36, **41**, *41*, 73, 75, **136**, 156ff, 201, *241*, 243
Nitrous acid
  role in ozone generation, 9
Nitroxyl radicals, 144f, 213, 217ff
trans,cis-2,6-Nonadienal, 70, **83**, 106
Nonylphenol, 56ff, **111**, 125
NO$_x$, in ozone generation, 9
N-oxide, see aminoxide
NTA, see nitrilotriacetic acid
NVP, see nevaripine

O

OCD, see Organic carbon detection
Octanal, **170**
1-Octanol, **170**
4-Octylphenol, 56f, **111**, 125
17β-Oestradiol, Oestradiol (E2), 56, 58f, 60, **111**
Oestriol, 112
Oestrogen receptor, 56, 58
Oestrogenicity
  as water quality parameter, 62ff, 76
  elimination of, by hydroxyl radical, 58f
  elimination of, by ozone, 56, 58f, 60
  elimination of, by chlorine, bromine, chlorine dioxide, ferrate(VI), 59
Oestrone (E1), 56, 58ff, **111**
Oestrone anion, **111**
Olefins, 81ff
Organic carbon detection (OCD), 31f
Orotic acid, **83**
Oseltamivir acid, 62, **84**, 104, 151
Oxalate ion, **170**, 208, 211
Oxalic acid, **170**, 211
Oxamyl, see vydate
N-Oxide, see aminoxide
Oxygen
  as radical scavenger, 215, 220
  excited states of, 11
  O–O bond length in, 222
  oxygen-centered radicals, 218
  production of ozone, from, 9
  reaction with carbon-centred radicals, **233**
  reduction potential of, 18
  solubility of (in water), 10

Index

Oxygen atoms, reactions of, 45f
Oxyl radicals, see alkoxyl radicals
Ozone
    absorption coefficients of, in UV/Vis region, 10ff
    activation energies of reactions of, 20, 165, 170, 174f, 179
    advanced oxidation processes, based on, 42ff
    alkalinity, effect on stability of, 25ff
    application in drinking water, 65ff
    application in wastewater, 65, 72
    application in the urban water cycle, 77ff
    as reducing agent, 191, 193, 208
    boiling point of, 8
    bond angle of, 8
    bond length of, 7
    cost of generation of, 8, 74
    decay of, 27f
    decomposition of, by •OH, 9, 217
    detection of, 1, 7
    determination of concentration of, 12ff
    dipole moment of, 7
    discovery of, 1ff
    disinfection by, 49ff
    effect on COD of, 33
    electron transfer to, 85, 117, 122, 171, 192
    energy for production of, 73f
    explosion threshold of, 8
    exposure, 29, 51, 55
    formation of, 2, 8f
    formation of assimilable organic carbon, from, 33f
    formation of hydroxyl radicals by, in wastewater, 23f, 40
    generation of, 8f
    H-abstraction by, 171ff
    Henry constant of, 8
    history of, 1ff, 65ff
    hydride transfer to, 171ff
    hydroxyl radical yield, from, 39f
    insertion by, 171ff
    kinetics of, measurement of, 14ff
    kinetics of decay, in wastewater, 29f
    maximum daily allowance in air of, 7
    melting point of, 8
    micropollutant elimination, by, 37ff
    mitigation of disinfection by-product formation from, 35f
    modeling decay of, 28
    molecular weight of, 7
    mutations, induced by, 50, 52
    name of, origin of the, 2
    quenching of, 16
    reactivity pK in reactions of, 20ff
    reduction potential of, 18
    smell of, 1, 2, 8
    solubility in water, 9f
    stability of, in drinking water, 23
    stability of, in groundwater, 25
    stability of, in lake water, 25, 43f
    stability of, in spring water, 25
    stability of, in wastewater, 23, 29f
    stability of, in water, 9f, 19, 26
    toxicity of, 7
    toxicity after application of, 55ff, 62
    UV photolysis of, 42, 45f
    UV spectrum of, 10
Ozonide
    decay of, via free radicals, 88
    formation of, 84f, 114ff
    reactions of, 84ff, 91ff, 98f
    origin of name of, 1
Ozonide radical anion ($O_3^{•−}$)
    as precursor of hydroxyl radical, 23
    decay of, 189
    formation of, 23, 117, 141f, 214

P

Parabene, 57
Paracetamol, **112**, 124
Partial oxidation, 87f
Patulin, 106
pCBA, see *p*-Chlorobenzoic acid
Penciclovir, 103f
Penicillin G, 62, **162**, 166f
Penicillin G-(*R*)-sulfoxide, 166
Penicillin G-(*S*)-sulfoxide, 166
Penicillum, 127
Pentabromophenol, **112**, 228
    singlet oxygen formation from, in reaction with ozone, 118
2,3,3′,5,6-Pentachlorobiphenyl, **112**
Pentachlorophenol, **112**, 228
    singlet oxygen formation from, in reaction with ozone, 118
Pentafluorophenol, 228
2-Pentanone, **170**
1-Penten-3-one, 70, **84**, 106
PER, see tetrachloroethene
Perbromate, 194
Percarbonate, 27
Perchlorate ion, **186**, 192, 222

Perfluorooctanoic acid (PFOA), *241*, 248
Perfluorooctanesulfonic acid (PFOS), 241, 248
Periodate ion, **186**
Permanganate
 formation of in reaction of Mn(II) with ozone, 208f
 in COD determination, 33
 oxidation of microcystin-LR by, 105
 reaction with carbamazepine, 102
 reaction with indigo trisulfonate, 14, 209
Peroxidase, 1
Peroxone process, 42ff, 201, 230
 hydroxyl radical yield of, 189
 mechanism of, 188ff
Peroxodisulfate, 174
Peroxyl radicals, **213**
 addition to C–C double bonds, 235
 formation of, 24, 215, 220, 233
 H-abstraction by, 234
 $HO_2^{\bullet}/O_2^{\bullet-}$ elimination of, 27, 234
 in COD reduction, 33
 reaction of ozone with, 217f
 reactions of, 33, 152, 225, 234ff, 246f
 reduction potential of, 233
Peroxynitrite, 196f
PFOA, see perfluorooctanoic acid
PFOS, see perfluorooctanesulfonic acid
Phenazone, 155f
Phenol, **112**
 consumption by ozone, 15, 17, 121
 endocrine disrupting compounds containing, 56ff
 halophenol formation from, 35, 66
 Hammett-type correlation for reaction with ozone, 115
 pH dependence of reaction with ozone, 20f
 products from its reaction with ozone, 121f
 reactions of halophenols with hydroxyl radical, 228
 reduction potential of, 228
 singlet oxygen formation from, in reaction with ozone, 118
Phenol radical cation, 228
Phenols
 formation of, by reaction of ozone with DOM, 24
 reactions with hydroxyl radical, 24
Phenoxyl radical
 formation of, 118, 121f
 reactions of, 122, 125, 216f
1-Phenoxy-2-propanol, **112**
1-Phenoxy-3-propanol, **136**
Phenylalanine, **136**
o-Phenylenediamine, 143
Phenytoin, 156

Phosphoric acid, Phosphate, **186**
Phosphorus, ozone formation by autoxidation of, 2
Photolysis
 of acetone, 177
 of hydrogen peroxide, 46, 174
 of NDMA, 73, 243
 of ozone, 42, 45f
 of tetrachloroethene, 247
 of peroxodisulfate, 174
 of triclosan, 62
Photosynthesis inhibition test, 63
Phytotoxicity, 62, 76
Pivalate ion, **170**
Pivalic acid, **170**
pK
 pK_a value of acrylic acid derived ozonide, 94
 pK_a value of aniline radical cation, 145
 pK_a value of H_xO_y compounds, 183
 pK_a value of HOBr, 191
 pK_a of HOCl, 191
 pK_a value of HOI, 192, 195
 pK_a value of hydrogen peroxide, 188
 pK_a value of phenol radical cation, 228
 pK_a value of protonated acetone, 175
 of halogen species, 191f
 reactivity pK of, in reaction with ozone, 16, 20f
 see also tables of ozone rate constants for further
   pK_a values
Plasma, chemistry of, in ozone generation, 8f
Plug-flow reactor, 51, 202
Polycyclic aromatic compounds, 129f
Popper, K. R., 4
Post-chlorination, 35f
Powdered activated carbon, see activated carbon
Primidone, 156, **156**
Progesterone, **84**, 103, 106f
Proline, **136**
Prometon, **136**
Promoter concept, 27f
Propachlor, **136**, 155
Propanal, **170**
1-Propanol, **170**
2-Propanol, **170**, 172f, 176ff
 activation energy of ozone reaction, 20, 179
 Arrhenius plot for ozone reaction with, 179
 products from ozonolysis, 176
 reactions with hydroxyl radical, 228
Propene, **84**, 87
Propionate ion, **170**
Propionic acid, **170**

Propranolol, **136**, 141, 149
Propyl acetate, **170**
Propylamine, **136**
Protozoa, disinfection of, 49ff
Pyridine, **136**, 137, 145f
Pyridine-2-carboxlic acid,144
Pyridine-*N*-oxide, 145
Pyrimidine nucleobases, 53, 99ff
Pyrrole, 46f

Q

Quantitative structure activity relationship (QSAR), 113, 115, 127
Quantum-chemical calculations, 86ff, 113ff, 139f, 143, 145f, 162ff, 171, 175, 187, 192f, 208, 214ff, 236
Quantum yield, 45f, 247
Quenching of ozone
  by activated carbon, 202
  by buten-3-ol, 16
  by indigo trisulfonic acid, 13f, 16
  by Mn^{2+}, 12
Quinoline, **136**, 146
Quinolinic acid, 146
Quinones, 37, 119ff
Quinonimine, 152
QSAR, see quantitative structure activity relationship

R

Radical cation, 13, 18, 117, 141f, 145, 152, 219, 226, 228f
Radicals, 213ff
  reaction with ozone of carbon-centered, 213, 215f
  reaction with ozone of halogen-centered, 220ff
  reaction with ozone of nitrogen-centered, 219f
  reaction with ozone of oxygen-centered, 213, 217f
  reaction with ozone of reducing, 213, 214f
  reaction with ozone of sulfur-centred, 219f
Rate constants
  apparent, pH dependence of, 20
  compilation of activation energies, for ozone, 20
  compilation of hydroxyl radical, 26, 41, 241
  compilation of ozone, 41
  compilation of, for inactivation of micro-organisms, 51
  determination of, by Hammet-type correlations, 115
  determination of, for hydoxyl radical, 229f
  determination of, for ozone, 14ff, 17f
RBV, see ribavirin
R_{ct} value, 28f, 231
  definition of, 29
RDX, see hexahydro-1,3,5-trinitro-1,3,5-triazine
Reactive absorption, 16f

Reactivity pK, 20
  of anatoxin-a in reaction with ozone, 105
  of bicarbonate in reaction with hydroxyl radical, 26f
  of metoprolol in reaction with ozone, 21f
  of phenol in reaction with ozone, 21
Reactor hydraulics, 51, 55
Reclamation of wastewater, 75f
Reduction potential
  of 1,4-benzoquinone, 122
  of carbon dioxide radical anion ($CO_2^{\bullet-}$), 214
  of catechol, 122
  of Co^{3+}, 207
  of hydrated electron (e_{aq}^-), 214
  of hydrogen atom ($^{\bullet}H$), 214
  of hydrogen peroxide, 18
  of hydroquinone, 122
  of hydroxyl radical, 18, 225, 228
  of oxygen species, 18, 214, 225
  of ozone, 18, 214
  of peroxyl radical, 122
  of phenol, 122, 228
  of phenoxyl radicals, 228
  singlet oxygen, 18
  of superoxide radical ($O_2^{\bullet-}$), 214, 233
  of *N,N,N',N'*-tetramethyl-*p*-phenylenediamine, 13
Repair, of genomic damage, 49f
Resorcinol, **112**
Retionic receptor test, 63
Reverse osmosis (RO), 75, 77
Ribavirin (RBV), 151
RNA, 52f, 55
RO, see reverse osmosis
Röntgen, W. K., 2
Rotavirus, **51**, 52
Roxithromycin, 61, **136**, 147
Russell reaction, 232, 235

S

Salicylic acid, **112**
Sand filtration, 35, 63f, 65f, 68, 72f
Scavenging capacity, 28, 230
Scavenging of hydroxyl radicals, 26, 29
Scavenging rate, for hydroxyl radicals, 39ff, 75, 232
Schönbein, C. F., 1ff, 7f, 13f, 205, 208, 209
SEC, see Size exclusion chromatography
Selenium(III), **206**, 209
Serine, **136**
Siemens, W. von, 2f, 8
Silver(I), **205**, 209ff
Simazine, **137**

Singlet oxygen
   adducts as intermediates in formation of, 187
   formation of, in decay of hydroperoxides, 125
   formation of, in hydroxylation reactions, 24, 117f
   from ozone photolysis, 45
   heavy atom effect in formation of, 187
   in ozone generation, 9
   in ozone reactions with amines, 140f
   in ozone reactions with aromatic compounds, 108, 117ff
   in ozone reactions with catechol, 124
   in ozone reactions with hydroquinone, 123
   in ozone reactions with halide ions, 187, 190, 193
   in ozone reactions with muconic acid, 97f
   in ozone reactions with nucleobases, 53f, 100ff
   in ozone reactions with phenol, 121
   in ozone reactions with sulfur-containing compounds, 161ff
   in ozone reaction with azide, 196f
   in ozone reaction with lead, 208
   in ozone reaction with nitrite, 195f
   in reaction of hypobromite with hydrogen peroxide, 199
   reduction potential of, 18
Size exclusion chromatography (SEC), 31f
Smell
   of iron, 7
   of ozone, 1f, 7
SMX, see sulfamethoxazole Solar radiation, disinfection by, 49
Solubility (in water)
   oxygen of, 10
   ozone of, 8, 10
Sorbic acid, **84**
Source control, 74
Specific UV absorbance (SUVA), 31, 36f
Spectinomycin, 136f, 149f
Standard Gibbs free energy (ΔG^0), 98, 114, 117, 143, 153, 164, 171ff, 190, 193, 214, 216ff, 221f, 236
Stavudine, 103
Stumm, W., 3
Succinate ion, **170**
Succinic acid, **170**
Sucrose, **170**
Sulfamethoxazole (SMX), 39, 73f, **137**, 150
Sulfate ion, **186**
Sulfate radical ($SO_4^{\cdot-}$), 174, 248
Sulfides, 163ff, 166f, 197f, 226
Sulfinic peracid, 197f
Sulfite ion, **186**, 198

Sulfonamide, 137f, 150
Sulfur dioxide, 92, 157f, **186**, 198
Superoxide radical ($O_2^{\cdot-}$), **213**
   bimolecular decay of, 235f
   formation of, 118, 234
   in drinking water and wastewater ozonation, 24
   reaction with 1,4-benzoquinone of, 122
   reaction with ozone of, 24, 214f
   reduction potential of, 18, 214, 233
SUVA, see Specific UV absorbance
Suwannee River humic/fulvic acid, 26, 37

T
Tamiflu, see oseltamivir acid
Taste and odour, 35, 65f, 70f, 77, 106, 130, 146, 195, 202, 209, 242
TBA, see 2,4,6-tribromoanisole
TCA, see 2,4,6-trichloroanisole
TCC, see total cell count
TCPP, see tris-(2-chloroisopropyl) phosphate
TCEP, see tris-(2-chloroethyl) phosphate
TEMPO, 145, **213**, 218f
Tertiary butanol, see 2-methyl-2-propanol
Testosterone, 106
Tetrachloroethene (PER), **84**, 102, *241*, 245ff
Tetracycline, 103, **112**, 126, 155
Tetrahydrofuran, **170**
Tetramethylethene, **84**, 85f, 90
1,1,1′,1′-Tetramethyl-4,4′(methylene-di-p-phenylene) disemicarbazide (TMDS), 156
N,N,N',N'-Tetramethyl-p-phenylenediamine (TMPD)
   formation of hydroxyl radical in reaction of ozone with, 13, 143
   reaction with peroxyl radicals, 13
   reaction with ozone, 13
   singlet oxygen formation from, in reaction with ozone, 118, 140
Tetranitromethane, 46, 119, **137**
Thallium(I), 226f
Terephthalic acid, 231
Thiocarbamate herbicides, 167
Thiols, 162f
Thiosulfate ion, **186**
Thiyl radicals, 220
   1,2-H shift of, 220
   reaction of ozone with, 220
   reaction with O_2 of, 220
THM, see trihalomethane
Three-electron bond, 220, 226
Threonine, **137**

Thymidine, 53, **84**, 99ff
5′-Thymidylic acid, **84**
Thymine, 53, **84**, 99ff, *229*
Tin(II), **206**, 211ff
TMDS, see 1,1,1′,1′-tetramethyl- 4,4′(methylene-di-*p*-phenylene)disemicarbazide
TMPD, see *N,N,N′,N′*-Tetramethyl-*p*-phenylenediamine
TnBP, see tri-*n*-butyl phosphate
TOC, see total organic carbon
Tolidine, 12
Toluene, **112**, 226
Tolylfluanide, 156f
Tonalide (AHTN), **112**, 116, 128
Total cell count/concentration (TCC), 52, 68f, 72f
Total organic carbon (TOC), 32
Toxicity
   of cyanotoxins, elimination by ozone, 105
   of diuron, elimination by ozone, 155
   of ozone, 7
   of ozone-induced transformation products, 55ff, 64
   of ozone-treated wastewater, 62ff, 76
   removal of, by post sand filtration, 64, 72
   test systems for, 62ff
Tramadol, **137**, 148
Treatment
   biological, 65ff
   of drinking water with membranes, 67f
   of drinking water with ozone, 65ff
   trains including ozone, 65ff, 68f
   of wastewater with ozone, 38, 65ff, 72, 76
2,4,6-Tribromoanisole (TBA), 71, **112**, 130
Tri-*n*-butyl phosphate (TnBP), **41**, *41*, **241**, 242
Tributyl tin, 58, **206**, 211f
2,4,6-Trichloroanisole (TCA), 71, **112**, 130
1,2,3-Trichlorobenzene, **112**
TRI, see trichloroethene
Trichloroethene (TRI), **84**, 90, *241*, 245, 248
Trichloronitromethane (Chloropicrin), 35
2,4,5-Trichlorophenol, **112**
2,4,6-Trichlorophenol, **112**
Triclosan, 62, **112**, 115f, 124
   consumption by ozone, 17
Triethylamine, **137**, 140f
Trihalomethane (THM), 35, 66, 201
2,4,6-Triiodophenol, **112**, 228
   singlet oxygen formation from, in reaction with ozone, 118
Trimethoprim, **137**
1,3,5-Trimethoxybenzene, **112**, 114, 116f
   scheme for ozone reactions of, 119ff
   singlet oxygen formation from, in reaction with ozone, 118
3,4,5-Trimethoxytoluene, **112**
Trimethylamine, **137**, 139ff, 143
1,2,4-Trimethylbenzene, **112**
1,3,5-Trimethylbenzene, **112**
2,4,6-Trimethylphenol
   singlet oxygen formation from, in reaction with ozone, 118
Triphenyl tin, 58
Tris-(2-chloroethyl) phosphate (TCEP), **39**, *39*, 41
Tris-(2-chloroisopropyl) phosphate (TCPP), **41**, *41*, *241*, 242
Tryptophan, **137**
Tylosin, 61, **84**, 103, **137**, 148
Tyrosine
   singlet oxygen formation from, in reaction with ozone, 118

U
UF, see ultrafiltration
Ultrafiltration (UF), 67ff, 71, 75
Ultrasound, 230
Uracil, 53, **84**, 101, 105, 234
Urban water cycle, 77f
Urine, 74, 77ff
UV disinfection
   DNA lesions, 49
   history of, 2ff, 65
   in multibarrier treatment, 69
   in the Mülheim process, 68
   in water reclamation, 75
UV photolysis of ozone, 45f
UV/Vis absorbance of DOM, 36f
UV/Vis absorption coefficients
   of DPD radical cation, 13
   of halogen compounds of indigo, 14
   of ozone, 10ff

V
Valine, **137**
Vancomycin, **112**, 126, 155
Vinyl acetate, **84**, 91
Vinyl bromide, **84**, 90
Vinyl chloride, **84**, 86f, 90f, 102
Vinylidene carbonate, **84**, 91f
Vinyl phenyl sulfonate, **84**, 91f
Vinyl phosphonic acid, **84**
Vinyl sulfonate ion, **84**, 91f

Virus, inactivation of, 49ff, 52f
Vitellogenin (VTG), 58f
VTG, see Vitellogenin
Vydate (Oxamyl), **162**, 167

W

Wastewater
  antimicrobial compounds in, 62
  AOC formation in, after ozonation, 34
  disinfection of, by ozone, 55, 73
  elimination of micropollutants by ozonation, 38f
  endocrine disrupting compounds in, 59f
  formation of hydroxyl radicals by ozone in, 24ff
  kinetics of ozone decay in, 29ff
  ozone treatment of, 65f, 72f
  reuse of, 75f
  size exclusion chromatography–organic carbon detection of, before and after ozonation, 31f
  toxicity tests with, 62ff
  UV/Vis spectrum of, 36
  yield of hydroxyl radicals in, upon ozonation, 40, 43
Water, **186**
Wöhler, F., 1
Wurster's blue, 13

X

X-ray contrast media, 129, *241*, 243
m-Xylene, **112**
o-Xylene, **112**
p-Xylene, **112**

Y

Yeast estrogen screen (YES), 58ff, 64
YES, see Yeast (o)estrogen screen
Yolk-sac larvae test, 59

Z

Zidovudine, 103
ZnDTPA^{3-}, **134**